Henning F. Harmuth

Transmission of Information by Orthogonal Functions

Second Edition

Springer-Verlag New York Heidelberg Berlin 1972

Dr. HENNING F. HARMUTH
Consulting Engineer and Adjunct Associate Professor
Department of Electrical Engineering
The Catholic University of America, Washington, D. C.

With 210 Figures

ISBN 0-387-05512-6 2nd edition Springer-Verlag New York Heidelberg Berlin
ISBN 3-540-05512-6 2nd edition Springer-Verlag Berlin Heidelberg New York

ISBN 0-387-04842-1 1st edition Springer-Verlag New York Heidelberg Berlin
ISBN 3-540-04842-1 1st edition Springer-Verlag Berlin Heidelberg New York

Library of Congress Catalog Card Number 73-166 296

To my Teacher

EUGEN SKUDRZYK

PREFACE

The orthogonality of functions has been exploited in communications since its very beginning. Conscious and extensive use was made of it by Kotel'nikov in theoretical work in 1947. Ten years later a considerable number of people were working in this field. However, little experimental use could be made of the theoretical results before the arrival of solid state operational amplifiers and integrated circuits.

The advantages of Walsh functions, which are emphasized in this book, were recognized independently by several scientists in the early sixties. Among them were E. Gibbs, K. Henderson, F. Ohnsorg, G. Sandy and E. Vandivere, whose work was not published until many years later.

Somewhat more than half the illustrations in this second edition were not contained in the first edition and this reflects the changes in contents. The most striking difference between the two editions is the progress toward practical applications made in the intervening three years. However, it may turn out that the most important change is one that appears rather theoretical on the surface and that concerns shift-invariant features strongly connected with sine-cosine functions. These functions are projections of the exponential function which, in turn, is the character group of the real numbers. The topology of the real numbers is generally accepted to be the same as that of time or a one-dimensional space, and this is the basis for a variety of claims that sinusoidal functions are unique and superior to all others.

Time may well have the same topology as the real numbers, but we will never know. We can only make a finite number of measurements and there must be a finite distance between them, either in time or in space. The topology of time or space which permits experimental verification is thus that of a finite number of integers. Small as this theoretical point might appear, it provides enough uncertainty to admit at least denumerably many systems of functions to applications that were previously believed to be reserved for sine-cosine functions. Asynchronous filters and mobile radio communication based on nonsinusoidal functions are carried in this book all the way to block circuit diagrams, primarily to help demonstrate this point.

A new theory is not generally accepted unless its advantages can be demonstrated convincingly. In engineering, a convincing demonstration means working equipment. The long years between the dawn of a theoretical conception and working equipment can only be bridged in a country that tolerates unsinusoidal activities and with the support of many fellow scientists. The author wants to express his gratitude to some of those who helped him most: F. A. Fischer, Darmstadt; M. Scholz, Bonn; H. Schlicke, Milwaukee; G. Lochs, Innsbruck; F. H. Lange, Rostock; H. Lueg, Aachen; J. D. Lee, Washington; J. H. Rosenbloom, Washington; L. B. Wetzel, Washington; C. M. Herzfeld, New York. Furthermore he wants to thank the International Telephone and Telegraph Company, Electro-Physics Laboratories, for the financial support of his work. Last but not least, thanks are due to Ms. A. Navon, Washington, Ms. M. Haggard, San Jose, and Mr. T. Frank, Washington, for the proofreading.

Washington, D.C., February 1972

Henning F. Harmuth

CONTENTS

Equations are numbered consecutively within each of the sections 1.1.1 to 7.3.5. Reference to an equation from a different section is made by writing the number of the section in front of the number of the equation, e.g., (5.2.3-1) for (1) in section 5.2.3.

INTRODUCTION

Historical Background and Motivation for the Use of Nonsinusoidal Functions

During the nineteenth century, the most important functions for communications were block pulses. Voltage and current pulses could be generated by mechanical switches, amplified by relays and detected by a variety of magneto-mechanical devices. Sine-cosine functions and the exponential function were well known and so was Fourier analysis, although in a somewhat rudimentary form. Almost no practical use could be made of this knowledge with the technology available at that time. Hertz used the exponential function to obtain his famous solution of Maxwell's equations for dipole radiation but he was never able to produce sinusoidal waves[1]. His experiments with wave propagation were done with what we would call colored noise today.

Toward the end of the nineteenth century, more practical means to implement capacitances than Leyden jars and metallic spheres were found. The implementation of inductances through the use of coils had been known long before. Practical resonant circuits for the separation of sinusoidal electromagnetic waves of different frequencies could thus be built around the turn of the century. Low-pass and band-pass filters using coils and capacitors were introduced in 1915 and a large new field for the application of sinusoidal functions was opened. Speaking more generally, the usefulness of sinusoidal functions in communications is intimately related to the availability of linear, time invariant circuit components in a practical form.

A major breakthrough of technology was provided by the introduction of semiconductors. The switch was added as a linear, time variable component of extreme usefulness. Furthermore, the previous emphasis on fewer circuit components and, especially, the exclusion of active components such as tubes became obsolete. This change opened the way for the use of nonsinusoidal functions.

There were two schools of thought from the beginning. The one tried to find functions that were best suited for a given problem from the mathematical point of view. The parabolic cylinder functions were found to

[1] The word "sinusoidal" seems to have been introduced to communication engineering by A. Graham Bell in 1876. *See* pages 9 and 10 of [17] and page 192 of [18].

compress the signal energy into the smallest section of the time-frequency domain. This result was already known to Wiener in the early thirties. In the fifties, prolate spheroidal functions were found to be almost as good and separable by time sampling. Not much came of either system of functions since the practical difficulties were too great to justify their use.

The second school of thought tried to find functions that led to equipment easily implemented by the available technology. This approach will yield, from the mathematical point of view, optimal solutions for given problems by coincidence only. However, simplification of equipment is in itself a desirable goal for the engineer. In addition, inherent simplicity makes it possible to develop equipment that would otherwise be too complicated.

As an example of the last statement, consider television. Black-and-white TV pictures are signals with two space variables and the time variable. No filters on the basis of sine-cosine functions ever became known for such signals, despite the high degree of sophistication of filters for time signals. The so-called Walsh functions, on the other hand, made it possible to design such filters, and they have already been implemented in their simplest form.

The search for useful functions led first to the system of Walsh functions, which assume the two values $+1$ and -1 only. The basic discussion of theory and equipment is usually presented for this system of functions. When it comes to practical equipment design, other systems of two- or three-valued functions are often used; however, it would be impractical to constantly refer to all these other functions. A similar situation exists in time division multiplexing which is generally discussed in terms of block pulses while actually a variety of other pulse shapes is used.

There were unforeseen but happy developments when the applications of Walsh functions in communications were investigated. First, it was found that Walsh functions were excellently suited for the analysis of sampled signals and the design of equipment for such signals. Second, the heavy emphasis on time as the variable in communications was found to be closely connected to sinusoidal functions but the connection did not carry over to Walsh functions. As an example, consider a tunable generator for sinusoidal functions. Probably every reader will think of a generator for *time* variable sinusoidal functions. Indeed, it is hard to imagine a generator that produces *space* variable sinusoidal functions that can be tuned, e.g., from 20 cycles per meter to 20,000 cycles per meter. It is one of the primary tasks of science to recognize and overcome such impediments to thinking.

Walsh functions are presently the most important example of non-sinusoidal functions in communications. They have been used for the transposition of conductors in open wire lines for more than 60 years. Rademacher functions, which are a subset of the Walsh functions, were used for this purpose toward the end of the 19th century [1]. The complete system of Walsh functions seems to have been found around 1900 by J. A. Barrett[1]. The transposition of conductors according to Barrett's scheme was standard practice in 1923, when J. L. Walsh introduced them into mathematics [3, 4, 6]. Communications engineers and mathematicians were not aware of this common usage until very recently [5]. Individual Walsh functions have been used for a much longer time. They may be found on ancient temples[2], and the pattern of the checker board used for playing checkers or chess turns out to be a two-dimensional Walsh function.

Orthogonal Functions, Walsh Functions and Other Basic Mathematical Concepts

Two functions $f(j, \theta)$ and $f(k, \theta)$ are called orthogonal in the interval $-\frac{1}{2} \leqq \theta < \frac{1}{2}$ if the integral $\int_{-1/2}^{1/2} f(j, \theta) f(k, \theta)\, d\theta$ is zero for $j \neq k$. They are orthogonal and normal or orthonormal if the integral equals 1 for $j = k$. The best known system of orthonormal functions is the system of sine and cosine functions shown in Fig. 1, with the normalized time $\theta = t/T$ as variable. These functions are used in Fourier series expansions and it is easily verified that the integral of the product of any two functions is zero.

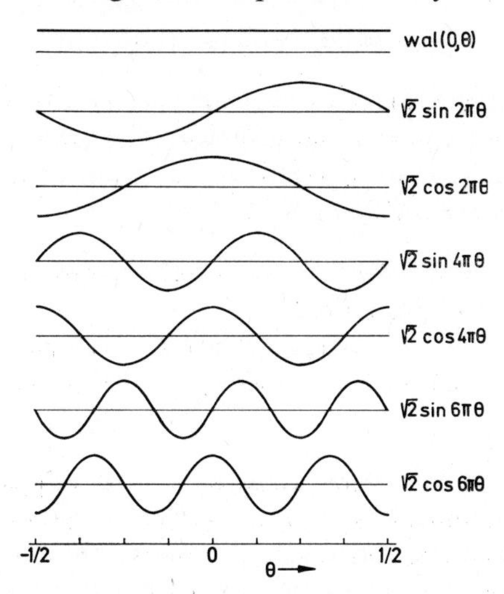

Fig. 1. Orthonormal sine and cosine elements.

[1] John A. Barrett is mentioned by Fowle in 1905 as inventor of the transposition of conductors according to Walsh functions; *see* particularly page 675 of [2].

[2] Rademacher functions are conspicuous all around the outside walls of the main palace at Mitla near Oaxaca, Mexico [7]. A Walsh function cal$(3/8, x)$ appears less conspicuously above the entrance.

Walsh functions are another example of orthonormal functions. Fig. 2 shows the first 64 of them. The notation $\mathrm{sal}(i, \theta)$ and $\mathrm{cal}(i, \theta)$ is used for these functions. The letters *s* and *c* allude to sine and cosine functions, to which the respective Walsh functions are closely related; the letters *al* are derived from the name Walsh. The functions $\mathrm{cal}(i, \theta)$ are even functions like $\sqrt{(2)}\cos 2\pi i\theta$, the functions $\mathrm{sal}(i, \theta)$ are odd functions like $\sqrt{(2)}\sin 2\pi i\theta$.

In Fig. 1 the parameter $i = 1, 2, \ldots$ in $\sqrt{(2)}\sin 2\pi i\theta$ and $\sqrt{(2)}\cos 2\pi i\theta$ gives the number of oscillations in the interval $-\frac{1}{2} \leqq \theta < \frac{1}{2}$, which is the normalized frequency $i = fT$. One may interpret i as "one half the number of zero crossings per unit time," rather than as "oscillations per unit time" [8, 9, 10]. The zero crossing at the left side, $\theta = -\frac{1}{2}$, but not the one at the right side, $\theta = +\frac{1}{2}$, of the time interval is counted for sine functions.

The parameter i also equals one half the number of zero crossings in the interval $-\frac{1}{2} \leqq \theta < \frac{1}{2}$ for the Walsh functions in Fig. 2. In contrast to sine-cosine functions, the sign changes are, in general, not equidistant. If, unlike in Fig. 2, i is not an integer, then it equals "one half the average number of zero crossings per unit time." The term "normalized sequency" has been introduced for i, and $\phi = i/T$ is called the nonnormalized sequency, which is measured in zps:

sequency in zps = $\frac{1}{2}$ (average number of *z*ero crossings *p*er *s*econd)

The concepts of period of oscillation $\tau = 1/f$ and wavelength $\lambda = v/f$ are connected with frequency. Substitution of sequency ϕ for frequency f leads to the more general concepts of "average period of oscillation" $\tau = 1/\phi$ and "average wavelength" $\lambda = v/\phi$:

average period of oscillation = average separation in time of the zero crossings multiplied by 2,
average wavelength = average separation in space of the zero crossings, multiplied by 2; v is the velocity of propagationof a zero crossing.

One of the most important features of sine and cosine functions is that almost all time functions used in communications can be represented by a superposition of sine and cosine functions, for which Fourier analysis is the mathematical tool. The transition from time to frequency functions is a result of this analysis. This is often taken so much for granted by the communications engineer that he instinctively sees a superposition of sine and cosine functions in the output voltage of a microphone or a teletype transmitter. Actually, the representation of a time function by sine and cosine functions is only one among many possible ones. Complete systems of orthogonal functions generally permit series expansions that correspond to the Fourier series. For instance, expansions into series of Bessel functions are much used in communications. There are also transforms corresponding to the Fourier transform for many systems of functions. Hence, one may see a superposition of Legendre polynomials, parabolic cylinder functions, etc., in the output voltage of a microphone.

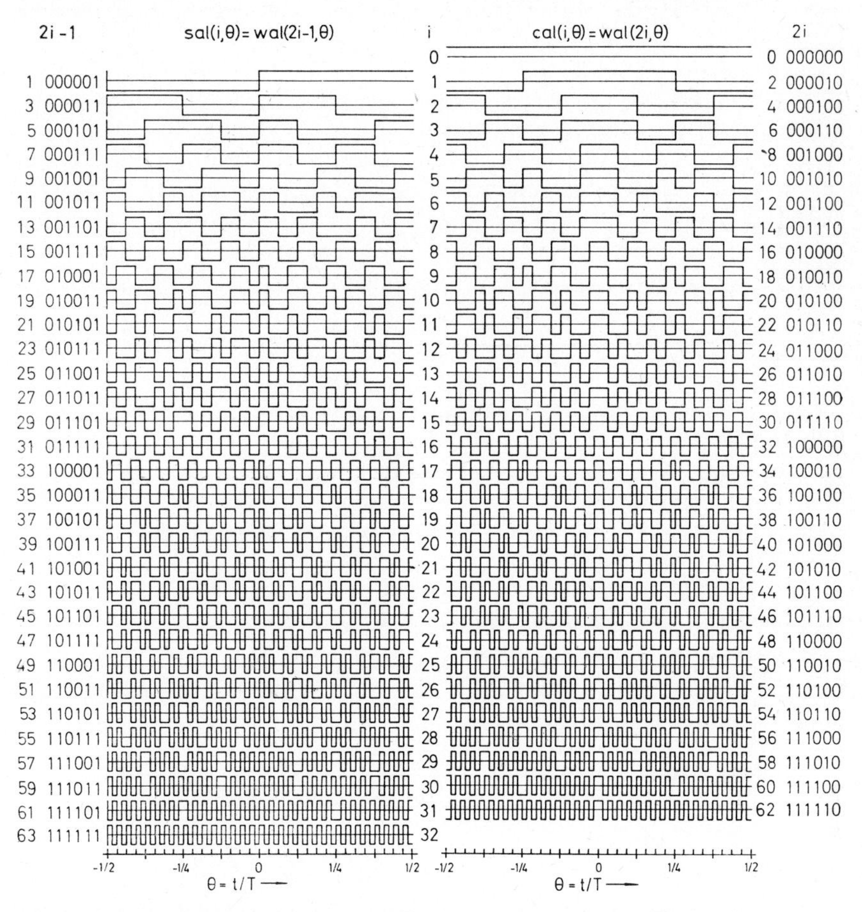

Fig. 2. Orthogonal Walsh elements wal$(0, \theta)$, sal(i, θ), cal(i, θ). The numbers $2i$ and $2i-1$ are given in decimal and binary notation; i is the normalized sequency.

The transition from the system of sine-cosine functions to general systems of orthogonal functions brings simplifications as well as complications to the mathematical theory of communication. One may, e.g., avoid the troublesome fact that any signal occupies an infinite section of the time-frequency-domain by substituting a time-function-domain. Any time-limited signal composed of a limited number of orthogonal functions occupies a finite section of this time-function-domain.

Filtering of Time and Space Signals

The acid test of any theory in engineering is its practical application. Filters are the area of application in which Walsh and similar functions have made the biggest progress toward practical use[1]. Filters for time signals have been developed that use the correlation principle and are implemented by multipliers, integrators, adders, and storage circuits. Furthermore, filters using the resonance principle have been implemented with the usual coils and capacitors supplemented by switches. It is worth noting that filters based on nonsinusoidal functions do not necessarily have to be synchronized in any way to the signal: There are filters that require synchronization but there are others that do not; they are referred to as asynchronous filters.

One of the most interesting new developments are filters for signals with space variables. There has been a need for such filters ever since television was introduced and such recent developments as the picturephone and picture transmission from deep space probes have increased this need. The usual theory of filters for time variable signals cannot be used for spatial filters. Digital computers have been applied very successfully to this problem by Alexandridis, Andrews, Carl, Kabrisky, Kennedy, Parkyn, Pratt, Wintz and others. These computer techniques are discussed in an excellent book by Andrews and are not dealt with here [11]. Analog circuits based on the computer techniques can be implemented by the usual means of electrical engineering and experimental equipment for the bandwidth reduction of TV signals has been developed succesfully by Enomoto[2], Hatori[3], Miyata[4], Shibata[5], Taki[3] and others in Japan.

Not surprisingly, binary functions lead to simple digital filters. Two varieties are known, one based on the generalized Fourier analysis the other on difference equations.

Direct and Carrier Transmission of Signals

The jumps between $+1$ and -1 of the Walsh functions make them appear to be poorly suited for the transmission through existing communication channels. However, one should keep in mind that a signal consisting of $2n$ independent amplitudes per second requires a frequency bandwidth of n hertz according to the sampling theorem, regardless of how those independent amplitudes were generated. A frequency bandpass or low-pass filter can transform signals consisting of a superposition

[1] Certain applications for which Walsh-Fourier analysis is used as mathematical tool while the implementation is done by computers are comparably or more advanced. *See*, for instance, the problems at the end of section 1.2.8.

[2] Tokyo Institute of Technology. — [3] Tokyo University. — [4] Hitachi Ltd., Central Research Lab., Tokyo. — [5] Kokusai Denshin Denwa Co. Ltd., Research and Development Lab., Tokyo.

of Walsh functions into a frequency band limited signal, while a sample-and-hold circuit will reconvert it into its original shape.

One of the most promising areas of application of Walsh functions is in multiplexing. Several experimental multiplexing systems have been developed that make use of the rather different technology[1]. The emphasis has been lately on the design of systems that are compatible with existing frequency or time multiplexing equipment. Several solutions to this problem have been found: a method for the transmission of digital signals matched to the requirements of the switched telephone plant; a mixed sequency-frequency method for the multiplexing of the usual 12 telephone channels into a group; and a digital multiplexing method that is compatible with digital cables but avoids the poor use of the amplifier power at low activity factors that is characteristic for time multiplexing.

Most existing communication channels favor sine-cosine functions. This is so because of technological advantages and not because of laws of nature. Cables or open wire lines that could not, nor need not, transmit sinusoidal functions have always existed. The telegraph lines of the 19th century, using electromechanical relays as amplifiers, were such lines, and they have recently made a comback as digital cables. Radio channels are allocated according to the frequency of sinusoidal functions by human laws. It will be shown in some detail that they could be allocated just as well according to the sequency of Walsh functions, at least on paper.

Nonsinusoidal Electromagnetic Waves

So far the investigation of nonsinusoidal electromagnetic waves has been restricted essentially to Walsh waves. The primary objective of their study at the present stage of development is the discovery of useful differences between sinusoidal and Walsh waves. The different technology for the implementation of equipment is evidently of great interest. Antennas designed for the radiation of Walsh waves hold the promise of being greatly reduced in size. The design of receivers, both for asynchronous and synchronous demodulation, is known and these contain all the characteristic features of the receivers for sinusoidal waves but use semiconductor circuits instead of coils and capacitors[2].

The mathematical differences lead to a variety of phenomena that go beyond technology. First, the sum of two sine functions with the same frequency but arbitrary phase and amplitude yields a sine function of the same frequency. This feature is the basis for interference phenomena

[1] Experimental sequency multiplex equipment has been developed in Canada, England, Netherlands, Switzerland, USA and West Germany [21].

[2] The first experimental project on radiation, propagation and reception of electromagnetic Walsh waves was started at Stanford Research Institute, Menlo Park, Calif., in September 1971.

ranging from multipath transmission to radiation diagrams of antennas to shaping of radar return signals by the target, and so forth. Second, the derivative of a sinusoidal function is again a sinusoidal function. As a result, sinusoidal electromagnetic waves have the same shape in the near zone and the wave zone; furthermore, the waves of multipole radiation have the same shape as waves of dipole radiation. Walsh waves sum very differently from sinusoidal waves and their shape is changed by differentiation. Hence, effects depending on summation and differentiation are very different from what one is used to from sinusoidal waves.

The general form of a sine function $V \sin(2\pi f t + \alpha)$ contains the parameters amplitude V, frequency f and phase angle α. The general form of a Walsh function $V \operatorname{sal}(\phi T, t/T + t_0/T)$ contains the parameters amplitude V, sequency ϕ, delay t_0 and time base T. The normalized delay, t_0/T, corresponds to the phase angle. The time base T is an additional parameter and it accounts for a major part of the differences in the applications of sine-cosine and Walsh functions. For instance, the Doppler effect changes the sequency exactly the way it changes the frequency. However, the time base is changed so that the product ϕT remains unchanged and it permits a correct identification of the Walsh waves up to relative velocities of the order of half the velocity of light.

Most problems concerning Walsh waves can presently be answered in terms of geometric optic only, since wave optic is a sine wave optic. On the other hand, there is little doubt that nonsinusoidal electromagnetic waves are a challenging field for basic research. The generation of nonsinusoidal radio waves implies that such waves can be generated in the region of visible light, and this leads ultimately to the question of why white light should be decomposed into sinusoidal functions[1].

Statistical Theory of Communication

Fourier analysis forces us to assume that a time signal is infinitely long, or has an infinite bandwidth or is infinite in time and frequency. This is particularly troublesome in the statistical theory of communication which requires at least one more infinite quantity. The substitution of general systems of orthogonal functions for sine-cosine functions and a time-function-domain for the time-frequency-domain brings a well-needed simplification.

Beyond mathematical simplification, Walsh functions provide us with some very practical advantages. A nonlinear compressor will not generate new Walsh functions if a signal consisting of a superposition of Walsh functions is passed through it; the compressor will merely redistribute

[1] Theoretical and experimental work on optical spectroscopy based on Walsh functions or Hadamard matrices was done in England [16] and the United States [14, 15, 19]. A scintillation counter for nuclear spectroscopy was developed in the Soviet Union [20].

the energy among the functions already present[1]. Jain pointed out that crosstalk in a multiplex system due to nonlinearities, e.g., the hard limiter in a satellite repeater, can be avoided by using Walsh functions as carriers [13].

In coding theory, Walsh functions have been known under the name Reed-Muller codes for about 20 years, and they are among the distinguished few codes that were actually used. Despite this early use in coding, the statistical theory of communication using general systems of orthogonal functions and Walsh functions in particular is in its infancy.

---

[1] The usefulness of Walsh functions for nonlinear processes was recognized by Weiser in the early sixties but his untimely death delayed the use of his work for several years [12].

1. MATHEMATICAL FOUNDATIONS

1.1 Orthogonal Functions

1.1.1 Orthogonality and Linear Independence

A system $\{f(j, x)\}$ of real and almost everywhere nonvanishing functions $f(0, x), f(1, x), \ldots$ is called orthogonal in the interval $x_0 \leqq x \leqq x_1$ if the following condition holds true:

$$\int_{x_0}^{x_1} f(j, x) f(k, x)\, dx = X_j\, \delta_{jk}$$

$$\delta_{jk} = 1 \quad \text{for} \quad j = k, \qquad \delta_{jk} = 0 \quad \text{for} \quad j \neq k \tag{1}$$

The functions are called orthogonal and normalized if the constant X_j is equal 1. The two terms are usually reduced to the single term orthonormal or orthonormalized.

A nonnormalized system of orthogonal functions may always be normalized. For instance, the system $\{X_j^{-1/2} f(j, x)\}$ is normalized, if X_j of Eq. (1) is not equal 1. Systems of orthogonal functions are special cases of systems of linearly independent functions. A system $\{f(j, x)\}$ of m functions is called linearly dependent, if the equation

$$\sum_{j=0}^{m-1} c(j) f(j, x) \equiv 0 \tag{2}$$

is satisfied for all values of x without all constants $c(j)$ being zero. The functions $f(j, x)$ are called linearly independent, if Eq. (2) is not satisfied. Functions of an orthogonal system are always linearly independent, since multiplication of Eq. (2) by $f(k, x)$ and integration of the products in the interval $x_0 \leqq x \leqq x_1$ yields $c(k) = 0$ for each constant $c(j)$.

A system $\{g(j, x)\}$ of m linearly independent functions can always be transformed into a system $\{f(j, x)\}$ of m orthogonal functions. One may write the following equations:

$$\left.\begin{aligned} f(0,x) &= c_{00}\,g(0,x) \\ f(1,x) &= c_{10}\,g(0,x) + c_{11}\,g(1,x) \\ f(2,x) &= c_{20}\,g(0,x) + c_{21}\,g(1,x) + c_{22}\,g(2,x) \\ &\text{etc.} \end{aligned}\right\} \tag{3}$$

Substitution of the $f(j, x)$ into Eq. (1) yields just enough equations for determination of the constants c_{pq}:

$$\begin{aligned} &\int_{x_0}^{x_1} f^2(0,x)\,dx = X_0 \\ &\int_{x_0}^{x_1} f^2(1,x)\,dx = X_1, \quad \int_{x_0}^{x_1} f(0,x)f(1,x)\,dx = 0 \\ &\int_{x_0}^{x_1} f^2(2,x)\,dx = X_2, \quad \int_{x_0}^{x_1} f(0,x)f(2,x)\,dx = 0, \quad \int_{x_0}^{x_1} f(1,x)f(2,x)\,dx = 0 \\ &\text{etc.} \end{aligned} \tag{4}$$

The coefficients $X_0, X_1, \ldots$ are arbitrary. They are 1 for normalized systems. It follows from Eq. (2) that Eq. (4) actually yields values for the coefficients c_{pq} as only a system $\{g(j, x)\}$ of linearly dependent functions could satisfy Eq. (4) identically.

Figures 1 to 3 show examples of orthogonal functions. The independent variable is the normalized time $\theta = t/T$. The functions of Fig. 1 are orthonormal in the interval $-\frac{1}{2} \leqq \theta \leqq \frac{1}{2}$; they will be referred to as sine and cosine elements. One may divide them into even functions $f_c(i, \theta)$, odd functions $f_s(i, \theta)$ and the constant 1 or $\mathrm{wal}(0, \theta)$:

$$\left.\begin{aligned} f(j,\theta) &= f_c(i,\theta) = \sqrt{(2)}\cos 2\pi\, i\,\theta && -\tfrac{1}{2} \leqq \theta \leqq \tfrac{1}{2} \\ &= f_s(i,\theta) = \sqrt{(2)}\sin 2\pi\, i\,\theta \\ &= \mathrm{wal}(0,\theta) = 1 \\ &= \text{undefined} && \theta < -\tfrac{1}{2},\ \theta > +\tfrac{1}{2} \end{aligned}\right\} \tag{5}$$

The term *element* is used to emphasize that a function is defined in a finite interval only and is undefined outside. The term *gulse* is used to emphasize that a function is identical zero outside a finite interval. Continuation of the sine and cosine elements of Fig. 1 outside of the interval $-\frac{1}{2} \leqq \theta \leqq \frac{1}{2}$ by $f(j, \theta) = 0$ yields the sine and cosine pulses; periodic continuation, on the other hand, yields the periodic sine and cosine functions.

It is easy to see, that the condition (1) for orthogonality is satisfied for sine and cosine elements:

$$\int_{-1/2}^{1/2} 1 \sqrt{(2)} \sin 2\pi\, i\, \theta\, d\theta = \int_{-1/2}^{1/2} 1 \sqrt{(2)} \cos 2\pi\, i\, \theta\, d\theta = 0$$

$$\int_{-1/2}^{1/2} \sqrt{(2)} \sin 2\pi\, i\, \theta \cdot \sqrt{(2)} \sin 2\pi\, k\, \theta\, d\theta$$

$$= \int_{-1/2}^{1/2} \sqrt{(2)} \cos 2\pi\, i\, \theta \cdot \sqrt{(2)} \cos 2\pi\, k\, \theta\, d\theta = \delta_{ik}$$

$$\int_{-1/2}^{1/2} \sqrt{(2)} \sin 2\pi\, i\, \theta \cdot \sqrt{(2)} \cos 2\pi\, k\, \theta\, d\theta = 0$$

$$\int_{-1/2}^{1/2} 1 \cdot 1\, d\theta = 1$$

Figure 2 shows the orthonormal system of Walsh functions or—more exactly—Walsh elements, consisting of a constant $\mathrm{wal}(0, \theta)$, even functions $\mathrm{cal}(i, \theta)$ and odd functions $\mathrm{sal}(i, \theta)$. These functions jump back and forth between $+1$ and -1. Consider the product of the first two functions. It is equal -1 in the interval $-\frac{1}{2} \leqq \theta < 0$ and $+1$ in the interval $0 \leqq \theta < < +\frac{1}{2}$. The integral of these products has the following value:

$$\int_{-1/2}^{1/2} \mathrm{wal}(0, \theta)\, \mathrm{wal}(1, \theta)\, d\theta = \int_{-1/2}^{0} (+1)(-1)\, d\theta + \int_{0}^{1/2} (+1)(+1)\, d\theta$$

$$= -\tfrac{1}{2} + \tfrac{1}{2} = 0$$

The product of the second and third element yields $+1$ in the intervals $-\frac{1}{2} \leqq \theta < -\frac{1}{4}$ and $0 \leqq \theta < +\frac{1}{4}$, and -1 in the intervals $-\frac{1}{4} \leqq \theta < 0$ and $+\frac{1}{4} \leqq \theta < +\frac{1}{2}$. The integral of these products again yields zero:

$$\int_{-1/2}^{1/2} \mathrm{wal}(1, \theta)\, \mathrm{wal}(2, \theta)\, d\theta = \int_{-1/2}^{-1/4} (-1)(-1)\, d\theta + \int_{-1/4}^{0} (-1)(+1)\, d\theta +$$

$$+ \int_{0}^{1/4} (+1)(+1)\, d\theta + \int_{1/4}^{1/2} (+1)(-1)\, d\theta = 0$$

One may easily verify that the integral of the product of any two functions is equal zero. A function multiplied with itself yields the products $(+1)(+1)$ or $(-1)(-1)$. Hence, these products have the value 1 in the whole interval $-\frac{1}{2} \leqq \theta < +\frac{1}{2}$ and their integral is 1. The Walsh functions are thus orthonormal.

Figure 3 shows a particularly simple system of orthogonal functions. Evidently, the product between any two functions vanishes and the integrals of the products must vanish too. For normalization the amplitudes of the functions must be $\sqrt{(5)}$.

An example of a linearly independent but not orthogonal system of functions are Bernoulli's polynomials $B_j(x)$ [4, 5]:

$$B_0(x) = 1, \quad B_1(x) = x - \tfrac{1}{2}, \quad B_2(x) = x^2 - x + \tfrac{1}{6}$$
$$B_3(x) = x^3 - \tfrac{3}{2}x^2 + \tfrac{1}{2}x, \quad B_4(x) = x^4 - 2x^3 + x^2 - \tfrac{1}{30}$$

$B_j(x)$ is a polynomial of order j. The condition

$$\sum_{j=0}^{m} c(j)\, B_j(x) \equiv 0$$

can be satisfied for all values of x only if $c(m)\, x^m$ is zero. This implies $c(m) = 0$. Now $c(m-1)\, B_{m-1}(x)$ is the highest term in the sum and the same reasoning can be applied to it. This proves the linear independence of the Bernoulli polynomials.

One may see from Fig. 4 without calculation that the Bernoulli polynomials are not orthogonal. For orthogonalization in the interval $-1 \leqq x < +1$ one may substitute them for $g(j, x)$ in Eq. (3):

$$P_0(x) = B_0(x) = 1$$
$$P_1(x) = c_{10}\, B_0(x) + c_{11}\, B_1(x), \quad \text{etc.}$$

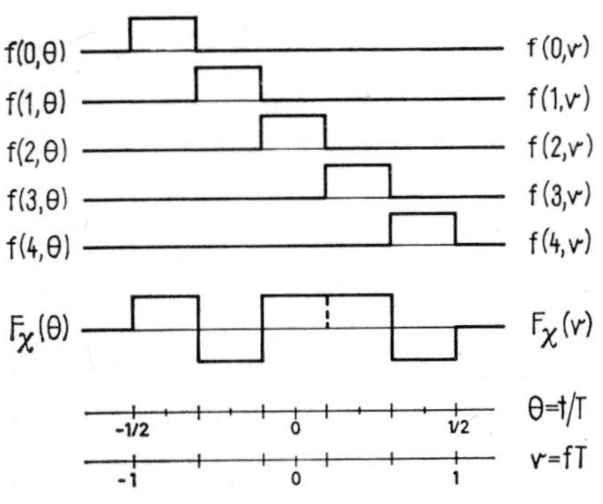

Fig. 3. Orthogonal block pulses $f(j, \theta)$ and $f(j, \nu)$.

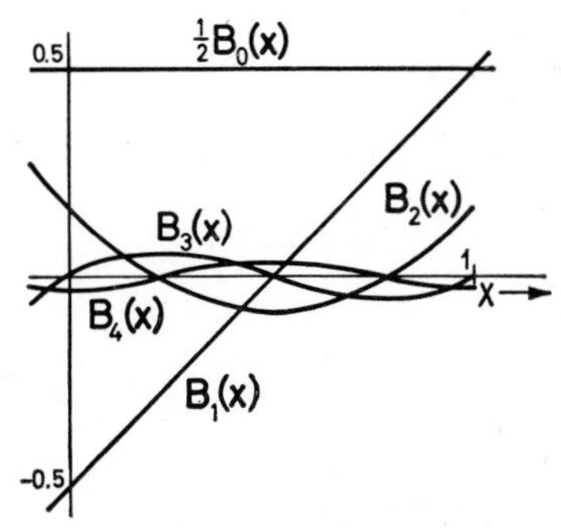

Fig. 4. Bernoulli polynomials.

Using the constants $X_j = 2/(2j + 1)$ one obtains from Eq. (4):

$$\int_{-1}^{1} 1\, dx = X_0 = 2$$

$$\int_{-1}^{1} [c_{10} + c_{11}(x - \tfrac{1}{2})]^2\, dx = X_1 = \tfrac{2}{3}, \quad \int_{-1}^{1} [c_{10} + c_{11}(x - \tfrac{1}{2})]\, dx = 0$$

The coefficients $c_{10} = \frac{1}{2}$, $c_{11} = 1$, etc. are readily obtained. The orthogonal polynomials $P_j(x)$ assume the following form:

$$P_0(x) = 1, \quad P_1(x) = x, \quad P_2(x) = -(3x^2 - 1)$$
$$P_3(x) = \tfrac{1}{2}(5x^3 - 3x), \quad P_4(x) = \tfrac{1}{8}(35x^4 - 30x^2 + 3)$$

These are the Legendre polynomials. $P_j(x)$ must be multiplied with $X_j^{-1/2} = (j + \frac{1}{2})^{1/2}$ for normalization. Figure 5 shows the first five polynomials.

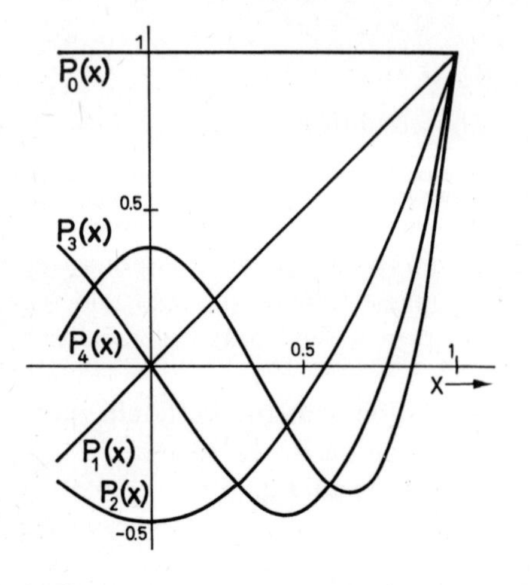

Fig. 5. Legendre polynomials.

Problems

111.1 The polynomials of Tschebyscheff, Hermite, Laguerre, Jacobi and Gegenbauer are not orthogonal but are made orthogonal by multiplication with a weighting function. State the weighting functions and the intervals of orthogonality [6].

111.2 Orthogonal functions are often encountered in the solution of differential equations. State the simplest differential equations that lead to sine-cosine functions, Legendre polynomials and Bessel functions.

111.3 Legendre polynomials are a special case of Legendre functions. Which differential equation defines Legendre functions?

1.1.2 Series Expansion by Orthogonal Functions

Let a function $F(x)$ be expanded in a series of the orthonormal system $\{f(j, x)\}$:

$$F(x) = \sum_{j=0}^{\infty} a(j) f(j, x) \tag{1}$$

The value of the coefficients $a(j)$ may be obtained by multiplying Eq. (1) by $f(k, x)$ and integrating the products in the interval of orthogonality $x_0 \leqq x < x_1$:

$$\int_{x_0}^{x_1} F(x) f(k, x)\, dx = a(k) \tag{2}$$

How well is $F(x)$ represented, if the coefficients $a(j)$ are determined by Eq. (2)? Let us assume a series $\sum b(j) f(j, x)$ having m terms yields a better representation. The criterion for "better" shall be the least mean square deviation Q of $F(x)$ from its representation:

$$Q = \int_{x_0}^{x_1} \left[F(x) - \sum_{j=0}^{m-1} b(j) f(j, x)\right]^2 dx$$

$$= \int_{x_0}^{x_1} F^2(x)\, dx - 2\sum_{j=0}^{m-1} b(j) \int_{x_0}^{x_1} F(x) f(j, x)\, dx + \int_{x_0}^{x_1} \left[\sum_{j=0}^{m-1} b(j) f(j, x)\right]^2 dx$$

Using Eq. (2) and the orthogonality of the functions $f(j, x)$ yields Q in the following form:

$$Q = \int_{x_0}^{x_1} F^2(x)\, dx - \sum_{j=0}^{m-1} a^2(j) + \sum_{j=0}^{m-1} [b(j) - a(j)]^2 \tag{3}$$

The last term vanishes for $b(j) = a(j)$ and the mean square deviation assumes its minimum.

The so-called Bessel inequality follows from Eq. (3):

$$\sum_{j=0}^{m-1} a^2(j) \leqq \sum_{j=0}^{\infty} a^2(j) \leqq \int_{x_0}^{x_1} F^2(x)\, dx \tag{4}$$

The upper limit of summation may be ∞ instead of $m - 1$, since the integral does not depend on m and must thus hold for any value of m.

The system $\{f(j, x)\}$ is called orthogonal, normalized and complete, if the mean square deviation Q converges to zero with increasing m for any function $F(x)$ that is quadratically integrable in the interval $x_0 \leqq x < x_1$:

$$\lim_{m\to\infty} \int_{x_0}^{x_1} \left[F(x) - \sum_{j=0}^{m-1} a(j) f(j, x)\right]^2 dx = 0 \tag{5}$$

The equality sign holds in this case in the Bessel inequality (4):

$$\sum_{j=0}^{\infty} a^2(j) = \int_{x_0}^{x_1} F^2(x)\, dx \tag{6}$$

Equation (6) is known as completeness theorem or Parseval's theorem. Its physical meaning is as follows: Let $F(x)$ represent a voltage as function of time across a unit resistance. The integral of $F^2(x)$ represents then the energy dissipated in the resistor. This energy equals, according to Eq. (6), the sum of the energy of the terms in the sum $\sum a(j) f(j, x)$. Putting it differently, the energy is the same whether the voltage is described by the time function $F(x)$ or its series expansion.

The system $\{f(j, x)\}$ is said to be closed[1], if there is no quadratically integrable function $F(x)$,

$$\int_{x_0}^{x_1} F^2(x)\, dx < \infty \tag{7}$$

[1] A complete orthonormal system is always closed. The inverse of this statement holds true if the integrals of this section are Lebesgue rather than Riemann integrals. The Riemann integral suffices for the major part of this book. Hence, integrable will mean Riemann integrable unless otherwise stated.

for which the equality

$$\int_{x_0}^{x_1} F(x) f(j, x)\, dx = 0 \tag{8}$$

is satisfied for all values of j.

Incomplete systems of orthogonal functions do not permit a convergent series expansion of all quadratically integrable functions. Nevertheless, they are of great practical interest. For instance, the output voltage of an ideal frequency low-pass filter may be represented exactly by an expansion in a series of the incomplete orthogonal system of $(\sin x)/x$ functions.

Whether a certain function $F(x)$ can be expanded in a series of a particular orthogonal system $\{f(j, k)\}$ cannot be told from such simple features of $F(x)$ as its continuity or boundedness[1] [5–7].

Problems

112.1 $F(x)$ represents a time-varying voltage across a resistor and is quadratically integrable. What is the physical meaning of this statement?

112.2 Can there be a signal in communications represented by a time-varying voltage that is not quadratically integrable?

112.3 A function $F(x)$ is zero for all rational values of x in the interval $0 < x < 1$ and 1 for all irrational values. The Lebesgue integral yields 1. Does a Riemann integral exist? Can you plot this function? Let $F(x)$ be 1 for all real values of x in the interval $0 < x < 1$. Can $F(x)$ now be plotted and Riemann-integrated?

1.1.3 Invariance of Orthogonality to Fourier Transformation

A time function $f(j, \theta)$ may be represented under certain conditions by two functions $a(j, \nu)$ and $b(j, \nu)$ by means of the Fourier transform:

$$f(j, \theta) = \int_{-\infty}^{\infty} [a(j, \nu) \cos 2\pi \nu \theta + b(j, \nu) \sin 2\pi \nu \theta]\, d\nu \tag{1}$$

$$a(j, \nu) = \int_{-\infty}^{\infty} f(j, \theta) \cos 2\pi \nu \theta\, d\theta$$

$$b(j, \nu) = \int_{-\infty}^{\infty} f(j, \theta) \sin 2\pi \nu \theta\, d\theta \qquad \theta = t/T,\ \nu = fT \tag{2}$$

It is advantageous for certain applications to replace the two functions $a(j, \nu)$ and $b(j, \nu)$ by a single function[2]:

$$g(j, \nu) = a(j, \nu) + b(j, \nu) \tag{3}$$

---

[1] For instance, the Fourier series of a continuous function does not have to converge in every point. A theorem due to Banach states that there are arbitrarily many orthogonal systems with the feature, that the orthogonal series of a continuously differentiable function diverges almost everywhere.

[2] Real notation is used for the Fourier transform to facilitate comparison with the formulas of the generalized Fourier transform derived later on.

It follows from Eq. (2) that $a(j, \nu)$ is an even and $b(j, \nu)$ an odd function of ν:

$$a(j, \nu) = a(j, -\nu), \qquad b(j, \nu) = -b(j, -\nu) \tag{4}$$

Equations (3) and (4) yield for $g(j, -\nu)$:

$$g(j, -\nu) = a(j, -\nu) + b(j, -\nu) = a(j, \nu) - b(j, \nu) \tag{5}$$

$a(j, \nu)$ and $b(j, \nu)$ may be regained from $g(j, \nu)$ by means of Eqs. (3) and (5):

$$a(j, \nu) = \tfrac{1}{2}[g(j, \nu) + g(j, -\nu)] \tag{6}$$

$$b(j, \nu) = \tfrac{1}{2}[g(j, \nu) - g(j, -\nu)]$$

Using the function $g(j, \nu)$ one may write Eqs. (1) and (2) in a more symmetric form:

$$f(j, \theta) = \int_{-\infty}^{\infty} g(j, \nu)\,(\cos 2\pi\,\nu\,\theta + \sin 2\pi\,\nu\,\theta)\,d\nu \tag{7}$$

$$g(j, \nu) = \int_{-\infty}^{\infty} f(j, \theta)\,(\cos 2\pi\,\nu\,\theta + \sin 2\pi\,\nu\,\theta)\,d\theta \tag{8}$$

The integrals of $b(j, \nu)\cos 2\pi\,\nu\,\theta$ and $a(j, \nu)\sin 2\pi\,\nu\,\theta$ in Eq. (7) vanish since $a(j, \nu)$ is an even and $b(j, \nu)$ is an odd function of ν.

Let $\{f(j, \theta)\}$ be a system orthonormal in the interval $-\frac{1}{2}\Theta \leqq \theta \leqq +\frac{1}{2}\Theta$ and zero outside. Θ may be finite or infinite. The functions $f(j, \theta)$ are Fourier transformable[1]. Their orthogonality integral,

$$\int_{-\infty}^{\infty} f(j, \theta) f(k, \theta)\,d\theta = \delta_{jk} \tag{9}$$

may be rewritten[2] using Eq. (7):

$$\left.\begin{aligned}
\int_{-\infty}^{\infty} f(j, \theta)\left[\int_{-\infty}^{\infty} g(k, \nu)\,(\cos 2\pi\,\nu\,\theta + \sin 2\pi\,\nu\,\theta)\,d\nu\right] d\theta &= \delta_{jk}\\
\int_{-\infty}^{\infty} g(k, \nu)\left[\int_{-\infty}^{\infty} f(j, \theta)\,(\cos 2\pi\,\nu\,\theta + \sin 2\pi\,\nu\,\theta)\,d\theta\right] d\nu &= \delta_{jk}\\
\int_{-\infty}^{\infty} g(j, \nu)\,g(k, \nu)\,d\nu &= \delta_{jk}
\end{aligned}\right\} \tag{10}$$

Hence, the Fourier transform of an orthonormal system $\{f(j, \theta)\}$ yields an orthonormal system $\{g(j, \nu)\}$.

---

[1] Orthonormality implies the existence of the Fourier transform and the inverse transform (Plancherel theorem).

[2] The integrations may be interchanged, since the integrands are absolutely integrable.

Substitution of

$$g(j,\nu) = a(j,\nu) + b(j,\nu), \qquad g(k,\nu) = a(k,\nu) + b(k,\nu)$$

into Eq. (10) yields it in terms of the notation $a(j,\nu)$, $b(j,\nu)$:

$$\int_{-\infty}^{\infty} g(j,\nu)\, g(k,\nu)\, d\nu = \int_{-\infty}^{\infty} [a(j,\nu) + b(j,\nu)]\,[a(k,\nu) + b(k,\nu)]\, d\nu$$

$$= \int_{-\infty}^{\infty} [a(j,\nu)\, a(k,\nu) + b(j,\nu)\, b(k,\nu)]\, d\nu = \delta_{jk}$$

Figure 6 shows as an example the Fourier transforms of sine and cosine pulses. These pulses are derived from the elements of Fig. 1 by continuing them identical to zero outside the interval $-\frac{1}{2} \leqq \theta \leqq +\frac{1}{2}$:

$$\left.\begin{aligned}
g(0,\nu) &= \int_{-1/2}^{1/2} 1(\cos 2\pi\nu\theta + \sin 2\pi\nu\theta)\, d\theta = \frac{\sin \pi\nu}{\pi\nu} \\
g_c(i,\nu) &= \int_{-1/2}^{1/2} \sqrt{(2)} \cos 2\pi i\theta (\cos 2\pi\nu\theta + \sin 2\pi\nu\theta)\, d\theta \\
&= \frac{1}{2}\sqrt{(2)} \left(\frac{\sin \pi(\nu - i)}{\pi(\nu - i)} + \frac{\sin \pi(\nu + i)}{\pi(\nu + i)} \right) \\
g_s(i,\nu) &= \int_{-1/2}^{1/2} \sqrt{(2)} \sin 2\pi i\theta (\cos 2\pi\nu\theta + \sin 2\pi\nu\theta)\, d\theta \\
&= \frac{1}{2}\sqrt{(2)} \left(\frac{\sin \pi(\nu - i)}{\pi(\nu - i)} - \frac{\sin \pi(\nu + i)}{\pi(\nu + i)} \right)
\end{aligned}\right\} \quad (11)$$

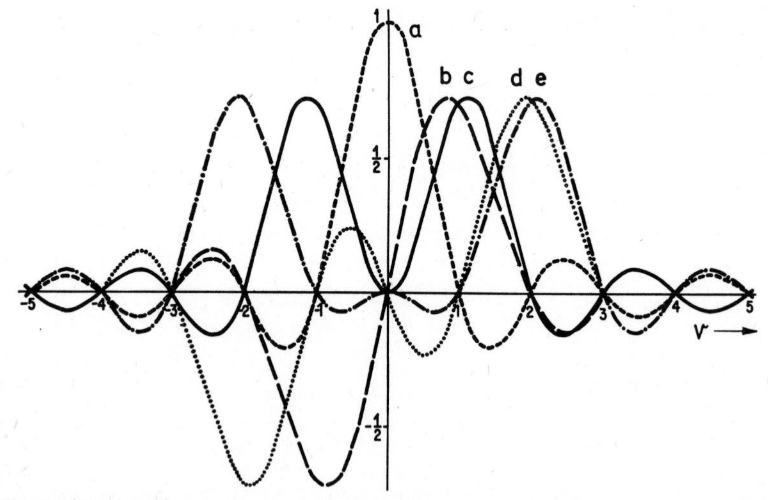

Fig. 6. Fourier transforms $g(j,\nu)$ of sine and cosine pulses according to Fig. 1. *a*, wal$(0,\theta)$; *b*, $\sqrt{(2)} \sin 2\pi\theta$; *c*, $\sqrt{(2)} \cos 2\pi\theta$; *d*, $\sqrt{(2)} \sin 4\pi\theta$; *e*, $\sqrt{(2)} \cos 4\pi\theta$.

Table 1 shows the Fourier transforms of Walsh pulses derived by continuing the elements $\mathrm{sal}(i, \theta)$ of Fig. 2 identical to zero outside the interval $-\frac{1}{2} \leqq \theta \leqq +\frac{1}{2}$:

$$g_s(i, \nu) = \int_{-1/2}^{1/2} \mathrm{sal}(i, \theta) \sqrt{(2)} \sin 2\pi \nu \theta \, d\theta$$

Table 1. *Fourier transforms of* $\mathrm{sal}(i, \theta)$, $i = 1, \ldots, 16$.

$$g_s(1, \nu) = +2\sqrt{(2)} \frac{1}{\pi \nu} \sin^2 \frac{\pi \nu}{2}$$

$$g_s(2, \nu) = -4\sqrt{(2)} \frac{1}{\pi \nu} \cos \frac{\pi \nu}{2} \sin^2 \frac{\pi \nu}{4}$$

$$g_s(3, \nu) = -8\sqrt{(2)} \frac{1}{\pi \nu} \sin \frac{\pi \nu}{2} \sin \frac{\pi \nu}{4} \sin^2 \frac{\pi \nu}{8}$$

$$g_s(4, \nu) = -8\sqrt{(2)} \frac{1}{\pi \nu} \cos \frac{\pi \nu}{2} \cos \frac{\pi \nu}{4} \sin^2 \frac{\pi \nu}{8}$$

$$g_s(5, \nu) = -16\sqrt{(2)} \frac{1}{\pi \nu} \sin \frac{\pi \nu}{2} \cos \frac{\pi \nu}{4} \sin \frac{\pi \nu}{8} \sin^2 \frac{\pi \nu}{16}$$

$$g_s(6, \nu) = +16\sqrt{(2)} \frac{1}{\pi \nu} \cos \frac{\pi \nu}{2} \sin \frac{\pi \nu}{4} \sin \frac{\pi \nu}{8} \sin^2 \frac{\pi \nu}{16}$$

$$g_s(7, \nu) = -16\sqrt{(2)} \frac{1}{\pi \nu} \sin \frac{\pi \nu}{2} \sin \frac{\pi \nu}{4} \cos \frac{\pi \nu}{8} \sin^2 \frac{\pi \nu}{16}$$

$$g_s(8, \nu) = -16\sqrt{(2)} \frac{1}{\pi \nu} \cos \frac{\pi \nu}{2} \cos \frac{\pi \nu}{4} \cos \frac{\pi \nu}{8} \sin^2 \frac{\pi \nu}{16}$$

$$g_s(9, \nu) = -32\sqrt{(2)} \frac{1}{\pi \nu} \sin \frac{\pi \nu}{2} \cos \frac{\pi \nu}{4} \cos \frac{\pi \nu}{8} \sin \frac{\pi \nu}{16} \sin^2 \frac{\pi \nu}{32}$$

$$g_s(10, \nu) = +32\sqrt{(2)} \frac{1}{\pi \nu} \cos \frac{\pi \nu}{2} \sin \frac{\pi \nu}{4} \cos \frac{\pi \nu}{8} \sin \frac{\pi \nu}{16} \sin^2 \frac{\pi \nu}{32}$$

$$g_s(11, \nu) = +32\sqrt{(2)} \frac{1}{\pi \nu} \sin \frac{\pi \nu}{2} \sin \frac{\pi \nu}{4} \sin \frac{\pi \nu}{8} \sin \frac{\pi \nu}{16} \sin^2 \frac{\pi \nu}{32}$$

$$g_s(12, \nu) = +32\sqrt{(2)} \frac{1}{\pi \nu} \cos \frac{\pi \nu}{2} \cos \frac{\pi \nu}{4} \sin \frac{\pi \nu}{8} \sin \frac{\pi \nu}{16} \sin^2 \frac{\pi \nu}{32}$$

$$g_s(13, \nu) = +32\sqrt{(2)} \frac{1}{\pi \nu} \sin \frac{\pi \nu}{2} \cos \frac{\pi \nu}{4} \sin \frac{\pi \nu}{8} \cos \frac{\pi \nu}{16} \sin^2 \frac{\pi \nu}{32}$$

$$g_s(14, \nu) = +32\sqrt{(2)} \frac{1}{\pi \nu} \cos \frac{\pi \nu}{2} \sin \frac{\pi \nu}{4} \sin \frac{\pi \nu}{8} \cos \frac{\pi \nu}{16} \sin^2 \frac{\pi \nu}{32}$$

$$g_s(15, \nu) = +32\sqrt{(2)} \frac{1}{\pi \nu} \sin \frac{\pi \nu}{2} \sin \frac{\pi \nu}{4} \cos \frac{\pi \nu}{8} \cos \frac{\pi \nu}{16} \sin^2 \frac{\pi \nu}{32}$$

$$g_s(16, \nu) = -32\sqrt{(2)} \frac{1}{\pi \nu} \cos \frac{\pi \nu}{2} \cos \frac{\pi \nu}{4} \cos \frac{\pi \nu}{8} \cos \frac{\pi \nu}{16} \sin^2 \frac{\pi \nu}{32}$$

Table 1 gives the Fourier series expansion of the periodic functions $\mathrm{sal}(i, \theta)$ if ν assumes the values $1, 2, \ldots$ only[1]. Figure 7 shows plots of the Fourier transforms of several Walsh pulses.

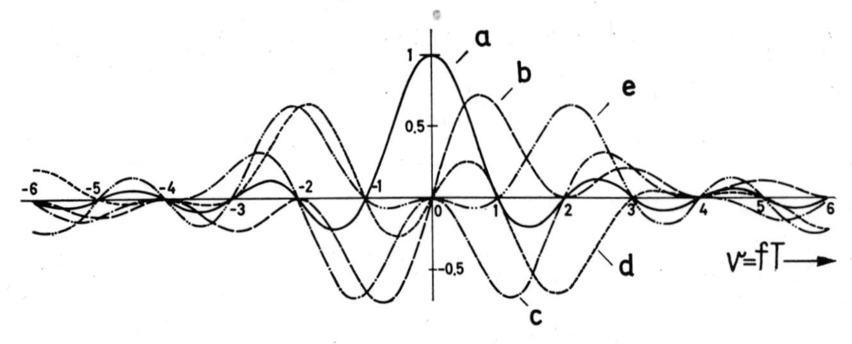

Fig. 7. Fourier transforms $g(j, \nu)$ of Walsh pulses according to Fig. 2. a, $\mathrm{wal}(0, \theta)$; b, $\mathrm{sal}(1, \theta)$; c, $-\mathrm{cal}(1, \theta)$; d, $-\mathrm{sal}(2, \theta)$; e, $\mathrm{cal}(2, \theta)$.

One may readily see from these examples that even time functions transform into even frequency functions and odd time functions transform into odd frequency functions. Negative values of the frequency have a perfectly valid physical meaning. The oscillation of frequency ν is a cosine oscillation with reference to $\theta = 0$, if the Fourier transform has the same value for $+\nu$ and $-\nu$; it is a sine oscillation, if the Fourier transform has the same absolute value but opposite sign for $+\nu$ and $-\nu$.

Figure 8 shows the Fourier transforms $g(j, \nu)$ of three block pulses of Fig. 3. They are no longer either even or odd[2].

Figure 9 shows a system of orthogonal sine and cosine pulses. They are time shifted compared with those of Fig. 1, so that all functions have jumps of equal magnitude at $\theta = -\frac{1}{2}$ and $\theta = +\frac{1}{2}$. Their Fourier transforms $g(j, \nu)$ are shown in Fig. 10:

$$g(j, \nu) = \frac{\sin \pi(\nu - k)}{\pi(\nu - k)}, \quad k = -\frac{1}{2} j \text{ for even } j$$
$$k = \frac{1}{2}(j + 1) \text{ for odd } j \tag{12}$$

---

[1] Compare this Table 1 with Table 1 of [7] showing the Walsh series expansion of sine and cosine functions. Plots of some of the functions in Table 1 without the weighting function $(\sin^2 x)/x$ were published by Filipowsky [9]. A closed formula for the Fourier series of Walsh functions is contained in an unpublished report by Simmons [10].

[2] The Fourier transforms of the various block pulses are different but their frequency power spectra are equal. The power spectrum is the Fourier transform of the auto-correlation function of a function, and not the Fourier transform of the function itself (Wiener-Khintchine theorem). The connection between Fourier transform, power spectrum and amplitude spectrum is discussed in section 1.3.2. *See* also [4].

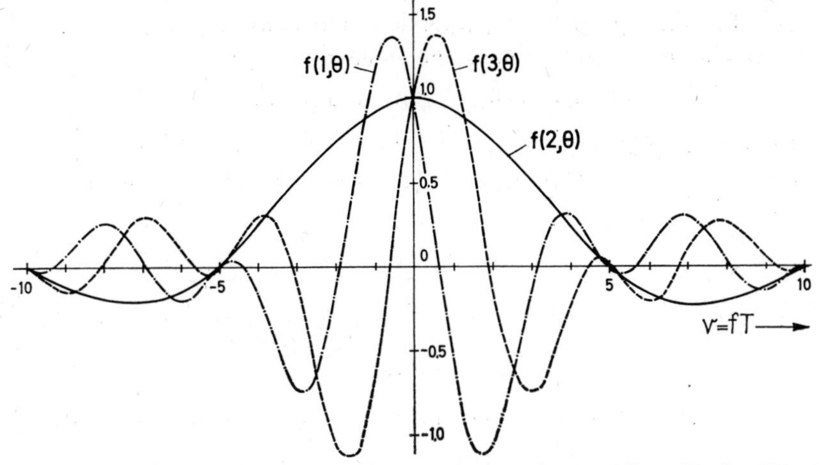

Fig. 8. Fourier transforms $g(j, \nu)$ of the block pulses $f(1, \theta)$, $f(2, \theta)$ and $f(3, \theta)$ of Fig. 3.

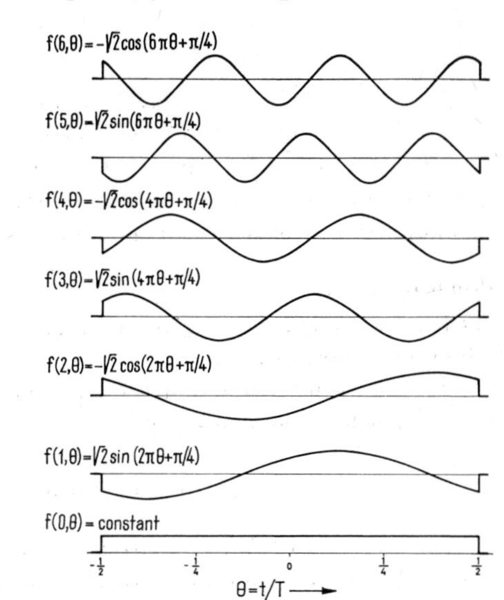

Fig. 9. Orthonormal system of sine and cosine pulses having jumps of equal height at $\theta = -\frac{1}{2}$ and $\theta = +\frac{1}{2}$.

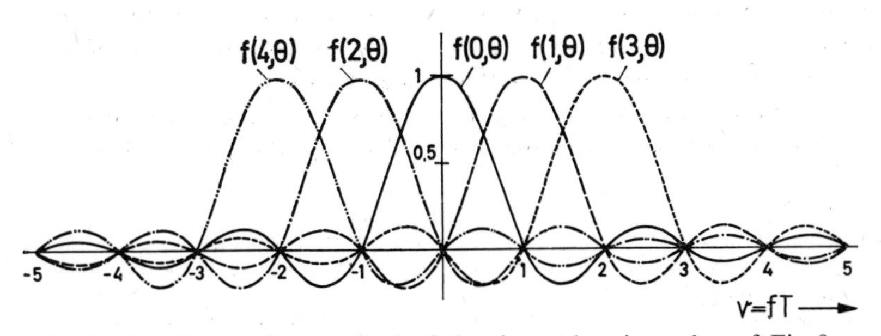

Fig. 10. Fourier transforms $g(j, \nu)$ of the sine and cosine pulses of Fig. 9.

The functions $\psi_j(\theta)$ of the parabolic cylinder shown in Fig. 11 and their Fourier transforms $g(j, \nu)$ have the same shape [5]:

$$\left.\begin{array}{c} f(j,\theta) = \psi_j(\theta), \quad g(0,\nu) = \psi_0(4\pi\nu) \\ g(2i,\nu) = (-1)^i \psi_{2i}(4\pi\nu), \quad g(2i+1,\nu) = -(-1)^i \psi_{2i-1}(4\pi\nu) \\ j = 0, 2i, 2i-1; \quad i = 1, 2, \ldots \end{array}\right\} \tag{13}$$

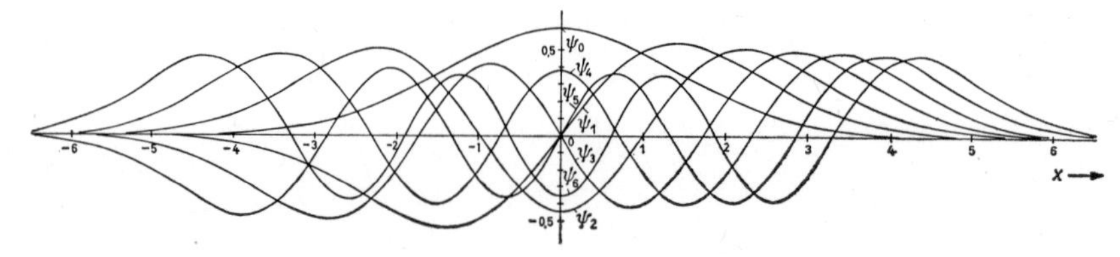

Fig. 11. The functions $\psi_j = \psi_j(\theta)$ or $\psi_j = \psi_j(4\pi\nu)$ of the parabolic cylinder.

$$\psi_j(x) = \frac{e^{-\frac{1}{4}x^2}}{\sqrt{(j!\sqrt{(2\pi)})}} \mathrm{He}_j(x); \quad \mathrm{He}_j(x) = e^{\frac{1}{2}x^2}\left(-\frac{d}{dx}\right)^j e^{-\frac{1}{2}x^2}$$

$x = \theta$ or $4\pi\nu$; $j = 0, 2i, 2i-1$; $i = 1, 2, \ldots$ $\mathrm{He}_j(x)$ = Hermite polynomials.

$\psi_j(\theta)$ decreases for large absolute values of θ proportionately to $\theta^j \exp(-\frac{1}{4}\theta^2)$ and $\psi_j(4\pi\nu)$ decreases for large absolute values of ν proportionately to $(4\pi\nu)^j \exp[-\frac{1}{4}(4\pi\nu)^2]$. Pulses with the shape of parabolic cylinder functions require a particularly small part of the time-frequency-domain for transmission of a certain percentage of their energy[1].

Problems

113.1 Use Table 1 to write the Fourier transform of sal(17, θ), sal(32, θ) and sal(33, θ); ignore the sign.

113.2 Use complex notation for the Fourier transform to derive Eq. (10).

113.3 Which differential equation leads to parabolic cylinder functions?

1.1.4 Walsh Functions

The Walsh functions wal(0, θ), sal(i, θ) and cal(i, θ) are of considerable interest in communications[2]. There is a close connection between sal and sine functions, as well as between cal and cosine functions. The letters s and c in sal and cal were chosen to indicate this connection, while the letters *al* are derived from the name Walsh.

For computational purposes it is sometimes more convenient to use sine and cosine functions, while at other times the exponential function is

[1] Pulses of the shape of parabolic cylinder functions use the time-frequency-domain theoretically "best". This has not been of much practical value so far, since sine-cosine pulses and pulses derived from sine-cosine pulses are almost as good, but much easier to generate and detect.

[2] The probably oldest use of Walsh functions in communications is for the transposition of conductors [18].

more convenient. A similar duality of notation exists for Walsh functions. A single function $\mathrm{wal}(j, \theta)$ may be defined instead of the three functions $\mathrm{wal}(0, \theta)$, $\mathrm{sal}(i, \theta)$ and $\mathrm{cal}(i, \theta)$:

$$\mathrm{wal}(2i, \theta) = \mathrm{cal}(i, \theta), \quad \mathrm{wal}(2i - 1, \theta) = \mathrm{sal}(i, \theta) \qquad i = 1, 2, \ldots \tag{1}$$

The functions $\mathrm{wal}(j, \theta)$ may be defined by the following difference equation[1,2]:

$$\left.\begin{array}{l} \mathrm{wal}(2j + p, \theta) \\ \quad = (-1)^{[j/2]+p}\{\mathrm{wal}[j, 2(\theta + \frac{1}{4})] + (-1)^{j+p}\,\mathrm{wal}[j, 2(\theta - \frac{1}{4})]\} \\ p = 0 \text{ or } 1; \quad j = 0, 1, 2, \ldots; \quad \mathrm{wal}(0, \theta) = 1 \quad \text{for} \quad -\frac{1}{2} \leqq \theta < \frac{1}{2}; \\ \qquad \mathrm{wal}(0, \theta) = 0 \quad \text{for} \quad \theta < -\frac{1}{2},\ \theta \geqq +\frac{1}{2} \end{array}\right\} \tag{2}$$

For explanation of this difference equation consider the function $\mathrm{wal}(j, \theta)$. The function $\mathrm{wal}(j, 2\theta)$ has the same shape, but is squeezed into the interval $-\frac{1}{4} \leqq \theta < +\frac{1}{4}$. $\mathrm{wal}[j, 2(\theta + \frac{1}{4})]$ is obtained by shifting $\mathrm{wal}(j, 2\theta)$ to the left into the interval $-\frac{1}{2} \leqq \theta < 0$, and $\mathrm{wal}[j, 2(\theta - \frac{1}{4})]$ is obtained by shifting $\mathrm{wal}(j, 2\theta)$ to the right into the interval $0 \leqq \theta < +\frac{1}{2}$.

As an example consider the cases $j = 0$, $p = 1$ and $j = 2$, $p = 1$. Using the values $[0/2] = 0$ and $[2/2] = 1$ one obtains:

$$\mathrm{wal}(1, \theta) = (-1)^{0+1}\{\mathrm{wal}[0, 2(\theta + \tfrac{1}{4})] + (-1)^{0+1}\,\mathrm{wal}[0, 2(\theta - \tfrac{1}{4})]\}$$
$$\mathrm{wal}(5, \theta) = (-1)^{1+1}\{\mathrm{wal}[2, 2(\theta + \tfrac{1}{4})] + (-1)^{2+1}\,\mathrm{wal}[2, 2(\theta + \tfrac{1}{4})]\}$$

It may be verified from Fig. 2 that $\mathrm{wal}(1, \theta) = \mathrm{sal}(1, \theta)$ is obtained from $\mathrm{wal}(0, \theta)$ by squeezing it to half its width, multiplying the function that is shifted to the left by -1, and the function that is shifted to the right by $+1$. $\mathrm{wal}(5, \theta) = \mathrm{sal}(3, \theta)$ is obtained by squeezing $\mathrm{wal}(2, \theta) = \mathrm{cal}(1, \theta)$ to half its width, multiplying the function that is shifted to the left by $+1$ and the function that is shifted to the right by -1.

The product of two Walsh functions yields another Walsh function:

$$\mathrm{wal}(h, \theta)\,\mathrm{wal}(k, \theta) = \mathrm{wal}(r, \theta)$$

This relation may readily be proved by writing the difference equation for $\mathrm{wal}(h, \theta)$ and $\mathrm{wal}(k, \theta)$, and multiplying them with each other. It turns out that the product $\mathrm{wal}(h, \theta)\,\mathrm{wal}(k, \theta)$ satisfies a difference equation of the same form as Eq. (2).

---

[1] Walsh functions are usually defined by products of Rademacher functions. This definition has many advantages but does not yield the Walsh functions ordered by the number of sign changes as does the difference equation. This order is important for the generalization of frequency in section 1.3.1. Rademacher functions are the functions $\mathrm{sal}(1, \theta)$, $\mathrm{sal}(2, \theta)$, $\mathrm{sal}(4, \theta)$, ... in Fig. 2.

[2] $[j/2]$ means the largest integer smaller or equal $\frac{1}{2}j$.

The determination of the value of r from the difference equation is somewhat cumbersome. The result is that r equals the modulo 2 sum of h and k:

$$\mathrm{wal}(h, \theta)\,\mathrm{wal}(k, \theta) = \mathrm{wal}(h \oplus k, \theta) \tag{3}$$

The sign $\oplus$ stands for an addition modulo 2. k and h are written as binary numbers and added according to the rules $0 \oplus 1 = 1 \oplus 0 = 1$, $0 \oplus 0 = 1 \oplus 1 = 0$ (no carry). Addition modulo 2 is what a half adder does in binary digital computers. As an example, consider the multiplication of $\mathrm{wal}(6, \theta)$ and $\mathrm{wal}(12, \theta)$. Using binary numbers for 6 and 12 one obtains 10 for the sum $6 \oplus 12$.

$$\begin{array}{rl} 0110 \ldots & 6 \\ \oplus\ \underline{1100} \ldots & 12 \\ 1010 \ldots & 10 \end{array}$$

It may be verified from Fig. 2 that the product $\mathrm{wal}(6, \theta)\,\mathrm{wal}(12, \theta)$ equals $\mathrm{wal}(10, \theta)$.

The product of a Walsh function with itself yields $\mathrm{wal}(0, \theta)$, since only the products $(+1)(+1)$ and $(-1)(-1)$ occur.

$$\mathrm{wal}(j, \theta)\,\mathrm{wal}(j, \theta) = \mathrm{wal}(0, \theta) \tag{4}$$

$$j \oplus j = 0$$

The product of $\mathrm{wal}(j, \theta)$ with $\mathrm{wal}(0, \theta)$ leaves $\mathrm{wal}(j, \theta)$ unchanged:

$$\mathrm{wal}(j, \theta)\,\mathrm{wal}(0, \theta) = \mathrm{wal}(j, \theta) \tag{5}$$

$$j \oplus 0 = j$$

The multiplication of Walsh functions is associative since only products of $+1$ and -1 occur:

$$[\mathrm{wal}(h, \theta)\,\mathrm{wal}(j, \theta)]\,\mathrm{wal}(k, \theta) = \mathrm{wal}(h, \theta)\,[\mathrm{wal}(j, \theta)\,\mathrm{wal}(k, \theta)] \tag{6}$$

Walsh functions form a group with respect to multiplication. Equation (3) shows that the product of two functions yields again a Walsh function; the inverse element is defined by Eq. (4) and is equal to the element itself; the unit element is $\mathrm{wal}(0, \theta)$ according to Eq. (5); the associative law is shown to hold by Eq. (6). The group of Walsh functions is an Abelian or commutative group since the factors in Eqs. (3), (4) and (5) may be commuted. Mathematically speaking, the group of Walsh functions is isomorphic to the discrete dyadic group.

To determine the number of elements in a group and its subgroups, consider what numbers can occur if two numbers, k and h, that are both smaller or equal $2^s - 1$, are added modulo 2. k and h are written as binary numbers:

$$\left.\begin{aligned} h &= p_{s-1}\cdot 2^{s-1} + p_{s-2}\cdot 2^{s-2} + \cdots + p_1\cdot 2^1 + p_0\cdot 2^0 \leqq 2^s - 1 \\ k &= q_{s-1}\cdot 2^{s-1} + q_{s-2}\cdot 2^{s-2} + \cdots + q_1\cdot 2^1 + q_0\cdot 2^0 \leqq 2^s - 1 \\ & p_0, \ldots, p_{s-1}, q_0, \ldots, q_{s-1} = 0 \text{ or } 1 \end{aligned}\right\} \quad (7)$$

The modulo 2 sum of h and k yields:

$$h \oplus k = (p_{s-1} \oplus q_{s-1})\cdot 2^{s-1} + \cdots + (p_0 \oplus q_0)\cdot 2^0 \quad (8)$$

The smallest number occurs if all the factors in front of the powers of 2 are zero. This number is obtained for $h = k$ and equals 0. The largest number is obtained if all these factors are 1; the resulting number,

$$2^{s-1} + 2^{s-2} + \cdots + 2^1 + 2^0 = 2^s - 1,$$

is obtained for $h = (2^s - 1) \oplus k$. This means, that in binary notation k has zeros where h has ones and vice versa. A group thus contains the Walsh functions $\text{wal}(0, \theta)$ to $\text{wal}(2^s - 1, \theta)$, a total of 2^s functions. Subgroups contain the functions $\text{wal}(0, \theta)$ to $\text{wal}(2^r - 1, \theta)$, $0 \leqq r < s$. These are all subgroups. Since a subgroup contains 2^r elements it has $2^s/2^r = 2^{s-r}$ cosets. Evidently, powers of 2 play an important role for Walsh functions.

Using Eq. (1) one may rewrite the multiplication theorem (3) of the Walsh functions as follows:

$$\left.\begin{aligned} \text{cal}(i, \theta)\,\text{cal}(k, \theta) &= \text{cal}(i \oplus k, \theta) \\ \text{sal}(i, \theta)\,\text{cal}(k, \theta) &= \text{sal}\{[k \oplus (i-1)] + 1, \theta\} \\ \text{cal}(i, \theta)\,\text{sal}(k, \theta) &= \text{sal}\{[i \oplus (k-1)] + 1, \theta\} \\ \text{sal}(i, \theta)\,\text{sal}(k, \theta) &= \text{cal}[(i-1) \oplus (k-1), \theta] \\ \text{cal}(0, \theta) &\equiv \text{wal}(0, \theta) \end{aligned}\right\} \quad (9)$$

The sine and cosine functions $\sin 2\pi i\theta$ and $\cos 2\pi i\theta$ are orthogonal in the interval $-\frac{1}{2} \leqq \theta < +\frac{1}{2}$. This is the system required for a Fourier series expansion. The Fourier transform requires the system $\{\sin 2\pi\nu\theta, \cos 2\pi\nu\theta\}$ which is orthogonal in the whole interval $-\infty < \theta < +\infty$. Note that i is an integer and thus denumerable, while ν is a real number and thus nondenumerable.

The system of Walsh functions orthogonal and complete in the whole interval $-\infty < \theta < +\infty$ is denoted by $\{\text{sal}(\mu, \theta), \text{cal}(\mu, \theta)\}$, where μ is a real number. It will be shown later on that this system may be obtained by "stretching" $\text{sal}(i, \theta)$ and $\text{cal}(i, \theta)$ just as the system $\{\sin 2\pi\nu\theta, \cos 2\pi\nu\theta\}$ can be obtained by stretching $\sin 2\pi i\theta$ and $\cos 2\pi i\theta$. Another definition due to Pichler[1] starts from the periodically continued functions $\mathrm{sal}(1,\theta)$ and $\mathrm{cal}(1,\theta)$. From them one may define the subset of the Walsh functions known as Rademacher functions [8, 9]:

$$\mathrm{cal}(2^k,\theta) = \mathrm{cal}(1, 2^k\theta), \quad \mathrm{sal}(2^k,\theta) = \mathrm{sal}(1, 2^k\theta) \tag{10}$$
$$k = \pm 1, \pm 2, \ldots; \; -\infty < \theta < +\infty$$

Let now μ be written as a binary number:

$$\mu = \sum_{s=-\infty}^{\infty} \mu_{-s} \cdot 2^{-s} = \ldots \mu_2 \cdot 2^2 + \mu_1 \cdot 2^1 + \mu_0 \cdot 2^0 + \mu_{-1} \cdot 2^{-1} + \mu_{-2} \cdot 2^{-2} \ldots$$

μ_{-s} is either 1 or 0. μ is called dyadic rational, if the sum has a finite number of terms. This means, there must be at most a finite number of binary digits to the right of the binary point. $\mathrm{cal}(\mu,\theta)$ and $\mathrm{sal}(\mu,\theta)$ are then defined as follows:

$$\left.\begin{aligned}
&\mathrm{cal}(\mu,\theta) = \prod_{s=-\infty}^{\infty} \mathrm{cal}(\mu_{-s} \cdot 2^{-s},\theta), \quad -\infty < \theta < +\infty \\
&\mathrm{sal}(\mu,\theta) = \begin{cases} -\mathrm{cal}(\mu,\theta), & -\infty < \theta < 0 \\ +\mathrm{cal}(\mu,\theta), & 0 < \theta < \infty \end{cases} \quad \mu = \text{dyadic irrational} \\
&\mathrm{sal}(\mu,\theta) = \mathrm{cal}(g \cdot 2^{-M},\theta)\,\mathrm{sal}(2^{-M},\theta), \quad -\infty < \theta < \infty \\
&\qquad g = \text{even number}; \quad \mu = (g+1)/2^M = \text{dyadic rational}
\end{aligned}\right\} \tag{11}$$

$\mathrm{cal}(\mu,\theta)$ and $\mathrm{sal}(\mu,\theta)$ are shown in Figs. 12 and 13 for the intervals $-4 < \mu < +4$ and $-3 < \theta < +3$. Black areas indicate the value $+1$, white areas the value -1. By drawing a line parallel to the θ-axis one obtains $\mathrm{cal}(\mu,\theta)$ or $\mathrm{sal}(\mu,\theta)$ as function of θ for a certain value of μ. Vice versa, a line parallel to the μ-axis shows the values of $\mathrm{cal}(\mu,\theta)$ or $\mathrm{sal}(\mu,\theta)$ as function of μ for a certain value of θ.

The following additional formulas are important for computations with Walsh functions:

$$\left.\begin{aligned}
\mathrm{wal}(\mu,\theta) &= \mathrm{wal}(0,\theta), && 0 \leqq \mu < 1 \\
\mathrm{cal}(\mu,\theta) &= \mathrm{cal}(i,\theta), && i \leqq \mu < i+1 \qquad -\tfrac{1}{2} \leqq \theta < +\tfrac{1}{2} \\
\mathrm{sal}(\mu,\theta) &= \mathrm{sal}(i,\theta), && i-1 < \mu \leqq i
\end{aligned}\right\} \tag{12}$$

$$\begin{aligned}
\mathrm{cal}(\mu,\theta \oplus \theta') &= \mathrm{cal}(\mu,\theta)\,\mathrm{cal}(\mu,\theta') \\
\mathrm{sal}(\mu,\theta \oplus \theta') &= \mathrm{sal}(\mu,\theta)\,\mathrm{sal}(\mu,\theta')
\end{aligned} \tag{13}$$

[1] The nondenumerable system of Walsh functions required for the Walsh-Fourier transform is due to Fine [12], who also first pointed out the existence of such a transform. The correct mathematical theory of the Walsh-Fourier transform using sal and cal functions, which are somewhat different from the system used by Fine, is due to Pichler [9]. A term like Fine or Pichler transform appears fair as well as shorter than the cumbersone term Walsh-Fourier transform. Mathematicians use this term, because the Walsh-Fourier transform is a special case of the general Fourier transforms on topologic groups, published by Vilenkin two years after Fine's paper [22].

Since θ and θ' may be positive or negative one has to extend the definition of addition modulo 2 to negative numbers $-a$ and $-b$:

$$\begin{aligned} (-a) \oplus (-b) &= a \oplus b \\ (-a) \oplus b = a \oplus (-b) &= -(a \oplus b) \end{aligned} \tag{14}$$

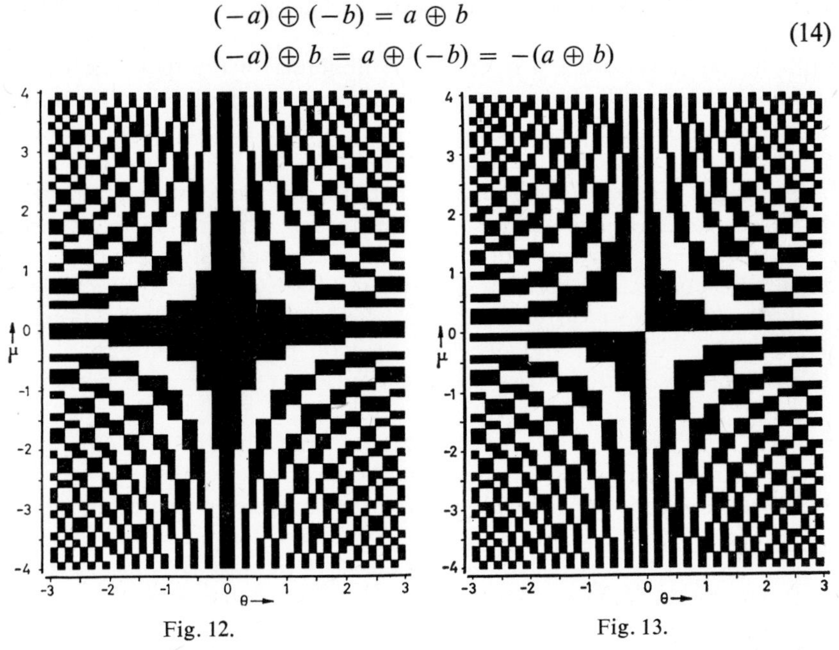

Fig. 12. Fig. 13.

Fig. 12. The function cal(μ, θ) in the interval $-3 < \theta < +3$, $-4 < \mu < +4$. A function, e.g., cal$(1.5, \theta)$, is obtained by drawing a line at $\mu = 1.5$ parallel to the θ-axis. cal$(1.5, \theta)$ is $+1$ where this line runs through a black area and -1 where it runs through a white area. At borders between black and white areas use the value holding for the absolutely larger μ. The function cal$(\mu, 1.5)$ is obtained by drawing a line at $\theta = 1.5$ parallel to the μ-axis and proceeding accordingly.

Fig. 13. The functions sal(μ, θ) in the interval $-3 < \theta < +3$, $-4 < \mu < +4$. The values $+1$ and -1 of the functions are obtained by drawing lines as explained in the caption of Fig. 12. At borders between black and white areas use the value holding for the absolutely smaller μ or θ. There are no functions sal$(0, \theta)$ or sal$(\mu, 0)$.

μ is equal to one half the average number of sign changes of cal(μ, θ) or sal(μ, θ) in a time interval of duration 1. This may easily be verified for the periodic functions cal(i, θ) and sal(i, θ) by counting the sign changes in Fig. 2. cal(μ, θ) and sal(μ, θ) are not periodic if μ is not dyadic rational, but the interpretation of μ as one half the average number of sign changes per time interval of duration 1 still holds true.

If an arbitrarily small section of a sine function is known, the function is known everywhere. This feature is frequently expressed by saying that sinusoidal functions transmit information at the rate zero. Walsh functions are quite different in this respect. Assume that a measurement has yielded the value $+1$ for a Walsh function in the interval $-\frac{1}{2} \leqq \theta < +\frac{1}{2}$. It follows from Figs. 12 and 13 that it must be a function cal(μ, θ) with μ in the interval $0 \leqq \mu < 1$. Let an additional measurement in the interval $\frac{1}{2} \leqq \theta < 1$ yield -1; the value of μ is thus restricted to the smaller interval $\frac{1}{2} \leqq \mu < 1$ according to Fig. 12. A further measurement yields, e.g., -1 for the interval $1 \leqq \theta < 1.5$ and $+1$ for the interval $1.5 \leqq \theta < 2$; this restricts μ to the still smaller interval $0.5 \leqq \mu < 0.75$. A doubling of the time interval $\Delta\theta$ required for measurement successively halves the interval $\Delta\mu$ within which the sequency μ remains undetermined. The product $\Delta\theta\,\Delta\mu$ remains constant and may be interpreted as the uncertainty relation for Walsh functions[1]. The transmission rate of information is not zero, since more information about the exact value of μ is obtained with increasing observation interval $\Delta\theta$.

A few words may be added for the mathematically inclined reader about the connection between the systems $\{\text{wal}(0, \theta), \text{sal}(i, \theta), \text{cal}(i, \theta)\}$ and $\{1, \sqrt{(2)} \sin 2\pi\, i\, \theta, \sqrt{(2)} \cos 2\pi\, i\, \theta\}$. Both are orthonormal systems in Hilbert space $L_2(0, 1)$ and one may base on both of them very similar theories of the Fourier series and the Fourier transform. The reason for this is that both may be derived from character groups. The system of circular functions $\{\cos k\,x, \sin k\,x\}$ is derived from the group $\{e^{ikx}\}$, which is the character group of the topologic group of real numbers. The system of Walsh functions may be derived from the character group of the dyadic group; the dyadic group is the topologic group derived from the set of binary representations of the real numbers. The most striking difference between the functions—continuity of circular functions and discontinuity of Walsh functions—is caused by the different topology of the real numbers and the dyadic group [8, 11, 12, 20].

An example showing the close connection between the functions $\text{sal}(i, \theta)$ and $\sin 2\pi\, i\, \theta$ is given in Fig. 14. Short and long distances between the zero crossings alternate in such a way as to give a strikingly good fit. Note that for $i \leqq 10$ the Walsh function is $+1$ or -1 whenever the sine function is $+1$ or -1 but this does not always hold true for larger values of i.

Problems

114.1 A system of functions cannot form a finite group unless the magnitude of all amplitudes is 1. Why can they not be larger or smaller?

114.2 Which real values can the amplitudes of a system of functions assume if the system is to form a group?

114.3 Can there be a system of real three-valued functions that forms a group?

[1] Care should be taken not to confuse the uncertainty relations of transforms with that of quantum mechanics. The quantum mechanical uncertainty relation is an axiom that connects the accuracy of measurement of the location and the momentum of a particle, which are independent physical quantities. The uncertainty relations of transforms are mathematical deductions that connect the accuracy with which a function is known in the original domain with the resulting accuracy in the transform domain. The function and its transform are not independent, they are usually connected by one-to-one mapping. The mathematical formalism is sometimes the same for different fields of science but a quantum mechanical interpretation of a result requires that a quantum mechanical axiom is introduced first.

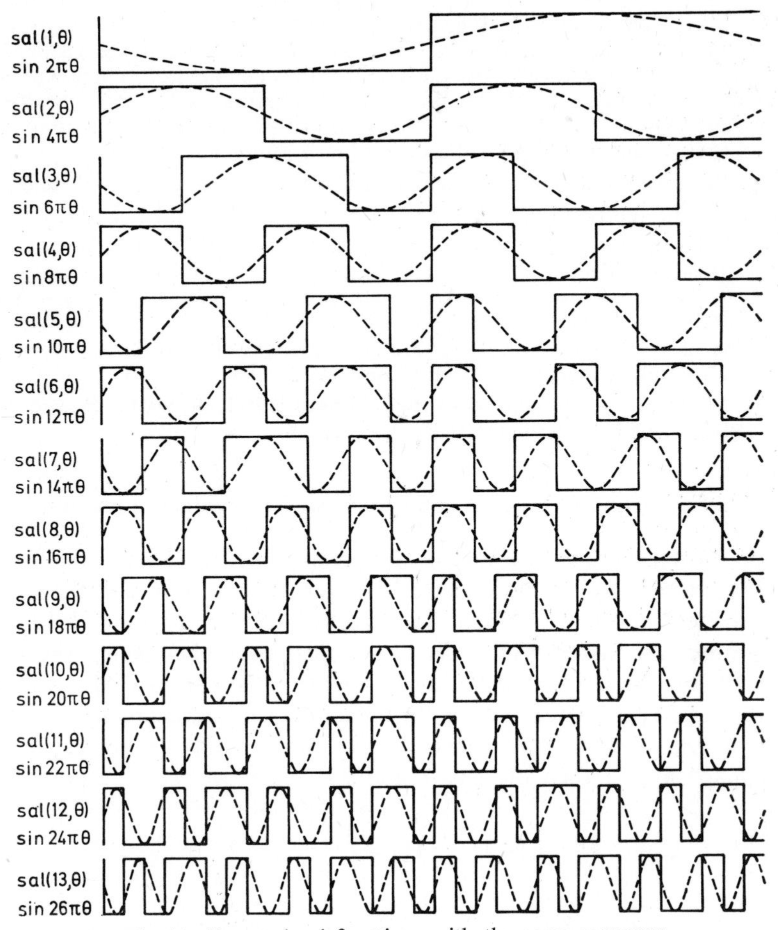

Fig. 14. Sine and sal functions with the same sequency.

114.4 The exponential functions $e^{i2k\pi/m}$, $k = 0, 1, \ldots, m$ = positive integer, form a group under multiplication. State inverse and unit element. Is the group commutative? How many elements are there? Find a subgroup when m is even. Can there be a subgroup when m is odd, or when m is a prime number?

114.5 Compute the first 8 coefficients of a Walsh series expansion of the function $F(\theta) = \theta,\ -\frac{1}{2} \leqq \theta < \frac{1}{2}$.

114.6 Using Fig. 12 plot the following functions in the interval $-3 < \theta < +3$ and see in which interval they are equal: $\mathrm{cal}(1, \theta)$, $\mathrm{cal}(1.5, \theta)$, $\mathrm{cal}(1.25, \theta)$, $\mathrm{cal}(1.75, \theta)$.

114.7 The functions $\mathrm{wal}(2^n - 1, \theta)$, $n > 0$, are called Rademacher functions. All Walsh functions can be written as products of Rademacher functions. A Walsh function has the multiplicity s if it is obtained by multiplication of s Rademacher functions. E.g., $\mathrm{wal}(10, \theta) = \mathrm{wal}(15, \theta)\,\mathrm{wal}(7, \theta)\,\mathrm{wal}(3, \theta)\,\mathrm{wal}(1, \theta)$ has the multiplicity $s = 4$ (observe: $10 = 15 \oplus 7 \oplus 3 \oplus 1$). Plot the multiplicity s of $\mathrm{wal}(j, \theta)$ as function of j for $j = 0, \ldots, 31$.

114.8 Plot a step function with 8 steps of equal width in the interval $-\frac{1}{2} \leqq \theta < \frac{1}{2}$ and amplitudes $+1, +2, +3, +1, -1, +1, +1, -1$. Compute all coefficients of its Walsh series expansion. Let the step function be amplitude clipped at ± 2. Compute the coefficients of the Walsh series expansion. Are any new terms generated? How many terms unequal to zero can exist for a step function having 2^n steps of equal width in the interval $-\frac{1}{2} \leqq \theta < \frac{1}{2}$?

114.9 Does the product of $2n$ Rademacher functions yield a cal or a sal function?

114.10 Is the multiplicity of a sal function an even or odd number?

114.11 Plot the derivatives of the functions $\mathrm{wal}(31, \theta)$, $\mathrm{wal}(15, \theta)$ and $\mathrm{wal}(7, \theta)$, which consist of positive and negative Dirac functions. Plot the derivative of $\mathrm{wal}(k, \theta)$ for $k = 31 \oplus 15$, $31 \oplus 7$ and $31 \oplus 15 \oplus 7$. State the connection between modulo 2 addition and the way the derivative of $\mathrm{wal}(k, \theta)$ may be obtained from the derivatives of $\mathrm{wal}(31, \theta)$, $\mathrm{wal}(15, \theta)$ and $\mathrm{wal}(7, \theta)$. Is the sequency of a Walsh function changed by differentiation?

114.12 Compare the derivatives of the functions $\mathrm{wal}(1, \theta)$, $\mathrm{wal}(2, \theta)$, $\mathrm{wal}(4, \theta)$ and $\mathrm{wal}(8, \theta)$ with that of $\mathrm{wal}(k, \theta)$, $k < 16$.

1.1.5 Hadamard Matrices and General Two-Valued Functions

Let $\mathbf{M}$ denote a square matrix with real elements. The inverse matrix $\mathbf{M}^{-1}$ is defined by the product $\mathbf{M}\,\mathbf{M}^{-1} = \mathbf{1}$, where $\mathbf{1}$ is a unit matrix with elements 1 along the main diagonal and elements 0 everywhere else. The transposed matrix $\mathbf{M}^*$ of $\mathbf{M}$ is obtained by interchanging the rows and columns. $\mathbf{M}$ is called an orthogonal matrix if the inverse $\mathbf{M}^{-1}$ and the transposed $\mathbf{M}^*$ are equal except for a factor[1]. A Hadamard matrix $\mathbf{H}$ is an orthogonal matrix with elements $+1$ and -1 only. There exists one Hadamard matrix each of rank 1 and 2:

$$\mathbf{H}_1 = [+1], \quad \mathbf{H}_2 = \begin{bmatrix} +1 & +1 \\ +1 & -1 \end{bmatrix} \tag{1}$$

Two non-trivially different matrices exist with rank 4:

$$\mathbf{H}_{41} = \begin{bmatrix} +1 & +1 & +1 & +1 \\ -1 & -1 & +1 & +1 \\ -1 & +1 & +1 & -1 \\ +1 & -1 & +1 & -1 \end{bmatrix}, \quad \mathbf{H}_{42} = \begin{bmatrix} +1 & +1 & +1 & -1 \\ +1 & +1 & -1 & +1 \\ +1 & -1 & +1 & +1 \\ -1 & +1 & +1 & +1 \end{bmatrix} \tag{2}$$

Matrices of higher rank can be obtained by Kronecker products[2]. It is usual to write + and − instead of $+1$ and -1 for the matrix elements.

---

[1] Many authors require that this factor must be one, but we shall use the term orthogonal and normalized for this case.

[2] The Kronecker product $\mathbf{S} \times \mathbf{K}$ of two matrices is obtained by multiplying $\mathbf{S}$ with each element k_{ij} of $\mathbf{K}$ and substituting the multiplied matrices $k_{ij}\,\mathbf{S}$ for the elements k_{ij} of $\mathbf{K}$. The Kronecker product $\mathbf{H}_2 \times \mathbf{H}_2$ is shown in Eq. (3), other examples are given in section 5.4.2.

Using this notation the Kronecker product of $\mathbf{H}_2$ with itself yields:

$$\mathbf{H}_2 \times \mathbf{H}_2 = \begin{bmatrix} \mathbf{H}_2 & \mathbf{H}_2 \\ \mathbf{H}_2 & -\mathbf{H}_2 \end{bmatrix} = \begin{bmatrix} + & + & + & + \\ + & - & + & - \\ + & + & - & - \\ + & - & - & + \end{bmatrix} \tag{3}$$

This matrix is equal to $\mathbf{H}_{41}$ except for a different ordering of the rows.

The rank of a Hadamard matrix with rank higher than 2 must be an integer multiple of 4. With one exception at least one Hadamard matrix is known for all possible ranks up to 200. Certain Hadamard matrices of rank 2^n are related to Walsh functions. This relationship is readily recognized from Fig. 15 for the rank 4 and 8.

A process that generates complete systems of two-valued functions from Hadamard matrices will be explained with reference to Fig. 16a. The

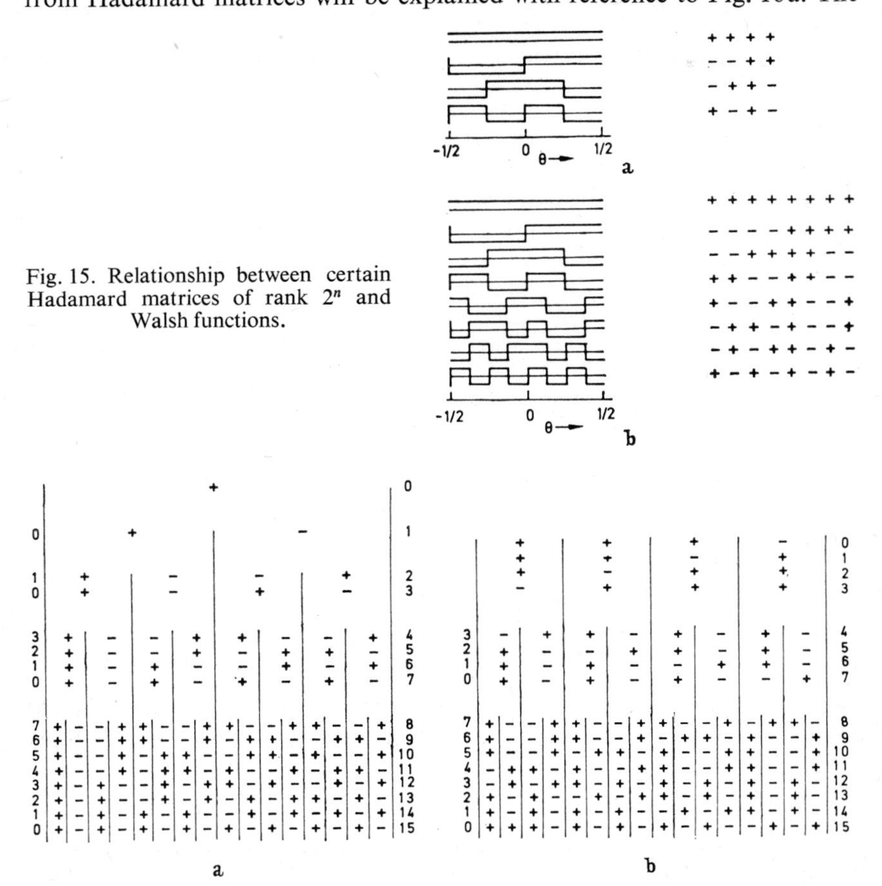

Fig. 15. Relationship between certain Hadamard matrices of rank 2^n and Walsh functions.

Fig. 16. Derivation of complete systems of two-valued orthogonal functions from Hadamard matrices by means of the alternating process.

system of Walsh functions in the interval (0, 1) is obtained from the Hadamard matrix $\mathbf{H}_1 = (+)$ in row 0 by dividing the interval in half, and writing + in the left half and − in the right half; this yields row 1. Rows 2 and 3 are obtained by dividing the intervals of row 1 in half; writing + − in the intervals where row 1 shows +, and − + where row 1 shows − yields row 2. Writing + − where row 0 shows + yields row 3. The process may be continued ad infinitum. Fig. 16a shows on the right the number of the row and on the left the number of the row from which it was derived.

Figure 16b shows the same process applied to the Hadamard matrix $\mathbf{H}_{42}$. Again the number of the row is shown on the right and the number of the row from which it was derived on the left. Figure 17 shows the plotted functions.

The Kronecker product also permits the generation of new functions. However, the two functions represented by $\mathbf{H}_2$ in Eq. (1) are represented by rows 1 and 3 in Eq. (3). Hence, the Kronecker product not only adds new functions but also changes the ordering of the ones already obtained.

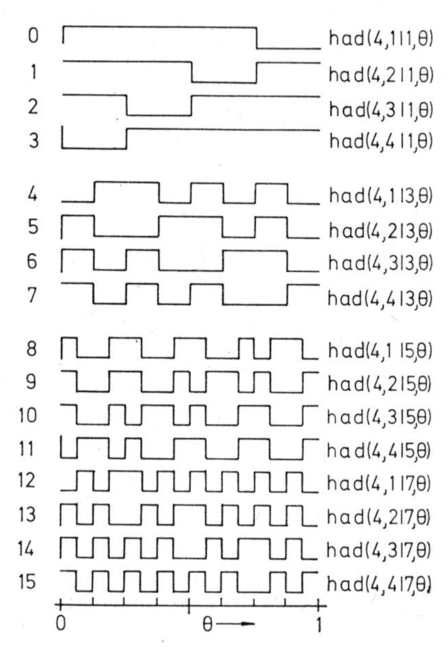

Fig. 17. Example of functions generated according to Fig. 16b.

The "alternating" process of Fig. 16 only adds new functions. Crittenden has generalized this process and proved the completeness of the systems of functions generated [1].

A Hadamard matrix of rank 12 was reported by Paley [2]. It is of interest in communications since it is usual to multiplex twelve telephony channels into a group. Figure 18 shows the orthogonal functions associated with this matrix. The notation pal(j, θ) is used in honor of Paley. It may readily be verified that the product of two Paley functions does not necessarily yield a function of the system. Hence, these functions do not form

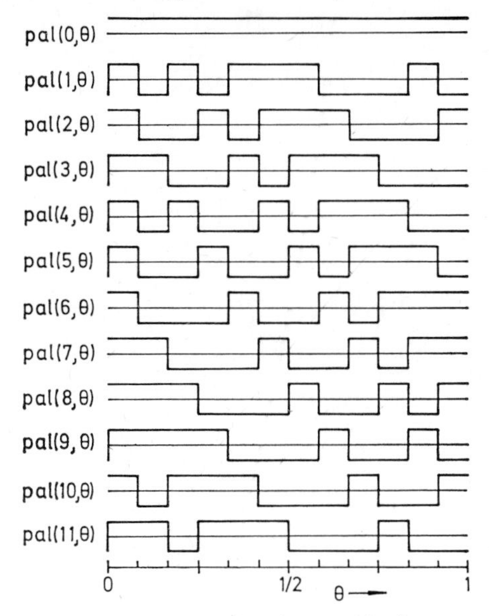

Fig. 18. Paley functions pal(j, θ).

a group as the Walsh functions do. However, the product of a function with itself yields the constant pal$(0, \theta)$:

$$\mathrm{pal}(j, \theta)\,\mathrm{pal}(j, \theta) = \mathrm{pal}(0, \theta). \tag{4}$$

Problems

115.1 Does a relation like Eq. (4) hold for systems of two-valued functions other than the Walsh and Paley functions?

115.2 The alternating process is based on the Hadamard matrix $\mathbf{H}_2$. Use the matrix $\mathbf{H}_{42}$ instead of $\mathbf{H}_2$ to derive functions analogous to Fig. 16.

115.3 Extend the system of Paley functions in Fig. 18 from twelve to twenty-four functions with the help of Figs. 16 and 17.

115.4 Assume the Paley functions of Fig. 18 are available as time varying voltages. What is the time variation of the voltage that produces the functions pal$(12, \theta)$ to pal$(23, \theta)$ according to problem 115.3? What is the time variation of the voltage that produces pal$(24, \theta)$ to pal$(47, \theta)$ from pal$(0, \theta)$ to pal$(23, \theta)$?

1.1.6 Haar Functions and General Three-Valued Functions

Haar functions assume the values $+1$, 0 and -1 multiplied by powers of $\sqrt{(2)}$. The first few functions are shown in Fig. 19. har$(0, 0, \theta)$ and har$(0, 1, \theta)$ are identical to the first two Walsh functions wal$(0, \theta)$ and

$\mathrm{wal}(1, \theta)$. The function $\mathrm{har}(0, 1, \theta)$ is then squeezed into half the interval and shifted to yield $\mathrm{har}(1, 1, \theta)$ and $\mathrm{har}(1, 2, \theta)$. Each one of the new functions is again squeezed and shifted, and so on ad infinitum.

Let a function $F(\theta)$ defined in the interval $(-\frac{1}{2}, +\frac{1}{2})$ be expanded into a series of Haar functions. The expansion coefficients of the first two functions $\mathrm{har}(0, 0, \theta)$ and $\mathrm{har}(0, 1, \theta)$ are affected by all values of $F(\theta)$.

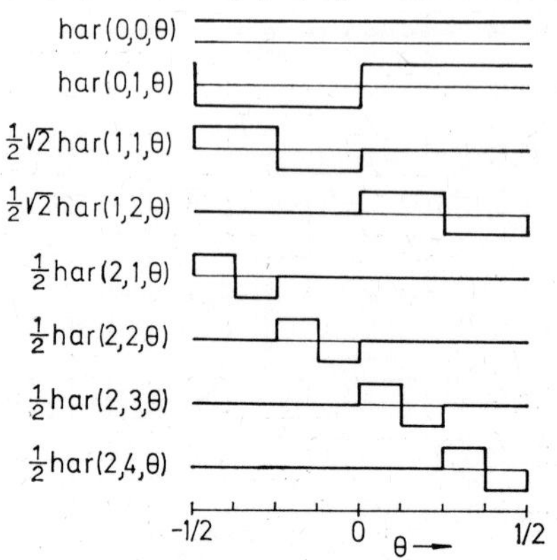

Fig. 19. Haar functions.

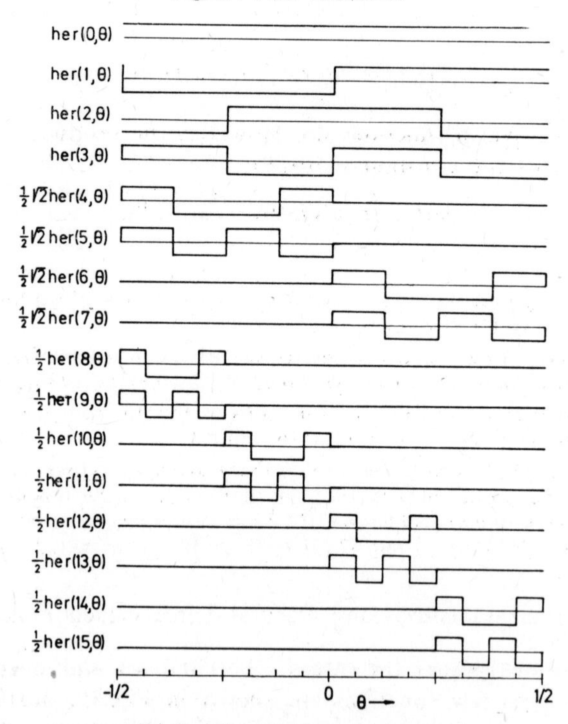

Fig. 20. The functions $\mathrm{her}(j, \theta)$.

The coefficients of the other Haar functions are affected by fractions $\frac{1}{2}, \frac{1}{4}, \frac{1}{8}, \ldots$ of the values of $F(\theta)$ only. These functions help represent certain local sections of $F(\theta)$ and are therefore referred to as local functions. $\mathrm{har}(0, 0, \theta)$ and $\mathrm{har}(0, 1, \theta)$ are called global functions. The engineering interest in this difference stems from the fact that a better representation of certain sections of a function is often desired. For instance, the most important parts of a TV picture are usually in the center and a better representation of the center sections is thus desirable than of the edge sections.

The system of Walsh functions consists of global functions only, while the Haar system contains two global functions only. One may generate systems with global functions between these extremes. Figure 20 shows a system using the first four Walsh functions. The Haar process of squeezing to one half and shifting is then applied to the functions $\mathrm{her}(2, \theta)$ and $\mathrm{her}(3, \theta)$. The signs are chosen so that all functions are non-negative for $+0$.

Another method to generate systems of three-valued functions is shown in Fig. 21. Again the first four Walsh functions are used. The Haar process

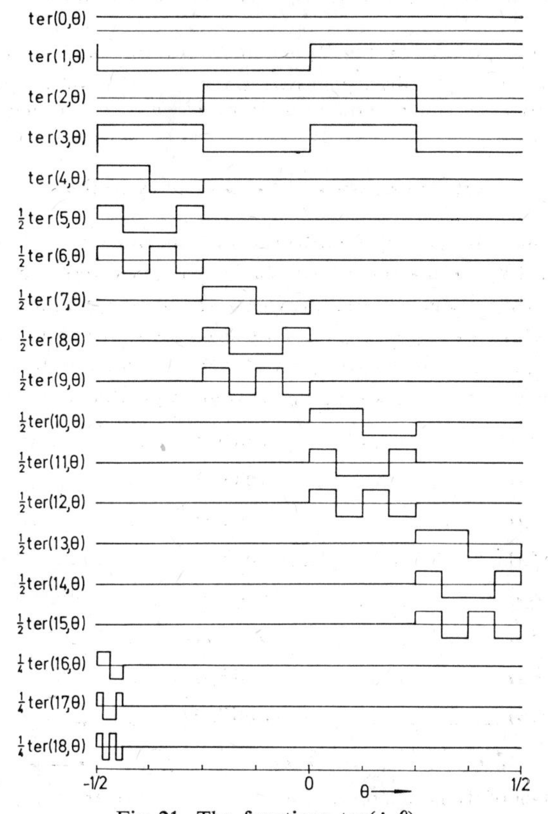

Fig. 21. The functions $\mathrm{ter}(j, \theta)$.

of squeezing and shifting is applied to the three functions ter$(1, \theta)$, ter$(2, \theta)$ and ter$(3, \theta)$. The squeezing in Fig. 21 is done by substituting 4θ for θ, while 2θ was substituted in Fig. 20.

The processes shown in Figs. 20 and 21 permit one to choose 2^n Walsh functions for a global representation. One then still has a choice of n generating processes for the local functions. For $2^n = 2^1$ one has only the Haar functions, for $2^n = 2^2$ one has the processes of Figs. 20 and 21, for $2^n = 2^3$ one can squeeze the functions by the substitution of 2θ, 4θ or 8θ for θ; etc.

Note that the amplitudes of the functions in Fig. 21 are $0, \pm 1, \pm 2, \pm 4, \ldots$ Multiplications by powers of 2 require only a shift of the binary point in digital computers and are fast to perform. Additional multiplications by $\sqrt{(2)}$ are required for the functions of Fig. 19 and Fig. 20 if the normalization is to be retained.

Problems

116.1 Can one apply the processes of Figs. 20 and 21 to the first four functions of Fig. 17?

116.2 Can one apply the processes of Figs. 20 and 21 to the first eight functions of Fig. 17?

116.3 Apply the process of Fig. 21 to the Paley functions of Fig. 18.

116.4 The human eye resolves pictures in the center (yellow spot) better than further away from the center. Which functions in Fig. 21 could one use to produce such an effect?

1.1.7 Functions with Several Variables

Many signals in communications are functions of several variables. For instance, TV signals have two space variables and the time variable. The functions discussed in the previous sections are readily extended to several variables but some changes of notations are indicated.

Figure 22 defines the notation blo(k, x) for block pulses. Block pulses with two and three variables may be written as products blo(k, x) blo(m, y)

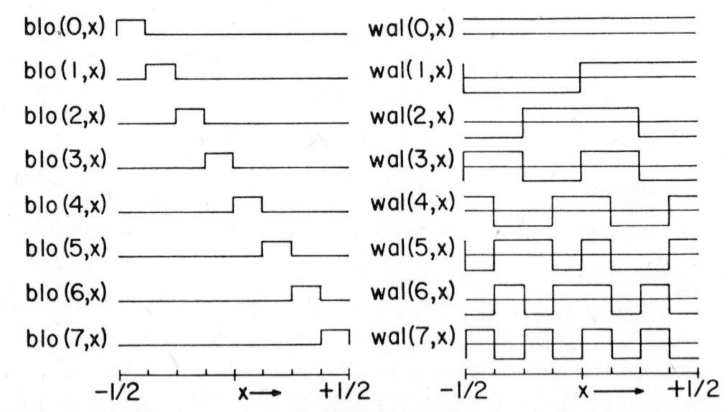

Fig. 22. Definition of the notation used for block pulses and Walsh functions with space variables.

and $\mathrm{blo}(k, x)\,\mathrm{blo}(m, y)\,\mathrm{blo}(n, z)$. The notation $\mathrm{wal}(k, x)$ defined by Figs. 2 and 22 will be used for Walsh functions instead of $\mathrm{wal}(0, x)$, $\mathrm{sal}(i, x)$ and $\mathrm{cal}(i, x)$. Sine-cosine functions with more than one variable have never played a great role in communications and there is thus little incentive to keep stressing the similarity between Walsh and sine-cosine functions. The notation $\mathrm{wal}(k, x)$ permits the definition of Walsh functions with two or three variables by one product $\mathrm{wal}(k, x)\,\mathrm{wal}(m, y)$ or $\mathrm{wal}(k, x)\,\mathrm{wal}(m, y)\,\mathrm{wal}(n, z)$ rather than by 9 or 27 products.

Figure 23 shows block pulses with two variables in the interval $-\frac{1}{2} \leqq x < \frac{1}{2}$, $-\frac{1}{2} \leqq y < \frac{1}{2}$ for $k, m = 0, \ldots, 7$. The function $\mathrm{blo}(k, x) \times \mathrm{blo}(m, y)$ is found at the intersection of the column denoted $\mathrm{blo}(k, x)$

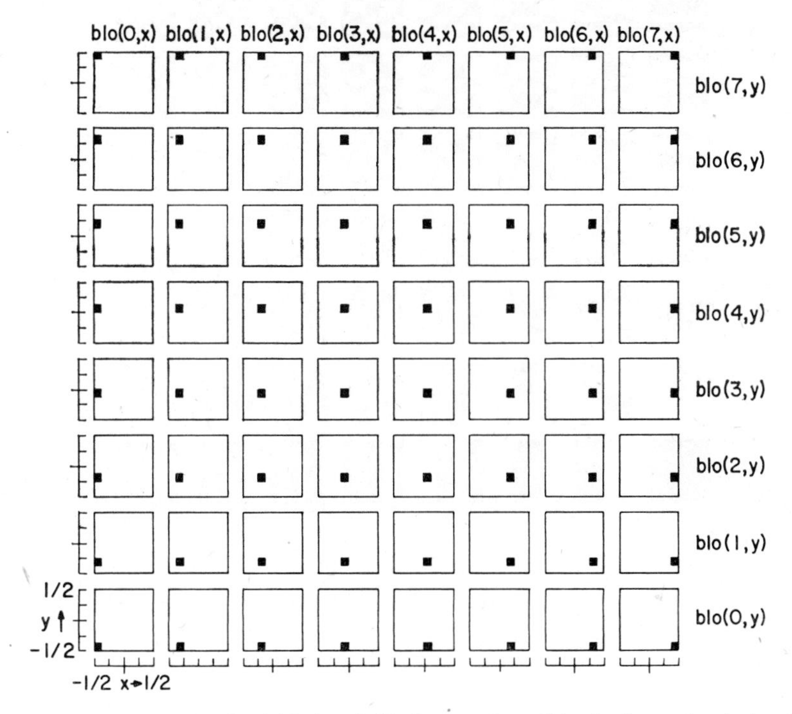

Fig. 23. Block pulses $\mathrm{blo}(k, x)\,\mathrm{blo}(m, y)$ for $k, m = 0 \ldots 7$ in the interval $-\frac{1}{2} < x < \frac{1}{2}$, $-\frac{1}{2} < y < \frac{1}{2}$. Black areas represent $+1$, white areas 0.

and the row denoted $\mathrm{blo}(m, y)$. Black areas represent the value $+1$, white areas the value 0. The black area moves from left to right as k increases and from bottom to top as m increases. This corresponds to the movement of the illuminated spot on a TV tube that is scanned from left to right and from bottom to top.

Figure 24 shows Walsh functions with two variables in the interval $-\frac{1}{2} \leqq x < \frac{1}{2}$, $-\frac{1}{2} \leqq y < \frac{1}{2}$ for $k, m = 0, \ldots, 7$. As before, the function wal(k, x) wal(m, y) is located at the intersection of the column denoted wal(k, x) and the row wal(m, y). Black areas represent again the value $+1$, but white areas represent now the value -1. The function wal$(0, x) \times$ $\times$ wal$(0, y)$ is $+1$ everywhere and its integral over the interval $-\frac{1}{2} \leqq x <$ $< \frac{1}{2}$, $-\frac{1}{2} \leqq y < \frac{1}{2}$ equals $+1$. All other functions are $+1$ as often as -1 and their integrals yield zero[1].

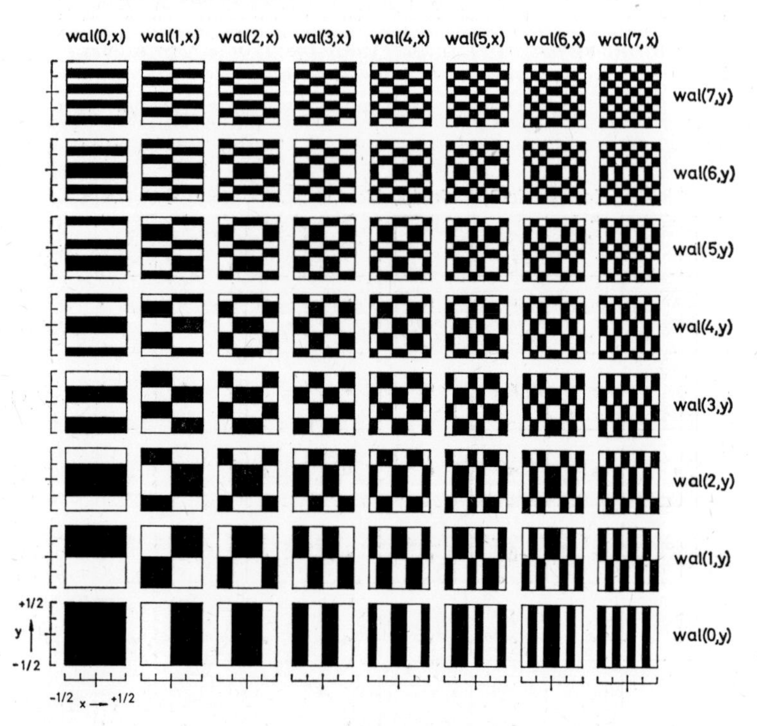

Fig. 24. Walsh functions wal(k, x) wal(m, y) for $k, m = 0 \ldots 7$ in the interval $-\frac{1}{2} < x < \frac{1}{2}$, $-\frac{1}{2} < y < \frac{1}{2}$. Black areas represent $+1$, white areas -1.

Consider the expansion of a function $F(x, y)$ into series of Walsh functions wal(k, x) wal(m, y):

$$F(x, y) = \sum_{k=0}^{\infty} \sum_{m=0}^{\infty} a(k, m) \operatorname{wal}(k, x) \operatorname{wal}(m, y) \tag{1}$$

$$a(k, m) = \int_{-1/2}^{1/2} \int_{-1/2}^{1/2} F(x, y) \operatorname{wal}(k, x) \operatorname{wal}(m, y) \, dx \, dy \tag{2}$$

A typical example for $F(x, y)$ is a black-and-white photograph and the terms picture, image or two-dimensional signal may thus be used for

[1] This representation of two-dimensional Walsh functions seems to have been used first in [1].

$F(x, y)$. The term $a(0,0)\,\mathrm{wal}(0,x)\,\mathrm{wal}(0,y)$ in the series expansion represents the average value of $F(x, y)$ or – in easily understandable terminology – the average brightness or the dc component of the image $F(x, y)$. The other terms $a(k,m)\,\mathrm{wal}(k,x)\,\mathrm{wal}(m,y)$ may then be said to represent ac components.

Figure 25 shows the functions of Fig. 17 for two variables. Note that there are 16 functions that would represent the average value of a function $F(x, y)$ in a series expansion while only the function $\mathrm{wal}(0,x)\,\mathrm{wal}(0,y)$ would do this in the case of Walsh functions.

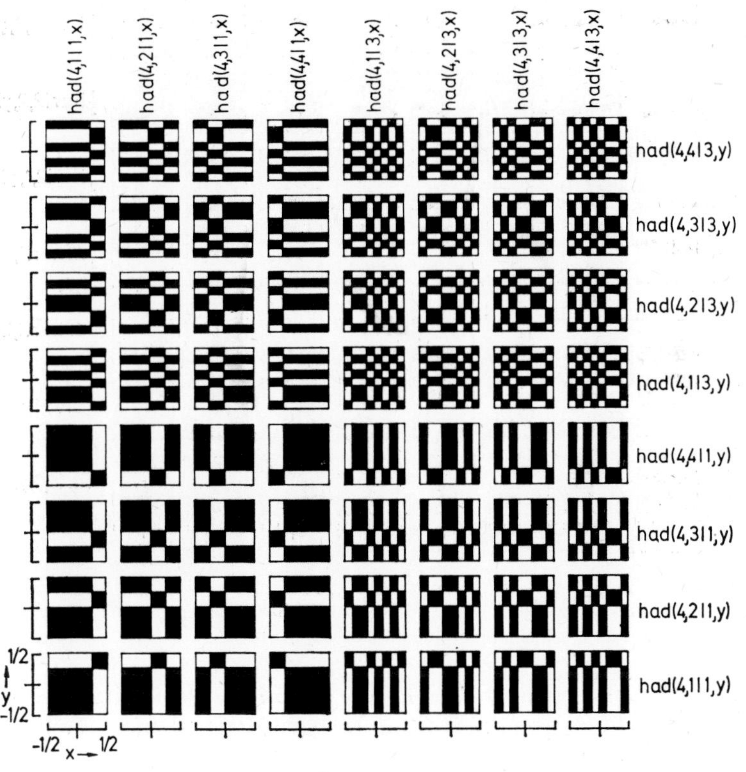

Fig. 25. The functions $\mathrm{had}(4,k \mid K,x)\,\mathrm{had}(4,m \mid M,y)$ for $k, m = 0 \ldots 4$ and $K, M = 1, 3$ in the interval $-\frac{1}{2} < x < \frac{1}{2}$, $-\frac{1}{2} < y < \frac{1}{2}$. Black areas represent $+1$, white areas -1.

Figure 26 shows Haar functions with two variables. Black and white areas again represent the values $+1$ and -1 while grey areas represent the value 0. Note that this representation shows the Haar functions multiplied by certain coefficients. For instance, the function $\frac{1}{4}\mathrm{har}(2,2,x) \times \times\, \mathrm{har}(2,2,y)$ rather than $\mathrm{har}(2,2,x)\,\mathrm{har}(2,2,y)$ is found at the intersection of the column denoted $\frac{1}{2}\mathrm{har}(2,2,x)$ and the row denoted

$\frac{1}{2}\text{har}(2, 2, y)$. Four of the functions are global, and the others local in an extension of the terminology introduced in section 1.1.6.

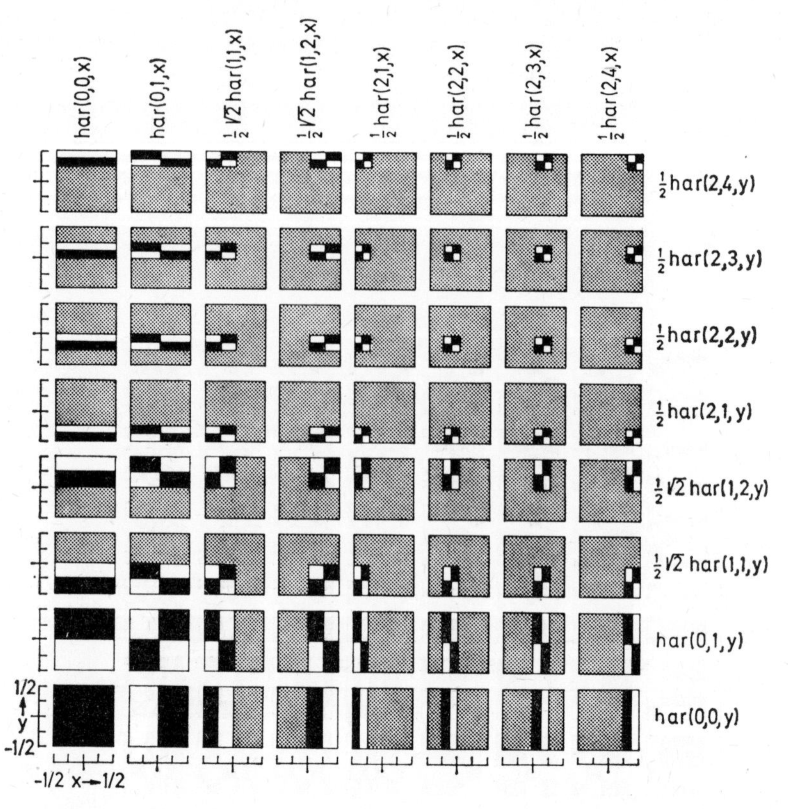

Fig. 26. Haar functions $\text{har}(K, k, x)\,\text{har}(M, m, y)$ for $K, M = 0, 1, 2$ and $k, m = 0 \ldots 4$ in the interval $-\frac{1}{2} < x < \frac{1}{2}$, $-\frac{1}{2} < y < \frac{1}{2}$. Black areas represent $+1$, gray areas 0, and white areas -1.

Figure 27 shows the functions of Fig. 20 for two variables. Black, grey and white areas represent again the vuales $+1$, 0 and -1. There are now 16 global functions.

The representation of the three-dimensional function $\text{wal}(k, x) \times \times \text{wal}(m, y)\,\text{wal}(n, x)$ requires black and white volumes rather than black and white areas for the representation of the values $+1$ and -1. Figure 28 shows the 64 functions for $k, m, n = 0, \ldots, 4$ in the interval $-\frac{1}{2} \leqq x < \frac{1}{2}$, $-\frac{1}{2} \leqq y < \frac{1}{2}$, $-\frac{1}{2} \leqq z < \frac{1}{2}$. Let us first consider the specific function $\text{wal}(2, x)\,\text{wal}(3, y)\,\text{wal}(1, z)$ which is shown in more detail in Fig. 29. It consists of 64 small cubes which are black or white inside and outside[1]. The function is $+1$ for a coordinate x, y, z inside a black

[1] A red tomato is red inside and outside while a red apple is red on the outside only.

cube and -1 for a coordinate inside a white cube. The four layers of cubes on the right hand side show how the 64 small cubes are arranged inside the large cube.

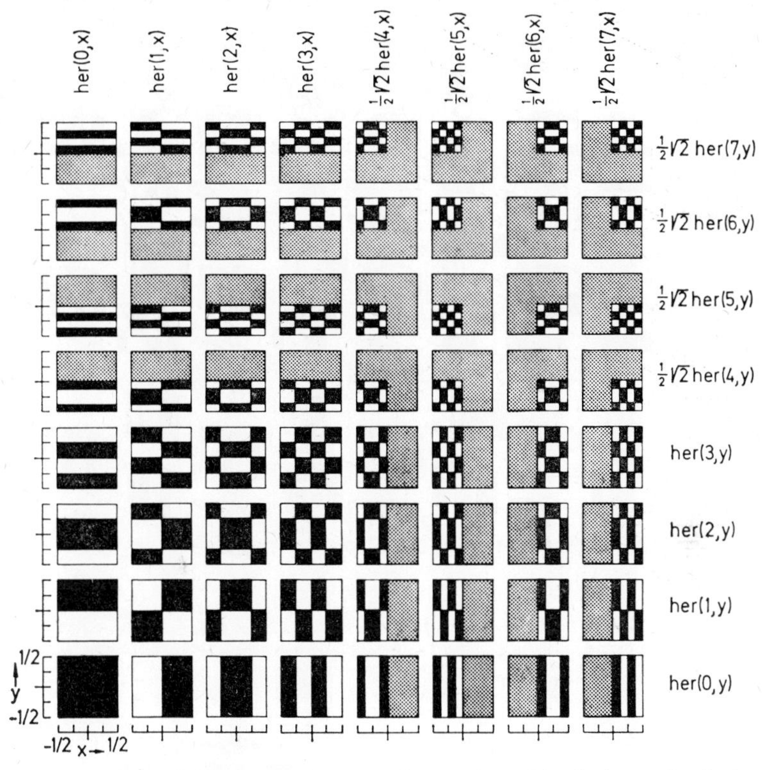

Fig. 27. The functions her(k, x) her(m, y) for $k, m = 0 \ldots 7$ in the interval $-\frac{1}{2} < x < \frac{1}{2}$, $-\frac{1}{2} < y < \frac{1}{2}$. Black areas represent $+1$, gray areas 0, and white areas -1.

Let us now return to Fig. 28. Sixty-four large cubes like the one of Fig. 29 are arranged in a cubic pattern just as the squares in Fig. 24 are arranged in a square pattern. The dashed lines are intended to make this three-dimensional arrangement more apparent. Furthermore, scales emphasize the 16 large cubes in the front plane. The function wal$(k, x) \times$ $\times$ wal(m, y) wal(n, z) is found at the intersection of the x-row denoted wal(k, x), the y-row denoted wal(m, y) and the z-column denoted wal(n, z). Note that the terms x-row and y-row have to be used since the terms row and column suffice for the description of square patterns but a third term for the description of cubic patterns is lacking.

Since a representation of functions with three variables according to Fig. 28 is not very lucid, only examples of a specific Haar function and a specific block pulse with three variables are shown in Figs. 30 and 31.

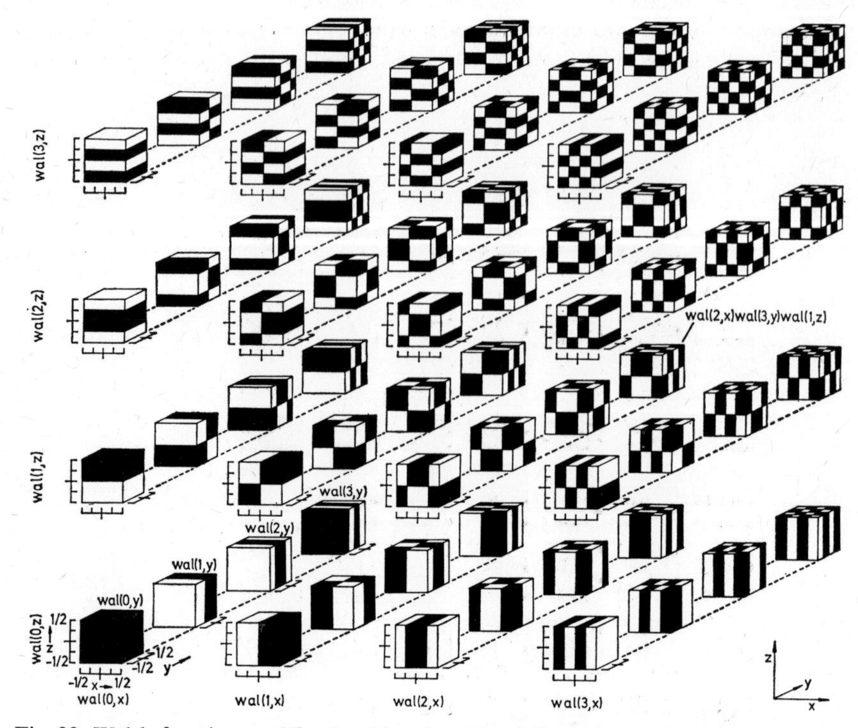

Fig. 28. Walsh functions $\mathrm{wal}(k, x)\,\mathrm{wal}(m, y)\,\mathrm{wal}(n, z)$ for $k, m, n = 0 \ldots 3$ in the interval $-\frac{1}{2} < x < \frac{1}{2}$, $-\frac{1}{2} < y < \frac{1}{2}$, $-\frac{1}{2} < z < \frac{1}{2}$. Black volumes represent $+1$, white volumes -1.

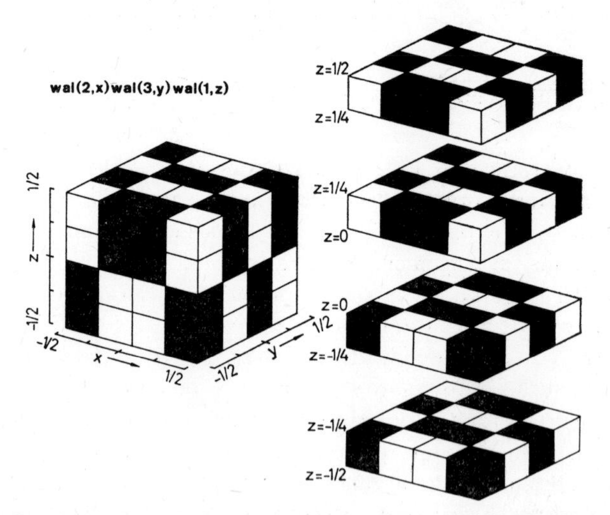

Fig. 29. The function $\mathrm{wal}(2, x)\,\mathrm{wal}(3, y)\,\mathrm{wal}(1, z)$. Black volumes represent $+1$, white volumes -1.

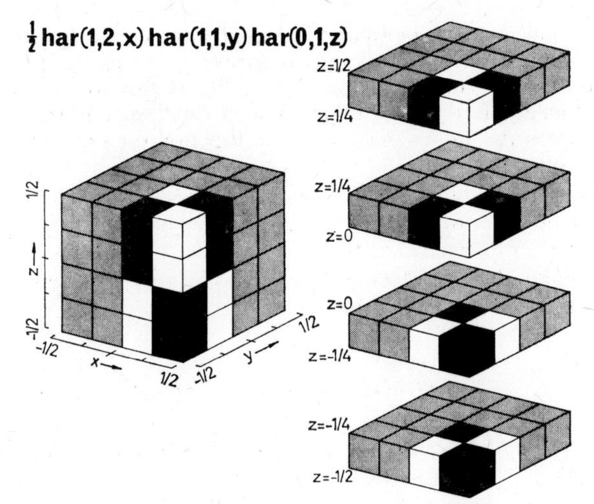

Fig. 30. The function $\frac{1}{2}$har(1, 2, x) har(1, 1, y) har(0, 1, z). Black volumes represent +1, gray volumes 0, and white volumes −1.

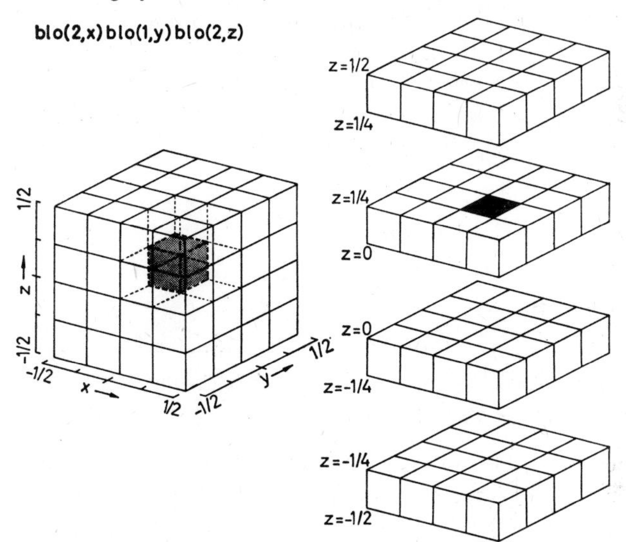

Fig. 31. The block pulse blo(2, x) blo(1, y) blo(2, z). The function is +1 inside the black volume and zero elsewhere. (The notation differs somewhat from that of Fig. 22.)

Problems

117.1 Where is the function wal(3, x) wal(3, y) wal(3, z) located in Fig. 28?

117.2 Draw the function wal(1, x) wal(2, y) wal(3, z) of Fig. 28 in the manner of Fig. 29.

117.3 Draw the function $\frac{1}{2}$ har(0, 1, x) har(1, 1, y) har(1, 2, z) in the manner of Fig. 30.

117.4 What is the difference between the block pulse blo(2, x) used in Fig. 31 and the block pulse blo(2, x) in Fig. 22?

117.5 The arrangement of the small black and white cubes inside the large cube in Fig. 29 is uniquely determined by the arrangement of the black and white squares on the visible surface of the large cube. Is this true for all the Walsh functions in Fig. 28? Is this true for all Walsh functions? Is there a similar relationship between the values of an analytic function on a closed surface and in the enclosed volume?

117.6 Is a Haar function with three variables defined by its values on the surface of a cube? Compare the functions $\mathrm{har}(2,2,x)\,\mathrm{har}(2,2,y)\,\mathrm{har}(2,2,z)$ and $\mathrm{har}(2,3,x)\,\mathrm{har}(2,3,y)\,\mathrm{har}(2,3,z)$.

117.7 How many different block pulses according to Fig. 31 exist that are zero on the surface like the block pulse shown?

1.2 Generalized Fourier Analysis

1.2.1 Transition from Fourier Series to Fourier Transform

The Fourier transform belongs to the basic knowledge of every communications engineer. Its derivation from the Fourier series is shown here in a special way that will facilitate understanding of the more general transition from orthogonal series to orthogonal transforms[1].

Consider the orthonormal system $\{f(j,\theta)\}$ of sine and cosine elements, the first few of which are shown in Fig. 1. The elements $f(j,\theta)$ are divided into even elements $f_c(i,\theta)$, odd elements $f_s(i,\theta)$ and the constant $f(0,\theta)$:

$$f(j,\theta)=\begin{cases} f(0,\theta)=\mathrm{wal}(0,\theta)=1 & -\frac{1}{2}\leqq\theta<+\frac{1}{2}\\ f_c(i,\theta)=\sqrt{(2)}\cos 2\pi i\theta & \\ f_s(i,\theta)=\sqrt{(2)}\sin 2\pi i\theta & \\ \text{undefined} & \theta<-\frac{1}{2},\ \theta\geqq+\frac{1}{2}\end{cases} \tag{1}$$

$$\theta=t/T;\qquad i=1,2,\ldots$$

Sine and cosine elements may be continued periodically outside the interval $-\frac{1}{2}\leqq\theta<+\frac{1}{2}$ to obtain the periodic sine and cosine functions:

$$f(j,\theta)\begin{cases} f(0,\theta)=1 & -\infty<\theta<+\infty\\ f_c(i,\theta)=\sqrt{(2)}\cos 2\pi i\theta & \\ f_s(i,\theta)=\sqrt{(2)}\sin 2\pi i\theta & \end{cases} \tag{2}$$

Periodic continuation of a function defined in a finite interval is one way to extend the interval of definition. Consider a function $F(\theta)$ defined in the interval $-\frac{1}{2}\leqq\theta<\frac{1}{2}$. An example is the triangular function shown on top of Fig. 32a. If conditions required for convergence are satisfied, one may expand $F(\theta)$ into a series of the orthonormal system $\{f(j,\theta)\}$ defined

[1] The transition from the Fourier series to the Fourier transform has mainly tutorial value. A mathematical correct transition without an additional assumption is not possible, since the Fourier series uses a system of denumerable functions but the Fourier transform one of non-denumerable functions. A corresponding remark applies to the transition from orthogonal series to the generalized Fourier transforms in section 1.2.2.

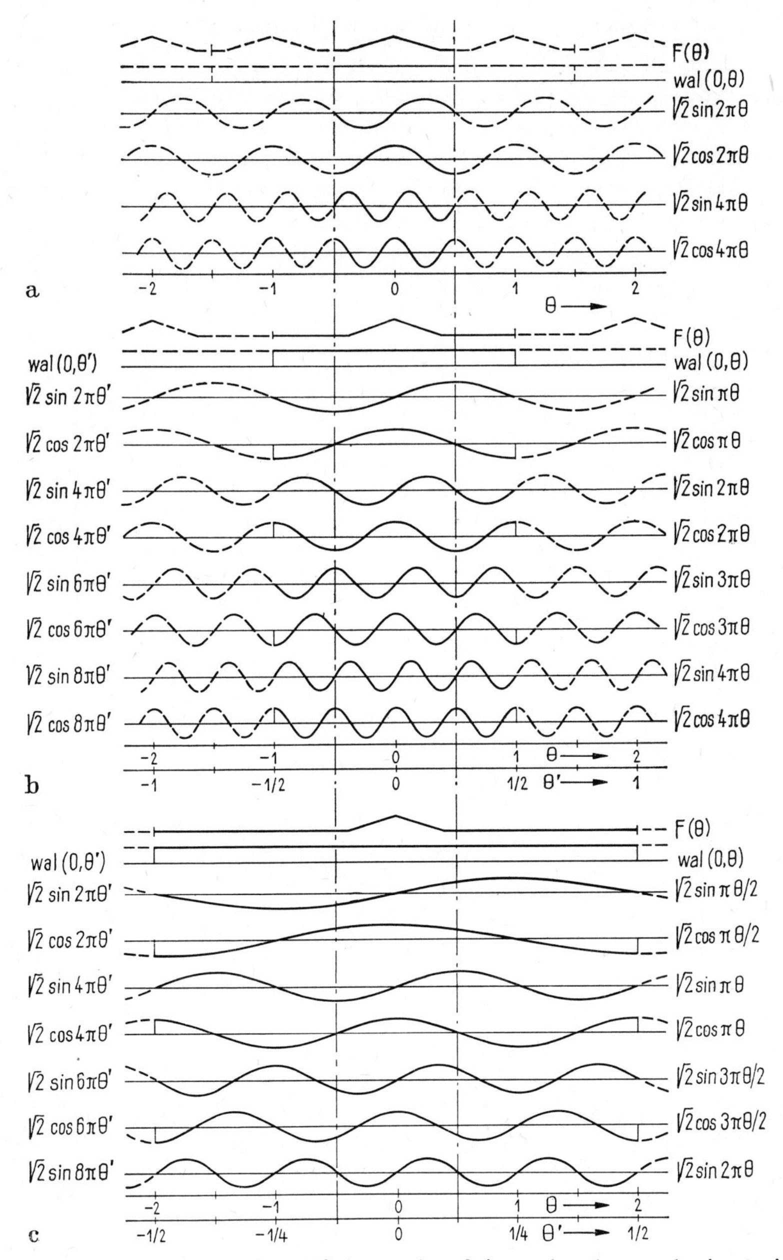

Fig. 32. Expansion of a function $F(\theta)$ in a series of sine-cosine elements having various intervals of orthogonality.

(a) $-\frac{1}{2} \leqq \theta < \frac{1}{2}$, $\{\text{wal}(0, \theta), \sqrt{(2)} \cos 2\pi i \theta, \sqrt{(2)} \sin 2\pi i \theta\}$

(b) $-1 \leqq \theta < 1$, $\{\text{wal}(0, \frac{1}{2}\theta), \sqrt{(2)} \cos 2\pi(\frac{1}{2}i) \theta, \sqrt{(2)} \sin 2\pi(\frac{1}{2}i) \theta\}$

(c) $-2 \leqq \theta < 2$, $\{\text{wal}(0, \frac{1}{4}\theta), \sqrt{(2)} \cos 2\pi(\frac{1}{4}i) \theta, \sqrt{(2)} \sin 2\pi(\frac{1}{4}i) \theta\}$

in the same interval as $F(\theta)$. The triangular function of Fig. 32a is expanded into a series of sine and cosine elements. If the triangular function is continued outside its interval of definition, one must continue the sine and cosine elements in the same way; two of the possible ways are particularly important: Periodic continuation of the triangular function requires periodic continuation of the sine and cosine elements. Hence, the periodic triangular function of Fig. 32a is expanded in a series of the periodic sine and cosine functions. If, on the other hand, the triangular function is continued by $F(\theta) \equiv 0$ outside the interval $-\frac{1}{2} \leqq \theta < \frac{1}{2}$, it has to be expanded in a series of sine and cosine pulses, which are zero outside that interval.

Let $F(\theta)$ be expanded in a series of sine and cosine elements:

$$\left.\begin{aligned}
F(\theta) &= a(0)\, f(0,\theta) + \sqrt{(2)} \sum_{i=1}^{\infty} [a_c(i) \cos 2\pi\, i\, \theta + a_s(i) \sin 2\pi\, i\, \theta] \\
a(0) &= \int_{-1/2}^{1/2} F(\theta) f(0,\theta)\, d\theta = \int_{-1/2}^{1/2} F(\theta)\, d\theta \\
a_c(i) &= \sqrt{(2)} \int_{-1/2}^{1/2} F(\theta) \cos 2\pi\, i\, \theta\, d\theta \\
a_s(i) &= \sqrt{(2)} \int_{-1/2}^{1/2} F(\theta) \sin 2\pi\, i\, \theta\, d\theta
\end{aligned}\right\} \quad (3)$$

The coefficients $a(0)$ and $a_c(i)$ are plotted for the triangular function of Fig. 32a in Fig. 33a. All coefficients $a_s(i)$ are zero, since the triangular function is an even function.

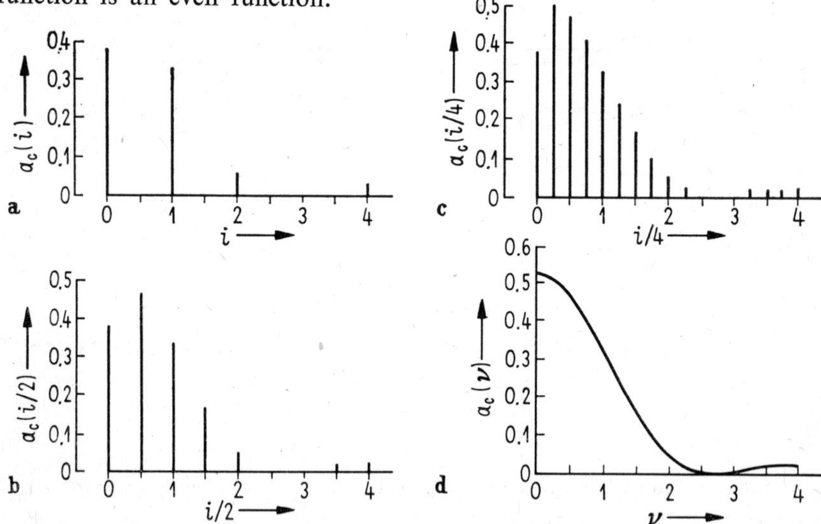

Fig. 33. Coefficients of the expansion of the triangular function $F(\theta)$ in a series of sine and cosine elements according to Fig. 32. $a_c(\nu)$ denotes the limit curve for the elements stretched by a factor $\xi \to \infty$.

Let the variable θ on the right hand side of Eq. (3) be replaced by the new variable θ':

$$\theta' = \theta/\xi, \ \xi > 1 \tag{4}$$

This substitution "stretches" the elements $\sqrt{(2)}\sin 2\pi i\theta$, $\sqrt{(2)}\cos 2\pi i\theta$ and $f(0,\theta)$ by a factor ξ. The new interval of orthogonality is $-\xi/2 \leqq \theta < \xi/2$. The orthogonal system of the stretched elements $\sqrt{(2)}\sin 2\pi i\theta'$, $\sqrt{(2)}\cos 2\pi i\theta'$ and $f(0,\theta')$ is not normalized, since these functions have the same amplitude as the original elements but are ξ-times as wide. The integral over the square of the stretched functions yields ξ rather than 1. Hence, the stretched functions have to be multiplied by $\xi^{-1/2}$ to retain normalization.

$F(\theta)$ is not stretched, but is continued into the interval $-\xi/2 \leqq \theta < -\frac{1}{2}$ and $\frac{1}{2} \leqq \theta < \xi/2$ by $F(\theta) \equiv 0$. This continuation of $F(\theta)$ and the stretching of $f(0,\theta)$, $\sqrt{(2)}\cos 2\pi i\theta$ and $\sqrt{(2)}\sin 2\pi i\theta$ is shown for $\xi = 2$ and $\xi = 4$ in Figs. 32b and c.

The expansion of $F(\theta)$ in a series of the stretched elements has the following form:

$$F(\theta) = \frac{1}{\sqrt{(\xi)}}\Big\{a(\xi,0)f(0,\theta') + \sqrt{(2)}\sum_{i=1}^{\infty}[a_c(\xi,i)\cos 2\pi i\theta' + a_s(\xi,i)\sin 2\pi i\theta']\Big\} \tag{5}$$

The factor ξ^{-1} contained in θ' may be combined with the factor i in the argument $2\pi i\theta'$. This is trivial for sine and cosine functions but it may be used as a point of departure for the generalization of the Fourier transform:

$$\left.\begin{aligned} \cos 2\pi i\theta' &= \cos 2\pi i(\theta/\xi) = \cos 2\pi(i/\xi)\theta \\ \sin 2\pi i\theta' &= \sin 2\pi i(\theta/\xi) = \sin 2\pi(i/\xi)\theta \\ f(0,\theta') &= f(0,\theta/\xi) \quad\ = f(0/\xi,\theta) \end{aligned}\right\} \tag{6}$$

The notation $f(0/\xi,\theta)$ is strictly formal and is of no consequence.

The series expansion of $F(\theta)$ assumes the following form:

$$\left.\begin{aligned} &F(\theta) = \frac{1}{\sqrt{(\xi)}}\Big\{a(\xi,0)f\Big(\frac{0}{\xi},\theta\Big) + \sqrt{(2)}\sum_{i/\xi=1/\xi}^{\infty}\Big[a_c(\xi,i)\cos 2\pi\frac{i}{\xi}\theta + a_s(\xi,i)\sin 2\pi\frac{i}{\xi}\theta\Big]\Big\} \\ &a_c(\xi,i) = \sqrt{\Big(\frac{2}{\xi}\Big)}\int_{-\xi/2}^{\xi/2}F(\theta)\cos 2\pi\frac{i}{\xi}\theta\,d\theta, \quad a(\xi,0) = \sqrt{\Big(\frac{1}{\xi}\Big)}\int_{-\xi/2}^{\xi/2}F(\theta)\,d\theta \\ &a_s(\xi,i) = \sqrt{\Big(\frac{2}{\xi}\Big)}\int_{-\xi/2}^{\xi/2}F(\theta)\sin 2\pi\frac{i}{\xi}\theta\,d\theta \end{aligned}\right\} \tag{7}$$

Introduction of new constants,

$$a_c\left(\frac{i}{\xi}\right) = \sqrt{(\xi)}\, a_c(\xi, i), \quad a_s\left(\frac{i}{\xi}\right) = \sqrt{(\xi)}\, a_s(\xi, i),$$

$$a\left(\frac{0}{\xi}\right) = \sqrt{(\xi)}\, a(\xi, 0), \tag{8}$$

yields

$$F(\theta) = \frac{1}{\xi}\left\{a\left(\frac{0}{\xi}\right) f\left(\frac{0}{\xi}, \theta\right) + \right.$$

$$\left. + \sqrt{(2)} \sum_{i/\xi = 1/\xi}^{\infty} \left[a_c\left(\frac{i}{\xi}\right) \cos 2\pi \frac{i}{\xi}\theta + a_s\left(\frac{i}{\xi}\right) \sin 2\pi \frac{i}{\xi}\theta\right]\right\} \tag{9}$$

$a(0/\xi)$ and $a_c(i/\xi)$ are plotted for $\xi = 2$ and $\xi = 4$ in Fig. 33b and c; they hold for the expansion of $F(\theta)$ in a series of the sine and cosine elements of Fig. 32b and c.

Let ξ increase to infinity; i/ξ shall remain constant:

$$\lim_{i,\xi \to \infty} a_c\left(\frac{i}{\xi}\right) = a_c(\nu), \quad \lim_{i,\xi \to \infty} a_s\left(\frac{i}{\xi}\right) = a_s(\nu) \tag{10}$$

$$i/\xi = \nu = fT$$

i may be any integer number. i as well as i/ξ are denumerable. ν, on the other hand, must be allowed to be any non-negative real number and thus be non-denumerable or some of the following integrals would be zero. Hence, the Fourier series contains denumerably many orthogonal functions, but the Fourier transform contains non-denumerably many.

The limits $a_c(\nu)$ and $a_s(\nu)$ follow readily from Eqs. (6) and (7):

$$a_c(\nu) = \lim_{\xi \to \infty} \sqrt{(2)} \int_{-\xi/2}^{\xi/2} F(\theta) \cos 2\pi \frac{i}{\xi}\theta\, d\theta = \sqrt{(2)} \int_{-\infty}^{\infty} F(\theta) \cos 2\pi\, \nu\theta\, d\theta$$

$$a_s(\nu) = \lim_{\xi \to \infty} \sqrt{(2)} \int_{-\xi/2}^{\xi/2} F(\theta) \sin 2\pi \frac{i}{\xi}\theta\, d\theta = \sqrt{(2)} \int_{-\infty}^{\infty} F(\theta) \sin 2\pi\, \nu\, \theta\, d\theta \tag{11}$$

In order to find an integral representation for $F(\theta)$, consider a certain value $\theta = \theta_0$. Equation (9) yields $F(\theta_0)$ as a sum of denumerably many terms, which may be plotted along the numbers axis at the points i/ξ as shown in Fig. 34. The distance between the plotted terms is equal to $1/\xi$. Hence, the sum of the terms multiplied by $1/\xi$ as given by Eq. (9) is equal to the area under the step function of Fig. 34. Using Eq. (9), one may approximate this area arbitrarily close for sufficiently large values of ξ by the following integral:

$$F(\theta) = \sqrt{(2)} \int_0^{\infty} [a_c(\nu) \cos 2\pi\, \nu\, \theta + a_s(\nu) \sin 2\pi\, \nu\, \theta]\, d\nu \tag{12}$$

The lower limit of the integral is zero, because the lower limit of the sum in Eq. (9) approaches zero. The first term of the sum of Eq. (9) may be neglected, since it contributes arbitrarily little for large values of ξ. The variable ν in Eq. (12) must assume the values of all real positive numbers

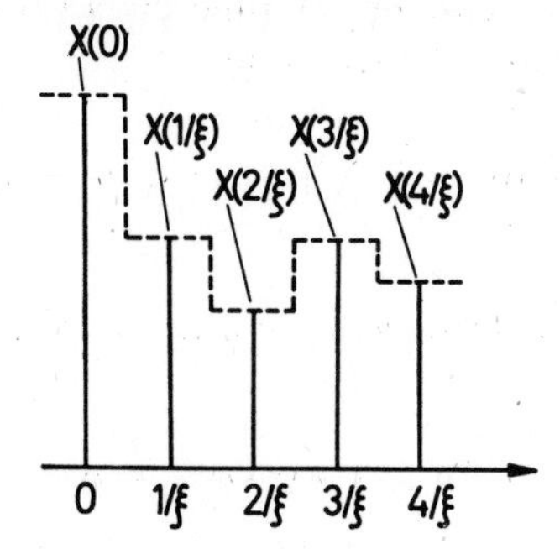

Fig. 34. Transition from Fourier series to Fourier transform.

$X(0) = a\left(\frac{0}{\xi}\right) f\left(\frac{0}{\xi}, \theta_0\right), \quad X\left(\frac{i}{\xi}\right) = a_c\left(\frac{i}{\xi}\right)\sqrt{(2)}\cos 2\pi \frac{i}{\xi}\theta + a_s\left(\frac{i}{\xi}\right)\sqrt{(2)}\sin 2\pi \frac{i}{\xi}\theta$

and not only of denumerably many of them, or the integral could not be interpreted as a Riemann integral.

Equation (11) shows that $a_c(\nu)$ is an even and $a_s(\nu)$ is an odd function of ν. Hence, Eq. (12) may be rewritten into the following form:

$$\begin{aligned} F(\theta) &= \int_{-\infty}^{\infty} [A(\nu)\cos 2\pi\nu\theta + B(\nu)\sin 2\pi\nu\theta]\,d\nu \\ A(\nu) &= \tfrac{1}{2}\sqrt{(2)}\,a_c(\nu), \quad B(\nu) = \tfrac{1}{2}\sqrt{(2)}\,a_s(\nu) \end{aligned} \tag{13}$$

$a_s(\nu)$ is identically zero for the triangular function of Fig. 32; $a_c(\nu)$ is plotted in Fig. 33d according to the following formula:

$$a_c(\nu) = 2\sqrt{(2)}\int_0^{3/8}\left(1 - \frac{8}{3}\theta\right)\cos 2\pi\nu\theta\,d\theta = \frac{3}{8}\sqrt{(2)}\left(\frac{\sin 3\pi\nu/8}{3\pi\nu/8}\right)^2$$

Problem

121.1 $a_c(0)$ has the same value in Fig. 33a to c but not in Fig. 33d. What is the reason and how much larger is $a_c(0)$ in Fig. 33d?

1.2.2 Generalized Fourier Transform[1]

Consider a system of functions $\{f(0,\theta), f_c(i,\theta), f_s(i,\theta)\}$ orthonormalized in the interval $-\frac{1}{2}\Theta \leq \theta < \frac{1}{2}\Theta$. The subscript c indicates an even function and the subscript s an odd function. Θ may be finite or infinite. Hence,

[1] For other generalizations *see* [1, 2].

the results will be applicable to functions having an infinite interval of orthogonality, such as the functions of the parabolic cylinder. Let all functions $f_c(i, \theta)$ be nonnegative for $\theta = 0$, and let all functions $f_s(i, \theta)$ cross from negative to positive values at $\theta = 0$. The functions do not have to be continuous or differentiable. A function $F(\theta)$ defined in the interval $-\frac{1}{2}\Theta \leqq \theta < \frac{1}{2}\Theta$ is expanded in a series:

$$\left.\begin{aligned} F(\theta) &= a(0) f(0, \theta) + \sum_{i=1}^{\infty} [a_c(i) f_c(i, \theta) + a_s(i) f_s(i, \theta)] \\ a_c(i) &= \int_{-\Theta/2}^{\Theta/2} F(\theta) f_c(i, \theta)\, d\theta, \quad a_s(i) = \int_{-\Theta/2}^{\Theta/2} F(\theta) f_s(i, \theta)\, d\theta \\ a(0) &= \int_{-\Theta/2}^{\Theta/2} F(\theta) f(0, \theta)\, d\theta \end{aligned}\right\} \tag{1}$$

θ is replaced[1] by θ' in the functions $f(0, \theta)$, $f_c(i, \theta)$ and $f_s(i, \theta)$:

$$\theta' = \theta/y, \quad y = y(\xi) > 1, \quad \lim_{\xi \to \infty} y(\xi) = \infty \tag{2}$$

The expansion of $F(\theta)$ in a series of the stretched functions is obtained in analogy to Eq. (1.2.1-5):

$$F(\theta) = \frac{1}{\sqrt{(y)}} \left\{ a(\xi, 0) f(0, \theta') + \sum_{i=1}^{\infty} [a_c(\xi, i) f_c(i, \theta') + a_s(\xi, i) f_s(i, \theta')] \right\} \tag{3}$$

The stretched functions are orthonormal in the interval $-\frac{1}{2} y\, \Theta \leqq \theta < \frac{1}{2} y\, \Theta$. $F(\theta)$ is continued by $F(\theta) \equiv 0$ into the intervals $-\frac{1}{2} y\, \Theta \leqq \theta < -\frac{1}{2}\Theta$ and $\frac{1}{2}\Theta \leqq \theta < \frac{1}{2} y\, \Theta$.

The factor $1/y$ is combined with i so that θ instead of θ' may be written on the right hand side of Eq. (3). $2\pi\, i(\theta/\xi)$ had been replaced trivially by $2\pi(i/\xi)\,\theta$ in Eq. (1.2.1-6); since i and θ are not necessarily connected as product in $f_c(i, \theta)$ and $f_s(i, \theta)$ the following substitutions must be considered purely formal until proved otherwise. In particular, i/ξ should be considered a symbol rather than a fraction:

$$\left.\begin{aligned} f_c(i, \theta') &= f_c(i, \theta/y) = f_c(i/\xi, \theta) \\ f_s(i, \theta') &= f_s(i, \theta/y) = f_s(i/\xi, \theta) \\ f(0, \theta') &= f(i, \theta/y) \ = f(0/\xi, \theta) \end{aligned}\right\} \tag{4}$$

[1] The method used applies to a large class of systems of functions. Exact mathematical proofs can be obtained without excessive mathematical requirements for individual systems of functions only. For instance, the results of this section seem to apply for dyadic rational values of $i/\xi = \mu$ only in the case of Walsh functions; in reality they apply to all real values of i/ξ.

The series expansion of $F(\theta)$ assumes the following form:

$$\left.\begin{aligned}
F(\theta) &= \frac{1}{\sqrt{(y)}}\left\{a(\xi,0)f\left(\frac{0}{\xi},\theta\right)+\sum_{i/\xi=1/\xi}^{\infty}\left[a_c(\xi,i)f_c\left(\frac{i}{\xi},\theta\right)+\right.\right.\\
&\qquad\qquad\left.\left.+\,a_s(\xi,i)f_s\left(\frac{i}{\xi},\theta\right)\right]\right\}\\
a_c(\xi,i) &= \frac{1}{\sqrt{(y)}}\int_{-y\Theta/2}^{y\Theta/2} F(\theta)f_c\left(\frac{i}{\xi},\theta\right)d\theta,\\
a(\xi,0) &= \frac{1}{\sqrt{(y)}}\int_{-y\Theta/2}^{y\Theta/2} F(\theta)\,d\theta\\
a_s(\xi,i) &= \frac{1}{\sqrt{(y)}}\int_{-y\Theta/2}^{y\Theta/2} F(\theta)f_s\left(\frac{1}{\xi},\theta\right)d\theta
\end{aligned}\right\} \quad (5)$$

New coefficients are introduced:

$$a_c\left(\frac{i}{\xi}\right)=\sqrt{(y)}\,a_c(\xi,i),\quad a_s\left(\frac{i}{\xi}\right)=\sqrt{(y)}\,a_s(\xi,i),\quad a\left(\frac{0}{\xi}\right)=\sqrt{(y)}\,a(\xi,0) \quad (6)$$

In order to make Eqs. (5) and (6) more than a formal notation, one must demand that the coefficients $a_c(i/\xi)$ or $a_s(i/\xi)$ have either the same value for all values of i and ξ, as long as $i/\xi=\mu$ is constant, or that they converge[1] toward a limit for large values of i and ξ:

$$\lim_{i,\xi\to\infty} a_c\left(\frac{i}{\xi}\right)=a_c(\mu),\quad \lim_{i,\xi\to\infty} a_s\left(\frac{i}{\xi}\right)=a_s(\mu),\quad \frac{i}{\xi}=\mu \quad (7)$$

Again, one has to postulate that μ is a nonnegative real number and thus is non-denumerable, while i or i/ξ is denumerable.

The limits in Eq. (7) exist, if $f_c(i/\xi,\theta)$ and $f_s(i/\xi,\theta)$ approach limit functions $f_c(\mu,\theta)$ and $f_s(\mu,\theta)$ that are defined as follows[2]:

$$\left.\begin{aligned}
\lim_{i,\xi\to\infty}\int_{-y\Theta/2}^{y\Theta/2} F(\theta)f_c\left(\frac{i}{\xi},\theta\right)d\theta &= \lim_{\xi\to\infty}\int_{-y\Theta/2}^{y\Theta/2} F(\theta)f_c(\mu,\theta)\,d\theta\\
\lim_{i,\xi\to\infty}\int_{-y\Theta/2}^{y\Theta/2} F(\theta)f_s\left(\frac{i}{\xi},\theta\right)d\theta &= \lim_{\xi\to\infty}\int_{-y\Theta/2}^{y\Theta/2} F(\theta)f_s(\mu,\theta)\,d\theta\\
y &= y(\xi)
\end{aligned}\right\} \quad (8)$$

---

[1] The left hand limit shall be taken, if left and right hand limit differ.

[2] The integrals shall represent Cauchy's principal value. They must hold for all quadratically integrable functions $F(\theta)$.

The functions $f_c(i/\xi, \theta)$ and $f_s(i/\xi, \theta)$ converge in the interval $-\frac{1}{2} y \Theta \leq \leq \theta < \frac{1}{2} y \Theta$ to the limit functions $f_c(\mu, \theta)$ and $f_s(\mu, \theta)$. This type of convergence is called *weak convergence* [3].

It follows from Eq. (5) to (8):

$$a_c(\mu) = \lim_{\xi \to \infty} \int_{-y\Theta/2}^{y\Theta/2} F(\theta) f_c\left(\frac{i}{\xi}, \theta\right) d\theta$$
$$a_s(\mu) = \lim_{\xi \to \infty} \int_{-y\Theta/2}^{y\Theta/2} F(\theta) f_s\left(\frac{i}{\xi}, \theta\right) d\theta \tag{9}$$

Let $F(\theta)$ be a function that vanishes outside a finite interval. Equations (9) reduce to the following simplified form:

$$a_c(\mu) = \int_{-\infty}^{\infty} F(\theta) f_c(\mu, \theta)\, d\theta, \quad a_s(\mu) = \int_{-\infty}^{\infty} F(\theta) f_s(\mu, \theta)\, d\theta \tag{10}$$

In order to find an integral representation for $F(\theta)$, consider a certain value $\theta = \theta_0$. Equation (5) yields $F(\theta_0)$ as a sum of denumerably many terms which may be plotted along the numbers axis at the points $i/y = i/y(\xi)$ instead of i/ξ as in Fig. 34. The distance between the plotted terms is $1/y$. Hence, the sum of the terms multiplied by $1/y$ as given by Eq. (5) is equal to the area under a step function. This area may be represented by an integral, if ξ and thus $y(\xi)$ grow beyond all bounds:

$$F(\theta) = \int_{0}^{\infty} [a_c(\mu) f_c(\mu, \theta) + a_s(\mu) f_s(\mu, \theta)]\, d\theta \tag{11}$$

$a_c(\mu)$ and $a_s(\mu)$ are called the generalized Fourier transform of $F(\theta)$ for the functions $f_c(\mu, \theta)$ and $f_s(\mu, \theta)$. Equation (11) is an integral representation of $F(\theta)$ or its generalized inverse Fourier transform. Whether these integrals actually exist cannot be stated without specifying the functions $f_c(\mu . \theta)$ and $f_s(\mu, \theta)$ more closely. The variable μ plays the same role as the variable ν in the usual Fourier transform. Hence, μ is called a generalized —and normalized—frequency.

$f_c(i, \theta)$ and $f_s(i, \theta)$ are defined for positive integers i only. Hence, $f_c(\mu, \theta)$ and $f_s(\mu, \theta)$ are defined for non-negative real numbers μ only. One may extend the definitions to negative real numbers:

$$f_c(\mu, \theta) = f_c(-\mu, \theta), \quad f_s(\mu, \theta) = -f_s(-\mu, \theta) \tag{12}$$

$f_c(\mu, \theta)$ is an even function of θ as well as of μ, and $f_s(\mu, \theta)$ is an odd function of θ as well as of μ.

Equations (9) and (10) show that $a_c(\mu)$ is an even and $a_s(\mu)$ is an odd function of μ. Hence, Eq. (11) may be brought into the form of

Eq. (1.2.1-13):

$$F(\theta) = \int_{-\infty}^{\infty} [A(\mu) f_c(\mu, \theta) + B(\mu) f_s(\mu, \theta)] \, d\mu \tag{13}$$

$$A(\mu) = \tfrac{1}{2} a_c(\mu), \quad B(\mu) = \tfrac{1}{2} a_s(\mu)$$

Problem

122.1 The weak convergence in Eq. (8) does not define the functions $f_c(\mu, \theta)$ and $f_s(\mu, \theta)$ completely. In which way may these functions vary and still satisfy Eq. (8)?

1.2.3 Invariance of Orthogonality to the Generalized Fourier Transform

Consider the function $G(\mu)$:

$$G(\mu) = \sqrt{(2)}\,[A(\mu) + B(\mu)] = \tfrac{1}{2}\sqrt{(2)}\,[a_c(\mu) + a_s(\mu)] \tag{1}$$

Since $A(\mu)$ is even and $B(\mu)$ is odd, one obtains for $G(-\mu)$:

$$G(-\mu) = \sqrt{(2)}\,[A(-\mu) + B(-\mu)] = \sqrt{(2)}\,[A(\mu) - B(\mu)]$$

$A(\mu)$ and $B(\mu)$ may be regained from $G(\mu)$:

$$A(\mu) = \tfrac{1}{4}\sqrt{(2)}\,[G(\mu) + G(-\mu)], \quad B(\mu) = \tfrac{1}{4}\sqrt{(2)}\,[G(\mu) - G(-\mu)] \tag{2}$$

Using $G(\mu)$ one may rewrite Eqs. (1.2.2-10) and (1.2.2-13) into the form of Eqs. (1.1.3-7) and (1.1.3-8):

$$F(\theta) = \tfrac{1}{2}\sqrt{(2)} \int_{-\infty}^{\infty} G(\mu)\,[f_c(\mu, \theta) + f_s(\mu, \theta)] \, d\mu \tag{3}$$

$$G(\mu) = \tfrac{1}{2}\sqrt{(2)} \int_{-\infty}^{\infty} F(\theta)\,[f_c(\mu, \theta) + f_s(\mu, \theta)] \, d\theta \tag{4}$$

Use is made in Eq. (3) of the fact that the integrals of $A(\mu) f_s(\mu, \theta)$ and $B(\mu) f_c(\mu, \theta)$ vanish.

Consider a system $\{f(j, \theta)\}$ of orthonormal functions that vanish outside a finite interval:

$$\int_{-\infty}^{\infty} f(j, \theta) f(k, \theta) \, d\theta = \delta_{jk} \tag{5}$$

Let $g(j, \mu)$ denote the generalized Fourier transform of $f(j, \theta)$. It follows from Eq. (4)

$$g(j, \mu) = \tfrac{1}{2}\sqrt{(2)} \int_{-\infty}^{\infty} f(j, \theta)\,[f_c(\mu, \theta) + f_s(\mu, \theta)] \, d\theta \tag{6}$$

Equation (5) may be transformed as follows:

$$\left.\begin{aligned}\int_{-\infty}^{\infty} f(j,\theta\left\{\tfrac{1}{2}\sqrt{(2)}\int_{-\infty}^{\infty} g(k,\mu)[f_c(\mu,\theta)+f_s(\mu,\theta)]\,d\mu\right\}d\theta &= \delta_{jk}\\ \int_{-\infty}^{\infty} g(k,\mu)\left\{\tfrac{1}{2}\sqrt{(2)}\int_{-\infty}^{\infty} f(j,\theta)[f_c(\mu,\theta)+f_s(\mu,\theta)]\,d\theta\right\}d\mu &= \delta_{jk}\\ \int_{-\infty}^{\infty} g(k,\mu)\,g(j,\mu)\,d\mu &= \delta_{jk}\end{aligned}\right\} \qquad (7)$$

An orthogonal system $\{f(j,\theta)\}$ that vanishes outside a finite interval is transformed by the generalized Fourier transform into an orthogonal system $\{g(j,\mu)\}$.

1.2.4 Examples of the Generalized Fourier Transform

Consider the generalized Fourier transform of the triangular function of Fig. 35 for Legendre polynomials [1]:

$$\mathrm{P}_0(x) = 1, \quad \mathrm{P}_1(x) = x, \quad \mathrm{P}_2(x) = \tfrac{1}{2}(3x^2-1), \quad \text{etc.}$$

The interval of orthogonality is $-1 \leqq x < +1$. $x = 2\theta$ is substituted and the following transformations are made:

$$\left.\begin{aligned} f(0,\theta) &= \mathrm{P}_0(2\theta)\\ f_c(i,\theta) &= \mathrm{P}_c(i,\theta) = \quad (-1)^i(4i+1)^{1/2}\,\mathrm{P}_{2i}(2\theta)\\ f_s(i,\theta) &= \mathrm{P}_s(i,\theta) = -(-1)^i(4i-1)^{1/2}\,\mathrm{P}_{2i-1}(2\theta)\\ & \qquad i = 1,2,\ldots\end{aligned}\right\} \qquad (1)$$

The system $\{f(0,\theta), \mathrm{P}_c(i,\theta), \mathrm{P}_s(i,\theta)\}$ is orthonormal in the interval $-\frac{1}{2} \leqq \theta < +\frac{1}{2}$. All functions $\mathrm{P}_c(i,\theta)$ are positive for $\theta = 0$, and all functions $\mathrm{P}_s(i,\theta)$ have a positive differential quotient. Written explicitly, the first few polynomials read as follows:

$$\begin{aligned} &f(0,\theta) = 1, \quad \mathrm{P}_s(1,\theta) = 2\sqrt{(3)}\theta, \quad \mathrm{P}_c(1,\theta) = -\tfrac{1}{2}\sqrt{(5)}(12\theta^2-1)\\ &\mathrm{P}_s(2,\theta) = -\sqrt{(7)}(20\theta^3-3\theta), \quad \mathrm{P}_c(2,\theta) = \tfrac{1}{8}\sqrt{(9)}(560\theta^4-120\theta^2+3)\end{aligned} \qquad (2)$$

The coefficients $a_c(i)$ and $a_s(i)$ for Fig. 35a may be readily computed:

$$\begin{aligned} a_c(i) &= \int_{-1/2}^{1/2} F(\theta)\,\mathrm{P}_c(i,\theta)\,d\theta = 2\int_0^{3/8}(1-\tfrac{8}{3}\theta)\,\mathrm{P}_c(i,\theta)\,d\theta\\ a_s(i) &= \int_{-1/2}^{1/2} F(\theta)\,\mathrm{P}_s(i,\theta)\,d\theta = 0, \quad a(0) = 2\int_0^{3/8}(1-\tfrac{8}{3}\theta)\,d\theta\end{aligned} \qquad (3)$$

$a_c(i)$ and $a(0)$ are plotted in Fig. 36a.

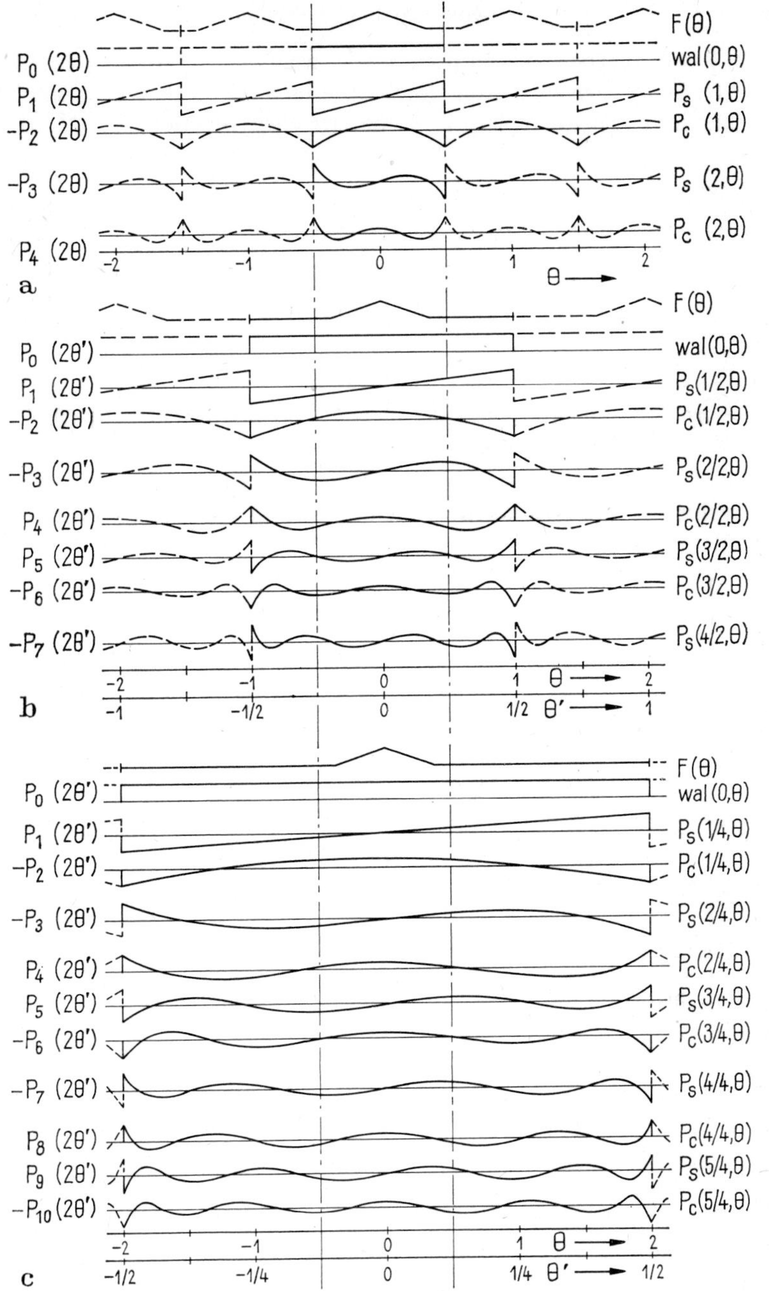

Fig. 35. Expansion of a function $F(\theta)$ in a series of Legendre polynomials having various intervals of orthogonality.

(a) $-\frac{1}{2} \leqq \theta < \frac{1}{2}$, $\{\mathrm{wal}(0, \theta), P_c(i, \theta), P_s(i, \theta)\}$

(b) $-1 \leqq \theta < 1$, $\{\mathrm{wal}(0, \theta), P_c(i/2, \theta), P_s(i/2, \theta)\}$

(c) $-2 \leqq \theta < 2$, $\{\mathrm{wal}(0, \theta), P_c(i/4, \theta), P_s(i/4, \theta)\}$

Let θ in Eq. (2) be replaced by $\theta' = \theta/y$, where $y = y(\xi) = \xi = 2$. $P_c(i, \theta)$ and $P_s(i, \theta)$ are stretched over double the interval as shown in Fig. 35b. The functions (2) are replaced by the stretched functions $P_c(i/2, \theta)$ and $P_s(i/2, \theta)$:

$$\left.\begin{aligned} P_s(1/2, \theta) &= P_s(1, \theta/2) = 2\sqrt{(3)}(\tfrac{1}{2}\theta) \\ P_c(1/2, \theta) &= P_c(1, \theta/2) = -\tfrac{1}{2}\sqrt{(5)}[12(\tfrac{1}{2}\theta)^2 - 1] \\ P_s(2/2, \theta) &= P_s(2, \theta/2) = -\sqrt{(7)}[20(\tfrac{1}{2}\theta)^3 - 3(\tfrac{1}{2}\theta)] \\ P_c(2/2, \theta) &= P_c(2, \theta/2) = \tfrac{1}{8}\sqrt{(9)}[560(\tfrac{1}{2}\theta)^4 - 120(\tfrac{1}{2}\theta)^2 + 3] \end{aligned}\right\} \quad (4)$$

The coefficients $a_c(i/2)$ have the following value:

$$a_c(i/2) = \int_{-1}^{+1} F(\theta)\, P_c(i/2, \theta)\, d\theta = 2\int_{0}^{3/8} (1 - \tfrac{8}{3}\theta)\, P_c(i/2, \theta)\, d\theta \quad (5)$$

Values of $a_c(i/2)$ are plotted in Fig. 36b. They do not have exactly the same values as the coefficients $a_c(i)$ of Fig. 36a since, e.g., $P_c(2/2, \theta)$ is not equal $P_c(1, \theta)$.

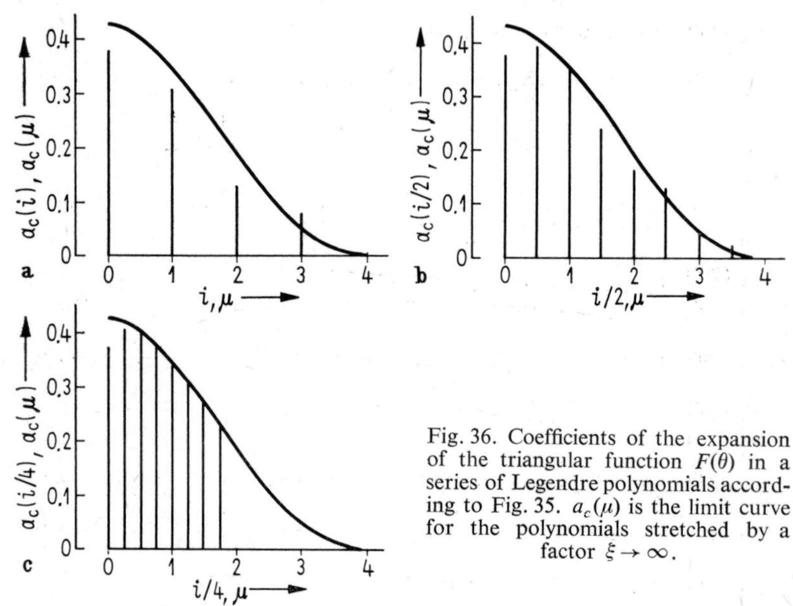

Fig. 36. Coefficients of the expansion of the triangular function $F(\theta)$ in a series of Legendre polynomials according to Fig. 35. $a_c(\mu)$ is the limit curve for the polynomials stretched by a factor $\xi \to \infty$.

Let the functions (2) be stretched over four times the interval by the substitution $\theta' = \theta/y$, where $y = y(\xi) = \xi = 4$ as shown in Fig. 35c:

$$\left.\begin{aligned} P_s(1/4, \theta) &= P_s(1, \theta/4) = 2\sqrt{(3)}(\tfrac{1}{4}\theta) \\ P_c(1/4, \theta) &= P_c(1, \theta/4) = -\tfrac{1}{2}\sqrt{(5)}[12(\tfrac{1}{4}\theta)^2 - 1] \\ P_s(2/4, \theta) &= P_s(2, \theta/4) = -\sqrt{(7)}[20(\tfrac{1}{4}\theta)^3 - 3(\tfrac{1}{4}\theta)] \\ P_c(2/4, \theta) &= P_c(2, \theta/4) = \tfrac{1}{8}\sqrt{(9)}[560(\tfrac{1}{4}\theta)^4 - 120(\tfrac{1}{4}\theta)^2 + 3] \end{aligned}\right\} \quad (6)$$

Some coefficients $a_c(i/4)$ are plotted in Fig. 36c:

$$a_c(i/4) = \int_{-2}^{+2} F(\theta)\, \mathrm{P}_c(i/4, \theta)\, d\theta = 2 \int_{0}^{3/8} (1 - \tfrac{8}{3}\theta)\, \mathrm{P}_c(i/4, \theta)\, d\theta \qquad (7)$$

In order to compute the limit $a_c(i/\xi)$ for large values of i and ξ, one needs $\mathrm{P}_c(i/\xi, \theta) = \mathrm{P}_c(i, \theta/\xi)$ for large values of i and small values of θ/ξ. An asymptotic series for Legendre polynomials $\mathrm{P}_j(x)$ is known that holds for large values of j and for small values of x:

$$\mathrm{P}_j(x) \cong \frac{\sqrt{(2)}}{\sqrt{[\pi j \sqrt{(1-x^2)}]}} \left\{ \left(1 - \frac{1}{4j} \sin\left[\left(j + \frac{1}{2}\right) \cos^{-1} x + \frac{1}{4}\pi\right] + \right.\right.$$
$$\left.\left. - \frac{x}{8j\sqrt{(1-x^2)}} \cos\left[\left(j + \frac{1}{2}\right) \cos^{-1} x + \frac{1}{4}\pi\right]\right)\right\} \qquad (8)$$

Using Eq. (1) one obtains:

$$\mathrm{P}_c(i, \theta/\xi) \cong \frac{2}{\sqrt{(\pi)}} \cos 4 \frac{i}{\xi} \theta$$

The limit functions $\mathrm{P}_c(\mu, \theta)$ and $a_c(\mu)$ follow for $\xi \to \infty$:

$$\mathrm{P}_c(\mu, \theta) = \frac{2}{\sqrt{(\pi)}} \cos 4\mu\, \theta \qquad (9)$$

$$a_c(\mu) = \int_{-\infty}^{\infty} F(\theta)\, \mathrm{P}_c(\mu, \theta)\, d\theta = \frac{4}{\sqrt{(\pi)}} \int_{0}^{3/8} \left(1 - \frac{8}{3}\theta\right) \cos 4\mu\, \theta\, d\theta$$
$$= \frac{3}{4\sqrt{(\pi)}} \left(\frac{\sin 3\mu/4}{3\mu/4}\right)^2 \qquad (10)$$

$a_c(\mu)$ is the generalized Fourier transform of the triangular function of Fig. 35 for Legendre polynomials. It is plotted in Fig. 36a to c. One may readily see how the coefficients $a_c(i)$, $a_c(i/2)$ and $a_c(i/4)$ converge to $a_c(\mu)$.

$a_c(\nu)$ in Fig. 33 and $a_c(\mu)$ in Fig. 36 are equal except for scale factors. One may see from the differential equation of Legendre polynomials that this is generally so:

$$(1 - x^2)\, z'' - 2x\, z' + j(j+1)\, z = 0; \quad j = 0, 1, 2, \ldots \qquad (11)$$

This equation reduces for small values of x and large values of j to the differential equation of sine and cosine functions:

$$z'' + j^2 z = 0 \qquad (12)$$

Hence, the generalization of the Fourier transform is mainly of interest for systems of orthogonal functions that are not defined by such differential equations, which are reduced by stretching to the one of sine and cosine functions. Since Walsh functions are defined by a difference rather than a differential equation, they may be expected to yield a more rewarding result than Legendre polynomials. The generalization of the Fourier

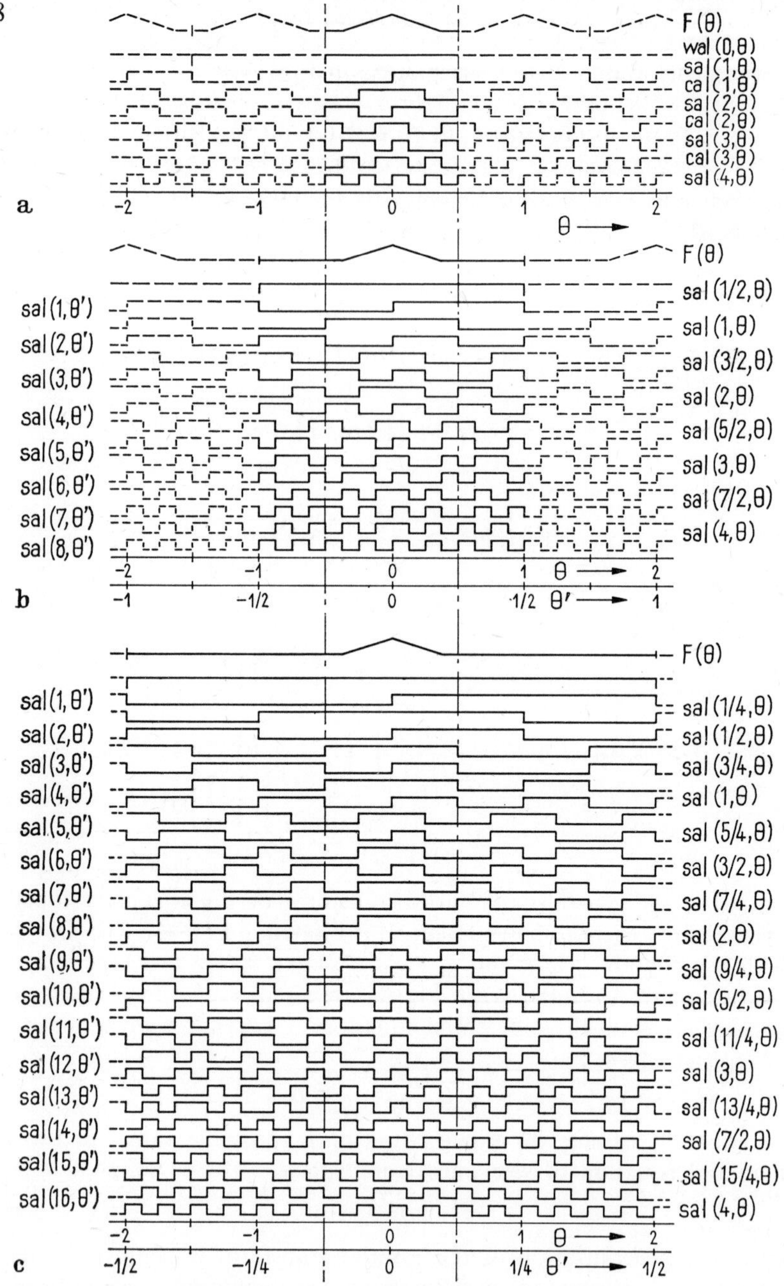

Fig. 37. Expansion of a function $F(\theta)$ in a series of Walsh elements having various intervals of orthogonality.

(a) $-\frac{1}{2} \leqq \theta < \frac{1}{2}$, $\{\mathrm{wal}(0, \theta), \mathrm{cal}(i, \theta), \mathrm{sal}(i, \theta)\}$

(b) $-1 \leqq \theta < 1$, $\{\mathrm{wal}(0, \theta), \mathrm{cal}(i/2, \theta), \mathrm{sal}(i/2, \theta)\}$

(c) $-2 \leqq \theta < 2$, $\{\mathrm{wal}(0, \theta), \mathrm{cal}(i/4, \theta), \mathrm{sal}(i/4, \theta)\}$

transform to the Walsh-Fourier transform is due to Fine. However, Fine did not distinguish between even and odd functions. This distinction is important for the applications of Walsh-Fourier analysis to communications. The mathematically rigorous theory for Walsh functions separated into even and odd functions—that is cal and sal functions—is due to Pichler [2].

Let the functions $f(0, \theta), f_c(i, \theta)$ and $f_s(i, \theta)$ represent Walsh functions:

$$f(0, \theta) = \text{wal}(0, \theta), \quad f_c(i, \theta) = \text{cal}(i, \theta), \quad f_s(i, \theta) = \text{sal}(i, \theta) \tag{13}$$

The triangular function of Fig. 37a yields the coefficients

$$a(0) = \int_{-1/2}^{1/2} F(\theta)\, \text{wal}(0, \theta)\, d\theta = 2 \int_0^{3/8} (1 - \tfrac{8}{3}\theta) d\theta,$$

$$a_c(i) = 2 \int_0^{3/8} (1 - \tfrac{8}{3}\theta)\, \text{cal}(i, \theta)\, d\theta, \qquad a_s(i) = 0$$

Fig. 38a shows some values of $a(0)$ and $a_c(i)$.

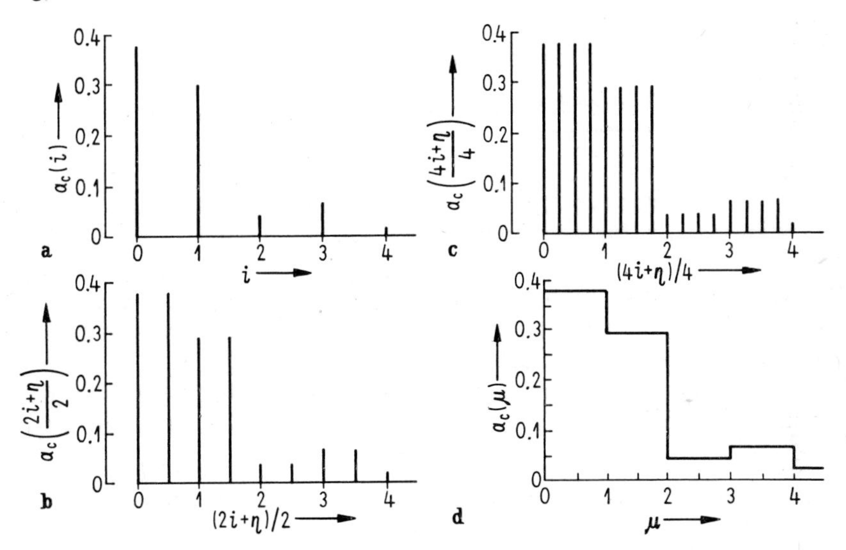

Fig. 38. Coefficients of the expansion of the triangular function $F(\theta)$ in a series of Walsh elements according to Fig. 37. $a_c(\mu)$ is the limit curve for the elements stretched by a factor $\xi \to \infty$.

With $y(\xi) = \xi$ one obtains $\text{cal}(i/\xi, \theta) = \text{cal}(i, \theta/\xi)$ and $\text{sal}(i/\xi, \theta) = \text{sal}(i, \theta/\xi)$. Inspection of Fig. 37a to c shows that $\text{cal}(2i/2, \theta)$ and $\text{cal}(4i/4, \theta)$ are equal to the function $\text{cal}(i, \theta)$ continued periodically over double or four times the original interval of definition. This result may also be inferred readily from the difference equation (1.1.4-2). Hence, it holds in the interval $-\frac{1}{2} \leqq \theta < \frac{1}{2}$:

$$\begin{gathered} \operatorname{cal}(i, \theta) = \operatorname{cal}(\xi\, i, \theta/\xi) = \operatorname{cal}(\xi\, i/\xi, \theta) \\ \xi = 2^r, \quad r = 1, 2, \ldots; \quad -\tfrac{1}{2} \leqq \theta < \tfrac{1}{2} \end{gathered}$$

Inspection of Fig. 37a to c shows further that the following relations hold in the interval $-\frac{1}{2} \leqq \theta < \frac{1}{2}$:

$$\begin{aligned} \operatorname{cal}(i, \theta) &= \operatorname{cal}(2i/2, \theta) = \operatorname{cal}[(2i+1)/2, \theta] \\ &= \operatorname{cal}(4i/4, \theta) = \operatorname{cal}[(4i+1)/4, \theta] = \operatorname{cal}[(4i+2)/4, \theta] \\ &\qquad = \operatorname{cal}[(4i+3)/4, \theta] \\ &= \operatorname{cal}[(\xi\, i + \eta)/\xi, \theta] \quad \eta = 0, 1, \ldots, \xi - 1; \ \xi = 2^r \end{aligned} \tag{14}$$

Substituting $(\xi\, i + \eta)/\xi = \mu$, $i \leqq \mu < i + 1$, one obtains:

$$\left.\begin{aligned} &\operatorname{cal}(\mu, \theta) = \operatorname{cal}(i, \theta), \quad i \leqq \mu < i + 1, \quad i = 1, 2, \ldots \\ &\operatorname{cal}(\mu, \theta) = \operatorname{wal}(0, \theta), \quad 0 \leqq \mu < 1 \\ &\qquad -\tfrac{1}{2} \leqq \theta < \tfrac{1}{2} \end{aligned}\right\} \tag{15}$$

Corresponding relations are obtained for $\operatorname{sal}(\mu, \theta)$:

$$\begin{gathered} \operatorname{sal}(\mu, \theta) = \operatorname{sal}(i, \theta), \quad i - 1 < \mu \leqq i, \quad i = 1, 2, \ldots \\ -\tfrac{1}{2} \leqq \theta < \tfrac{1}{2} \end{gathered} \tag{16}$$

The limit functions $\operatorname{cal}(\mu, \theta)$ and $\operatorname{sal}(\mu, \theta)$ have been derived here in an heuristic manner for the interval $-\frac{1}{2} \leqq \theta < \frac{1}{2}$. Pichler has obtained $\operatorname{cal}(\mu, \theta)$ and $\operatorname{sal}(\mu, \theta)$ in a mathematically rigorous way for the whole interval $-\infty < \theta < \infty$, but his proofs require a very good command of mathematics. Figures 12 and 13 show a very ingenious representation of the functions $\operatorname{cal}(\mu, \theta)$ and $\operatorname{sal}(\mu, \theta)$ found by him [2].

Functions that are identical in the interval $-\frac{1}{2} < \theta < \frac{1}{2}$ yield the same expansion coefficients for $F(\theta)$. Hence, one obtains for $a_c(\mu)$ and $a_s(\mu)$:

$$\left.\begin{aligned} &a_c(\mu) = a_c(i) = \int_{-1/2}^{1/2} F(\theta) \operatorname{cal}(i, \theta)\, d\theta \qquad i \leqq \mu < i+1 \\ &a_s(\mu) = a_s(i) = \int_{-1/2}^{1/2} F(\theta) \operatorname{sal}(i, \theta)\, d\theta \qquad i - 1 < \mu \leqq i \\ &a_c(\mu) = a(0) = \int_{-1/2}^{1/2} F(\theta)\, d\theta \qquad 0 \leqq \mu < 1 \\ &-\tfrac{1}{2} \leqq \theta < \tfrac{1}{2}; \ i = 1, 2, \ldots \end{aligned}\right\} \tag{17}$$

$a_c[(2i + \eta)/2]$, $a_c[(4i + \eta)/4]$ and the limit $a_c(\mu)$ are shown in Fig. 38b to d for the triangular function of Fig. 37. The computation of the functions

$a_c(\mu)$ and $a_s(\mu)$ is very simple for Walsh functions, since one only has to compute the coefficients $a(0)$, $a_c(i)$ and $a_s(i)$ and plot these values from 0 to 1, from i to $i+1$, or from i to $i-1$ to obtain $a_c(\mu)$ and $a_s(\mu)$ in the intervals $0 \leqq \mu < 1$, $i \leqq \mu < i+1$ or $i-1 < \mu \leqq i$.

Problems

124.1 Try to generalize the Fourier transform for the parabolic cylinder functions of Fig. 11. Use $y = \sqrt{(\xi)}$ [1].

124.2 $a_c(0/\xi)$ in Fig. 36 differs from the value of the limit curve. Why is that so and how large is the difference?

1.2.5 Fast Walsh-Fourier Transform

The time required to obtain the Fourier transform may be drastically reduced by means of a method known as fast Fourier transform. A corresponding fast Walsh-Fourier transform was found by Green [1] and generalized by Welch [2, 3]. Kane, Andrews and Pratt have used a two-dimensional fast Walsh-Fourier transform for the compression of information of pictures [4]. Whelchel and Guinn have used it for signal classification [5]. The form presented here distinguishes between even and odd functions and lists them according to the number of sign changes as in Fig. 2.

Consider a function $F(\theta)$ in some interval. Let this interval be divided into 2^n equally wide subintervals. For illustration, the special case $2^3 = 8$ will be discussed. The average values of $F(\theta)$ in the 8 intervals of length 1/8 are denoted by $8A, 8B, \ldots, 8H$. $F(\theta)$ is thus represented by a step function that is a least mean square fit of $F(\theta)$ for this number of intervals. The Walsh-Fourier transforms $a_c(\mu)$ and $a_s(\mu)$ of these step functions may be obtained from the average values $A, B, \ldots, H$ with the help of Fig. 2:

$$+A+B+C+D+E+F+G+H = a(0) = a_c(\mu), \qquad 0 \leqq \mu < 1$$

$$-A-B-C-D+E+F+G+H = a_s(1) = a_s(\mu), \qquad 0 < \mu \leqq 1$$

$$-A-B+C+D+E+F-G-H = a_c(1) = a_c(\mu), \qquad 1 \leqq \mu < 2$$

$$+A+B-C-D+E+F-G-H = a_s(2) = a_s(\mu), \qquad 1 < \mu \leqq 2$$

$$+A-B-C+D+E-F-G+H = a_c(2) = a_c(\mu), \qquad 2 \leqq \mu < 3$$

$$-A+B+C-D+E-F-G+H = a_s(3) = a_s(\mu), \qquad 2 < \mu \leqq 3$$

$$-A+B-C+D+E-F+G-H = a_c(3) = a_c(\mu), \qquad 3 \leqq \mu < 4$$

$$+A-B+C-D+E-F+G-H = a_s(4) = a_s(\mu), \qquad 3 < \mu \leqq 4$$

There are $2^3(2^3-1) = 56$ or generally $2^n(2^n-1)$ additions necessary to obtain the 2^n coefficients $a_c(\mu)$ and $a_s(\mu)$. The fast Walsh-Fourier transform requires $2^n n$ additions only. Note that the Walsh-Fourier transform does not require multiplications, which are time consuming in the case of the fast Fourier transform[1].

For an explanation of the fast Walsh-Fourier transform refer to Table 2. Column 0 lists the 8 amplitude samples $A, B, \ldots, H$ together with a more general notation $s^{0,0}_{k,0}$. Column 1 lists sums and differences

Table 2. *Fast Walsh-Fourier transform*

0	1	2	3
$s^{0,0}_{0,0}=A$	$s^{0,0}_{0,1}=+(s^{0,0}_{0,0}+s^{0,0}_{1,0})$ $=+A+B$	$s^{0,0}_{0,2}=+(s^{0,0}_{0,1}+s^{0,0}_{1,1})$ $=+A+B+$ $+C+D$	$s^{0,0}_{0,3}=+(s^{0,0}_{0,2}+s^{0,0}_{1,2})$ $=+A+B+C+D+E+$ $+F+G+H$
$s^{0,0}_{1,0}=B$	$s^{0,1}_{0,1}=-(s^{0,0}_{0,0}-s^{0,0}_{1,0})$ $=-A+B$	$s^{0,1}_{0,2}=-(s^{0,0}_{0,1}-s^{0,0}_{1,1})$ $=-A-B+$ $+C+D$	$s^{0,1}_{0,3}=-(s^{0,0}_{0,2}-s^{0,0}_{1,2})$ $=-A-B-C-D+E+$ $+F+G+H$
$s^{0,0}_{2,0}=C$	$s^{0,0}_{1,1}=+(s^{0,0}_{2,0}+s^{0,0}_{3,0})$ $=+C+D$	$s^{1,0}_{0,2}=+(s^{0,1}_{0,1}-s^{0,1}_{1,1})$ $=-A+B+$ $+C-D$	$s^{1,0}_{0,3}=+(s^{0,1}_{0,2}-s^{0,1}_{1,2})$ $=-A-B+C+D+E+$ $+F-G-H$
$s^{0,0}_{3,0}=D$	$s^{0,1}_{1,1}=-(s^{0,0}_{2,0}-s^{0,0}_{3,0})$ $=-C+D$	$s^{1,1}_{0,2}=-(s^{0,1}_{0,1}+s^{0,1}_{1,1})$ $=+A-B+$ $+C-D$	$s^{1,1}_{0,3}=-(s^{0,1}_{0,2}+s^{0,1}_{1,2})$ $=+A+B-C-D+E+$ $+F-G-H$
$s^{0,0}_{4,0}=E$	$s^{0,0}_{2,1}=+(s^{0,0}_{4,0}+s^{0,0}_{5,0})$ $=+E+F$	$s^{0,0}_{1,2}=+(s^{0,0}_{2,1}+s^{0,0}_{3,1})$ $=+E+F+$ $+G+H$	$s^{2,0}_{0,3}=-(s^{1,0}_{0,2}+s^{1,0}_{1,2})$ $=+A-B-C+D+E-$ $-F-G+H$
$s^{0,0}_{5,0}=F$	$s^{0,1}_{2,1}=-(s^{0,0}_{4,0}-s^{0,0}_{5,0})$ $=-E+F$	$s^{0,1}_{1,2}=-(s^{0,0}_{2,1}-s^{0,0}_{3,1})$ $=-E-F+$ $+G+H$	$s^{2,1}_{0,3}=+(s^{1,0}_{0,2}-s^{1,0}_{1,2})$ $=-A+B+C-D+E-$ $-F-G+H$
$s^{0,0}_{6,0}=G$	$s^{0,0}_{3,1}=+(s^{0,0}_{6,0}+s^{0,0}_{7,0})$ $=+G+H$	$s^{1,0}_{1,2}=+(s^{0,1}_{2,1}-s^{0,1}_{3,1})$ $=-E+F+$ $+G-H$	$s^{3,0}_{0,3}=-(s^{1,1}_{0,2}-s^{1,1}_{1,2})$ $=-A+B-C+D+E-$ $-F+G-H$
$s^{0,0}_{7,0}=H$	$s^{0,1}_{3,1}=-(s^{0,0}_{6,0}-s^{0,0}_{7,0})$ $=-G+H$	$s^{1,1}_{1,2}=-(s^{0,1}_{2,1}+s^{0,1}_{3,1})$ $=+E-F+$ $+G-H$	$s^{3,1}_{0,3}=+(s^{1,1}_{0,2}+s^{1,1}_{1,2})$ $=+A-B+C-D+E-$ $-F+G-H$

[1] Care must be taken not to misinterprete the meaning of *fast transform*. The fast transforms are only fast if additions and multiplications are performed one after the other. A circuit that produces the sum of 4, 8, ... numbers can compute faster by means of the ordinary Walsh-Fourier transform or by a method intermediate between ordinary and fast transform than by a circuit producing the sum of two numbers by means of the fast Walsh-Fourier transform.

of each two of the samples, again together with a more general notation. Sums and differences of column 1 are shown in column 2, while column 3 shows sums and differences of column 2. The general notation $s_{k,m}^{j,p}$ shows in each case, which terms of the previous column are added or subtracted. The third column yields the Walsh-Fourier coefficients $a(0)$, $a_c(i)$ and $a_s(i)$.

The fast Walsh-Fourier transform can be represented by a recurrence formula or difference equation that follows from that of the Walsh functions (1.1.4-2):

$$s_{k,m}^{j,p} = (-1)^{[j/2]+p} \, [s_{k,m-1}^{[j/2],x} + (-1)^{j+p} \, s_{k+1,m-1}^{[j/2],x}] \tag{1}$$

$[j/2]$ = largest integer smaller or equal $\frac{1}{2}j$

$x = 0$ for $j =$ even, $x = 1$ for $j =$ odd

$k = 0, 1, \ldots, 2^{n-m} - 1$; $m = 0, 1, \ldots, n$; $p = 0$ or 1; $j = 0, \ldots, 2^{m-1} - 1$; 2^n = number of amplitude samples

As an example consider the term for $j = 3$, $p = 1$, $k = 0$, $m = 3$. It follows with $[j/2] = [3/2] = 1$ and $x = 1$:

$$s_{0,3}^{3,1} = (-1)^2 \, [s_{0,2}^{1,1} + (-1)^4 \, s_{1,2}^{1,1}]$$

This is identical with the term in the lower right corner of Table 2.

The quantities $[j/2]$ and x may be produced in a binary computer as follows: Let j be represented by a binary number. Division by 2 shifts the binary point by one place. The number to the left of the binary point is $[j/2]$, the number to the right is x. Example:

$j = 27 = 11011$, $\quad \frac{1}{2}j = 13.5 = 1101.1$, $\quad [j/2] = 13 = 11011$, $\quad x = 1$

The computation starts with the 2^n terms $s_{k,0}^{0,0}$, $k = 0, \ldots, 2^n - 1$. It follows from $[j/2] = [0/2] = 0$ that the terms $s_{k,1}^{0,p}$, $k = 0, \ldots, 2^{n-1} - 1$, can be computed. These are the terms in the second column of Table 2. Further terms with $[j/2] = [1/2] = 0$ cannot be computed, since this would require terms $s_{k,0}^{0,1}$ while only terms with $x = 0$ are available.

The terms $s_{k,1}^{0,p}$ permit the computation of the 2^{n-1} terms $s_{k,2}^{0,p}$ and the 2^{n-1} terms $s_{k,2}^{1,p}$ since x may be zero or 1, and j may thus be 0 or 1, both values yielding $[j/2] = 0$.

The fast inverse Walsh-Fourier transform is obtained by computing the coefficients $A, B, \ldots, H$ from the coefficients $a(0), a_s(1), \ldots, a_s(4)$. This may be done by inverting the recursion formula for the $s_{k,m}^{j,p}$. One obtains from the sum and difference of $s_{k,m}^{j,0}$ and $s_{k,m}^{j,1}$ the following two recursion formulas:

$$s_{k+1,m-1}^{[j/2],x} = \tfrac{1}{2}(-1)^{j+[j/2]} \, (s_{k,m}^{j,0} + s_{k,m}^{j,1})$$

$$s_{k,m-1}^{[j/2],x} = \tfrac{1}{2}(-1)^{[j/2]} \, (s_{k,m}^{j,0} - s_{k,m}^{j,1})$$

Both may be written together in one formula:

$$s_{k+p,m-1}^{[j/2],x} = \tfrac{1}{2}(-1)^{pj+[j/2]}\left[s_{k,m}^{j,0} + (-1)^p s_{k,m}^{j,1}\right] \quad (2)$$

$p = 0$ or 1; $x = 0$ for $j =$ even, $x = 1$ for $j =$ odd;

$[j/2] =$ largest integer smaller or equal $\frac{1}{2}j$.

Problems

125.1 Develop a fast Walsh transform for a function $F(x, y)$ represented by 4×4 values as shown in Fig. 78. Discuss the saving over the ordinary Walsh transform for a function represented by $2^n \times 2^n$ values rather than 4×4.

125.2 Do the same for a function $F(x, y, z)$ defined by $4 \times 4 \times 4$ values.

1.2.6 Fast Haar-Fourier Transform

Let us denote the coefficients of a Haar-Fourier expansion of $F(\theta)$ by $a(J,j)$:

$$a(J,j) = \int_{-1/2}^{1/2} F(\theta)\,\mathrm{har}(J,j,\theta)\,d\theta \quad (1)$$

Using the notation of section 1.2.5 and the Haar functions of Fig. 19 one obtains the following values for $a(J,j)$:

$$\begin{aligned}
+A + B + C + D + E + F + G + H &= a(0,0)\\
-A - B - C - D + E + F + G + H &= a(0,1)\\
+A + B - C - D \qquad\qquad\qquad\qquad &= \tfrac{1}{2}\sqrt{(2)}\,a(1,1)\\
+E + F - G - H &= \tfrac{1}{2}\sqrt{(2)}\,a(1,2)\\
+A - B \qquad\qquad\qquad\qquad &= \tfrac{1}{2}a(2,1)\\
+C - D \qquad\qquad\qquad &= \tfrac{1}{2}a(2,2)\\
+E - F \qquad\quad &= \tfrac{1}{2}a(2,3)\\
+G - H &= \tfrac{1}{2}a(2,4)
\end{aligned}$$

There are $2(2^3 - 1) + 2^1(2^2 - 1) + 2^2(2^1 - 1) = 24$ additions and subtractions necessary to obtain the 2^3 Haar coefficients multiplied by powers of $\sqrt{(2)}$. Using 2^n values instead of the 2^3 values $A, \ldots, H$ one requires $2(2^n - 1) + 2^1(2^{n-1} - 1) + \cdots + 2^r(2^{n-r} - 1) + \cdots + 2^{n-1}(2^1 - 1) = 2^n n$ additions and subtractions. The number of operations for the ordinary Haar-Fourier transform is thus the same as for the fast Walsh-Fourier transform if one ignores the multiplications by powers of $\sqrt{(2)}$.

Andrews was first to point out the existence of a fast Haar-Fourier transform [1]. Its principle is shown for the eight values $A, \ldots, H$ in Table 3. A comparison of Tables 2 and 3 shows immediately what savings in computations can be obtained. The fast Walsh-Fourier transform requires 8 + 8 + 8 additions and subtractions while the fast Haar-Fourier transform requires 8 + 4 + 2 only. In the general case of 2^n values instead of 2^3 one needs $2^n + 2^{n-1} + \cdots + 2 = 2(2^n - 1)$ additions and subtractions.

Table 3. *Fast Haar-Fourier transform*

0	1	2	3
A	$A+B$	$(A+B)+(C+D)$	$[(A+B)+(C+D)]+$ $+[(E+F)+(G+H)]=a(0,0)$
B	$C+D$	$(E+F)+(G+H)$	$-[(A+B)+(C+D)]+$ $+[(E+F)+(G+H)]=a(0,1)$
C	$E+F$	$(A+B)-(C+D)$	$=\frac{1}{2}\sqrt{(2)}a(1,1)$
D	$G+H$	$(E+F)-(G+H)$	$=\frac{1}{2}\sqrt{(2)}a(1,2)$
E	$A-B$		$=\frac{1}{2}a(2,1)$
F	$C-D$		$=\frac{1}{2}a(2,2)$
G	$E-F$		$=\frac{1}{2}a(2,3)$
H	$G-H$		$=\frac{1}{2}a(2,4)$

The multiplications by powers of $\sqrt{(2)}$ have so far been ignored. The even powers require a shift of the binary point only and this takes little time in a binary computer. About half the coefficients require an additional multiplication by $\sqrt{(2)}$. This multiplication may often be avoided by using nonnormalized Haar functions. The use of nonnormalized functions is a drawback when noise is involved but $\sqrt{(2)}$ is so close to 1 that this drawback is not too important. The main objection to Haar functions is that they do not seem to give equally good results in image processing as Walsh functions[1]. Pearl has introduced a performance measure for comparing the effectiveness of various systems of functions for image representation [2].

Problems

126.1 Show that the fast transform for the functions her(j, θ) of Fig. 20 requires more additions and subtractions than the fast Haar transform but fewer than the fast Walsh transform.

126.2 Derive the fast transform for the functions ter(j, θ) of Fig. 21 and discuss their advantages and drawbacks (multiplications required, use of 4^n instead of 2^n functions).

126.3 Develop a fast Haar transform for the functions of two variables shown in Fig. 26 and compare it with the fast Walsh transform of problem 125.1.

---

[1] A very promising way to reduce computing time and equipment complexity for transformations is shown by the *logical Walsh-Fourier transform* discussed in section 4.2.3. The idea is to replace addition and subtraction by reversible algebraic operations that require no carry, for instance by the addition modulo 2.

1.2.7 Generalized Laplace Transform

The Laplace transform $X(\sigma, \nu)$ of a time function $F(\theta)$ and its inverse may be written as follows:

$$X(\sigma, \nu) = \int_0^\infty F(\theta)\, e^{-\sigma\theta}\, e^{-i2\pi\nu\theta}\, d\theta \tag{1}$$

$$F(\theta) = e^{\sigma\theta} \int_{-\infty}^{\infty} X(\sigma, \nu)\, e^{+i2\pi\nu\theta}\, d\nu \tag{2}$$

It is apparent that the Laplace transform of $F(\theta)$ may be considered to be a Fourier transform of $F(\theta)\, e^{-\sigma\theta}$. The factor $e^{-\sigma\theta}$ makes functions $F(\theta)$ Fourier transformable that are not quadratically integrable. The generalized Laplace transform in real notation follows from this remark from Eqs. (1) and (2):

$$\left.\begin{aligned} a_s(\sigma, \nu) &= \int_0^\infty F(\theta)\, e^{-\sigma\theta} f_s(\nu, \theta)\, d\theta \\ a_c(\sigma, \nu) &= \int_0^\infty F(\theta)\, e^{-\sigma\theta} f_c(\nu, \theta)\, d\theta \\ F(\theta) &= e^{\sigma\theta} \int_{-\infty}^{\infty} [a_c(\sigma, \nu) f_c(\nu, \theta) + a_s(\sigma, \nu) f_s(\nu, \theta)]\, d\nu \end{aligned}\right\} \tag{3}$$

The integrals in Eq. (3) do not have the lower limit $-\infty$ as do the integrals of the generalized Fourier transform, since the factor $e^{-\sigma\theta}$ might make them divergent. $F(\theta)$ must vanish sufficiently fast for large negative values of θ. The usual assumption $F(\theta) \equiv 0$ for $\theta < 0$ is used here.

1.2.8 Dyadic Differentiation, Integration and Correlation

Signals in communications may be continuous functions such as the output voltage of a microphone. Other signals are represented by discrete samples. Digital equipment always uses signals in sampled form. The mathematical tool of calculus is well suited for continuous signals but it does not fit the requirements of sampled signals very well. For instance, differentiation of a polynomial is extremely easy on paper but rather cumbersome for a computer. Gibbs has undertaken to develop mathematical methods specifically for sampled signals or sampled data systems. This effort leads to Walsh functions if one considers 2^n equally spaced signal samples, since all possible signals can be represented exactly by a Walsh series expansion containing the 2^n functions $\mathrm{wal}(0, x), \ldots, \mathrm{wal}(2^n - 1, x)$, where x stands for a space or time variable[1]. The mathematical theory of

[1] The important point seems to be the assumption of a finite number of samples, which is closer to reality than the assumption of a denumerable or nondenumerable number of samples. The calculus of finite differences uses a denumerable number of

Gibbs is still in its infancy and it may eventually turn out as one of the most important contributions to sequency theory. A short introduction is attempted here.

Consider the differentiation of the exponential function:

$$\frac{d}{d\theta} \mathrm{e}^{i\,2\pi j\theta} = i\,2\pi j\,\mathrm{e}^{i\,2\pi j\theta} \tag{1}$$

This relation may be used to *define* the process of differentiation. The close relationship between exponential functions and Walsh functions mentioned at the end of section 1.1.4 suggests the definition of a new operator which will be denoted by $g/g\,\theta$ analogous to the differential operator $d/d\theta$:

$$\frac{g}{g\,\theta} \operatorname{wal}(j, \theta) = j \operatorname{wal}(j, \theta) \tag{2}$$

The process defined by Eq. (2) is variously called logical differentiation, dyadic differentiation or Gibbs differentiation[1] (the last expression explains the notation $g/g\,\theta$ used for the operator).

Consider the logical differentiation of a function represented by a finite Walsh series:

$$F(\theta) = \sum_{j=0}^{2^n-1} a(j) \operatorname{wal}(j, \theta) \tag{3}$$

The logical derivative[2] $g\,F(\theta)/g\,\theta = F^{(\prime)}(\theta)$ follows from Eq. (2):

$$F^{(\prime)}(\theta) = \sum_{j=0}^{2^n-1} j\,a(j) \operatorname{wal}(j, \theta) \tag{4}$$

The process of integration may also be defined with the help of the exponential function and then carried over to Walsh functions[3]:

$$\int \mathrm{e}^{i\,2\pi j\theta}\,d\theta = \frac{1}{i2\pi j} \mathrm{e}^{i\,2\pi j\theta} \tag{5}$$

$$\oint \operatorname{wal}(j, \theta)\,g\,\theta = \frac{1}{j} \operatorname{wal}(j, \theta) \tag{6}$$

---

samples, e.g., when used as in section 2.2.1, or a non-denumerable number, e.g., when the gamma function is defined by a difference equation. Any mathematical method that assumes more than a finite number of samples and sample values is a mathematical overkill if used to analize or predict the results of measurements. Concepts like convergence, completeness, Riemann or Lebesgue integrability, introduced in section 1.1.2, will no longer be required once the mathematical methods for finite numbers of independent signal samples are sufficiently developed.

[1] The definition of the logical derivative is to some extend arbitrary. The definition used here corresponds to the logical derivative of the second kind in Gibbs' terminology.

[2] The sign ⊕ is generally used to distinguish addition on the dyadic group from addition on the real numbers. This notation is extended here to differentiation, integration, convolution, etc. Parentheses are used instead of a circle for the logical derivative since a prime with circle was not available to the typesetter.

[3] The case $j = 0$ must be excluded. The integral of the exponential function is in this case θ while the logical integral does not exist. Only functions with mean value zero have a logical integral.

The logical integral of the function $F(\theta)$ in Eq. (3) becomes[1]:

$$\oint F(\theta)\, g\, \theta = \sum_{j=1}^{2^n-1} \frac{1}{j} a(j) \operatorname{wal}(j, \theta) \tag{7}$$

The exponential functions are eigenfunctions of the differential equation

$$y' - i 2\pi j y = 0, \quad y(1) = y(0), \quad y(0) = 1 \tag{8}$$

while the Walsh functions are eigenfunctions of the logical differential equation

$$y^{(\prime)} - j y = 0, \quad y(0) = 1 \tag{9}$$

as may readily be seen from Eqs. (1) and (2).

For example, consider the following function $F_m(\theta)$:

$$\begin{aligned} F_m(\theta) &= \frac{1}{4} \operatorname{wal}(2^1 - 1, \theta) - \frac{1}{2} \sum_{n=2}^{m} 2^{-n} \operatorname{wal}(2^n - 1, \theta) \\ &= \frac{1}{4} \operatorname{wal}(1, \theta) - \frac{1}{8} \operatorname{wal}(3, \theta) - \frac{1}{16} \operatorname{wal}(7, \theta) - \frac{1}{32} \operatorname{wal}(15, \theta) - \\ &\quad - \frac{1}{64} \operatorname{wal}(31, \theta) \ldots \end{aligned} \tag{10}$$

The staircase function F in Fig. 39a shows $F_4(\theta)$ which consists of the first four terms of the sum, while the staircase function F in Fig. 39b shows $F_5(\theta)$ which consists of the first five terms. The function F in Fig. 39c shows the infinite sum $F_\infty(\theta) = \theta$.

The first logical derivative of $F_m(\theta)$ follows from Eq. (2):

$$\frac{g}{g\,\theta} F_m(\theta) = \frac{1}{4} \operatorname{wal}(1, \theta) - \frac{1}{2} \sum_{n=2}^{m} \frac{2^n - 1}{2^n} \operatorname{wal}(2^n - 1, \theta) \tag{11}$$

$F_4^{(\prime)}(\theta)$ and $F_5^{(\prime)}(\theta)$ are shown by the dashed-dotted lines in Figs. 39a and b. The logical derivative of $F_\infty(\theta)$ diverges.

The logical integral of $F_m(\theta)$ is defined by the series

$$\oint F_m(\theta)\, g\, \theta = \frac{1}{4} \operatorname{wal}(1, \theta) - \frac{1}{2} \sum_{n=2}^{m} \frac{1}{2^n(2^n - 1)} \operatorname{wal}(2^n - 1, \theta). \tag{12}$$

The integral function of $F_4(\theta)$ is shown by the dashed line in Fig. 39a. The integral function of $F_5(\theta)$ is practically the same due to the rapid convergence of the series. The integral exists for $F_\infty(\theta) = \theta$ and is shown in Fig. 39c by dots. Dots rather than a line are used since the function has jumps at dyadically rational[2] values of θ. Some of these jumps are clearly visible.

---

[1] According to footnote 3, page 67, functions $F(\theta)$ are excluded for which $a(0)$ is not zero.

[2] A binary number 0.10111... is called dyadically rational if it contains a finite number of ones to the right of the binary point.

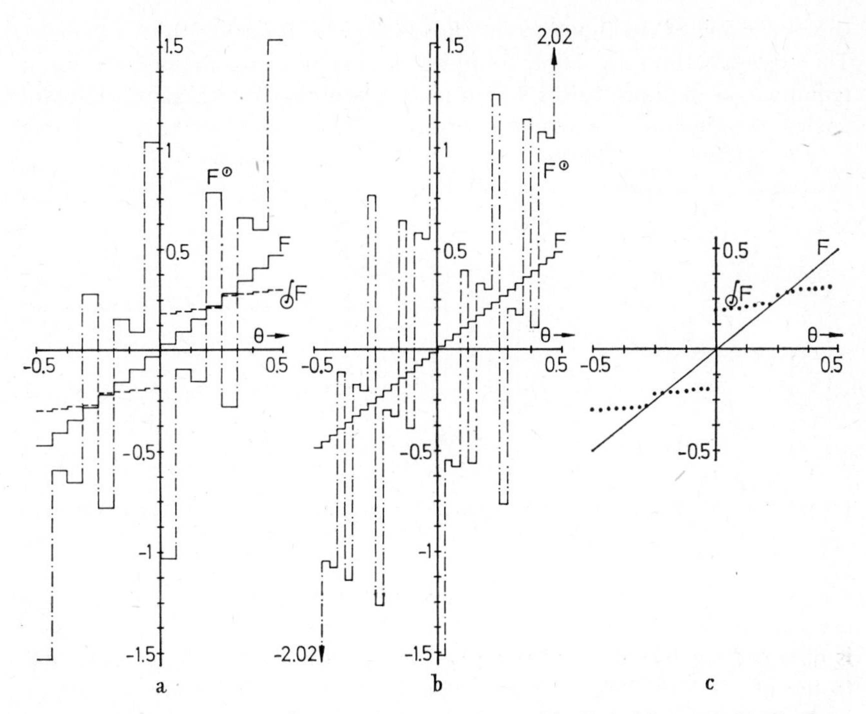

Fig. 39. (a) The function $F_4(\theta)$, its logical derivative and its logical integral. (b) The function $F_5(\theta)$ and its logical derivative. (c) The function $F_\infty(\theta) = \theta$ and its logical integral.

It is evident from our results that logical differentiation and integration are useful operations for step functions having 2^n steps of equal width. Ordinary differentiation is of no use for such functions while logical differentiation fails for such a simple continuous function as $F(\theta) = \theta$. Ordinary integration can be applied to the step functions and logical integration to the continuous function $F(\theta) = \theta$ in Fig. 39 since integration improves convergence according to Eqs. (5) and (6) while differentiation according to Eqs. (1) and (2) has the opposite effect.

The role of ordinary differentiation in science stems from its many useful applications. One cannot foresee whether a logical differential calculus will prove equally useful for sampled functions. A practical application was found by Pearl in the logical characteristic function, which is the Walsh transform of a statistical density function [6]. Pichler has used a logical differential equation for the definition of dyadic time invariant systems [7].

A few comments on the representation of sampled functions are required at this point. Figure 40 shows a staircase curve (*a*), Dirac samples (*b*) and dots (*c*). They all represent the same function. From the mathematical point of view the representation by dots is the most desirable one since it shows a value of the function for those values of θ for which it is defined. The representation by Dirac samples is appealing to engineers since it reminds one of short pulses which usually represent sampled functions in analog equipment. The representation by dots as well as that by Dirac

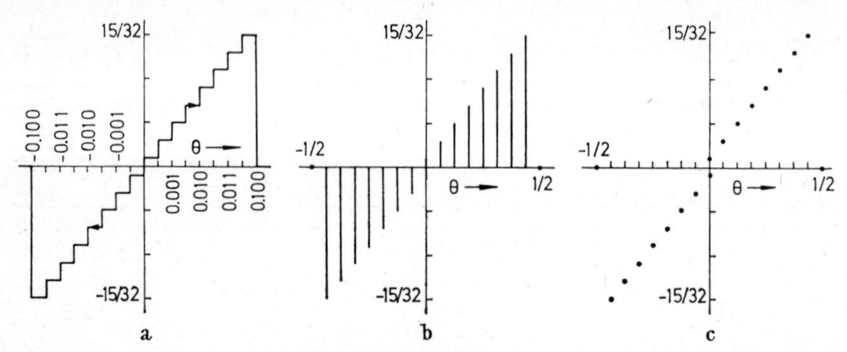

Fig. 40. Graphic representation of a sampled function $F_4(\theta)$ by a staircase curve (a), Dirac samples (b) and dots (c).

samples would be confusing in Fig. 39 and the staircase representation was used for this reason. One may readily see that the staircase curve in Fig. 40a is obtained for $\theta > 0$ by drawing in Fig. 40c a horizontal line from a dot to the abscissa of the next dot to the right; for $\theta < 0$ the horizontal line is drawn to the left. The arrows in Fig. 40a show these different ways of plotting for positive and negative values of θ. As a result one must be careful to distinguish between functional values for $\theta = +0$ and $\theta = -0$ as will soon become evident when dyadic correlation is investigated.

In Fourier analysis the auto-correlation function of a signal $F(\theta)$ is defined by the integral

$$K_{FF}(\theta_v) = \lim_{\Theta\to\infty} \frac{1}{\Theta} \int_{-\Theta/2}^{\Theta/2} F(\theta)\, F(\theta + \theta_v)\, d\theta. \tag{13}$$

In Walsh-Fourier analysis the following integral defines the dyadic auto-correlation function

$$D_{FF}(\theta_v) = \int_{-1/2}^{1/2} F(\theta)\, F(\theta \oplus \theta_v)\, d\theta = F(\theta) \odot F(\theta_v). \tag{14}$$

There is no need to distinguish between dyadic correlation and dyadic convolution since addition and subtraction modulo 2 are identical operations[1]. The dyadic correlation integral can be defined for the infinite interval just like the ordinary correlation integral in Eq. (13) [7]. However,

[1] Consider the two points +1 and −1 on the complex unit circle. Going clockwise (subtraction) or counterclockwise (addition) half the length of the circle brings one from −1 to +1 or from +1 to −1.

the dyadic correlation integral can also be defined for a finite interval which we chose to be $-\frac{1}{2} \leqq \theta < \frac{1}{2}$. This is the more useful integral in communications since any practical signal has a beginning and an end. If the signal $F(\theta)$ is zero outside the interval $-\frac{1}{2} \leqq \theta < \frac{1}{2}$ the dyadically shifted signal $F(\theta \oplus \theta_v)$ is zero outside the interval $-\frac{1}{2} \leqq \theta \oplus \theta_v < \frac{1}{2}$. If $\theta \oplus \theta_v$ were replaced in this condition and in Eq. (14) by $\theta + \theta_v$ the integral could yield nonzero values for $D_{FF}(\theta_v)$ in the interval $-1 < \theta_v < +1$, which is twice the range of θ. The addition modulo 2 assures that $D_{FF}(\theta_v)$ is zero outside the interval $-\frac{1}{2} \leqq \theta_v < \frac{1}{2}$. This is readily seen from Fig. 41 which shows the staircase function $F_4(\theta)$ of Fig. 40a and the dyadically shifted function $F_4(\theta \oplus \theta_v)$ for various values of θ_v.

For an explanation of Fig. 41 consider the value $F_4(\theta)$ in Fig. 40 at the point $\theta = 6/16$ or $\theta = 0.0110$ in binary notation. The value $F_4(0.0110)$

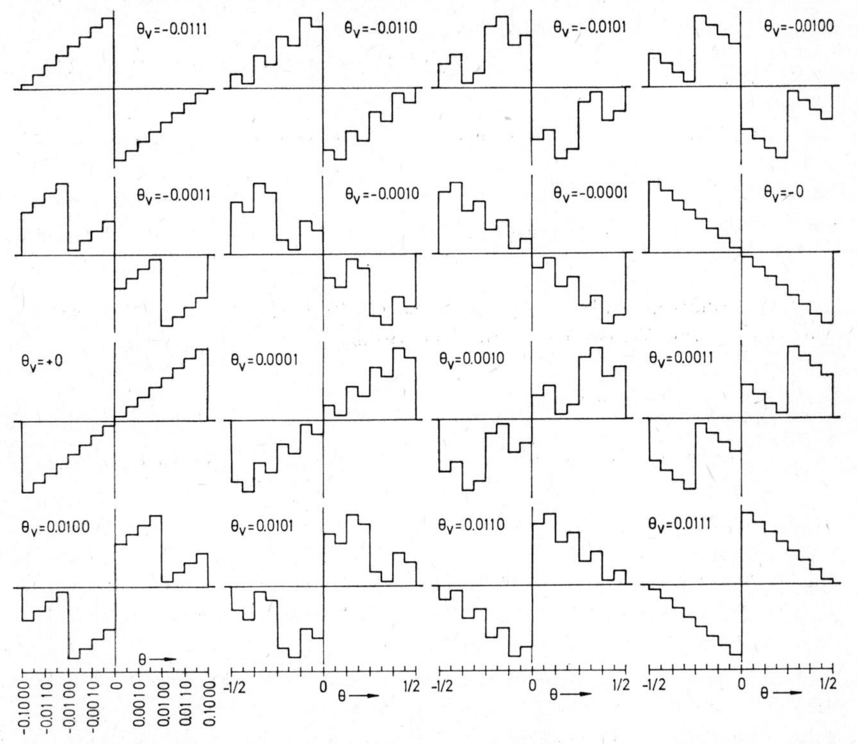

Fig. 41. The dyadically shifted staircase function $F_4(\theta \oplus \theta_v)$ for $\theta_v = -0.0111 = -7/16$, $-0.0110, \ldots, +0.0111$. $F_4(\theta)$ is shown by the curve denoted $\theta_v = +0$.

is plotted from this point to the right till the next larger value of θ is reached for which $F_4(\theta)$ is defined. Consider now the shift $\theta_v = 0.0100$ as shown in the lower left corner of Fig. 41. The point $\theta = 0.0110$ becomes $\theta \oplus \theta_v = 0.0110 \oplus 0.0100 = 0.0010$ and $F_4(0.0110)$ is plotted from $\theta = 0.0010$

to the right until the next larger value of θ is reached for which $F(\theta \oplus \theta_v)$ is defined.

The definitions for addition modulo 2 given in section 1.1.4 have to be used if θ or θ_v are negative. The modulo 2 addition of two numbers with equal signs yields a positive number, while the addition of numbers with different signs yields a negative number. Special cases occur when two numbers of equal absolute value are added: $(+a)\oplus(+a)=(-a)\oplus(-a) = +0$, $(+a) \oplus (-a) = (-a) \oplus (+a) = -0$. Note also that two different curves for $\theta_v = +0$ and $\theta_v = -0$ are shown in Fig. 41 since $a \oplus (+0)$ yields a but $a \oplus (-0)$ yields $-a$.

Using Fig. 41 and the definition of the dyadic auto-correlation function (14) one could compute the dyadic auto-correlation function of $F_4(\theta)$. There is, however, a faster method. In Fourier analysis it is usual to compute the auto-correlation function via the Fourier transform. The same technique can be used in Walsh-Fourier analysis.

Let the Walsh-Fourier series in Eq. (3) be substituted for $F(\theta)$ in Eq. (14):

$$D_{FF}(\theta_v) = \int_{-1/2}^{1/2} \sum_{j=0}^{2^n-1} a(j)\,\mathrm{wal}(j,\theta) \sum_{k=0}^{2^n-1} a(k)\mathrm{wal}(k,\theta \oplus \theta_v)\,d\theta \qquad (15)$$

$$a(j) = \int_{-1/2}^{1/2} F(\theta)\,\mathrm{wal}(j,\theta)\,d\theta \qquad (16)$$

Using the relation $\mathrm{wal}(k, \theta \oplus \theta_v) = \mathrm{wal}(k,\theta)\,\mathrm{wal}(k,\theta_v)$ and the orthonormality of the Walsh functions one obtains:

$$\begin{aligned} D_{FF}(\theta_v) &= \sum_{j=0}^{2^n-1} a^2(j)\,\mathrm{wal}(j,\theta_v) \\ &= a^2(0)\,\mathrm{wal}(0,\theta_v) + \sum_{i=1}^{2^{n-1}-1} a_c^2(i)\,\mathrm{cal}(i,\theta_v) + \sum_{i=1}^{2^n-1} a_s^2(i)\,\mathrm{sal}(i,\theta_v) \end{aligned} \qquad (17)$$

$$a_c(i) = a(2i), \qquad a_s(i) = a(2i-1)$$

The Walsh-Fourier transform (16) of $F(\theta)$ yields the coefficients $a(j)$. Squaring them and applying the inverse transform according to Eq. (17) yields the dyadic auto-correlation function.

The ordinary auto-correlation function is always symmetric but this is not necessarily so for the dyadic auto-correlation function since its series expansion (17) contains the skew-symmetric functions $\mathrm{sal}(i,\theta_v)$. Consider $D_{FF}(\theta_v)$ for $\theta_v > 0$. The functions $\mathrm{cal}(i,\theta_v)$ and $\mathrm{sal}(i+1,\theta_v)$ are identical for $\theta_v > 0$ and one may rewrite the sum (17):

$$\begin{aligned} D_{FF}(\theta_v) &= \sum_{i=1}^{2^{n-1}-1} [a_c^2(i) + a_s^2(i+1)]\,\mathrm{cal}(i,\theta_v) \\ \mathrm{cal}(0,\theta_v) &\equiv \mathrm{wal}(0,\theta_v), \qquad a_c(0) \equiv a(0), \qquad \theta_v > 0 \end{aligned} \qquad (18)$$

$a_c^2(i) + a_s^2(i+1)$ represents the average power of the terms $a_c(i)\,\text{cal}(i, \theta)$ *and* $a_s(i+1)\,\text{sal}(i+1, \theta)$. $D_{FF}(\theta_v)$ for $\theta_v > 0$ is thus the Walsh transform of the sequency power spectrum of $F(\theta)$. Without the restriction $\theta_v > 0$ the function $D_{FF}(\theta_v)$ is still a power spectrum, but it is the power spectrum $a^2(j)$ [7]. Its analog in Fourier analysis would be a frequency power spectrum separated into sinusoidal and cosinusoidal components.

The relationship between the dyadic correlation function and the sequency power spectrum was first reported by Pichler. Due to its similarity to the Wiener-Khintchine theorem it is referred to as the Pichler theorem.

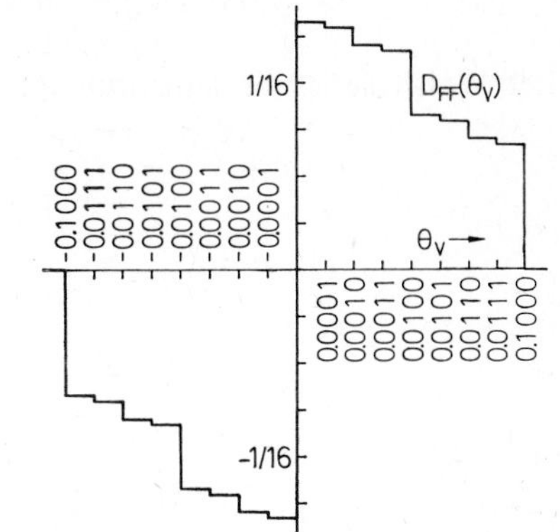

Fig. 42. The dyadic auto-correlation function $D_{FF}(\theta_v)$ of the staircase function $F_4(\theta)$.

Figure 42 shows the dyadic correlation function $F_4(\theta)$ defined by Eq. (10). Using Eqs. (16) and (17) one may write it without any computations:

$$D_{FF}(\theta_v) = 4^{-2}\text{wal}(1, \theta_v) + 8^{-2}\text{wal}(3, \theta_v) + 16^{-2}\text{wal}(7, \theta_v) + 32^{-2}\text{wal}(15, \theta_v)$$

Let us multiply Eq. (17) by $\text{wal}(j, \theta_v)$ and integrate:

$$\int_{-1/2}^{1/2} D_{FF}(\theta_v)\,\text{wal}(j, \theta_v)\,d\theta_v = a^2(j) \tag{19}$$

The meaning of this equation is the following: Let a dyadic auto-correlation function $D_{FF}(\theta_v)$ be specified by 2^n equidistant samples. Its Walsh transform yields 2^n coefficients $a^2(j)$. There are 2^{2^n} signals $F(\theta)$ that yield the same dyadic auto-correlation function due to the two possible signs of the square roots of the 2^n coefficients $a^2(j)$. All signals are defined by Eq. (3).

The dyadic correlation function (14) requires a finite time interval while the ordinary correlation function (13) requires an infinite interval. Since infinite intervals are an abstraction and never occur in engineering, the dyadic correlation function appears more desirable. However, a signal often consists of many more independent samples than can be stored in processing equipment. It is of interest to generalize the dyadic correlation function for this case. Gibbs defined such a function in 1967 but its properties have not been investigated in any detail and the implementation of a practical correlator is very difficult. We shall use a correlation function that can be produced by the asynchronous filter to be discussed in section 2.1.4.

Let $F(\theta)$ in Eq. (14) be replaced by $F(\theta + \theta')$, which is a signal of the same shape but shifted. The function $F(\theta \oplus \theta_v)$ in Eq. (14) remains unchanged. The new correlation function $E_{FF}(\theta', \theta_v)$ is thus defined as follows:

$$E_{FF}(\theta', \theta_v) = \int_{-1/2}^{1/2} F(\theta + \theta')\, F(\theta \oplus \theta_v)\, d\theta \tag{20}$$

Let Walsh series be substituted[1]:

$$\left.\begin{aligned} F(\theta + \theta') &= \sum_{j=0}^{2^n-1} a(j, \theta')\, \mathrm{wal}(j, \theta) \\ F(\theta \oplus \theta_v) &= \sum_{j=0}^{2^n-1} a(j, 0)\, \mathrm{wal}(j, \theta)\, \mathrm{wal}(j, \theta_v) \\ a(j, \theta') &= \int_{-1/2}^{1/2} F(\theta + \theta')\, \mathrm{wal}(j, \theta)\, d\theta \end{aligned}\right\} \tag{21}$$

One obtains:

$$E_{FF}(\theta', \theta_v) = \sum_{j=0}^{2^n-1} a(j, \theta')\, a(j, 0)\, \mathrm{wal}(j, \theta_v) \tag{22}$$

Figure 43 shows five correlation functions (22) for the function $F_4(\theta)$ of Fig. 40. As one would expect they all have their maximum at $\theta' = 0$. The use of the Walsh transform instead of the Fourier transform for the computation of the correlation function (22) affects of course only the speed of the computation and the complexity of the equipment but leaves the correlation function itself unchanged. The functions of Fig. 43 all look rather similar. The reason is that the coefficient $a(1) = a(1, 0)$ of $F_4(\theta)$ equals $\frac{1}{4}$ according to Eq. (10), while the other coefficients $a(3, 0)$, $a(7, 0)$ and $a(15, 0)$ equal $-1/8$, $-1/16$ and $-1/32$. If two or more coefficients are of approximately equal magnitude the correlation functions for different values of θ_v differ strongly and provide information about symmetry properties of the correlated function.

In many uses of correlation functions, such as the detection of radar return signals, one prefers very short correlation functions rather than the

---

[1] The use of Walsh series avoids multiplications if the coefficients $a(j, 0)$ are chosen to be $+1$ and -1 or $+2^n$ and -2^n, which means simpler equipment and faster computation. A further improvement is obtained by avoiding ordinary addition and subtraction. Instead of interpreting the values $+1$ and -1 of the Walsh functions by the 'operators' $+$ and $-$ one interprets them by the operators $\oplus$ and $\ominus$ (Searle operation, *see* section 4.2.3). These operators require no carries, full adders are replaced by half adders, etc. The use of these methods requires, however, a more general concept of orthogonal series expansion.

wide functions of Fig. 43, some of which have very flat maxima. It will be discussed in section 5.4.2 how correlation functions can be transformed into narrow pulses. A different method may be found by solving problem 128.6.

A circuit that can produce the correlation functions $E_{FF}(\theta', \theta_v)$ according to Eq. (22) is shown in Fig. 65. Let the signal $F(\theta)$ be applied to the input of the shift register R. The delay in each stage is 1/16. The output voltages of the stages 0, 1, ... represent the functions $F(\theta - 15/16)$,

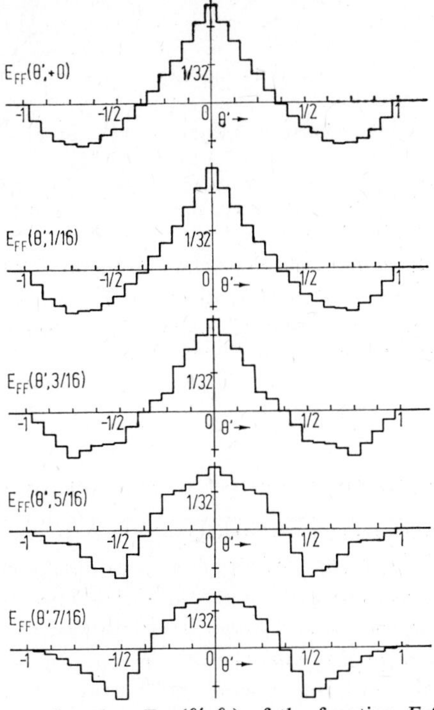

Fig. 43. The correlation function $E_{FF}(\theta', \theta_v)$ of the function $F_4(\theta)$ of Fig. 40 for five values of θ_v.

$F(\theta - 14/16), \ldots$ or generally $F(\theta + \theta')$. The output voltages of the summers S 0, S 1, ..., S 15 represent the Walsh-Fourier coefficients $a(0, \theta'), a(1, \theta'), \ldots, a(15, \theta')$. They may be multiplied by the coefficients $a(1, 0), \ldots, a(15, 0)$, which represent $F(\theta)$, using resistors and operational amplifiers. The output voltages of the summers SU 0, SU 1, ..., SU 15 then represent $E_{FF}(\theta', \theta_v)$ for $\theta_v = 0, 1/16, \ldots$

Problems

128.1 Write the r-th logical derivative $g^r/g\,\theta^r$ of wal(j, θ) and $F_m(\theta)$ of Eq. (10).

128.2 Plot the function $F_4(\theta)$ in Fig. 41 for a dyadic shift $\theta_v = +0.1000$ and $\theta_v = -0.1100$.

128.3 Expand the function $F_4(\theta \oplus 0.0101)$ in Fig. 41 into a Walsh series; determine the logical derivative $g/g\,\theta$ and plot it.

128.4 The dyadic auto-correlation function of Fig. 41 has amplitudes specified at 16 points. One might expect that there are 2^{16} signals that generate this auto-correlation function but there are only 2^4. What is the reason?

128.5 Substitute the Walsh functions $\mathrm{wal}(2^n, \theta)$ for $\mathrm{wal}(2^n - 1, \theta)$ in Eq. (10). Plot the sum of the first four terms, its logical derivative, the dyadically shifted functions according to Fig. 41 and the dyadic auto-correlation function.

128.6 A standard problem in radar is the elimination of the so-called sidelobes of structured radar signals (Sidelobes are the sections of the auto-correlation function that do not vanish outside a small interval around $\theta' = 0$). Show that the dyadic correlation function $E_{FF}(\theta', \theta_v)$ of Eq. (20) can be used to solve this problem. Show that the computation according to Eq. (22) leads to high speed digital circuit implementation.

128.7 Multiplication of the kernel of Eq. (13) by $\exp(-i\,2\pi\,\nu\,\theta)$ yields Woodward's two-dimensional auto-correlation function [13]. The square of its magnitude is called the ambiguity function and it is used to process radar signals with a Doppler shift [14]. Show that a sidelobe-free certainty function in the range-Doppler domain can be derived by linear, time-invariant processes by means of the four-dimensional dyadic correlation function $\int F(\theta + \theta')\,F(\theta \oplus \theta_v) \times$ $\times \exp\{-i\,2\pi[\nu'(\theta + \theta') - \nu_d\,(\theta \oplus \theta_v)]\}\,d\theta$, where $\theta = t/T$ and $\nu = fT$.

128.8 Apply the results of problems 128.6 and 128.7 to sidelobe suppression of one- and two-dimensional antenna arrays with 2^n and $2^n \times 2^n$ dipoles.

1.3 Generalized Frequency

1.3.1 Physical Interpretation of the Generalized Frequency

Frequency is a parameter that distinguishes the individual functions of the systems $\{\cos 2\pi f t\}$ or $\{\sin 2\pi f t\}$. Its usual physical interpretation is "number of cycles per unit of time." The normalized frequency $\nu = fT$ is interpreted as "number of cycles in a time interval of duration 1."

The generalized frequency may be interpreted as "average number of zero crossings per unit of time divided by 2" or as "average number of sign changes per unit of time divided by 2." The normalized generalized frequency μ is interpreted as "average number of zero crossings per time interval of duration 1 divided by 2." The generalized frequency has the dimension s^{-1}:

$$\mu = \phi\,T, \quad \phi = \mu/T \quad [s^{-1}] \tag{1}$$

The definition of the generalized frequency has been chosen so that it coincides with that of frequency if applied to sine and cosine functions. For instance, a sine oscillation with frequency 100 Hz has 100 cycles per second or 200 zero crossings per second. One half the number of zero crossings per second equals 100, which is the same number and dimension as that of the frequency[1]. The zero crossings of sine and cosine functions

[1] The number of sign changes per unit of time has been used to define an instantaneous frequency of frequency modulated sinusoidal oscillations [1, 2, 3].

are equally spaced but the definition of the generalized frequency makes it applicable to functions whose zero crossings are not equally spaced and which need not even be periodic.

It is useful to introduce the new term "sequency" for the generalized frequency ϕ. One reason is that the term generalized frequency is already used in connection with damped oscillations; another is that there are transversal waves in three-dimensional space which have a frequency as well as a sequency. The measure of sequency is "average number of zero crossings *per* *s*econd divided by 2," for which one may use the abbreviation zps.

Consider the Walsh functions $\mathrm{cal}(i, \theta)$ and $\mathrm{sal}(i, \theta)$ in Fig. 2. i equals one half the number of sign changes in the interval $-\frac{1}{2} \leqq \theta < \frac{1}{2}$ and $\phi = i/T$ is the sequency of the periodically continued functions. If the functions are stretched by a factor ξ they will have $2i$ sign changes in the interval $-\frac{1}{2}\xi \leqq \theta < \frac{1}{2}\xi$; $i/\xi = \mu$ will be one half the average[1] number of sign changes in an interval of duration 1.

Consider as a further example the periodically continued Legendre polynomials $\mathrm{P}_c(i, \theta)$ and $\mathrm{P}_s(i, \theta)$ of Fig. 35a. They have $2i$ sign changes in the interval $-\frac{1}{2} \leqq \theta < +\frac{1}{2}$. Stretching them by a factor ξ makes the duration of this interval equal ξ and $i/\xi = \mu$ becomes one half the average number of sign changes per time interval of duration 1.

Let the normalized variables ν and θ in $\sin 2\pi\,\nu\,\theta$ be replaced by the non-normalized variables $f = \nu/T$ and $t = \theta\,T$:

$$\sin 2\pi\,\nu\,\theta = \sin 2\pi (f\,T) \frac{t}{T} = \sin 2\pi\,f\,t \tag{2}$$

The time base T drops out. Sine and cosine functions contain the three parameters amplitude, frequency and phase angle. This is not so for complete systems of orthogonal functions which do not have sequency and time base connected by multiplication. Walsh functions $\mathrm{sal}(\mu, \theta)$ or Legendre polynomials $\mathrm{P}_s(\mu, \theta)$ have a comma between μ and θ. Hence, the substitutions $\phi = \mu/T$ and $t = \theta\,T$ yield:

$$\mathrm{sal}(\mu, \theta) = \mathrm{sal}(\phi\,T, t/T), \quad \mathrm{P}_s(\mu, \theta) = \mathrm{P}_s(\phi\,T, t/T) \tag{3}$$

These functions contain in their general form the four parameters amplitude V, sequency ϕ, delay t_0, and time base T:

$$V\,\mathrm{sal}\left(\phi\,T, \frac{t - t_0}{T}\right), \quad V\,\mathrm{P}_s\left(\phi\,T, \frac{t - t_0}{T}\right).$$

[1] The sequency of a periodic function equals one half the number of sign changes per period. The sequency of a non-periodic function equals one half the number of sign changes per unit of time, if this limit exists.

Problems

131.1 For a certain sequency there is one sine and one cosine function, or one sal and one cal function. How many different functions with an equal number of zero crossings exist in the following systems: a) parabolic cylinder functions; b) had functions of Fig. 17; c) pal functions of Fig. 18; d) $(\sin x)/x$ functions.

131.2 What measure frequency meters that "count cycles" in a certain time interval?

131.3 Why give frequency meters that use resonance more information than cycle counting frequency meters?

1.3.2 Power Spectrum, Amplitude Spectrum, Filtering of Signals

One may derive the frequency function $a_c^2(\nu) + a_s^2(\nu)$ from the Fourier transforms $a_c(\nu)$ and $a_s(\nu)$ of Eq. (1.2.1-11) and interpret it as frequency power spectrum. In analogy, one may interpret the sequency function $a_c^2(\mu) + a_s^2(\mu)$ derived from the generalized Fourier transforms $a_c(\mu)$ and $a_s(\mu)$ of Eqs. (1.2.2-9) and (1.2.2-10) as a sequency power spectrum.

Let Eq. (1.2.2-5) be squared and integrated using the notation of Eq. (1.2.2-6) for the coefficients:

$$\int_{-\infty}^{\infty} F^2(\theta)\, d\theta = \frac{1}{y^2} \int_{-\infty}^{\infty} \left\{ a\left(\frac{0}{\xi}\right) f\left(\frac{0}{\xi}, \theta\right) + \right.$$

$$\left. + \sum_{i/\xi = 1/\xi}^{\infty} \left[a_c\left(\frac{i}{\xi}\right) f_c\left(\frac{i}{\xi}, \theta\right) + a_s\left(\frac{i}{\xi}\right) f_s\left(\frac{i}{\xi}\right), \theta\right) \right] \right\}^2 d\theta \quad (1)$$

The integrals of the cross-products of different functions vanish due to the orthogonality of the functions. The integrals of $f^2(0/\xi, \theta)$, $f_c^2(i/\xi, \theta)$ and $f_s^2(i/\xi, \theta)$ multiplied by y^{-1} yield 1:

$$\int_{-\infty}^{\infty} F^2(\theta)\, d\theta = \frac{1}{y} \left\{ a^2\left(\frac{0}{\xi}\right) + \sum_{i/\xi = 1/\xi}^{\infty} \left[a_c^2\left(\frac{i}{\xi}\right) + a_s^2\left(\frac{i}{\xi}\right) \right] \right\} \quad (2)$$

The sum has the same form as that of Eq. (1.2.2-5). Hence, it may be interpreted as the area under a step function and the sum may be replaced by an integral for large values of ξ and $y = y(\xi)$:

$$\int_{-\infty}^{\infty} F^2(\theta)\, d\theta = \int_{0}^{\infty} [a_c^2(\mu) + a_s^2(\mu)]\, d\mu = \tfrac{1}{2} \int_{-\infty}^{\infty} [a_c^2(\mu) + a_s^2(\mu)]\, d\mu \quad (3)$$

Using non-normalized notation one obtains:

$$\int_{-\infty}^{\infty} F^2(t/T)\, dt = T \int_{0}^{\infty} [a_c^2(\phi T) + a_s^2(\phi T)]\, d(\phi T) \quad (4)$$

$T[a_c^2(\mu) + a_s^2(\mu)]\,d\mu$ is the energy of the components $a_c(\mu) f_c(\mu, \theta)$ to $a_c(\mu + d\mu) f_c(\mu + d\mu, \theta)$ and $a_s(\mu) f_s(\mu, \theta)$ to $a_s(\mu + d\mu) f_s(\mu + d\mu, \theta)$, if the integral of $F^2(t/T)$ is interpreted as the energy of the signal $F(\theta)$. Hence, $a_c^2(\mu) + a_s^2(\mu)$ has the dimension of power and may be interpreted as sequency power spectrum or sequency power density spectrum.

Using the function $G(\nu)$,

$$G(\nu) = A(\nu) + B(\nu) = \tfrac{1}{2}\sqrt{(2)}\,[a_c(\nu) + a_s(\nu)],$$

one may rewrite the frequency power spectrum $a_c^2(\nu) + a_s^2(\nu)$ into the following form:

$$a_c^2(\nu) + a_s^2(\nu) = 2[A^2(\nu) + B^2(\nu)] = G^2(\nu) + G^2(-\nu) \tag{5}$$

Use has been made of Eqs. (1.1.3-3), (1.1.3-6) and (1.2.1-12). The sequency power spectrum may be rewritten as follows:

$$a_c^2(\mu) + a_s^2(\mu) = 4[A^2(\mu) + B^2(\mu)] = G^2(\mu) + G^2(-\mu) \tag{6}$$

The square root $[a_c^2(\nu) + a_s^2(\nu)]^{1/2}$ may be interpreted as frequency amplitude spectrum. Such an interpretation is not possible for the square root $[a_c^2(\mu) + a_s^2(\mu)]^{1/2}$ of the sequency power spectrum, since a specific feature of sine and cosine functions is required for it. Using the relation

$$A \sin x + B \cos x = (A^2 + B^2)^{1/2} \cos\left(x - \tan^{-1}\frac{A}{B}\right) \tag{7}$$

one may rewrite Eq. (1.2.1-12) as follows:

$$F(\theta) = \sqrt{(2)} \int_0^\infty [a_c^2(\nu) + a_s^2(\nu)]^{1/2} \cos\left[2\pi\nu\theta - \tan^{-1}\frac{a_s(\nu)}{a_c(\nu)}\right] d\nu \tag{8}$$

The factor $[a_c^2(\nu) + a_s^2(\nu)]^{1/2}$ may be interpreted as frequency amplitude spectrum, since it represents the amplitude of the oscillation with frequency ν without regard to the phase angle $\tan^{-1} a_s(\nu)/a_c(\nu)$. Systems of functions that do not have an addition theorem like Eq. (7) do not permit this interpretation of the square root $[a_c^2(\mu) + a_s^2(\mu)]^{1/2}$. However, $a_c(\mu)$ and $a_s(\mu)$ are just like $a_c(\nu)$ and $a_s(\nu)$ the amplitude spectra of the even and odd part of the function $F(\theta)$.

Filters or, more generally, systems that change an input signal $F(\theta)$ into an output signal $F_0(\theta)$ may be described by operators. The concept of linear operators describing linear systems is of particular importance in connection with complete systems of orthogonal functions. Let Ω denote an operator and $\{f(j, \theta)\}$ a complete system of orthogonal functions. Application of Ω to a particular function or input signal $f(j, \theta)$ generates an output signal $g(j, \theta)$:

$$\Omega f(j, \theta) = g(j, \theta) \tag{9}$$

The operator Ω is called linear if the proportionality law and the superposition law hold for all functions of the system $\{f(j, \theta)\}$:

$$\Omega\, a(j) f(j, \theta) = a(j)\, \Omega f(j, \theta) \qquad \text{proportionality law} \tag{10}$$

$$\Omega \sum_{j=0}^{\infty} a(j) f(j, \theta) = \sum_{j=0}^{\infty} \Omega\, a(j) f(j, \theta) \quad \text{superposition law} \tag{11}$$

Ω may be a function of j and θ. If Ω depends on θ, the operator and the system it describes are linear and time-variable; otherwise they are linear and time-invariant. An example of a linear, time-variable system is the amplitude modulator. Let an input signal $F(\theta)$ be represented by the sum $\sum a(j) f(j, \theta)$ and the carrier by $h(k, \theta) = \Omega$. $h(k, \theta)$ may be, e.g., a sine carrier $\sqrt{(2)} \sin 2\pi\, k\, \theta$ or a Walsh carrier $\mathrm{wal}(k, \theta)$. Amplitude modulation with suppressed carrier yields:

$$F(\theta)\, h(k, \theta) = \Omega\, F(\theta) = \Omega \sum_{j=0}^{\infty} a(j) f(j, \theta) = \sum_{j=0}^{\infty} a(j)\, g(j, \theta) \tag{12}$$

$$g(j, \theta) = h(k, \theta) f(j, \theta)$$

It is best to use Walsh functions $\mathrm{wal}(j, \theta)$ for $f(j, \theta)$ if $h(k, \theta)$ is a Walsh carrier $\mathrm{wal}(k, \theta)$. One obtains for $g(j, \theta)$:

$$g(j, \theta) = \mathrm{wal}(k, \theta)\, \mathrm{wal}(j, \theta) = \mathrm{wal}(k \oplus j, \theta)$$

If $h(k, \theta)$ is a sine carrier $\sqrt{(2)} \sin 2\pi\, k\, \theta$ one should use the functions $f(0, \theta)$, $\sqrt{(2)} \sin 2\pi\, i\, \theta$ and $\sqrt{(2)} \cos 2\pi\, i\, \theta$ for the system $\{f(j, \theta)\}$. The functions $g(j, \theta)$ are then

$$\begin{aligned} g(0, \theta) &= \sqrt{(2)} \sin 2\pi\, k\, \theta \\ g(2i, \theta) &= \cos 2\pi(k - i)\, \theta - \cos 2\pi(k + i)\, \theta \\ g(2i - 1, \theta) &= \sin 2\pi(k - i)\, \theta + \sin 2\pi(k + i)\, \theta \end{aligned}$$

$$j = 0, 2i, 2i - 1;\ i = 1, 2, \ldots$$

The definition of linearity has changed during development of communication theory. First it was restricted to differential operators with constant coefficients, then to time invariable but not necessarily differential operators. The present definition does not require Ω to be a differential or time invariable operator. It has been used by mathematicians for a long time.

If a system is described by a linear operator Ω and if one is free to choose the system of functions $\{f(j, \theta)\}$, one may choose the system of eigenfunctions of Ω. Equation (9) assumes the following form in this case:

$$\Omega f(j, \theta) = b(j) f(j, \theta) \tag{13}$$

It is convenient to call $f(j, \theta)$ an eigenfunction of Ω even if $f(j, \theta)$ on the right hand side of Eq. (13) has to be replaced by the time shifted function $f[j, \theta - \theta(j)]$.

In the frequency theory of communication the electrical characteristics of filters are described by the frequency response of attenuation and phase shift. This description assumes that a voltage $V \cos 2\pi f t$ is applied to the input of a filter. The steady state voltage $V_c(f) \cos[2\pi f t + \alpha_c(f)]$ appears at the output. The frequency functions $-2 \log V_c(f)/V = -2 \log V_c(\nu)/V$ and $\alpha_c(f) = \alpha_c(\nu)$ are called frequency response of attenuation and phase shift. Let an input signal $F(\theta)$ have the Fourier transforms $a_c(\nu)$ and $a_s(\nu)$. The output signal $F_0(\theta)$ follows from Eq. (1.2.1-12):

$$F_0(\theta) = \sqrt{(2)} \int_0^\infty \{a_c(\nu) K_c(\nu) \cos[2\pi \nu \theta + \alpha_c(\nu)] + $$

$$+ a_s(\nu) K_c(\nu) \sin[2\pi \nu \theta + \alpha_c(\nu)]\} d\nu \tag{14}$$

$$K_c(\nu) = V_c(\nu)/V; \quad \nu = fT, \quad \theta = t/T$$

The description of filters by means of frequency response of attenuation and phase shift is eminently suited for telephone filters. Matched filters, on the other hand, are usually described by means of the pulse response. A voltage pulse of the shape of the Dirac function $\delta(\theta)$ is applied to the input and the shape of the output voltage $D(\theta)$ is determined. No reference to sine and cosine functions is required. Which system of functions is used for the description of a filter is strictly a matter of convenience.

Let the voltages $V f_c(\mu, \theta)$ and $V f_s(\mu, \theta)$ be applied to the input of a filter instead of $V \cos 2\pi f t$. The functions $f_c(\mu, \theta)$ and $f_s(\mu, \theta)$ are the same that occur in the generalized Fourier transform (1.2.2-10). The steady state voltages $V_c(\mu) f_c[\mu, \theta - \theta_c(\mu)]$ and $V_s(\mu) f_s[\mu, \theta - \theta_s(\mu)]$ shall occur at the filter output. Let $-2 \log V_c(\mu)/V$ and $-2 \log V_s(\mu)/V$ be called attenuation. $\theta_c(\mu)$ and $\theta_s(\mu)$ are called delay, since the term "phase shift" cannot be applied to functions other than sine and cosine. These simple relations between input and output voltage exist for filters consisting of coils and capacitors if $f_c(\mu, \theta)$ and $f_s(\mu, \theta)$ are sine and cosine functions. However, one may design filters that contain multipliers, integrators, storages, resistors and switches, which will attenuate and delay Walsh functions but will distort sine and cosine functions. Such filters are better described by Walsh functions than by sine-cosine functions.

Let a signal $F(\theta)$ have the generalized Fourier transforms $a_c(\mu)$ and $a_s(\mu)$. Let the steady state attenuation and delay be $-2 \log V_c(\mu)/V$, $-2 \log V_s(\mu)/V$ and $\theta_c(\mu), \theta_s(\mu)$. The output signal follows from Eq. (1.2.2-11):

$$F_0(\theta) = \int_0^\infty \{a_c(\mu) K_c(\mu) f_c[\mu, \theta - \theta_c(\mu)] + a_s(\mu) K_s(\mu) f_s[\mu, \theta - \theta_s(\mu)]\} d\mu$$

$$K_c(\mu) = V_c(\mu)/V, \quad K_s(\mu) = V_s(\mu)/V \tag{15}$$

Comparison of Eqs. (14) and (15) shows that only $K_c(\nu)$ and $\alpha_c(\nu)$ occurs in Eq. (14), but not $K_s(\nu)$ and $\alpha_s(\nu)$. Such terms would occur if frequency filters would distinguish between sine and cosine functions of the same frequency. The input voltage $V \sin 2\pi f t$ would then produce the output voltage $V_s(f) \sin[2\pi f t + \alpha_s(f)]$ rather than $V_c(f) \sin[2\pi f t + \alpha_c(f)]$. Such a distinction between sine and cosine requires some time-variable circuit element and can thus not occur in frequency filters which are linear and time-invariant. Filters based on sine and cosine pulses rather than on the periodic sine and cosine functions distinguish between sine and cosine. An example of such a filter will be given later on.

Problems

132.1 An operator transforms all sine functions into sal functions and all cosine functions into cal functions. Can this operator be linear? Must this operator be linear?

132.2 A more general definition of the superposition law replaces the functions $f(j, \theta)$ by their logarithm [1]. What restriction is imposed by this definition? Are there useful signals that satisfy this restriction?

1.3.3 Examples of Walsh-Fourier Transforms and Power Spectra

Figure 44 shows time functions $F(\theta)$, their Walsh-Fourier transforms $G(\mu)$, $a_c(\mu)$, $a_s(\mu)$ and their sequency power spectra $a_c^2(\mu) + a_s^2(\mu)$:

$$\left.\begin{aligned} & G(\mu) = \tfrac{1}{2}\sqrt{(2)} \int_{-\infty}^{\infty} F(\theta)\,[\mathrm{cal}(\mu,\theta) + \mathrm{sal}(\mu,\theta)]\,d\theta \\ & a_c(\mu) = \tfrac{1}{2}\sqrt{(2)}\,[G(\mu) + G(-\mu)], \quad a_s(\mu) = \tfrac{1}{2}\sqrt{(2)}\,[G(\mu) - G(-\mu)] \\ & a_c^2(\mu) + a_s^2(\mu) = G^2(\mu) + G^2(-\mu) \end{aligned}\right\} \quad (1)$$

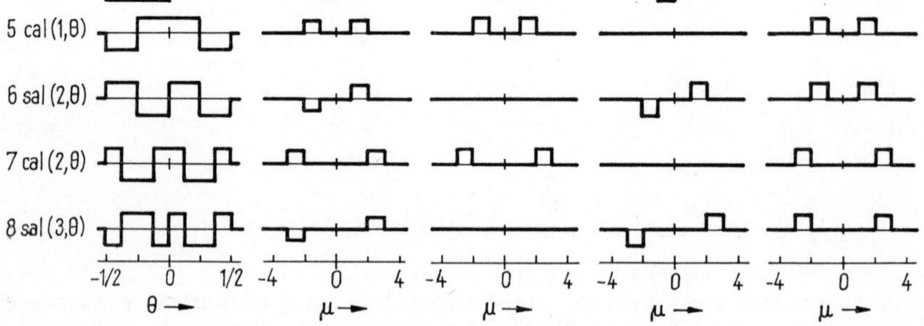

Fig. 44. Some time functions $F(\theta)$, their Walsh-Fourier transforms $G(\mu)$, $a_c(\mu)$, $a_s(\mu)$ and their sequency power spectra $a_s^2(\mu) + a_c^2(\mu) = G^2(\mu) + G^2(-\mu)$.

One may see that compression of the first block pulse by a power of 2 in the time-domain produces a proportionate stretching of the transform $G(\mu)$. The delta function $\delta(\theta)$ is obtained in the limit. Its transform $G(\mu)$ has a constant value in the whole interval $-\infty < \mu < \infty$.

One may further see, that the transform $G(\mu)$ of the Walsh pulses in lines $1, 4, 5, \ldots, 8$ are "sequency-limited". This is in contrast to the well known result of Fourier analysis that a time-limited function cannot have a frequency-limited Fourier transform. The Fourier transforms shown in Fig. 6 for the sine and cosine pulses according to Fig. 1 go on to infinity. Walsh-Fourier transform avoids the troublesome infinite time-bandwidth products of the ordinary Fourier analysis; bandwidth refers of course to sequency bandwidth in the case of Walsh-Fourier transform.

A class of time functions that are time and sequency-limited may be inferred from Fig. 44. The Walsh pulses $\text{cal}(i, \theta)$ and $\text{sal}(i, \theta)$ vanish outside the time interval $-\frac{1}{2} \leqq \theta < \frac{1}{2}$. Their Walsh-Fourier transforms vanish outside the sequency intervals $-(i+1) \leqq \mu \leqq +(i+1)$ or $-i \leqq \mu \leqq +i$. Hence, any time function $F(\theta)$ consisting of a finite number of Walsh pulses is time and sequency-limited:

$$F(\theta) = a(0)\,\text{wal}(0, \theta) + \sum_{i=1}^{l} [a_c(i)\,\text{cal}(i, \theta) + a_s(i)\,\text{sal}(i, \theta)] \quad (2)$$

$$\text{wal}(0, \theta) = \text{cal}(i, \theta) = \text{sal}(i, \theta) = 0 \quad \text{for} \quad |\theta| > \tfrac{1}{2}$$

Let $F(\theta)$ have the Walsh-Fourier transform $G(\mu)$. It holds:

$$F(\theta) = 0 \quad \text{for} \quad |\theta| > \tfrac{1}{2} \quad (3)$$

$$G(\mu) = 0 \quad \text{for} \quad |\mu| > l + 1$$

The orthogonality of a system of functions is invariant to the generalized Fourier transform and that includes the Walsh-Fourier transform. Hence, one may write $G(\mu)$ explicitly, if the coefficients $a(0)$, $a_c(i)$ and $a_s(i)$ of the expansion (2) are known. Let $g(0, \mu)$, $g_c(i, \mu)$ and $g_s(i, \mu)$ denote the Walsh-Fourier transforms of $\text{wal}(0, \theta)$, $\text{cal}(i, \theta)$ and $\text{sal}(i, \theta)$. One obtains the transform $G(\mu)$ of $F(\theta)$:

$$G(\mu) = a(0)\,g(0, \mu) + \sum_{i=1}^{l} [a_c(i)\,g_c(i, \mu) + a_s(i)\,g_s(i, \mu)] \quad (4)$$

The functions $g(0), g_s(1, \mu), g_c(1, \mu), \ldots, g_s(3, \mu)$ are shown in Fig. 44, second column, lines $1, 4, 5, \ldots, 8$. One may readily infer the shape of $g_c(i, \mu)$ and $g_s(i, \mu)$ for larger values of i.

Figure 45 shows the Walsh-Fourier transforms of sine and cosine pulses that vanish outside the interval $-\frac{1}{2} \leqq \theta < \frac{1}{2}$. One may readily see how the orthogonality of the transformed functions is preserved. Figure 46 shows the coefficients $a(0)$, $a_c(i)$ and $a_s(i)$ of the expansion of periodic sine and cosine functions in a series of periodic Walsh functions. The band

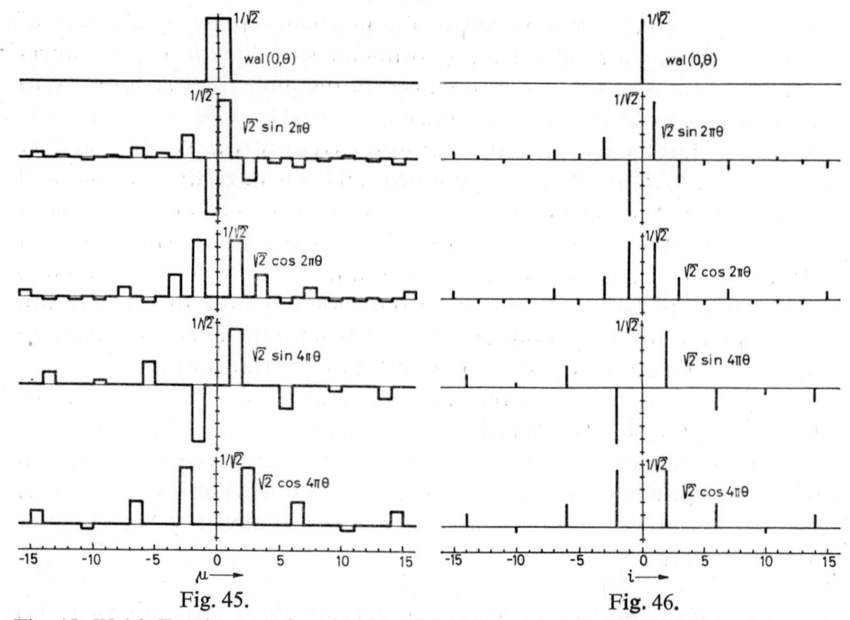

Fig. 45. Fig. 46.

Fig. 45. Walsh-Fourier transforms $G(\mu)$ of the sine and cosine pulses derived from the elements of Fig. 1.

Fig. 46. Coefficients of the expansion of the periodically continued sine and cosine elements of Fig. 1 in a series of periodic Walsh functions $\mathrm{cal}(i, \theta)$ and $\mathrm{sal}(i, \theta)$.

spectra of Fig. 45 are replaced by line spectra. The analogy to Fourier transform of a pulse and Fourier series of the corresponding periodic function is evident.

Figure 47 shows the frequency power spectra $a_c^2(\nu) + a_s^2(\nu) = G^2(\nu) + G^2(-\nu)$ for the first five sine and cosine pulses of Fig. 9 and the block

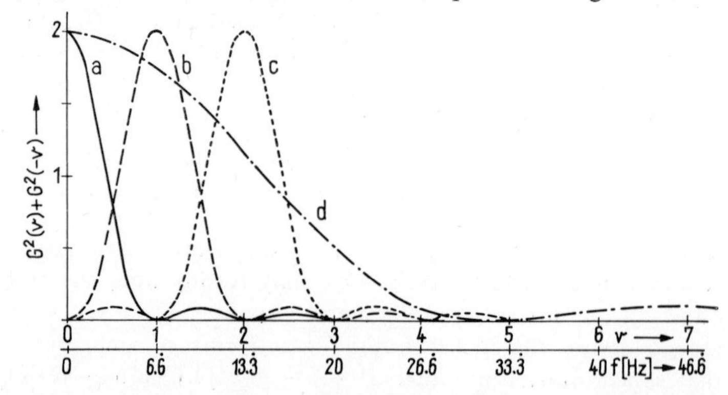

Fig. 47. Frequency power spectra $a_c^2(\nu) + a_s^2(\nu) = G^2(\nu) + G^2(-\nu)$ of the sine and cosine pulses of Fig. 9. (a) $f(0, \theta)$; (b) $f(1, \theta)$, $f(2, \theta)$; (c) $f(3, \theta)$, $f(4, \theta)$. Curve d is the frequency power spectrum of the block pulses of Fig. 3 if they have five times the energy of the block pulse of Fig. 9. The frequency scale in hertz holds for $T = 150$ ms.

pulses of Fig. 3. The area under the curves multiplied by T represents the energy of the signals. The curves in the whole interval $-\infty < \theta < \infty$ are obtained by continuing them as even functions into the interval $\nu < 0$. This continuation is of much less interest for power spectra than for the Fourier transform $G(\nu)$ or the Walsh-Fourier transform $G(\mu)$, since they are always even functions.

Problems

133.1 Plot the Walsh transforms $G(\mu)$, $a_c(\mu)$, $a_s(\mu)$ and the sequency power spectrum for the staircase function of Fig. 40a.

133.2 Do the same for the functions denoted $\theta_v = 0.0001$ and $\theta_v = 0.0010$ in Fig. 41.

1.3.4 Dyadic Time

The concept of dyadic time shifts was needed in connection with dyadic correlation and some of its ramifications will be discussed in this section. Consider the two binary numbers 00 and 01 located at the points -1 and $+1$ of the unit circle in the complex plane. Adding 01 modulo 2 to one of the two numbers produces the other. This is shown in Fig. 48 by the arrows pointing counter clockwise. Figure 48 shows further the numbers 10 and 11 located at the points -1 and $+1$ of a second unit circle. Again

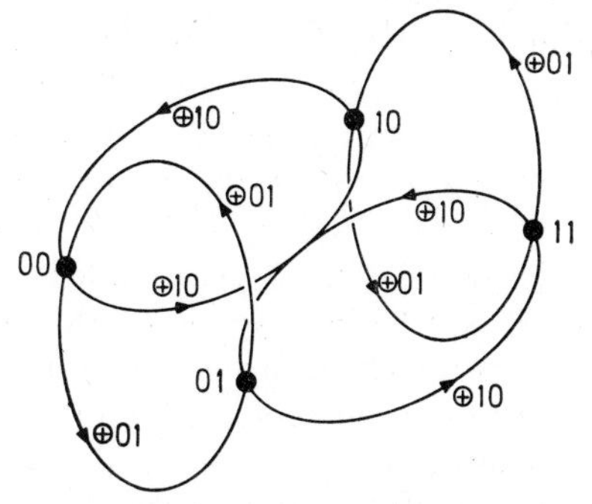

Fig. 48. Addition modulo 2 on the dyadic group with four elements 00, 01, 10 and 11.

the addition modulo 2 of 01 transforms one number into the other. Consider now the addition modulo 2 of 10. This process requires two unit circles located in a plane perpendicular to the previous ones. One may readily see that the points 10 and 01 are both one half circle length away from the point 00, while the distance between the points 00 and 11 is two half circle lengths. Distance measured in this way is called Hamming distance. Subtraction in Fig. 48 is done by advancing clockwise around one of the circles, which produces the same result as addition.

Our usual concept of time requires that the addition of a time interval to a certain time point yields a later time point. The process is equivalenl to adding a positive real number to another, which results in a larger reat number. One says that the topology of time is the same as that of real numbers.

Is it possible to think of time to behave like the points 00 and 01 in Fig. 48 when 01 is added modulo 2? Let us first reflect that most time measurements are done by a modulo addition: seconds and minutes are added modulo 60, hours modulo 12, days modulo 7. Only the years are added like the real numbers[1]. Consider now a primitive animal living in the tropics. It experiences day and night but knows nothing about seasons, clocks and calendars. All nights are equal and so are all days. Adding or subtracting a twelve hour period at any given moment will change day to night or night to day. The concept of time for this animal will be that of dyadic time. Of course, a new concept of time that distinguishes between past and future will dawn on it when it gets eaten by a bigger animal[2].

From the standpoint of physics the dyadic model of time plays a role comparable to the early models of non-Euclidean geometry. There may never be a use for it, but one cannot be sure since there are many animals in physics that have no need and no means to distinguish between past and future.

Things are much simpler in communications. If a sampled signal has been received and stored one can operate on it with any concept of time that yields useful results. Let us first replace the multi-dimensional model of the dyadic group in Fig. 48 by a model that requires two dimensions only. Figure 49 shows a street map of Dyadicville. There are several types of streets classified $A, B, \ldots$ according to increasing quality. All houses are located at the ends of class A streets. There are no street crossings, only intersections. A motorist travelling from the house 010111 near the upper left corner to the house 110101 in the upper middle measures the distance travelled as follows: Going east or north (to the right or to the top) he adds the distances travelled, but going west or south he subtracts them. Furthermore, due to the different quality of the streets he considers all streets $A, B, \ldots$ equally long and denotes their length by 1. His log reads as follows: $-\frac{1}{2}(A) - \frac{1}{2}(B) - \frac{1}{2}(C) + \frac{1}{2}(D) - \frac{1}{2}(E) + \frac{1}{2}(F) + \frac{1}{2}(F) + \frac{1}{2}(E) - \frac{1}{2}(D) + \frac{1}{2}(C) - \frac{1}{2}(B) + \frac{1}{2}(A) = -1(B) + 1(F)$. Adding the absolute values of the distances travelled on each class of street yields 2, which is the Hamming distance of the numbers 010111 and 110101. A comparison

[1] The Mayans extended the modulo addition in a complicated fashion to years. This troubled the last adherents of Maya culture on the island of Tayasal in Lake Petén, Guatemala, when their calendar was about to reach the end of a cycle. Reportedly, a Jesuit priest delivered them from this evil in 1697 by bringing the Christian calendar and the Spanish rule that came with it. This example shows that there is more than one concept of time one can strongly believe in, and that some concepts of time are better than others.

[2] This is the simplest manifestation of dyadic time. A more sophisticated one is presented in section 1.2.8. In wave propagation, ordinary addition creates an unsymmetry between space and time since a wave can propagate in the direction of a positive or negative space variable while the time variable is always increasing (positive). This unsymmetry does not occur if modulo 2 addition is used.

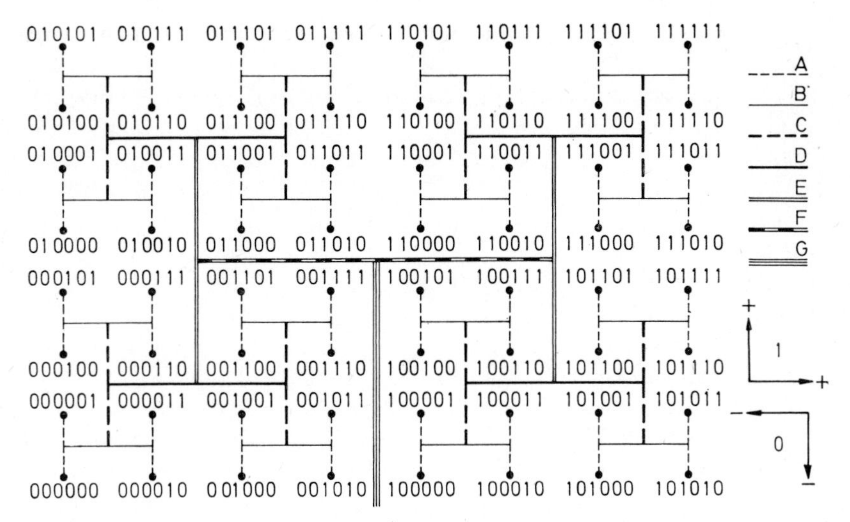

Fig. 49. Representation of the dyadic group by the street map of Dyadicville.

of the signs of the log with the number of the house at the start (010111) read from right to left (111010) and inverted (000101) showed the motorist that this number gave him travelling instructions for the first half of the way, while the number of the house he travelled to (110101) provided instructions for the second half.

Let a signal sample $F(\theta) = F(010111)$ be stored in the location 010111 of Dyadicville. A dyadic time shift $\theta \oplus 100010$ means that $F(010111)$ becomes $F(010111 \oplus 100010) = F(110101)$. Hence, the signal sample has to be transferred from location 010111 to location 110101. The street map of Dyadicville may thus be used as a wiring diagram for a shift register shifting in dyadic time[1].

Consider the dyadic shift of a function $F(\theta)$ by θ'. The new function is $F(\theta_d)$ with $\theta_d = \theta \oplus \theta'$. An example of a dyadically shifted stepfunction was given in Fig. 41. To shift a general function it is helpful to plot θ_d for various values of θ'. This is done in Fig. 50. Assume a shift by $\theta' = 0.1100$ as shown in the lower left corner. The value of the function $F(\theta_d)$ for $\theta_d = 0.1010$ equals the value of $F(\theta)$ for $\theta = 0.0110$.

Let us use Fig. 50 to shift the sine and cosine functions of Fig. 51 by $\theta' = 0.0101$. Since θ' has four digits to the right of the binary point the interval $0 \leqq \theta < 1$ is divided into $2^4 = 16$ subintervals, denoted by binary numbers 0.0000 to 0.1111. The plot for $\theta_d = \theta \oplus 0.0101$ in Fig. 50 shows that $\sin 2\pi\theta$ and $\cos 2\pi\theta$ in the interval $0 \leqq \theta < 0.0001$ are shifted to

[1] In a digital computer the ordinary (time) shifting of stored numbers is performed by adding a number S to the address of each stored number. Dyadic shifting is done by adding S modulo 2 to the addresses, which is a simpler and faster operation for a binary computer.

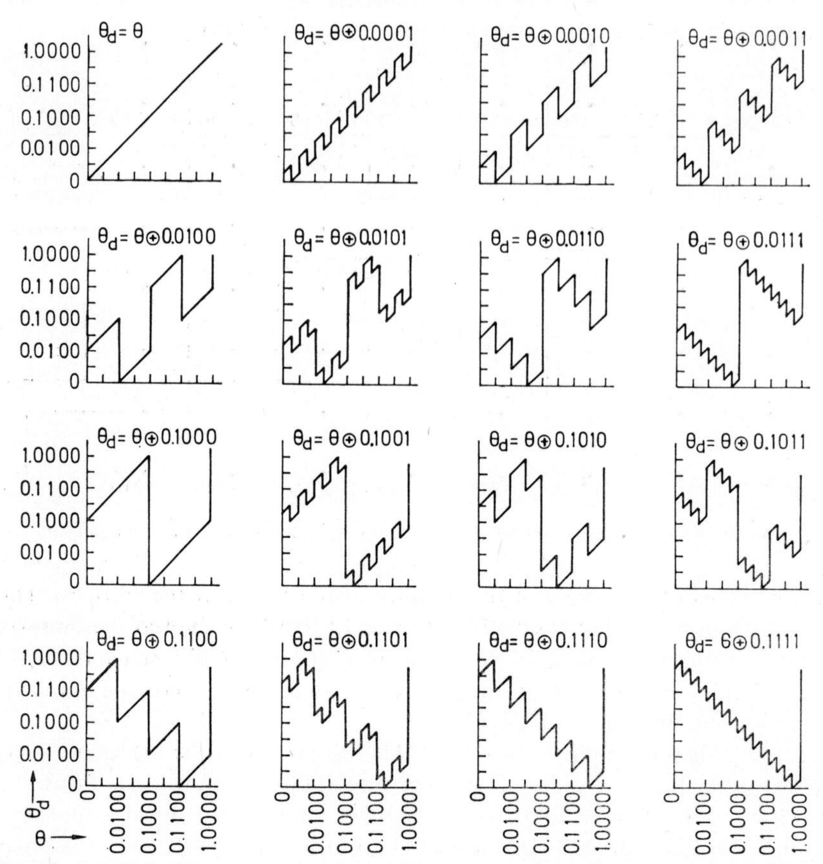

Fig. 50. Connection between the ordinary time θ and the dyadic time $\theta_d = \theta \oplus \theta'$ for various time shifts θ'.

the interval $0 \oplus 0.0101 \leqq \theta_d < 0 \oplus 0.0101 + 0.0001$. The numbers on the bottom of Fig. 51 show from which interval the plotted section of the function came. For instance, 0.0101 in the first interval of $\sin 2\pi(\theta \oplus 0.0101)$ and $\cos 2\pi(\theta \oplus 0.0101)$ indicate that the sections of the functions $\sin 2\pi\,\theta$ and $\cos 2\pi\,\theta$ in the interval 0.0101 are plotted.

Several features of dyadic shifting may readily be recognized from Fig. 51. Symmetry and skew-symmetry of the functions remain unchanged, just like in Fig. 41. The definite integral of the functions remains also unchanged. The orthogonality of $\sin 2\pi\,\theta$ and $\cos 2\pi\,\theta$ is preserved, and this will obviously hold for other orthogonal functions[1]. The definite integral of the square of a function will not be changed by a dyadic shift, implying the preservation of energy.

[1] The number of intervals must be finite or the concept of Riemann integration cannot be applied. There is no need to consider infinitely many intervals since this case cannot occur in communications.

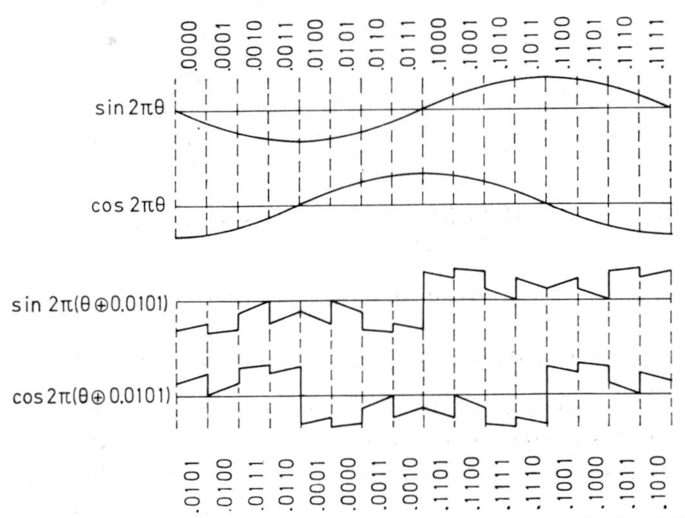

Fig. 51. Invariance of area, orthogonality and energy to a dyadic shift.

Problems

134.1 Write the motorist's log for a trip from house 111010 to 101010.

134.2 Draw the street map of Dyadicville when it contained 16 houses and number the houses, then enlarge Dyadicville to 32 houses.

134.3 Redraw Fig. 48 for a group that has its elements located at the points $+1$, $+i$, -1, $-i$ of the unit circle. Discuss addition and subtraction.

134.4 A binary counter advances according to the time θ in Fig. 50. How can one transform the output of the counter so that it advances according to the time θ_d?

134.5 A motorist in Dyadicville can proceed in two directions at each intersection. Ternarytown was laid out to have no intersections but street crossings only, which give a motorist three directions to proceed. Like in Dyadicville, the houses are all located at the ends of the lowest class streets. Draw the street map.

134.6 The school of Dyadicville was located at a place where it could be reached by the pupils from each house by an equal number of steps. Indicate the location. The physics teacher, interested in equalizing the travelling times of electromagnetic waves, devised the Dyadicville schoolhouse antenna for magnetic dipole radiation shown in Fig. 168.

2. SEQUENCY FILTERS FOR TIME AND SPACE SIGNALS

2.1 *Correlation Filters for Time Signals*

2.1.1 Generation of Time Variable Walsh Functions

Figure 52 shows a circuit for the generation of periodically repeated Walsh functions $\mathrm{cal}(i, \theta)$ and $\mathrm{sal}(i, \theta)$. This circuit is based on the multiplication theorem of the functions $\mathrm{wal}(j, \theta)$ as given by Eq. (1.1.4-3).

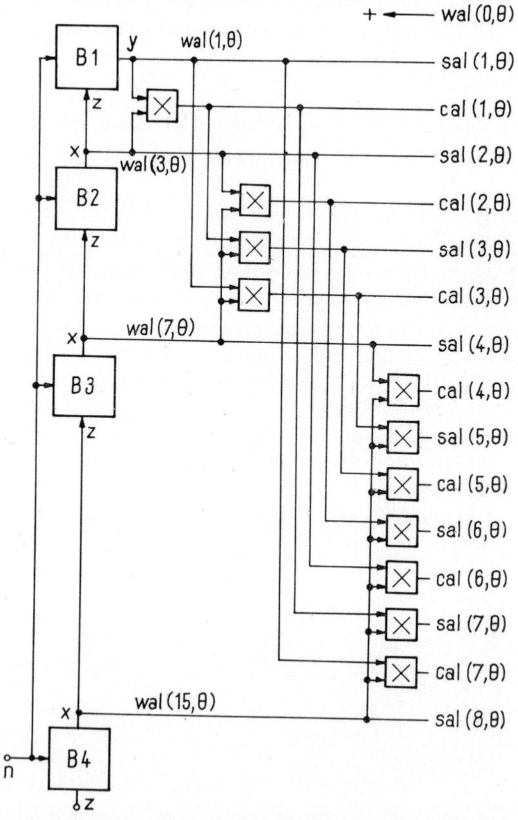

Fig. 52. Generator for time variable Walsh functions using products of Rademacher functions. B, binary counter stage; ×, multiplier = exclusive OR-gate; *z*, input for trigger pulses; *n*, input for reset pulses.

Binary counters B 1 to B 4 produce the functions $\text{wal}(1, \theta) = \text{sal}(1, \theta)$, $\text{wal}(3, \theta) = \text{sal}(2, \theta)$, $\text{wal}(7, \theta) = \text{sal}(4, \theta)$ and $\text{wal}(15, \theta) = \text{sal}(8, \theta)$. The multipliers shown in Fig. 52 produce from these Rademacher functions the complete system of Walsh functions. $\text{wal}(0, \theta)$ is a constant positive voltage.

The multipliers are gates having a truth table as shown in Table 4, since Walsh functions assume the values $+1$ and -1 only. Comparison of this truth table with that of the exclusive OR-gate or the half adder shows that the multipliers in Fig. 52 may be exclusive OR-gates, if an output 0 stands for a positive voltage $+V$ and an output 1 for a negative voltage $-V$.

Table 4. *Truth tables for a multiplier for two Walsh functions* (*a*) *and for an exclusive* OR-*gate* (*b*).

	a			*b*	
	+1	−1		0	1
+1	+1	−1	0	0	1
−1	−1	+1	1	1	0

Consider a Walsh function generator having 20 binary counters rather than 4 as shown in Fig. 52. A total of $2^{20} = 1\,048\,576$ different Walsh functions can be obtained. Nineteen exclusive OR-gates are required to produce any one of the 20^{20} possible functions. The accuracy of their sequency will depend on the trigger pulse generator which drives the binary counters. There are no drift or aging problems. It is worthwhile to compare the simplicity of such a generator with that of a frequency synthesizer delivering a million discrete sine functions. On the other hand, representative switching times of the fastest digital circuits are presently between 100 ps and 10 ns. This restricts the highest sequency of Walsh functions from 10^8 zps = 100 Mzps to 10^{10} zps = 10 Gzps at the present time. Sine waves with frequencies of 100 MHz to 10 GHz were produced decades ago.

The generator of Fig. 52 is not well suited for high switching speeds because of the accumulative delay in the exclusive OR-gates. The build-up of delays is avoided by the circuit shown in Fig. 53 [1]. There are again four counter stages B that produce the Rademacher functions $\text{wal}(1111, \theta)$ to $\text{wal}(1, \theta)$. They are parallel triggered for faster operation. The Rademacher functions are differentiated and yield the trigger pulses $\text{wal}'(1111, \theta)$ to $\text{wal}'(1, \theta)$ as shown in the pulse diagram of Fig. 54. The obtained negative trigger pulses[1] $\text{tri}(1000, \theta)$ to $\text{tri}(1, \theta)$ may or may not pass through

[1] The differentiated Rademacher functions do not have to be rectified to obtain these trigger pulses, since the following AND-gates will automatically suppress positive pulses.

the AND-gates A 1 to A 4, reach the OR-gate and trigger the flip-flop FF. Which pulses may pass through the AND-gates A 1 to A 4 is determined by the voltages supplied to the second input terminals of the gates. Fig. 53

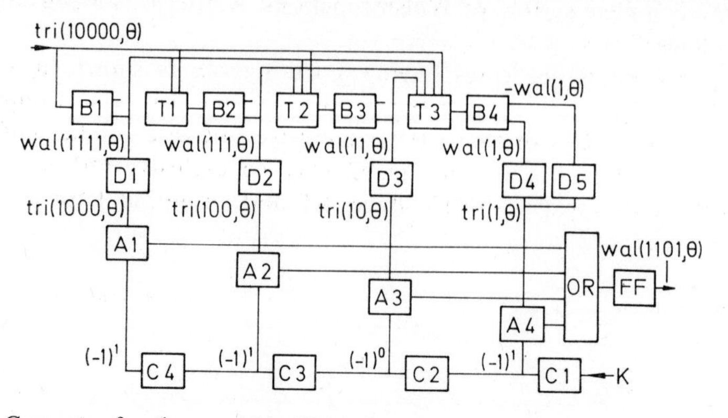

Fig. 53. Generator for time variable Walsh functions using derivatives of Rademacher functions. B, binary counter stage; T, trigger gate; D, differentiator; A, AND-gate; OR, OR-gate; FF, flip-flop; C, stage of a shift register or counter.

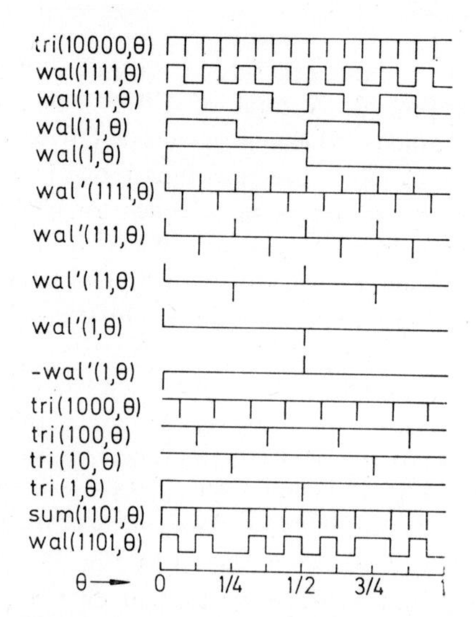

Fig. 54. Pulse diagram for the Walsh function generator of Fig. 53.

shows that these voltages are supplied by counter or shift register stages C. The number stored in these stages represents the parameter j of the output function wal(j, θ) in binary form.

Several other principles for the generation of Walsh functions are known [2, 3].

Problems

211.1 Draw a pulse diagram for Fig. 52.

211.2 How much time is approximately required to switch the circuit of Fig. 53 from a function wal(j, θ) to a different function wal(k, θ)? Compare this with fast switching frequency synthesizers.

211.3 Extend the circuit of Fig. 53 so that four Walsh functions can be produced simultaneously.

2.1.2 Sequency Low-Pass Filters

It has been shown in section 1.3.2 how a filter can be characterized by a system of functions $\{f_c(\mu, \theta), f_s(\mu, \theta)\}$ instead of sine and cosine functions [1–7]. Equation (1.3.2-15) had been obtained, representing the signal $F_0(\theta)$ at the filter output, if the signal $F(\theta)$ is applied to the input. Substitution of the system of Walsh functions $\{\text{cal}(\mu, \theta), \text{sal}(\mu, \theta)\}$ into Eq. (1.3.2-15) yields:

$$F_0(\theta) = \int_0^\infty \{a_c(\mu)\, K_c(\mu)\, \text{cal}[\mu, \theta - \theta_c(\mu)] + a_s(\mu)\, K_s(\mu)\, \text{sal}[\mu, \theta - \theta_s(\mu)]\}\, d\mu \qquad (1)$$

The following relations of section 1.2.4 are needed to derive filters from Eq. (1):

$$\left.\begin{aligned} \text{cal}(\mu, \theta) &= \text{wal}(0, \theta) & 0 \leqq \mu < 1 \\ \text{cal}(\mu, \theta) &= \text{cal}(i, \theta) & i \leqq \mu < i + 1 \\ \text{sal}(\mu, \theta) &= \text{sal}(i, \theta) & i - 1 < \mu \leqq i \end{aligned}\right\} \qquad (2)$$

$$-\tfrac{1}{2} \leqq \theta < \tfrac{1}{2}, \quad i = 1, 2, \ldots$$

Let a signal $G(\theta)$ be divided into time sections $-\frac{1}{2} \leqq \theta < \frac{1}{2}$, $\frac{1}{2} \leqq \theta < \frac{3}{2}, \ldots$, $F(\theta)$ denotes the section in the interval $-\frac{1}{2} \leqq \theta < \frac{1}{2}$. Such a division does not place any restrictions on the signal $G(\theta)$, but a synchronization signal is required from which the beginning and end of the intervals can be derived. The coefficients $a_c(\mu)$ and $a_s(\mu)$ of Eq. (1) may be computed for $F(\theta)$ with the help of Eq. (2):

$$\left.\begin{aligned} a_c(\mu) &= a(0) = \int_{-1/2}^{1/2} F(\theta)\, \text{wal}(0, \theta)\, d\theta & 0 \leqq \mu < 1 \\ a_c(\mu) &= a_c(i) = \int_{-1/2}^{1/2} F(\theta)\, \text{cal}(i, \theta)\, d\theta & i \leqq \mu < i + 1 \\ a_s(\mu) &= a_s(i) = \int_{-1/2}^{1/2} F(\theta)\, \text{sal}(i, \theta)\, d\theta & i - 1 < \mu \leqq i \end{aligned}\right\} \qquad (3)$$

The specific features of the Walsh functions make it possible to transform the representation of a time limited function $F(\theta)$ by an integral into a representation by a sum:

$$F(\theta) = \int_0^\infty [a_c(\mu)\,\mathrm{cal}(\mu, \theta) + a_s(\mu)\,\mathrm{sal}(\mu, \theta)]\,d\theta$$

$$= a(0)\,\mathrm{wal}(0, \theta) + \sum_{i=1}^{\infty} [a_c(i)\,\mathrm{cal}(i, \theta) + a_s(i)\,\mathrm{sal}(i, \theta)] \qquad (4)$$

The attenuation coefficients $K_c(\mu)$ and $K_s(\mu)$ as well as the delays $\theta_c(\mu)$ and $\theta_s(\mu)$ in Eq. (1) determine the filter. Attenuation coefficients and delays may be chosen freely within the limits of physical realization. The following choice is made in order to be able to represent the output signal by a sum rather than an integral:

$$\left.\begin{aligned} K_c(\mu) &= K(0), & \theta_c(\mu) &= \theta(0) & 0 &\leqq \mu < 1 \\ K_c(\mu) &= K_c(i), & \theta_c(\mu) &= \theta_c(i) & i &\leqq \mu < i + 1 \\ K_s(\mu) &= K_s(i), & \theta_s(\mu) &= \theta_s(i) & i - 1 &< \mu \leqq i \end{aligned}\right\} \qquad (5)$$

Equation (1) assumes the following form:

$$F_0(\theta) = a(0)\,K(0)\,\mathrm{wal}[0, \theta - \theta(0)] + \sum_{i=1}^{\infty} \{a_c(i)\,K_c(i)\,\mathrm{cal}[i, \theta - \theta_c(i)] +$$

$$+ a_s(i)\,K_s(i)\,\mathrm{sal}[i, \theta - \theta_s(i)]\} \qquad (6)$$

Let us consider a filter for which the following holds:

$$K(0) = 1, \quad \theta(0) = 1; \quad K_c(i) = K_s(i) = 0 \qquad (7)$$

$F_0(\theta)$ follows from Eqs. (1) to (6):

$$F_0(\theta) = \int_0^1 a_c(\mu)\mathrm{cal}(\mu, \theta - 1)\,d\mu = a(0)\,\mathrm{wal}(0, \theta - 1) \qquad (8)$$

The form of the integral suggests calling this a low-pass filter or, more precisely, a sequency low-pass filter, in order to distinguish it from the usual frequency low-pass filter. Its block diagram is shown in Fig. 55a, and a time diagram in Fig. 55b. An integrator I determines $a(0)$ according to Eq. (3). $a(0)$ can be sampled at the end of the interval $-\frac{1}{2} \leqq \theta < \frac{1}{2}$ at the integrator output by switch s_2. The integrator is then reset by switch s_1. $a(0)$ for the section of $G(\theta)$ in the interval $\frac{1}{2} \leqq \theta < \frac{3}{2}$ is obtained by integrating $G(\theta)$ during that time interval, etc. $a(0)$ must be multiplied by $\mathrm{wal}(0, \theta - 1)$ according to Eq. (8). $\mathrm{wal}(0, \theta)$ is a constant with value 1. Hence, $a(0)\,\mathrm{wal}(0, \theta - 1)$ is the voltage $a(0)$ obtained in the interval $-\frac{1}{2} \leqq \theta < \frac{1}{2}$, sampled at $\theta = +\frac{1}{2}$ and stored during the interval $\frac{1}{2} \leqq \theta < \frac{3}{2}$. A holding circuit H is shown in Fig. 55a, and a practical version of this sequency low-pass filter is shown in Fig. 55c.

For numerical values consider a frequency low-pass filter with 4 kHz cut-off frequency. A signal at the output of this filter has 8000 independent amplitudes per second. Hence, the steps of $G_0(\theta)$ in Fig. 55b must

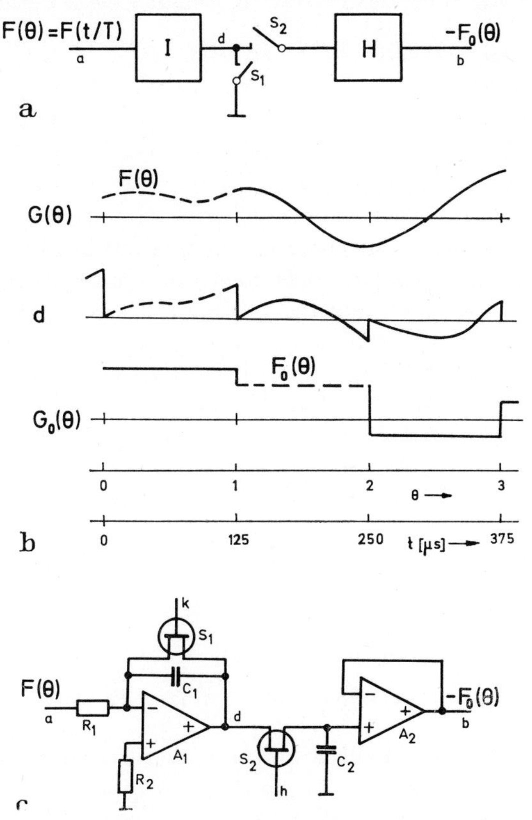

Fig. 55. Sequency low-pass filter. (a) block diagram, (b) time diagram, (c) practical circuit. I, integrator; H, holding circuit; A, operational amplifier.

be $T = 1/8000 = 125\ \mu s$ long; the cut-off sequency equals $\phi = 1/2T = 4000$ zps $= 4$ kzps. Use is made here of the sampling theorem of Walsh-Fourier analysis, which happens to be trivially simple for sequency low-pass filters [8–10].

Problems

212.1 The resetting time for the integrator in Fig. 55 is about 1% of the integration time. Show how this error can be avoided by using two integrators alternately.

212.2 The impedance of the field effect transistor s_2 in Fig. 55 is about 100 ohms when conducting. How large can C_2 be for a 4 kzps cut-off sequency? Would a reduction of the impedance to 10 ohms be desirable?

212.3 What is the purpose of the resistor R_2 in Fig. 55?

2.1.3 Sequency Band-Pass Filters

Let us derive simple sequency band-pass filters. The multiplication theorems of Walsh functions derived in section 1.1.4 are needed:

$$\left.\begin{aligned} \mathrm{cal}(i,\theta)\,\mathrm{cal}(k,\theta) &= \mathrm{cal}(i \oplus k,\theta) \\ \mathrm{sal}(i,\theta)\,\mathrm{cal}(k,\theta) &= \mathrm{sal}\{[k \oplus (i-1)]+1,\theta\} \\ \mathrm{sal}(i,\theta)\,\mathrm{sal}(k,\theta) &= \mathrm{cal}[(i-1)\oplus(k-1),\theta] \\ [\mathrm{cal}(0,\theta) &= \mathrm{wal}(0,\theta)] \end{aligned}\right\} \qquad (1)$$

The multiplication theorems (1) are very similar to those of sine and cosine functions, except that one term only stands on the right hand side instead of two terms for the sum and the difference of the frequencies. A consequence is that the modulation of a Walsh carrier by a signal yields a single (sequency) sideband modulation. This makes it possible to implement sequency band-pass filters by a principle well known but little used for frequency band-pass filters.

Let the signal $F(\theta)$ of Eq. (2.1.2-4) be "sequency shifted" by multiplication with $\mathrm{cal}(k,\theta)$. Using Eq. (1) one obtains:

$$\begin{aligned} F(\theta)\,\mathrm{cal}(k,\theta) = a(0)\,\mathrm{cal}(k,\theta) + \sum_{\substack{i=1\\ i\neq k}}^{\infty} \big[a_c(i)\,\mathrm{cal}(i\oplus k,\theta) &+ \\ + a_s(i)\,\mathrm{sal}\{[k\oplus(i-1)]+1,\theta\}\big] + a_c(k)\,\mathrm{wal}(0,\theta) &+ \\ + a_s(k)\,\mathrm{sal}\{[k\oplus(k-1)]+1,\theta\} & \end{aligned} \qquad (2)$$

Passing this signal through a low-pass filter described by Eq. (2.1.2-7) yields in analogy to Eq. (2.1.2-8) the output signal $F_{01}(\theta)$:

$$F_{01}(\theta) = a_c(k)\,\mathrm{wal}(0,\theta-1) = \int_k^{k+1} a_c(\mu)\,\mathrm{cal}(\mu,\theta)\,\mathrm{cal}(k,\theta-1)\,d\mu$$

$$\mathrm{cal}(k,\theta-1) \equiv \mathrm{cal}(k,\theta) \qquad (3)$$

Multiplication of $F_{01}(\theta)$ by $\mathrm{cal}(k,\theta-1)$ shifts the filtered signal to its original position in the sequency domain:

$$F_0(\theta) = F_{01}(\theta)\,\mathrm{cal}(k,\theta-1) = a_c(k)\,\mathrm{cal}(k,\theta-1) = \int_k^{k+1} a_c(\mu)\,\mathrm{cal}(\mu,\theta)\,d\mu$$

$$\mathrm{wal}(0,\theta-1)\,\mathrm{cal}(k,\theta-1) = \mathrm{cal}(k,\theta-1), \qquad \mathrm{cal}^2(k,\theta-1) \equiv 1 \qquad (4)$$

The last integral suggests the name sequency band-pass filter. For its practical implementation, one must put a multiplier in front of the sequency low-pass filter of Fig. 55 to perform the multiplication (2). A second multiplier after the sequency low-pass filter performs the multiplication (4). Figure 56 shows such a band-pass filter. The same function $\mathrm{cal}(k, \theta)$ is fed to both multipliers, since $\mathrm{cal}(k, \theta)$ has the period 1 and is thus identical

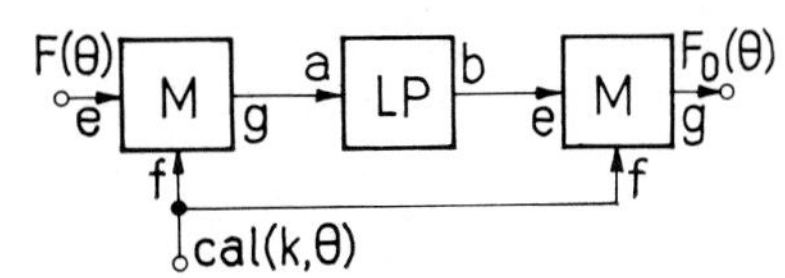

Fig. 56. Sequency band-pass filter. M, multiplier for Walsh functions; LP, sequency low-pass filter.

with $\mathrm{cal}(k, \theta - 1)$. Suitable multipliers are shown in section 3.2.1. Note that multiplication by a Walsh function means multiplication by $+1$ or -1 only; multiplication by $+1$ leaves a signal unchanged, multiplication by -1 reverses its amplitude.

Figure 57 shows attenuation and delay as functions of sequency for a sequency low-pass filter with $K(0) = 1$, and several band-pass filters. The coefficients $K_c(i)$ and $K_s(i)$ are zero, except for the values of i for

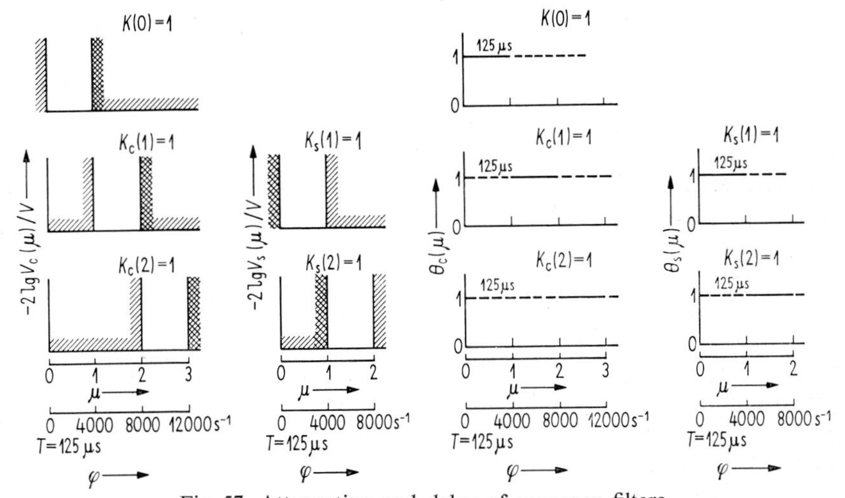

Fig. 57. Attenuation and delay of sequency filters.

which they are shown to be 1. The hatched areas at the band limits $\mu = i$ indicate that the function $\mathrm{cal}(i, \theta)$ or $\mathrm{sal}(i, \theta)$ passes through the filter; cross-hatched areas indicate that they do not pass.

The normalized bandwidth $\mu_2 - \mu_1 = i_2 - i_1 = \Delta\mu$ equals 1 for all filters of Fig. 57. Let us denote the quotient bandwidth/(lower band limit) $= \Delta\mu/\mu_1$ as relative bandwidth. The following relations may be seen to hold for $\Delta\mu/\mu_1$ from Fig. 57:

$$\Delta\mu/\mu_1 = 1 \quad \text{for} \quad K_c(1) = 1 \quad \text{or} \quad K_s(1) = 1$$

$$\Delta\mu/\mu_1 = \tfrac{1}{2} \quad \text{for} \quad K_c(2) = 1 \quad \text{or} \quad K_s(2) = 1$$

$$\Delta\mu/\mu_1 = 1/k \quad \text{for} \quad K_c(k) = 1 \quad \text{or} \quad K_s(k) = 1$$

The functions $\mathrm{cal}(k, \theta)$ or $\mathrm{sal}(k, \theta)$ fed into the sequency band-pass filter of Fig. 56 determine the relative bandwidth. Relative bandwidths smaller or equal 1 only can be achieved with this circuit.

The normalized sequency μ as well as the non-normalized sequency $\phi = \mu/T$ for $T = 125\ \mu s$ are plotted in Fig. 57. The values of ϕ show the channels that one would use in multiplexing telephone signals, if each signal may assume 8000 independent amplitudes per second.

The sequency response of attenuation and delay shown in Fig. 57 is idealized. At the present, in practical filters the root mean square deviation of the filter output voltage lies between 0.01 and 0.001. The mean square deviation is thus between 10^{-4} and 10^{-6}, which means the attenuation in the stop-bands of Fig. 57 is between -40 and -60 dB.

The infinitely steep filter edges shown in Fig. 57 also hold for practical filters. This result is startling to an engineer used to thinking in terms of frequency filters. However, sequency filters use switches that introduce a time quantization of the signal. Keeping this in mind, the discontinuous change of attenuation is not surprising. The discontinuity would disappear, if the Walsh functions of the whole interval $-\infty < \theta < +\infty$ were used rather than the sections in the interval $-\frac{1}{2} \leqq \theta < \frac{1}{2}$.

Attenuation and delay of sequency filters are constant in the pass-band according to Fig. 57. Hence, there are inherently no attenuation or delay distortions. Delay in the stop-bands is not defined for ideal filters, since no energy is passed. Real filters pass energy in the stop-bands. This delay is shown by dashed lines in Fig. 57.

Figure 58 shows a sine wave (a) at the input of several sequency filters and the resulting output signals. (b) is the output of a sequency low-pass filter with $K_c(0) = 1$. (c), (d) and (e) are the outputs of band-pass filters with $K_s(1) = 1$, $K_c(1) = 1$ and $K_s(2) = 1$. (f), (g), and (h) are the outputs obtained from several band-pass filters connected in parallel: $K_c(0) = K_s(1) = 1$ (f); $K_c(0) = K_s(1) = K_c(1) = 1$ (g); $K_c(0) = K_s(1) = K_c(1) = K_s(2) = 1$ (h).

Figure 59 shows sequency amplitude spectra of sinusoidal functions (a) of equal frequency and amplitude but various phases. The amplitude spectra $a_c(\mu) = a_c(\phi T)$ are shown by (b), the amplitude spectra $a_s(\mu) = a_s(\phi T)$ are shown by (c). The oscillograms (b) and (c) were obtained by sampling the output voltages of a bank of 16 sequency filters. Squaring and adding the traces (b) and (c) yields the sequency power spectra.

Band-pass filters according to Fig. 56 permit relative bandwidths $\Delta\mu/\mu_1 = 1, \frac{1}{2}, \frac{1}{3}, \ldots$ only. Figure 60 shows a band-pass filter for relative bandwidths $\Delta\mu/\mu_1 = 1, 2, 3, \ldots$. This circuit uses a low-pass filter LP 1 according to Fig. 55, which integrates the input signals over time intervals of duration T. A further low-pass filter LP 2 integrates over time intervals of duration $T/2$, or $T/3$, or $T/4, \ldots$. The output voltages of the two low-pass filters are shown in Fig. 60a and b; the integration period of LP 1 and LP 2 is chosen equal to $T/2$. The different delay times of LP 1 and LP 2 are compensated by the holding circuit SP. The difference of the

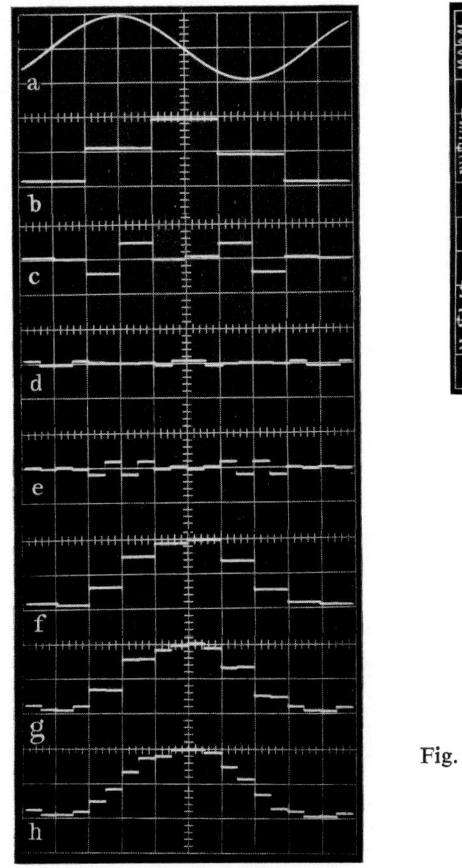

Fig. 58.

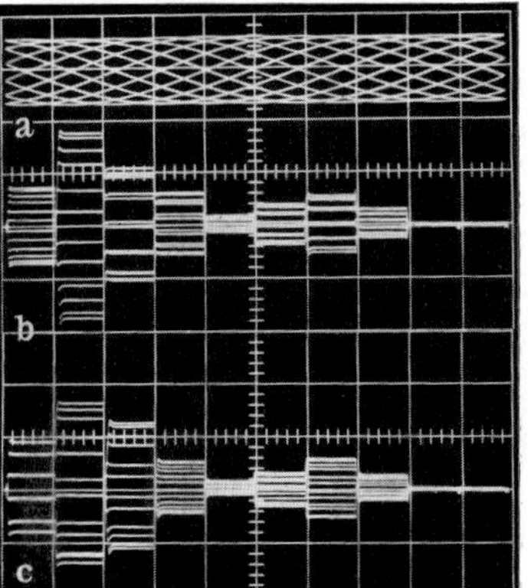

Fig. 59.

Fig. 58. Approximation of sinusoidal functions by Walsh functions. (*a*) sinusoidal function, frequency 250 Hz; (*b*) $a(0)\,\mathrm{wal}(0,\theta)$; (*c*) $a_s(1)\,\mathrm{sal}(1,\theta)$; (*d*) $a_c(1)\,\mathrm{cal}(1,\theta)$; (*e*) $a_s(2)\,\mathrm{sal}(2,\theta)$; (*f*) is the sum of *b* and *c*; (*g*) is the sum of *b*, *c* and *d*; (*h*) is the sum of *b*, *c*, *d* and *e*. Time base $T = 1$ ms; horizontal scale 0.5 ms/div. (Courtesy C. Boesswetter and W. Klein of Technische Hochschule Darmstadt.)

Fig. 59. Walsh-Fourier transforms of sinusoidal waves. (*a*) sinusoidal waves, frequency 1 kHz, various phases; horizontal scale 0.1 ms/div. (*b*) and (*c*) show Walsh-Fourier transforms $a_c(\phi\, T)$ and $a_s(\phi\, T)$ of *a*; time base $T = 1.6$ ms; horizontal scale 625 zps/div. (Courtesy C. Boesswetter and W. Klein of Technische Hochschule Darmstadt.)

voltages of Fig. 60a and b yields the output signal $F_0(\theta)$ of this wide band-pass filter.

There is another important difference between the band-pass filters of Fig. 60 and 56 besides the different relative bandwidth. The functions sal(μ, θ) as well as cal(μ, θ) may pass the filter of Fig. 60 in the pass-band, while only sal(μ, θ) or cal(μ, θ) may pass a filter according to Fig. 56.

A great variety of filters may be derived from the basic types discussed. Fig. 61 shows a sequency high-pass and a sequency band-stop filter derived

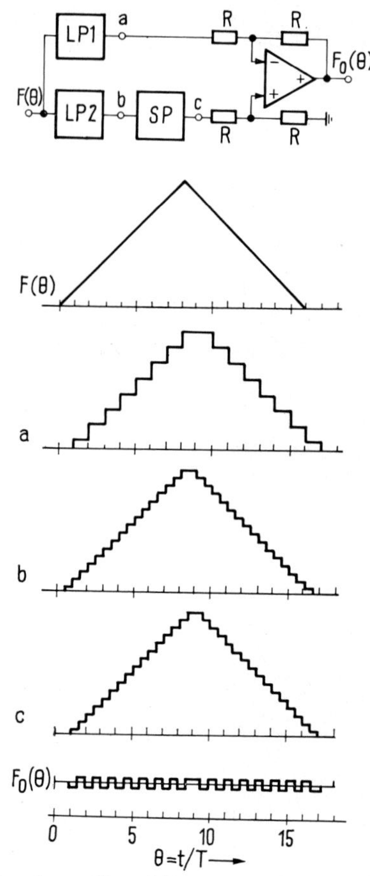

Fig. 60. Sequency wide band-pass filter. LP, sequency low-pass filter; SP, holding circuit.

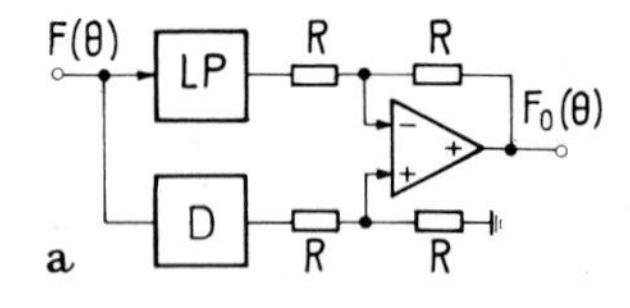

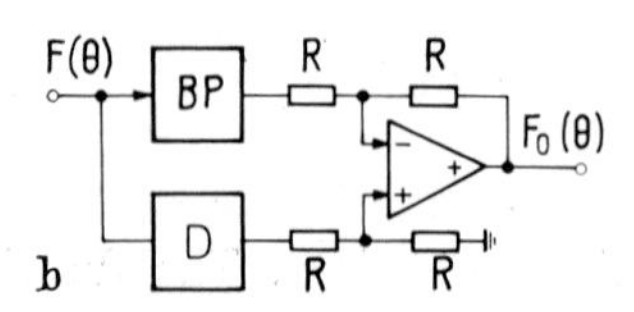

Fig. 61. Sequency high-pass filter (a) and sequency band-stop filter (b). LP, sequency low-pass filter; BP, sequency band-pass filter; D, delay circuit.

from low-pass and band-pass filters. Parallel connection of several band-pass filters according to Fig. 56 yields filters that let pass sal(μ, θ) as well as cal(μ, θ) and have relative bandwidths $\Delta\mu/\mu_1 = 1, \frac{1}{2}, \frac{1}{3}, \ldots$; on the other hand, one may obtain wide band-pass filters that let pass sal(μ, θ) or cal(μ, θ) only as shown by the oscillograms of Fig. 58f to h.

Problems

213.1 Let the function $F(\theta)$ of Fig. 60 be applied to the circuits of Fig. 61, and plot the output voltages.

213.2 The following method is used in commercial radio communication to preserve privacy: The voice band is divided into 5 frequency bands; modulation of 5 carriers permits the permutation of the bands; there are $5! = 120$ permutations; furthermore, upper and lower sideband modulation may be used for the permutations; this results in $5!\,2^5 = 3840$ keys. Design a corresponding system for sequency filters. Why can one use 16 and more filters? How many keys are obtained? How can the keys be changed by a binary key number supplied by a number generator (*see* section 2.1.1.)? How may one introduce amplitude quantization? Is the sequency or frequency bandwidth of the signal increased by this processing?

2.1.4 Multiplicity Filters and Asynchronous Filters

Figure 62 shows a very general block diagram of a sequency filter. The multipliers M 01, ..., M 0s and the integrators I 0, ..., I s produce the coefficients $a(j)$, $j = 0 \ldots s$:

$$a(j) = \int_{-1/2}^{1/2} F(\theta)\,\text{wal}(j, \theta)\,d\theta \tag{1}$$

Sample and hold circuits H 0, ..., H s store the voltages that represent these coefficients. The multipliers M 11, ..., M 1s produce the products

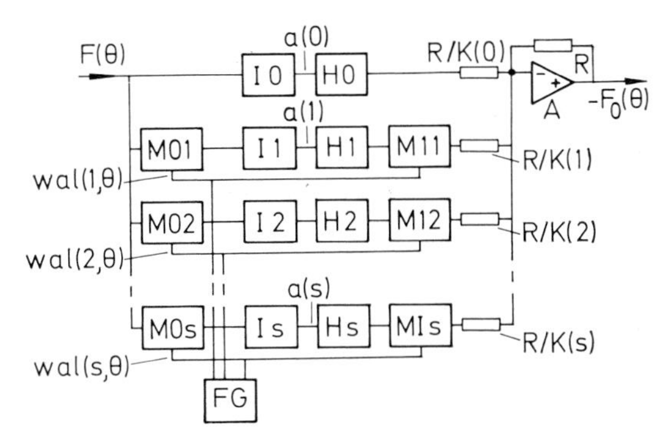

Fig. 62. General sequency filter using the correlation principle for functions of time. M, multiplexer; I, integrator; H, sampling and hold circuit; FG, Walsh function generator.

$a(j)\,\mathrm{wal}(j, \theta)$. The resistors $R/K(0), \ldots, R/K(s)$ together with the operational amplifier A implement the multiplications $K(j)\,a(j)\,\mathrm{wal}(j, \theta)$ and the summation of the $s+1$ terms:

$$-F_0(\theta) = \sum_{j=0}^{s} K(j)\,a(j)\,\mathrm{wal}(j, \theta) \tag{2}$$

It is evident that this block diagram is not restricted to filtering on the basis of Walsh functions, but that any other system of functions can be used if the function generator FG and the multipliers are modified.

Let us see how the circuit of Fig. 62 can be used to implement a so-called multiplicity filter. It has been mentioned previously that Walsh functions $\mathrm{wal}(j, \theta)$ are products of the Rademacher functions $\mathrm{wal}(2^n - 1, \theta)$. The number q of Rademacher functions in the product is called the multiplicity of the Walsh function. Figure 63 shows the multiplicity plotted as function of j. Note that even values of j yield cal functions while odd values yield sal functions. Circles are used in Fig. 63 for odd values of j and dots for even values.

A sequency low-pass filter would pass, e.g., all Walsh functions with $j \leqq 31$, which is a total of 32 functions. A multiplicity filter will pass functions according to their value of j and of q. Figure 63 shows as example the dashed line denoted $j \leqq 48$, $q \leqq 3$. There are 32 functions that will pass through such a filter, just as in the case of the low-pass filter.

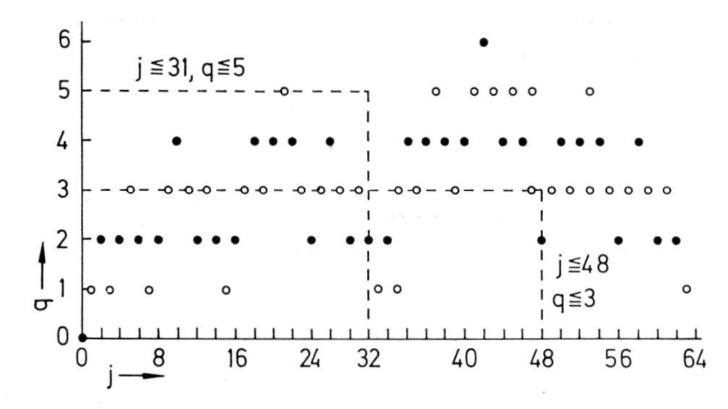

Fig. 63. Multiplicity q as function of j for Walsh functions $\mathrm{wal}(j, \theta)$; $\mathrm{wal}(2i, \theta) = \mathrm{cal}(i, \theta)$, $\mathrm{wal}(2i-1, \theta) = \mathrm{sal}(i, \theta)$. The dashed lines show band limits of multiplicity filters permitting 32 functions to pass.

As a matter of fact, the ordinary low-pass filter turns out to be a multiplicity filter in disguise. The dashed line denoted $j \leqq 31$, $q \leqq 5$ represents the ordinary low-pass filter, since the multiplicity of a function with $j \leqq 31$ cannot be larger than 5.

Multiplicity filters were first proposed by Liedl of Innsbruck University and developed by Roth [1]. The argument in their favor is as follows:

It is known that a polynomial with m terms approximates a function usually much better than a Fourier series with m terms. A polynomial of order q contains in its Walsh series expansion only functions with multiplicity q or less. This makes one suspect that Walsh functions with high multiplicity but low sequency contribute less to the approximation than such with low multiplicity but high sequency.

Multiplicity filters are very expensive for time signals but the principle carries over to filters for space signals and leads to more attractive implementations there.

It is often thought that filters based on functions other than sine-cosine require synchronization. This is not so. Asynchronous filters for Walsh functions have been known as long as synchronous filters [1]. Consider the sampled signal $F(\theta + \theta')$ in Fig. 64. It is shown as a step function with steps of equal width, but there is no difficulty applying the results of

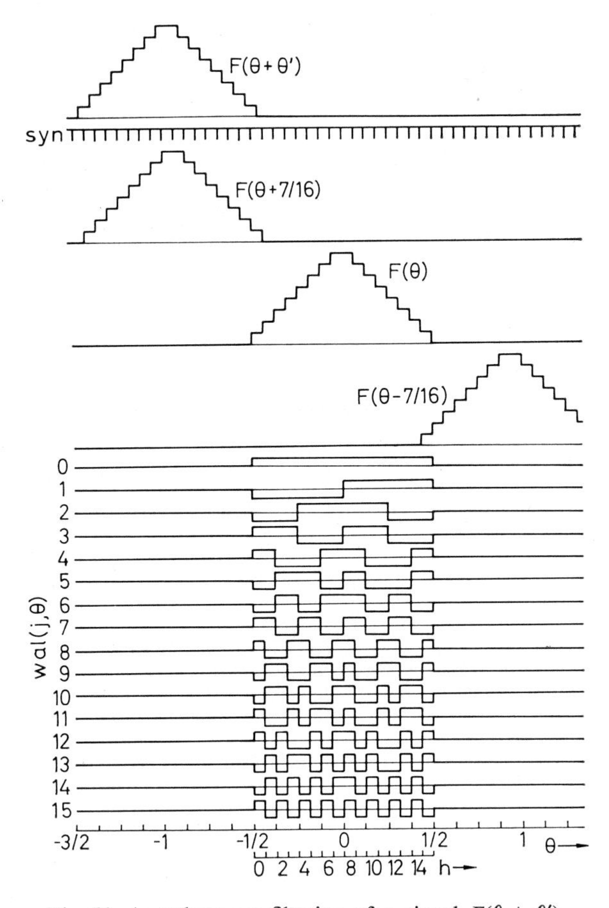

Fig. 64. Asynchronous filtering of a signal $F(\theta + \theta')$.

this section to a signal represented by Dirac samples. $F(\theta + \theta')$ may have any position in time. Synchronizing pulses syn sample $F(\theta + \theta')$ and shift it by a fraction of the width of a step. The signal is now synchronized with the filter. Let us point out once more that no synchronization is required between the original signal and the synchronization pulses.

The synchronized signal $F(\theta + 7/16)$ is decomposed by the functions $\text{wal}(j, \theta)$ shown in Fig. 64, and 16 coefficients $a(0), \ldots, a(15)$ are obtained:

$$a(j, 7/16) = \int_{-1/2}^{1/2} F(\theta + 7/16)\, \text{wal}(j, \theta)\, d\theta \tag{3}$$

The signal is then shifted by $-1/16$, and a second set of coefficients is obtained:

$$a(j, 6/16) = \int_{-1/2}^{1/2} F(\theta + 6/16)\, \text{wal}(j, \theta)\, d\theta \tag{4}$$

Again, the signal is shifted by $-1/16$ and decomposed. This process continues until the signal has been shifted by $-15/16$ and become $F(\theta - 7/16)$. The next and all succeeding shifts will yield zero for all coefficients $a(j, \theta')$.

Figure 65 shows a circuit that performs these decompositions. The signal is fed to the input of a 16 stage analog shift register R. The synchronization pulses syn of Fig. 64 are the shift pulses for the register. If

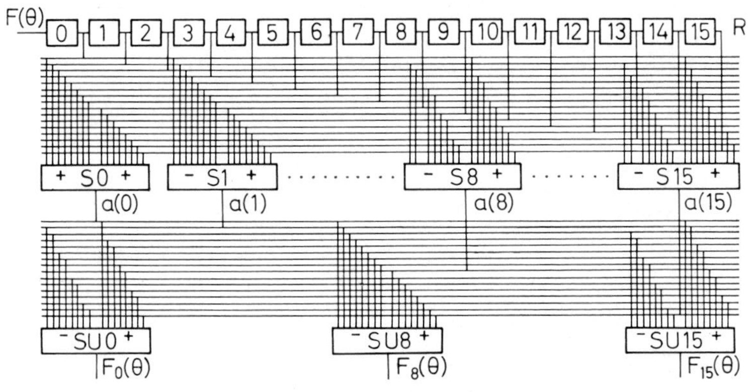

Fig. 65. Block diagram of an asynchronous filter according to Fig. 64. R, stages of an analog shift register; S, SU, summers with adding and subtracting input terminals.

$F(\theta + \theta')$ is fed to the input, the output voltages of the 16 register stages will successively represent $F(\theta + 7/16), F(\theta + 6/16), \ldots, F(\theta - 7/16)$. Summers S 0 to S 15 with adding (+) and subtracting (−) inputs are connected to the outputs of the register stages. The connections are made according to the positive and negative values of the Walsh functions in Fig. 64. The output voltages of the summers represent the coefficients $a(0) = a(0, \theta')$ to $a(15) = a(15, \theta')$.

Filtering is performed by multiplying the coefficients $a(j)$ by attenuation coefficients $K(j)$ and retransforming from the sequency to the time domain. Let us assume that all attenuation coefficients $K(0), \ldots, K(15)$ equal 1. The output voltages $a(0), \ldots, a(15)$ are then fed to the inputs of summers SU with adding and subtracting inputs. The voltages fed to SU 0 are added and subtracted according to the positive and negative signs of the 16 Walsh functions in the interval $h = 0$, or $-1/2 \leqq \theta < -1/2 + 1/16$, in Fig. 64. Similarly[1], adding and subtracting in SU 8 and SU 15 is done according to the signs in the intervals $h = 8$ and $h = 16$.

Figures 66a, d and g show the output signals $F_0(\theta)$, $F_8(\theta)$ and $F_{15}(\theta)$ for $K(0), \ldots, K(15) = 1$. Evidently, the filter acts as a delay line in this case.

Consider now the band-pass filter defined by $K(0) = 0, K(1), \ldots, K(15) = 1$. The output voltage of S 0 in Fig. 65 is not fed to the summers SU, but everything else remains unchanged. Figures 66b, e and h show the

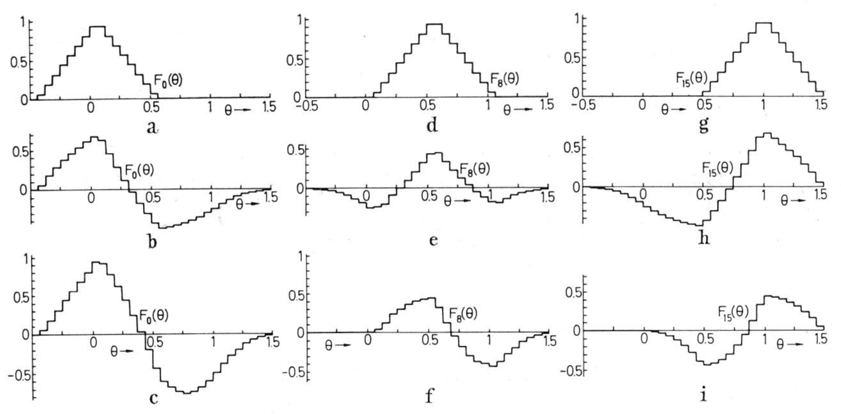

Fig. 66. Output signals $F_0(\theta)$, $F_8(\theta)$ and $F_{15}(\theta)$ of the filter of Fig. 65 if the signal $F(\theta + \theta')$ of Fig. 64 is applied to the input. (a), (d), (g): $K(0) \ldots K(15) = 1$; (b), (e), (h): $K(0) = 0$, $K(1) \ldots K(15) = 1$; (c), (f), (i): $K(0) = K(1) = 0$, $K(2) \ldots K(15) = 1$.

resulting output signals $F_0(\theta)$, $F_8(\theta)$ and $F_{15}(\theta)$. The output signals of another band-pass filter defined by $K(0) = K(1) = 0, K(2), \ldots, K(15) = 1$ are shown by Figs. 66c, f and i.

Problems

214.1 Compare the multiplicity of the Walsh functions in Fig. 63 with the binary representation of j in Fig. 1, and then plot the multiplicity for some values of j larger than 63.

214.2 What is the multiplicity of the Walsh functions in the series expansion of θ in section 1.2.8?

214.3 Plot the output signal $F_2(\theta)$ in analogy to Fig. 66.

---

[1] In terms of matrix notation the transformation from the time to the sequency domain is done by multiplication with a Hadamard matrix and the inverse transformation by multiplication with the transposed matrix.

2.2 Resonance Filters for Time Signals

2.2.1 Series and Parallel LCS Resonance Filters

Figure 67a shows a series resonance circuit with an inductor L and a capacitor C_2. A switch S, a capacitor C_1 and another switch s_0 have been added. The initial voltages across the capacitors are $v_1(0) = V_1$ and

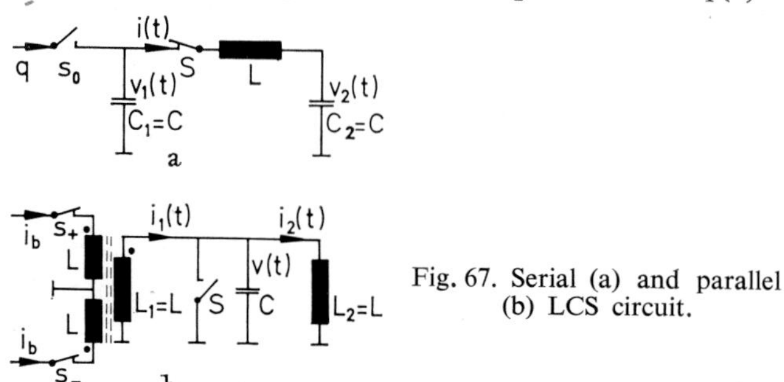

Fig. 67. Serial (a) and parallel (b) LCS circuit.

$v_2(0) = V_2$. The switch S is closed at the time $t = 0$. The current $i(t)$ and the voltages $v_1(t)$ and $v_2(t)$ are defined by the following equations:

$$\left.\begin{aligned} i(t) &= \frac{1}{Z}(V_1 - V_2)\sin\omega t \\ v_1(t) &= V_1 - \tfrac{1}{2}(V_1 - V_2)(1 - \cos\omega t) \\ v_2(t) &= V_2 + \tfrac{1}{2}(V_1 - V_2)(1 - \cos\omega t) \\ \omega &= \sqrt{(2/LC)}, \quad Z = \sqrt{(2L/C)} \end{aligned}\right\} \tag{1}$$

One obtains the following relations at the time $t = \pi/\omega$:

$$i(\pi/\omega) = 0, \quad v_1(\pi/\omega) = V_2, \quad v_2(\pi/\omega) = V_1 \tag{2}$$

The voltages across the capacitors at the times $t = 0$ and $t = \pi/\omega$ are interchanged by this circuit. To this point the circuit is identical with the resonant transfer switch by Haard and Svala [1–4].

Consider now the function $\mathrm{sal}(4, t/T)$ shown in line *1* of Fig. 68. It can be represented by the amplitude samples of line *2*. Let the switch s_0 in Fig. 67a add the charge $q = +V_1 C$ at the times of positive amplitude samples and the charge $q = -V_1 C$ at the times of negative amplitude samples to the charge in the capacitor C_1. The switch S shall always be closed as shown by line *3* of Fig. 68 (black = closed). Let the capacitors C_1 and C_2 of Fig. 67a be initially free of charge. The first positive charge $q = +V_1 C$ applied at the time $t = -T/2 + T/16$ according to Fig. 68 produces the voltage $V_1 = q/C$ across capacitor C_1. Line *4* of Fig. 68 shows $v_1(t)$ and $v_2(t)$ according to (1) for $\omega = 8\pi/T$. The voltage $v_1(t)$ becomes zero and $v_2(t)$ becomes V_1 at the time $t = -T/2 + 3T/16$. At this moment a negative charge $q = -V_1 C$ makes $v_1(t)$ jump to $-V_1$. During the following time interval of duration $\pi/\omega = T/8$ the voltages across C_1 and C_2 are interchanged, yielding $v_1(t) = +V_1$ and $v_2(t) = -V_1$ at the time $t = -T/2 + 5T/16$. At this moment, a positive charge $q = +V_1 C$ makes $v_1(t)$ jump from $+V_1$ to $+2V_1$ and the exchange of voltages begins again. Line *4* in Fig. 68 shows how $v_1(t)$ and $v_2(t)$ increase

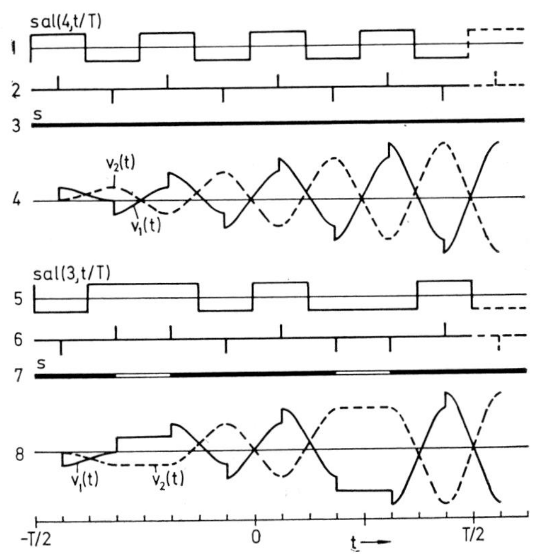

Fig. 68. Time diagrams for a LCS resonance filter tuned to the function applied to the input terminal.

proportionately with time. This increase is analogous to the increase of the voltage across the capacitor of a serial LC circuit fed from a current source by a sinusoidal current in resonance with the LC circuit. Since the Walsh function has been represented by the time samples of its amplitude, the current source had to be replaced by a *charge source*.

Let now the function sal$(3, t/T)$ be applied to the circuit of Fig. 67a. The switch S is operated according to the black (= closed) and white (= open) sections of line *7* in Fig. 68. The resulting voltages $v_1(t)$ and $v_2(t)$ in line *8* increase again proportionately with time. The circuit of Fig. 67a is thus a serial resonance circuit for time sampled Walsh functions.

Figure 69, lines *1* to *4*, shows the application of sal$(3, t/T)$ to the circuit of Fig. 67a, if the switch S is closed all the time and the circuit is thus tuned to sal$(4, t/T)$. The voltages do not increase with time. Similarly, lines *5* to *8* show the application of sal$(4, t/T)$ to a circuit tuned to sal$(3, t/T)$ by the switch S. Again, there is no increase of $v_1(t)$ and $v_2(t)$ proportionately with time.

Figure 70 shows the application of the eight functions wal$(0, t/T)$ to sal$(4, t/T)$ to a circuit tuned to sal$(3, t/T)$ by the switch S. For simplification, the voltages $v_1(t)$ and $v_2(t)$ between the sampling times $t = -T/2 + T/16, -T/2 + 3T/16, \ldots$ are shown by straight lines rather than the correct curved lines used in Figs. 68 and 69. One may see that only wal$(0, t/T)$ and sal$(3, t/T)$ produce voltages that increase proportionately with time.

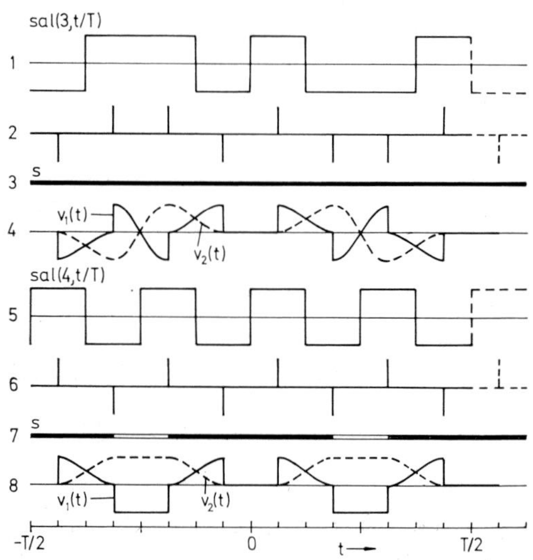

Fig. 69. Time diagrams for a LCS resonance filter *not* tuned to the function applied to its input terminal.

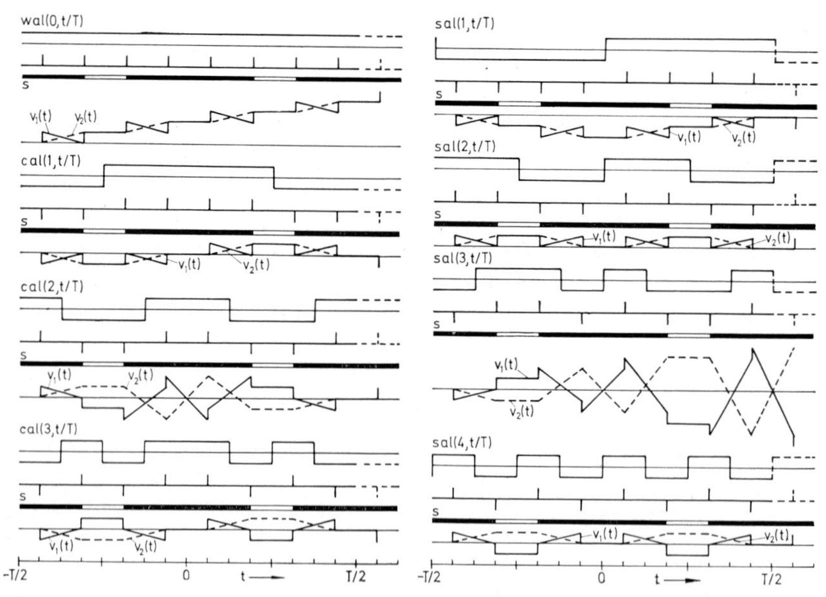

Fig. 70. Time diagrams for a LCS resonance filter tuned to sal$(3, t/T)$. The eight functions wal$(0, t/T)$ to sal$(4, t/T)$ are applied to its input terminal.

This is in complete analogy to the serial tuned LC circuit for sinusoidal functions: Feeding a dc current or a sinusoidal current of proper frequency from a current source to such a circuit produces a voltage across the capacitor that increases proportionately with time. Contrary to what one is used to from resonance circuits for sinusoidal functions, the circuit of Fig. 67a distinguishes between the functions $\mathrm{sal}(3, t/T)$ and $\mathrm{cal}(3, t/T)$ even though both have the same sequency. The reason is, of course, that the switch S introduces synchronization which is absent in resonance circuits for sinusoidal functions.

Figure 71 shows when the switch S has to be closed (= black) or open (= white) to cause the circuit of Fig. 67a to be in resonance with the respective Walsh function $\mathrm{sal}(1, \theta)$ to $\mathrm{sal}(4, \theta)$. One may readily see that

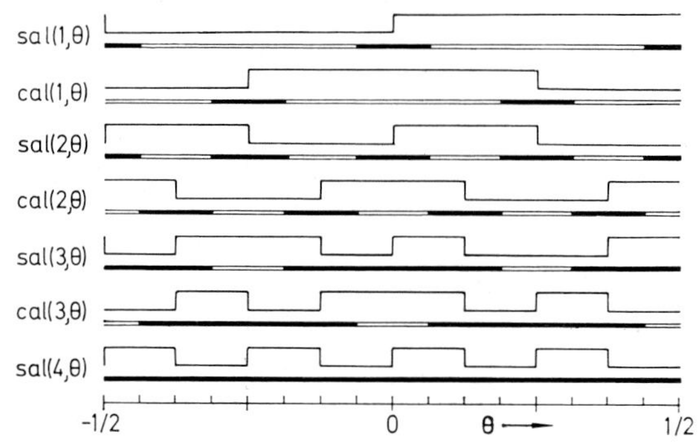

Fig. 71. Operating times of the switch S in Fig. 67 for tuning the filter to the functions $\mathrm{sal}(1, \theta)$ to $\mathrm{sal}(4, \theta)$. Black areas: S closed; white areas: S open.

S must be closed for intervals of duration $\Delta\theta = 1/8$ or $\Delta T = T/8$ whenever the Walsh function changes sign. For $\mathrm{sal}(3, \theta)$ and $\mathrm{cal}(3, \theta)$ some of these intervals are contiguous, while for $\mathrm{sal}(4, \theta)$ all these intervals are contiguous.

Figure 67b shows the parallel tuned LCS circuit for Walsh functions. Switch S, inductor L_2 and capacitor C are now in parallel rather than in series as S, L and C_2 are in Fig. 67a. The voltages $v_1(t)$ and $v_2(t)$ are replaced by the currents $i_1(t)$ and $i_2(t)$, while the current $i(t)$ is replaced by the voltage $v(t)$. A positive or negative magnetic flux must be added instead of a positive or negative charge. In principle, this can be done by driving the currents i_b from a current source through a transformer as shown. Opening the switch s_+ makes $i_1(t)$ jump by $+i_b$, while opening s_- makes $i_1(t)$ jump by $-i_b$.

The inductor L in Fig. 67a only serves to exchange the voltages across C_1 and C_2. The value of L is unimportant as long as it is small enough to allow the exchange of the voltages before the next amplitude sample is fed into the circuit. The switch S may be opened as soon as this exchange is performed and closed when the next charge q is fed into the capacitor C_1.

Evidently, the inductor in Fig. 67a may be replaced by a variety of circuits using operational amplifiers, switches and capacitors, but no inductors that will exchange the voltages across C_1 and C_2. An equivalent statement holds for the capacitor C in Fig. 67b.

Since the voltages $v_1(t)$ and $v_2(t)$ are only of interest at certain time points $t_n = (n + \frac{1}{2})\, T/8 = n\pi/\omega + T/16$ where n equals $0, \pm 1, \pm 2, \ldots$ according to Figs. 68 to 70, one may use difference equations for the analysis of the circuits of Fig. 67. Let us denote $v_1(t_n)$ by $v_1(n)$ and $v_2(t_n)$ by $v_2(n)$. Let the switch S in Fig. 67a always be closed and let s_0 always be open. According to Eq. (2) the voltages $v_1(n)$ and $v_2(n)$ are connected by the following two difference equations of first order:

$$v_1(n+1) = v_2(n), \qquad v_2(n+1) = v_1(n) \tag{3}$$

Let the switch s_0 be momentarily closed at the time t_n. The applied charge $q_n = C\, v_1'(n)$ increases the voltage across the capacitor C_1 from $v_1(n)$ to $v_1(n) + v_1'(n)$. The Eqs. (3) assume the following form for this case:

$$v_1(n+1) = v_2(n), \qquad v_2(n+1) = v_1(n) + v_1'(n) \tag{4}$$

Separation of the variables yields two difference equations of second order:

$$\begin{aligned} v_1(n+2) - v_1(n) &= v_1'(n) \\ v_2(n+2) - v_2(n) &= v_1'(n+1) \end{aligned} \tag{5}$$

The solutions of the homogeneous equations are:

$$\begin{aligned} v_1(n) &= V_{11}(+1)^n + V_{12}(-1)^n \\ v_2(n) &= V_{21}(+1)^n + V_{22}(-1)^n \end{aligned} \tag{6}$$

If the voltages $v_1'(n)$ represent the Walsh functions $\mathrm{wal}(0, t/T)$ or $\mathrm{sal}(4, t/T)$ by amplitude samples at the times t_n one obtains:

$$v_1'(n) = V(+1)^n \quad \text{or} \quad v_1'(n) = V(-1)^n \tag{7}$$

The inhomogeneous terms are thus in resonance with the solutions of the homogeneous equations.

The reader is referred to the literature for the solution of this type of difference equation [6–8]. The generalization of Eq. (4) to the case where the switch S is not closed all the time is straight forward and is omitted here.

Problems

221.1 Derive Eqs. (1) from the differential equation of the circuit of Fig. 67a.

221.2 Derive Eqs. (3) to (6) if the switch S is operated according to line 7 of Fig. 68.

221.3 Write $\mathrm{sal}(3, t/T)$ in the form of Eq. (7) and compare it with the solution of the homogeneous equation derived in problem 221.2.

2.2.2 Low-Pass LCS Resonance Filters

Consider the circuit of Fig. 72. Except for the switches S_1 and S_2 it is the basic low-pass filter section known from frequency theory. The initial

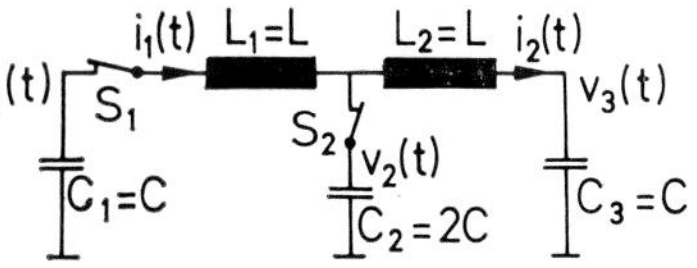

Fig. 72. Basic LCS low-pass filter.

voltages across the capacitors are $v_1(0) = V_1$, $v_2(0) = V_2$ and $v_3(0) = V_3$. The switches S_1 and S_2 are closed at the time $t = 0$. The currents $i_1(t)$, $i_2(t)$ and the voltages $v_1(t)$, $v_2(t)$, $v_3(t)$ are defined by the following equations:

$$\left.\begin{aligned}
i_1(t) &= \frac{1}{2\sqrt{(2)}Z}(V_1 - 2V_2 + V_3)\sin\sqrt{(2)}\omega t + \frac{1}{2Z}(V_1 - V_3)\sin\omega t \\
i_2(t) &= -\frac{1}{2\sqrt{(2)}Z}(V_1 - 2V_2 + V_3)\sin\sqrt{(2)}\omega t + \frac{1}{2Z}(V_1 - V_3)\sin\omega t \\
v_1(t) &= V_1 - \tfrac{1}{4}(V_1 - 2V_2 + V_3)(1 - \cos\sqrt{(2)}\omega t) - \\
&\qquad - \tfrac{1}{2}(V_1 - V_3)(1 - \cos\omega t) \\
v_2(t) &= V_2 + \tfrac{1}{4}(V_1 - 2V_2 + V_3)(1 - \cos\sqrt{(2)}\omega t) \\
v_3(t) &= V_3 - \tfrac{1}{4}(V_1 - 2V_2 + V_3)(1 - \cos\sqrt{(2)}\omega t) + \\
&\qquad + \tfrac{1}{2}(V_1 - V_3)(1 - \cos\omega t) \\
\omega &= 1/\sqrt{(LC)}, \qquad Z = \sqrt{(L/C)}
\end{aligned}\right\} \quad (1)$$

The currents $i_1(t)$ and $i_2(t)$ are equal at the time $t = \pi/\sqrt{(2)}\,\omega$. No current flows at that moment through switch S_2, and the switch may thus be opened. The voltage $v_2(t)$ and the currents $i_1(t) = i_2(t)$ have the following values at this time:

$$\begin{aligned}
v_2(\pi/\sqrt{(2)}\,\omega) &= \tfrac{1}{2}(V_1 + V_3) \\
i_1(\pi/\sqrt{(2)}\,\omega) &= \frac{1}{2Z}(V_1 - V_3)\sin\pi/\sqrt{(2)}
\end{aligned} \quad (2)$$

After opening switch S_2 the circuit of Fig. 72 is the same as that of Fig. 67a except that L is replaced by $2L$. Current and voltages for the time $t > \pi/\sqrt{(2)}\,\omega$ are defined by the following equations:

$$\left.\begin{aligned}
i_1(t) = i_2(t) &= \frac{1}{2Z}(V_1 - V_2)\sin\omega t \\
v_1(t) &= V_2 + \tfrac{1}{2}(V_1 - V_3)\cos\omega t \\
v_3(t) &= V_2 - \tfrac{1}{2}(V_1 - V_3)\cos\omega t
\end{aligned}\right\} \quad (3)$$

The current becomes zero at the time $t = \pi/\omega$ and the voltages assume the following values:

$$\left.\begin{aligned} v_1(\pi/\omega) &= V_2 - \tfrac{1}{2}(V_1 - V_3) \\ v_2(\pi/\omega) &= v_2(\pi/\sqrt{(2)}\,\omega) = \tfrac{1}{2}(V_1 + V_3) \\ v_3(\pi/\omega) &= V_2 + \tfrac{1}{2}(V_1 - V_3) \end{aligned}\right\} \quad (4)$$

In the preceding section the difference equations (2.2.1-3) were derived from Eq. (2.2.1-2) for the time points t_n. By analogy, one obtains from Eq. (4) the following difference equations if the switch S_2 is closed at the times $t_n = n\pi/\omega$ and opened at the times $n\pi/\omega + \pi/\sqrt{(2)}\,\omega$:

$$\left.\begin{aligned} v_1(n+1) &= v_2(n) - \tfrac{1}{2}[v_1(n) - v_3(n)] \\ v_2(n+1) &= \tfrac{1}{2}[v_1(n) + v_3(n)] \\ v_3(n+1) &= v_2(n) + \tfrac{1}{2}[v_1(n) - v_3(n)] \end{aligned}\right\} \quad (5)$$

If the charges $q_{1,n} = C v_1'(n)$ and $q_{3,n} = C v_3'(n)$ are applied to the capacitors C_1 and C_2 of Fig. 72 at the time t_n, one obtains the following equations instead of Eq. (5):

$$\left.\begin{aligned} v_1(n+1) &= v_2(n) - \tfrac{1}{2}[v_1(n) + v_1'(n) - v_3(n) - v_3'(n)] \\ v_2(n+1) &= \tfrac{1}{2}[v_1(n) + v_1'(n) + v_3(n) + v_3'(n)] \\ v_3(n+1) &= v_2(n) + \tfrac{1}{2}[v_1(n) + v_1'(n) - v_3(n) - v_3'(n)] \end{aligned}\right\} \quad (6)$$

Consider as the simplest case that switch S_1 of Fig. 72 is always closed. The separation of variables yields the following three difference equations of second order:

$$\left.\begin{aligned} v_1(n+2) - v_1(n) &= v_1'(n) - \tfrac{1}{2}[v_1'(n+1) - v_3'(n+1)] \\ v_2(n+2) - v_2(n) &= \tfrac{1}{2}[v_1'(n+1) + v_3'(n+1)] \\ v_3(n+2) - v_3(n) &= v_3'(n) + \tfrac{1}{2}[v_1'(n+1) - v_3'(n+1)] \end{aligned}\right\} \quad (7)$$

Comparison with Eq. (2.2.1-5) shows that the homogeneous equations are the same. The voltages $v_1'(n)$ and $v_3'(n+1)$ must be suitably chosen to avoid resonance between inhomogeneous terms and the solutions of the homogeneous equations. In other words, one must remove as much energy at the output terminal as is fed into the input terminal. This is accomplished by matched termination of the filter output in frequency theory. The general idea for obtaining resonance filters for Walsh functions seems fairly clear. However, particular solutions of the inhomogeneous equations (7) for specific functions $v_1'(n)$ and $v_3'(n)$ are of limited value. What is needed are particular solutions for classes of functions, but little progress has been made in this direction.

The inductors in Fig. 72 may readily be replaced by operational amplifiers, capacitors and switches since their function is strictly to transform the initial voltages V_1, V_2 and V_3 into the voltages $v_1(\pi/\omega)$, $v_2(\pi/\omega)$ and $v_3(\pi/\omega)$ according to Eq. (4). However, the difference equations (5) to (7) lead to a simple implementation by digital circuits and thus to a simple digital filter that is equivalent to the analog filter of Fig. 72, and this appears to be the more promising way to a practical use of LCS filters.

According to the results of the last two sections Walsh functions are excellently suited for filtering signals that are represented by amplitude samples at equidistant time points. The reason for this is that a simple sample-and-hold circuit can transform such amplitude samples into a step function with steps of equal width. Such step functions can be represented more easily by a superposition of Walsh functions than, e.g., by a superposition of sine-cosine or $(\sin x)/x$ functions. The emphasis on the Walsh functions is thus due to the fact that the present technology makes sample-and-hold circuits easier to implement than circuits that transform, for instance, an amplitude sample into a $(\sin x)/x$ function.

Problems

222.1 Derive Eqs. (1) from the differential equation of the circuit of Fig. 72.
222.2 Derive Eqs. (7) if the switch S_1 is operated according to line 7 of Fig. 68.
222.3 Assume $v_3'(n+1) = -v_1'(n)$ in Eq. (7), substitute a Walsh function for $v_1'(n)$, e.g., $v_1'(n) = V(-1)^n$, and find a particular set of solutions for Eq. (7).

2.2.3 Parametric Amplifiers

The amplification of sinusoidal oscillations by a periodic variation of an oscillating system has been investigated since the middle of the 19th century. The principle of this "parametric amplification" has been applied to sinusoidal electromagnetic oscillations since about 1915 [1]. Considerable interest was aroused when Manley and Rowe showed in 1956 that parametric amplifiers have very low noise figures [2].

Parametric amplifiers for sampled signals may readily be derived from the sequency resonance filters discussed in the two preceding sections. Let the charge Q_0 be in the capacitor C_1 of Fig. 73 and let C_1 have the capacity C. The voltage across C_1 is denoted by V_0:

$$V_0 = Q_0/C \tag{1}$$

The energy stored in C_1 equals W_0:

$$W_0 = \tfrac{1}{2} C V_0^2 = Q_0^2/2C \tag{2}$$

Let the capacity of C_1 be decreased from C to C/p, $p > 1$. Voltage V_0 and energy W_0 are transformed into V and W:

$$V = \frac{Q_0}{C/p} = p\, V_0$$

$$W = \frac{1}{2} \frac{C}{p} (p\, V_0)^2 = p\, W_0 \tag{3}$$

The energy W_0 is amplified by a factor p. However, this is a "one shot" amplification and it has to be shown how the process can be repeated. The principle is to transfer the energy W from capacitor C_1 to capacitor C_2,

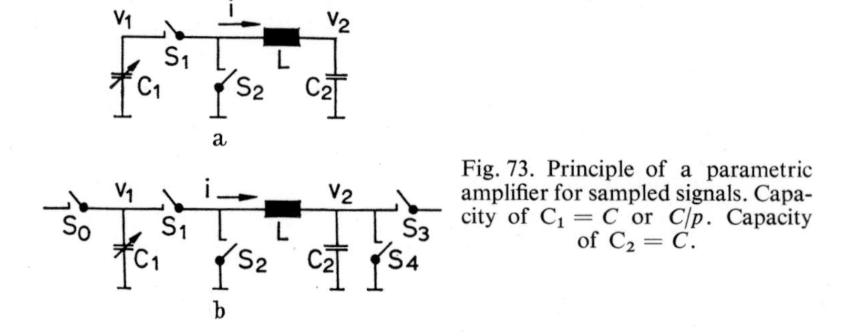

Fig. 73. Principle of a parametric amplifier for sampled signals. Capacity of $C_1 = C$ or C/p. Capacity of $C_2 = C$.

change the capacity of C_1 from C/p back to C, and transfer the energy stored in capacitor C_2 back into C_1. The capacity of C_1 may then be again reduced to C/p.

Let switch S_1 in Fig. 73a be closed when the initial voltage V_0 has been pumped to $V_1 = p V_0$. Current and voltages are defined by the following equations:

$$\left.\begin{aligned} i(t) &= \frac{V}{Z}\sin\omega t \\ v_1(t) &= V\left[1 - \frac{p}{p+1}(1 - \cos\omega t)\right] \\ v_2(t) &= V\frac{1}{p+1}(1 - \cos\omega t) \\ Z = \sqrt{(L/C_0)}, &\quad \omega = 1/\sqrt{(L C_0)}, \quad C_0 = C/(p+1) \end{aligned}\right\} \tag{4}$$

Plots of $v_1(t)$, $v_2(t)$ and $Z\,i(t)$ are shown for $p = 1, 2, 3$ in Fig. 74. Note that $Z\,i(t)$ is independent of p. The voltage $v_2(t)$ and thus the energy in capacitor C_2 reach their peak at the time $\omega t/2\pi = \frac{1}{2}$ or $t = \pi/\omega$. Unfortunately, the voltage $v_1(t)$ is not zero at this time, except for the case $p = 1$ when no amplification is achieved. To remedy the situation the switch S_2 is closed at the time t_1 when $v_1(t)$ becomes zero. t_1 follows from Eq. (4):

$$\cos\omega t_1 = -\frac{1}{p}, \qquad \pi/2 \leqq \omega t_1 \leqq \pi \tag{5}$$

t_1 varies from π/ω for $p = 1$ to $\pi/2\omega$ for $p = \infty$.

After S_2 has been closed, a current $i(t)$ will flow in Fig. 73a through S_2, L and C_2. This current will become zero at the time $t_2 > t_1$. All the energy $p W_0$ must be in C_2 at this time since no energy is left in C_1 or L.

The following equations define $i(t)$ and $v_2(t)$ in the interval $t_1 < t \leqq t_2$:

$$\left.\begin{aligned} i(t) &= \frac{V}{Z}\frac{p+1}{p}\left[\sqrt{(p-1)}\cos\omega_1(t-t_1) - \sin\omega_1(t-t_1)\right] \\ v_2(t) &= V\frac{1}{p}\left[\sqrt{(p-1)}\sin\omega_1(t-t_1) + \cos\omega_1(t-t_1)\right] \\ \omega_1 &= 1/\sqrt{(L\,C)} = \omega/\sqrt{(p+1)} \end{aligned}\right\} \quad (6)$$

The time t_2 follows from the condition $i(t_2) = 0$. Fig. 75 shows plots of current and voltages for $p = 1, 2, 3$, and ∞. The voltage $v_2(t_2)$ follows

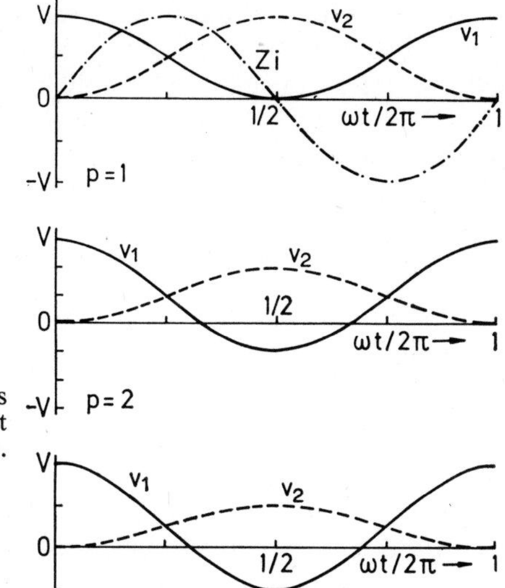

Fig. 74. Voltages and current as functions of time for the circuit of Fig. 73a if S_2 is always open.

from Eq. (6) or, more easily, from the preservation of energy:

$$v_2(t_2) = (2p\,W_0/C)^{1/2} = \sqrt{(p)}\,V_0 = V/\sqrt{(p)} \quad (7)$$

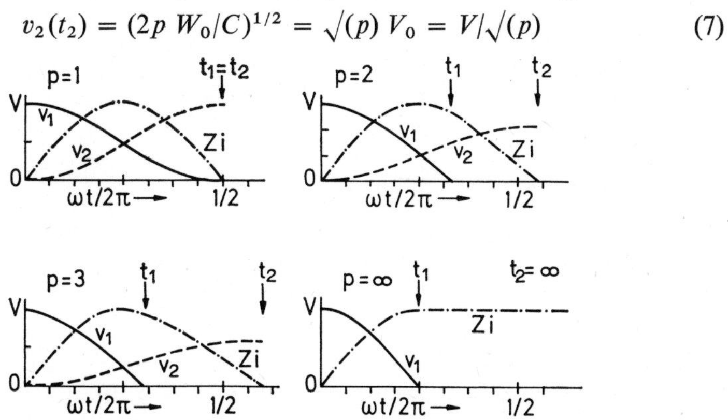

Fig. 75. Voltages and currents as functions of time for the circuit of Fig. 73a if S_2 is closed at the time t_1.

Let the capacity of C_1 be increased from C/p to C during the time interval $t_1 < t < t_2$ and let switch S_2 be opened at time t_2. The charge in C_2 will then flow back into C_1 in analogy to the plot $p = 1$ in Fig. 74. Voltage, charge and energy in C_1 at the time $t_2 + \pi/\omega$ are denoted V_1, Q_1 and W_1:

$$V_1 = \sqrt{(p)}\, V_0, \quad Q_1 = \sqrt{(p)}\, Q_0, \quad W_1 = p\, W_0 \tag{8}$$

A new cycle of operation may start. n cycles yield a voltage and charge amplification of $p^{n/2}$, and an energy amplification of p^n. Hence, the amplifier yields a digitally controlled exponential amplification even though no nonlinear circuit elements are used. Note that energy and charge are amplified, not power. A power amplification is achieved by sampling a signal k times per second and amplifying the energy of each sample. The efficiency of the amplifier is theoretically 100% since all the pumping energy is transformed into signal energy.

Figure 73b shows a parametric amplifier with input and output circuit. Its time diagram is shown by Fig. 76. Capacitor C_1 is charged through S_0 to a voltage V_0. Switch S_1 is open during this time. S_0 opens and S_1 closes

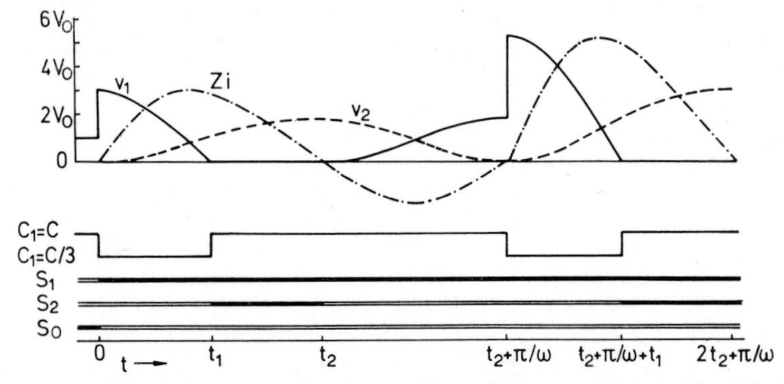

Fig. 76. Time diagram for a recycling parametric amplifier according to Fig. 73b (black = switch closed, white = switch open).

at $t = 0$. The capacitance of C_1 is changed from C to $C/3$ at the same moment, but this change does not have to be instantaneous. Switch S_2 is closed during the interval $t_1 < t < t_2$ and the capacitance of C_1 is changed from $C/3$ to C. During the time $t_2 < t < t_2 + \pi/\omega$ the energy flows back from C_2 to C_1. This time interval could be used for amplification if C_2 were made variable. The second pumping cycle starts at $t = t_2 + \pi/\omega$. The input energy amplified p^n times will be in C_2 at the time $n\,t_2 + (n - 1)\,\pi/\omega$. If S_1 and S_2 are open one may extract the amplified energy by closing S_3. Any residual energy can be removed from C_2 by closing S_4.

An experimental amplifier by Walker [3] using voltage variable capacitors in a bridge circuit for C_1 was not very successful because of the small capacitance and large leakage of these capacitors. There is, however, a very promising application in the mechanical amplification of an electric charge. Let capacitor C_1 in Fig. 73 be a mechanically variable capacitor, e.g., a rotary capacitor. Let the whole circuit be made superconducting. The amplifier will then be free of thermal noise. Careful construction should avoid contact potentials and Peltier effect. Since the inductivity L is fixed there should be little difficulty in suppressing noise caused by external magnetic fields. This is an advantage over the Ries-Satterthwaite amplifier, which amplifies, however, a current rather than a charge and is thus not directly comparable [4]. Since the capacitor C_1 and the switches can be operated mechanically there is no need to have any electromagnetic energy near the amplifier that could interfere with the signal energy. It appears that such an amplifier should be capable to amplify charges of the order of a single electron charge. The ratio p between largest and smallest capacity of a rotary capacitor can readily be made 10, yielding an energy amplification of 10 dB per pumping cycle. Allowing 10 ms operating time for the switches calls for a period of about 100 ms per pumping cycle, which means the energy can be amplified 100 dB per second. Using an amplifier with several stages rather than one recycling stage would increase the speed.

Problems

223.1 Design a parametric amplifier with three stages instead of one stage that is recycled three times.

223.2 Design a digitally controlled amplifier that amplifies a voltage $p^{n/2}$ times or attenuates it $p^{-n/2}$ times.

223.3 Design a bridge circuit that uses four voltage variable capacitors for C_1 in Fig. 73.

223.4 Why is it more desirable to make the capacitor C_1 than the inductor L variable?

2.3 Instantaneous Filters for Space Signals

2.3.1 Filters for Signals with One Space Variable

Let $F(\theta, x)$ be a signal with the time variable θ and the space variable x in the interval $-\frac{1}{2} \leqq \theta < \frac{1}{2}$, $-\frac{1}{2} \leqq x < \frac{1}{2}$. Its Walsh-Fourier transform is $a(j, k)$:

$$a(j, k) = \int_{-1/2}^{1/2} \int_{-1/2}^{1/2} F(\theta, x) \operatorname{wal}(j, \theta) \operatorname{wal}(k, x) \, d\theta \, dx \tag{1}$$

The inverse transform is defined by a double sum:

$$F(\theta, x) = \sum_{j=0}^{\infty} \sum_{k=0}^{\infty} a(j, k) \operatorname{wal}(j, \theta) \operatorname{wal}(k, x) \tag{2}$$

Space functions are usually encountered as sample functions. It is then convenient to replace the notation $F(\theta, x)$ by $F(\theta, h)$ with $h = 0, 1, \ldots, 2^n - 1$. The connection between the x and h scales is shown for $2^n = 8$

in Fig. 77. The integral in Eq. (1) assumes the following form for a function $F(\theta, h)$ that is sampled in space but continuous in time:

$$a(j,k) = \int_{-1/2}^{1/2} \sum_{h=0}^{2^n-1} F(\theta, h)\,\mathrm{wal}(j,\theta)\,\mathrm{wal}\left(k, \frac{h+\frac{1}{2}}{2^n} - \frac{1}{2}\right) d\theta \tag{3}$$

The inverse transform in Eq. (2) remains unchanged except that x is replaced by $(h + \frac{1}{2})/2^n - \frac{1}{2}$ and the upper limit of k becomes $2^n - 1$.

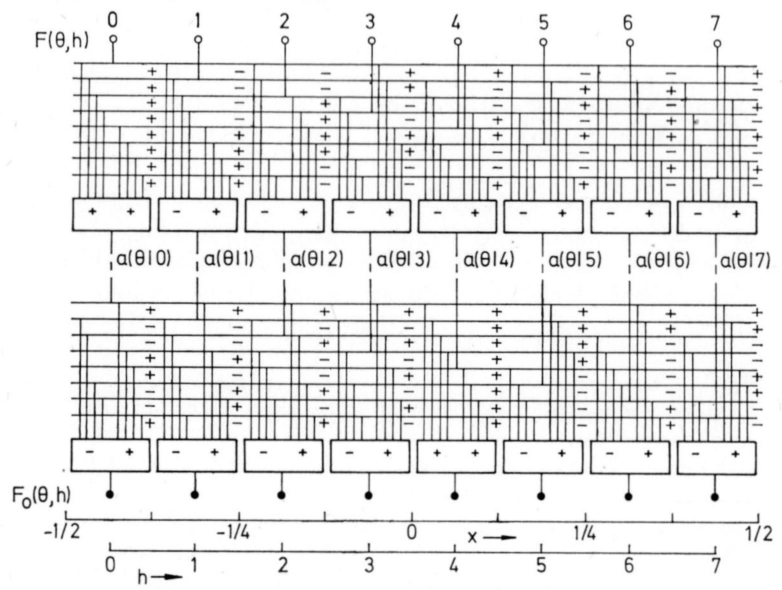

Fig. 77. Sequency filter using the correlation principle for functions with one space variable h assuming eight values.

The transformation in Eq. (3) and the inverse transformation may be performed in two steps:

$$a(\theta \mid k) = \sum_{h=0}^{2^n-1} F(\theta, h)\,\mathrm{wal}\left(k, \frac{h+\frac{1}{2}}{2^n} - \frac{1}{2}\right) \tag{4}$$

$$a(j,k) = \int_{-1/2}^{1/2} a(\theta \mid k)\,\mathrm{wal}(j,\theta)\,d\theta \tag{5}$$

$$a(\theta \mid k) = \sum_{j=0}^{\infty} a(j,k)\,\mathrm{wal}(j,\theta) \tag{6}$$

$$F(\theta, h) = \sum_{k=0}^{2^n-1} a(\theta \mid k)\,\mathrm{wal}\left(k, \frac{h+\frac{1}{2}}{2^n} - \frac{1}{2}\right) \tag{7}$$

Filtering is done by multiplying $a(j, k)$ by an attenuation coefficient $K(j, k)$. In the simplest cases $K(j, k)$ may be factored:

$$K(j,k) = G(j)\,H(k) \tag{8}$$

The time filtering may then be done as discussed in sections 2.1.2 and 2.1.3, and the transformation in Eqs. (5) and (6) may be ignored here. Space filtering is performed by multiplying $a(\theta \mid k)$ by $H(k)$ before or after time filtering, and then retransforming according to Eq. (7). A space shift could be incorporated by replacing $\text{wal}(k, x)$ by $\text{wal}[k, x - x(k)]$ during retransformation, but this has so far not been of interest.

Figure 77 shows a circuit for the space filtering. The input signal $F(\theta, h)$ is defined at $2^n = 8$ space points $h = 0, \ldots, 7$. The transformation in Eq. (4) is accomplished by feeding the eight time signals $F(\theta, 0), \ldots, F(\theta, 7)$ to the adding and substracting input terminals of eight summers. Note that the connections to adding and subtracting input terminals are made according to the functions $\text{wal}(0, x)$ to $\text{wal}(7, x)$ represented by the signs + and − written in vertical columns.

The inverse transformation in Eq. (7) of the attenuated coefficients $H(k)\, a(\theta \mid k)$ is again performed by summers with adding and subtracting input terminals. Note that now the signs + and − written in horizontal rows represent the functions $\text{wal}(0, x)$ to $\text{wal}(7, x)$. This interchange of rows and columns is equivalent to the use of a matrix for transformation, but of the transposed matrix for the inverse transformation in matrix algebra. From Figs. 12, 13 and 22 one may readily infer that the function $\text{wal}(j, \theta)$ or $\text{wal}(k, x)$ is plotted horizontally if θ or x are variables and j or k parameters, but vertically if j or k are variables and θ or x parameters.

Typical signals with the time variable and one space variable are produced by line reading equipment. For instance, a paper tape with symbols on it pulled past eight photocells will produce the function $F(\theta, h)$ of Fig. 77 as output voltages of the photocells.

Problems

231.1 Redraw the circuit diagram of Fig. 77 for filtering by Haar functions.

231.2 What effect has the possibility or impossibility of factoring $K(j, k)$ on the circuit of Fig. 77?

231.3 Write the transformation and retransformation of Fig. 77 in matrix notation.

2.3.2 Filters for Signals with Two Space Variables

The Walsh-Fourier transform and the inverse transform of a space signal $F(x, y)$ in the interval $-\frac{1}{2} \leq x < \frac{1}{2}$, $-\frac{1}{2} \leq y < \frac{1}{2}$ are defined by:

$$a(k, m) = \int_{-1/2}^{1/2} \int_{-1/2}^{1/2} F(x, y)\, \text{wal}(k, x)\, \text{wal}(m, y)\, dx\, dy \tag{1}$$

$$F(x, y) = \sum_{k=0}^{\infty} \sum_{m=0}^{\infty} a(k, m)\, \text{wal}(k, x)\, \text{wal}(m, y) \tag{2}$$

A filtered signal $F_0(x, y)$ is obtained by multiplying the transform $a(k, m)$ with attenuation coefficients $K(k, m)$:

$$F_0(x, y) = \sum_{k=0}^{\infty} \sum_{m=0}^{\infty} K(k, m)\, a(k, m)\, \mathrm{wal}(k, x)\, \mathrm{wal}(m, y) \tag{3}$$

No shifts of wal(k, x) and wal(m, y) to wal$[k, x - x(k)]$ and wal$[m, y - y(m)]$ will be considered.

Figure 78 shows a circuit that generates the Walsh-Fourier transform $a(k, m)$ of a space signal $F(x, y)$. The signal is represented by 4×4 voltages $a, b, \ldots, p$ in the x, y-plane as shown in the lower right corner.

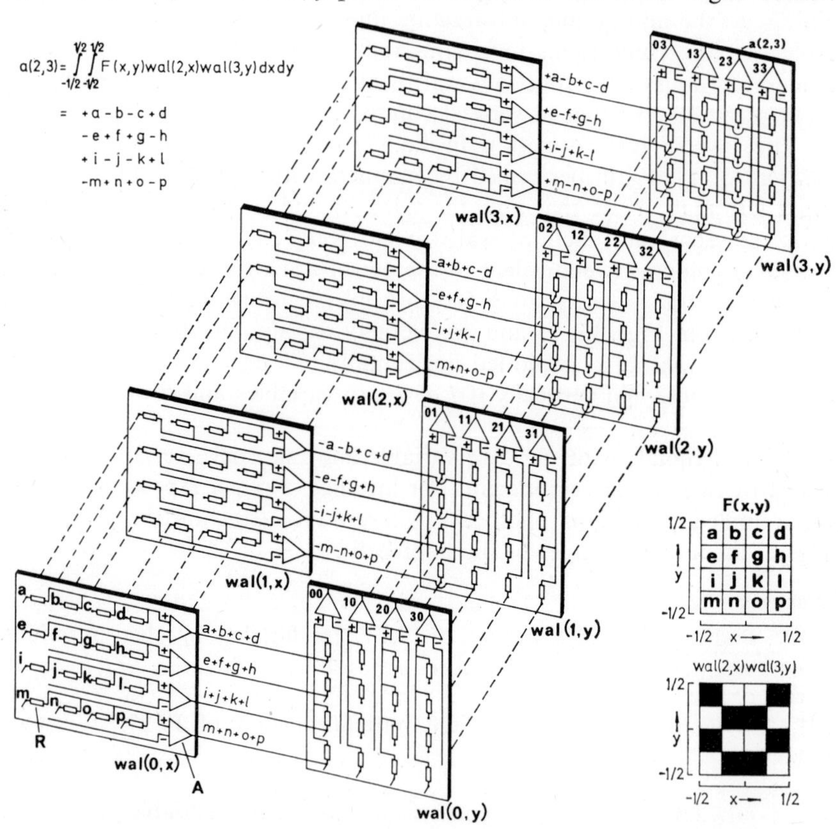

Fig. 78. Principle of a sequency filter for functions of two space variables x and y. The circuit shown transforms $F(x, y)$ into $a(k, m)$. A similar circuit is required for the inverse transformation.

Four printed circuit boards denoted by wal$(0, x)$ to wal$(3, x)$ are intersected by 4×4 wires; only seven of them are shown by dashed lines to avoid obscuring the picture. The voltages $a, \ldots, p$ are applied to these 4×4 wires. Resistors R feed them to the adding (+) or subtracting (−)

inputs of summing amplifiers A. At the outputs of the board wal$(0, x)$ occur the four voltages $a + b + c + d$, $e + f + g + h$, $i + j + k + l$ and $m + n + o + p$; the topmost output of the board wal$(1, x)$ yields the voltage $-a - b + c + d$, etc. Note that the resistors of each board are connected to the adding or subtracting input of the summing amplifier according to the positive and negative sign of the respective Walsh function: $++++$ represents wal$(0, x)$, $--++$ represents wal$(1, x)$, $-++-$ represents wal$(2, x)$ and $+-+-$ represents wal$(3, x)$.

A second set of four printed circuit cards is denoted by wal$(0, y)$ to wal$(3, y)$. These cards have the same circuitry as the cards wal$(0, x)$ to wal$(3, x)$ but are rotated 90°. Again, only seven of the sixteen intersecting wires are shown by dashed lines. The output voltages of the cards wal$(0, x)$ to wal$(3, x)$ are fed to these sixteen intersecting wires as shown. One may readily infer that the output voltage of the summing amplifier denoted by 00 is the sum $(a + b + c + d) + (e + f + g + h) + (i + j + k + l) + + (m + n + o + p)$ and thus represents the Walsh-Fourier coefficient $a(0, 0)$ except for an unimportant scaling factor. Generally, a summing amplifier denoted by $k\,m$ yields the Walsh-Fourier coefficient $a(k, m)$. For instance, $k\,m = 23$ yields $-(-a + b + c - d) + (-e + f + g - h) - - (-i + j + k - l) + (-m + n + o - p) = a(2, 3)$. This sum is written in a square in the upper left corner of Fig. 78. The positive and negative signs of the voltages $a, \ldots, p$ correspond to the values $+1$ (= black) and -1 (= white) of the Walsh function wal$(2, x)$ wal$(3, y)$ shown in the lower right corner.

The "half filter" of Fig. 78 produces the Walsh-Fourier transform $a(k, m)$ of the signal $F(x, y)$. An essentially equal circuit will do the inverse transformation and reproduce $F(x, y)$ from the transform $a(k, m)$. A filtering effect is obtained by modifying the voltages $a(k, m)$. Before we can enter into a discussion of this filtering we must develop a suitable and simple representation of the processes to be studied. When investigating the filtering of time signals one may plot the signal as a function of time, its Fourier transform[1] as a function of frequency, the attenuation and phase shift of the filter as a function of frequency, the Fourier transform of the filtered signal as a function of frequency, and finally the filtered signal as a function of time. All these plots can be done in a two-dimensional cartesian coordinate system.

The very same plots can be drawn for space signals $F(x, y)$ but one needs three-dimensional cartesian systems. Some simplification is required for their representation on two-dimensional paper. The first step is to replace the continuous function $F(x, y)$ by a sampled function $F(h, i)$. While $F(x, y)$ is defined for any value of x and y in an interval $-\frac{1}{2} \leqq x < \frac{1}{2}$,

---

[1] Usually one does not plot the Fourier transform but the frequency power spectrum, which does not contain phase information in terms of Fourier analysis. This procedure is permissable for telephony signals since the ear is very insensitive to phase information.

$-\frac{1}{2} \leqq y < \frac{1}{2}$, the function $F(h, i)$ is defined for non-negative integer values of h and i only. Such a sample function is shown in Fig. 79. This representation is still too cumbersome. Table 5 shows a better representation: the value of $F(h, i)$ is printed at the intersection of row h and column i. For instance, $F(6, 12)$ has the value $+3$.

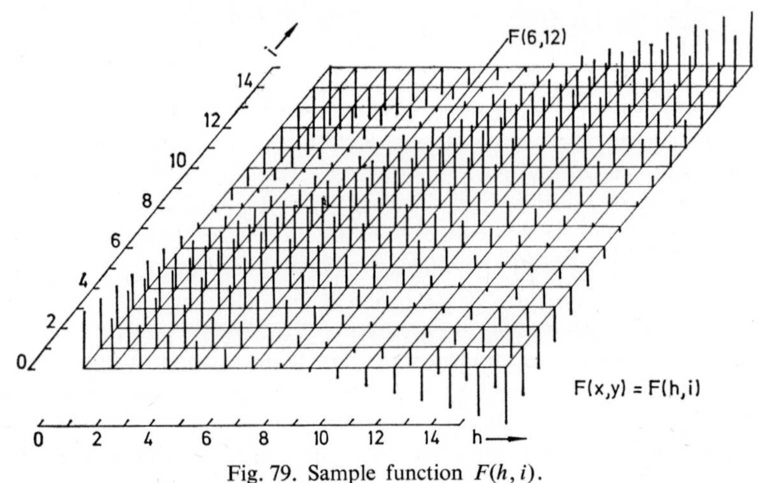

Fig. 79. Sample function $F(h, i)$.

The Walsh-Fourier transform $a(k, m)$ of $F(h, i)$ is represented by Table 6 and Fig. 80[1].

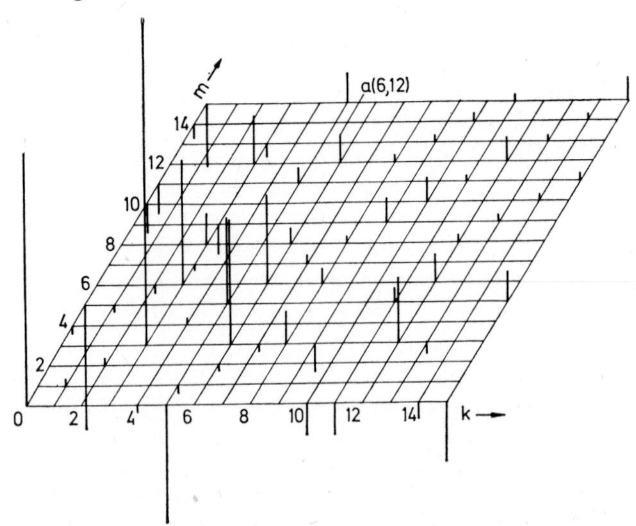

Fig. 80. Walsh-Fourier transform $a(k, m)$ of the sample function $F(h, i)$ of Fig. 79.

[1] The computations for Figs. 80 and 83 were done by H. C. Andrews of the University of Southern California.

Consider now the representation of the attenuation function $K(k, m)$ of the filter. Figure 81 shows such a function having the value $+1$ for $k + m \leqq 14$ and 0 for $k + m > 14$. It is evident that one can plot any attenuation function in this way or that one can represent $K(k, m)$ by a

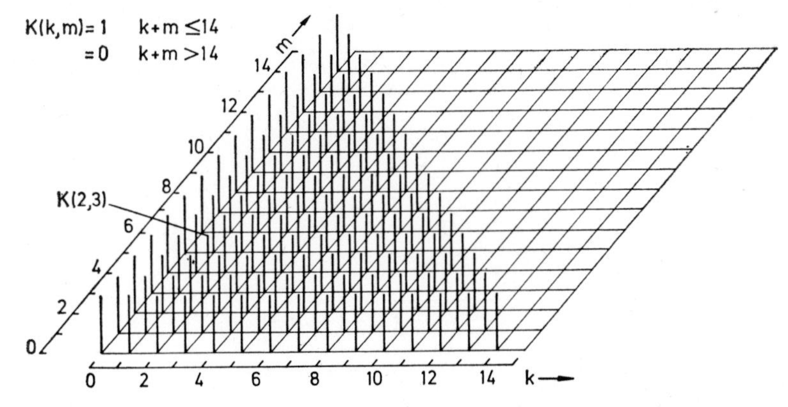

Fig. 81. Attenuation function $K(k, m)$ for a sequency low-pass filter.

table like Tables 5 and 6. This is shown in a further simplified form in Fig. 82. Instead of writing the values of $K(k, m)$ we only indicate by black dots where these values should be written. The line $k + m = 14$ is then

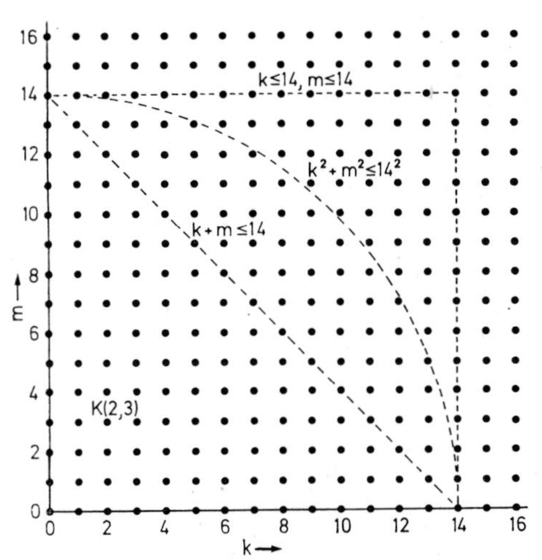

Fig. 82. The (k, m)-plane of the two-dimensional sequency domain with three band limits for low-pass filters.

drawn. The attenuation function of Fig. 81 may readily be visualized by imagining the value $+1$ at all those dots for which $k + m$ is smaller or equal 14 and the value 0 for all other values of $k + m$.

Table 5. *Representation of a two-dimensional sampled function $F(h, i)$*

i \ h	0	1	2	3	4	5	6	7	8	9	10	11	12	13	14	15
15	−15	−13	−11	−9	−7	−5	−3	−1	+1	+3	+5	+7	+9	+11	+13	+15
14	−13	−11	−9	−7	−5	−3	−1	+1	+3	+5	+7	+9	+11	+13	+15	+13
13	−11	−9	−7	−5	−3	−1	+1	+3	+5	+7	+9	+11	+13	+15	+13	+11
12	−9	−7	−5	−3	−1	+1	+3	+5	+7	+9	+11	+13	+15	+13	+11	+9
11	−7	−5	−3	−1	+1	+3	+5	+7	+9	+11	+13	+15	+13	+11	+9	+7
10	−5	−3	−1	+1	+3	+5	+7	+9	+11	+13	+15	+13	+11	+9	+7	+5
9	−3	−1	+1	+3	+5	+7	+9	+11	+13	+15	+13	+11	+9	+7	+5	+3
8	−1	+1	+3	+5	+7	+9	+11	+13	+15	+13	+11	+9	+7	+5	+3	+1
7	+1	+3	+5	+7	+9	+11	+13	+15	+13	+11	+9	+7	+5	+3	+1	−1
6	+3	+5	+7	+9	+11	+13	+15	+13	+11	+9	+7	+5	+3	+1	−1	−3
5	+5	+7	+9	+11	+13	+15	+13	+11	+9	+7	+5	+3	+1	−1	−3	−5
4	+7	+9	+11	+13	+15	+13	+11	+9	+7	+5	+3	+1	−1	−3	−5	−7
3	+9	+11	+13	+15	+13	+11	+9	+7	+5	+3	+1	−1	−3	−5	−7	−9
2	+11	+13	+15	+13	+11	+9	+7	+5	+3	+1	−1	−3	−5	−7	−9	−11
1	+13	+15	+13	+11	+9	+7	+5	+3	+1	−1	−3	−5	−7	−9	−11	−13
0	+15	+13	+11	+9	+7	+5	+3	+1	−1	−3	−5	−7	−9	−11	−13	−15

Table 6. *Walsh-Fourier transform $a(k, m)$ of the function $F(h, i)$ of Table 5*

m \ k	0	1	2	3	4	5	6	7	8	9	10	11	12	13	14	15
15	−16	0	0	0	0	+8	0	0	0	0	0	+2	0	0	0	+6
14	−4	0	0	0	0	0	0	0	0	0	+2	0	0	0	+2	0
13	0	0	0	−4	0	0	0	0	0	+2	0	0	0	+2	0	0
12	0	0	0	+16	0	0	+8	0	+2	0	0	0	+6	0	0	0
11	−8	0	0	0	0	+4	0	0	0	0	0	+2	0	0	0	+2
10	−8	0	0	0	0	0	0	0	0	0	+6	0	0	0	+2	0
9	0	0	0	−8	0	0	0	0	0	+6	0	0	0	+2	0	0
8	0	0	0	+8	0	0	+4	0	+2	0	0	0	+2	0	0	0
7	0	0	0	−2	0	0	0	+2	0	0	0	0	0	0	0	0
6	0	0	−2	+32	0	0	+22	0	+4	0	0	0	+8	0	0	0
5	−32	−2	0	0	0	+22	0	0	0	0	0	+4	0	0	0	+8
4	−2	0	0	0	+2	0	0	0	0	0	0	0	0	0	0	0
3	0	0	0	+86	0	0	+32	−2	+8	−8	0	0	+16	−4	0	0
2	0	0	+2	0	0	0	−2	0	0	0	0	0	0	0	0	0
1	0	+2	0	0	0	−2	0	0	0	0	0	0	0	0	0	0
0	+70	0	0	0	−2	−32	0	0	0	0	−8	−8	0	0	−4	−16

A filter defined by the attenuation function $K(k, m) = 1$ for $k + m \leq 14$ and $K(k, m) = 0$ for $k + m > 14$ is referred to as a low-pass filter. The all inclusive name would be "two-dimensional spatial attenuation low-pass filter." On the basis of Walsh and Haar functions or, generally speaking, on the basis of two- and three-valued functions one can build filters for signals with up to three space variables and the time variable, hence the term "two-dimensional spatial." The term "attenuation" is justified by the fact that one not only may have an attenuation function $K(k, m)$, but also a resolution function $R(k, m)$ which leads to "resolution low-pass filters." Indeed, it is the resolution filter which is mainly of interest for TV picture transmission at the present time.

The dashed line $k^2 + m^2 = 14^2$ in Fig. 82 defines another low-pass filter: $K(k, m)$ equals 1 for $k^2 + m^2 \leqq 14^2$ and 0 for $k^2 + m^2 > 14^2$. A third low-pass filter is defined by the condition $K(k, m) = 1$ for $k \leqq 14$, $m \leqq 14$ and $K(k, m) = 0$ for $k > 14, m > 14$.

The low-pass filters discussed so far let certain parts of the Walsh-Fourier transform $a(k, m)$ pass unchanged and suppressed others completely. The representation of Fig. 82 is, however, capable of handling more complicated cases. Consider, e.g., the attenuation function defined as follows[1]:

$$\begin{aligned}
K(k, m) &= 1 \quad \text{for} \quad k + m \leqq 14 \\
K(k, m) &= \tfrac{1}{2} \quad \text{for} \quad 14 < k + m, \quad k^2 + m^2 \leqq 14^2 \\
K(k, m) &= \tfrac{1}{4} \quad \text{for} \quad 14^2 \leqq k^2 + m^2, \quad k \leqq 14, m \leqq 14 \\
K(k, m) &= 0 \quad \text{for} \quad 14 < k, \quad 14 < m
\end{aligned}$$

The part of the transform $a(k, m)$ between the limits $k = 0$, $m = 0$ and $k + m = 14$ in Fig. 82 will pass this filter unchanged; the part between the limits $k + m = 14$ and $k^2 + m^2 = 14^2$ will be attenuated to half its amplitude, etc.

Let us now turn from attenuation filters to resolution filters. The output voltages $a(k, m)$ of the 16 operational amplifiers in Fig. 78 denoted by 00 to 33 shall be fed to 16 analog/digital converters. Let some of the voltages be converted into binary numbers with many digits and the other voltages into binary numbers with a few digits only. The voltages $a(k, m)$ are thus represented with different "resolution" and the resolution is a function of k and m.

Let us define a resolution function $R(k, m)$. The notation $R(k, m) = 5$ shall mean that the voltage $a(k, m)$ is represented by 5 binary digits. We can use Fig. 82 to define a resolution filter. For instance, a simple resolution low-pass filter is defined by:

$$\begin{aligned}
R(k, m) &= 5 \quad \text{for} \quad k + m \leqq 14 \\
R(k, m) &= 0 \quad \text{for} \quad k + m > 14
\end{aligned}$$

A more sophisticated resolution low-pass filter is obtained by the following definition:

$$\begin{aligned}
R(k, m) &= 5 \quad \text{for} \quad k + m \leqq 14 \\
R(k, m) &= 4 \quad \text{for} \quad 14 < k + m, \quad k^2 + m^2 \leqq 14^2 \\
R(k, m) &= 3 \quad \text{for} \quad 14^2 < k^2 + m^2, \quad k \leqq 14, m \leqq 14 \\
R(k, m) &= 0 \quad \text{for} \quad 14 < k, \quad 14 < m
\end{aligned}$$

[1] $K(k, m)$ is ambiguous for $k = 0$, $m = 14$ and $k = 14$, $m = 0$ but this is of no importance here.

In order to apply these results to TV pictures let i and h in Table 5 assume the values 0 to 511 rather than 0 to 15. This corresponds to a TV picture with 512 lines and 512^2 picture elements. k and m in Fig. 82 must also assume the values from 0 to 511. The implementation of a filter with 512^2 inputs would be a formidable task if done according to Fig. 78. There are two methods to simplify such a filter. First, one can decompose the

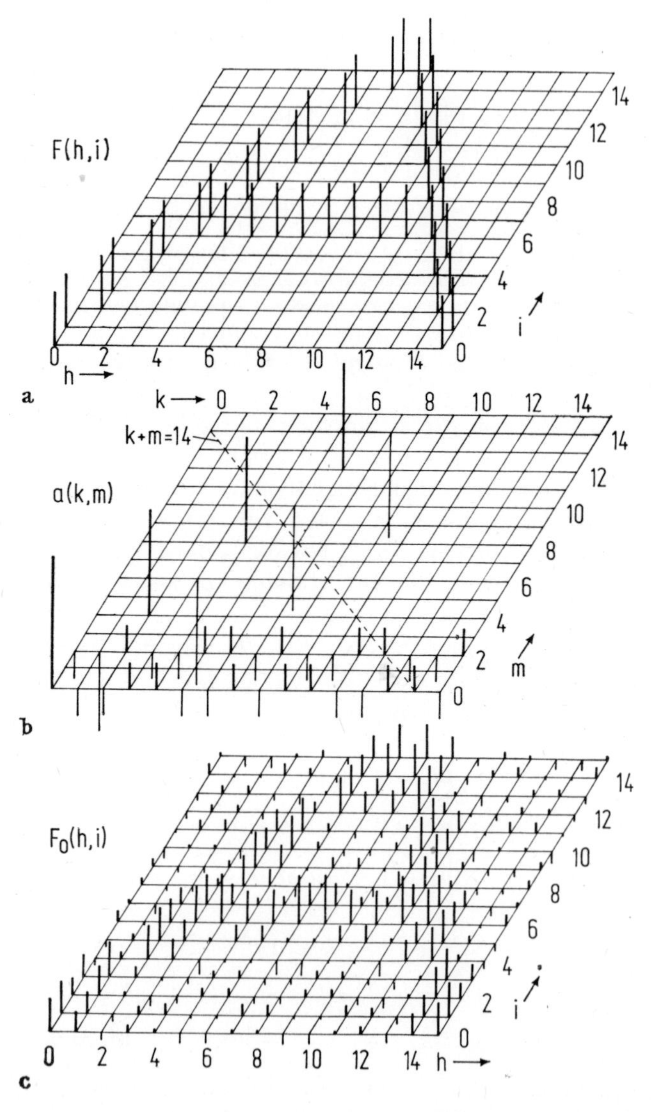

Fig. 83. The letter "A" as sample function $F(h, i)$, its Walsh-Fourier transform $a(k,m)$, and the retransformed function $F_0(h, i)$ after passing through a low-pass filter defined by Fig. 81.

square of 512^2 picture elements into smaller squares. For instance, a good choice would be 32 squares with 16^2 picture elements each. A filter with 16^2 inputs is then needed which transforms sequentially the 32 square picture sections[1]. Second, the circuit of Fig. 78 can be brought into a more practical form.

The distinction between attenuation and resolution filters does not vanish for binary space signals. Such a signal $F(h, i)$ consisting of a black letter A on a white background is shown in Fig. 83[2]. Also shown is its Walsh-Fourier transform $a(k, m)$ and the inverse transform $F_0(h, i)$ after filtering by a low-pass filter defined by $K(k, m) = 1$ for $k + m \leqq 14$, $K(k, m) = 0$ for $k + m > 14$.

Problems

232.1 Write the voltage at point 32 of the card $\text{wal}(2, y)$ in Fig. 78 in the way shown for point 23.

232.2 How many additions and subtractions are required for the Walsh-Fourier transformation of $F(h, i)$ in Fig. 79 if the ordinary and the fast transform are used?

232.3 A good low-pass filter for images of man-made objects is defined by $K(k, m) = 1$ for $k\,m \leqq r$, $K(k, m) = 0$ for $k\,m > r$. Plot this curve in Fig. 82 and explain why it yields better results than the three curves shown.

2.3.3 Practical Implementation of Filters for Signals with Two Space Variables

If the continuous function $F(x, y)$ is replaced by the sampled function $F(h, i)$ the Walsh-Fourier integral,

$$a(k, m) = \int_{-1/2}^{1/2} \int_{-1/2}^{1/2} F(x, y)\, \text{wal}(k, x)\, \text{wal}(m, y)\, dx\, dy \tag{1}$$

is replaced by a double sum:

$$a(k, m) = \sum_{i=0}^{2^n-1} \sum_{h=0}^{2^n-1} F(h, i)\, \text{wal}\left(k, \frac{h + \frac{1}{2}}{2^n} - \frac{1}{2}\right) \text{wal}\left(m, \frac{i + \frac{1}{2}}{2^n} - \frac{1}{2}\right) \tag{2}$$

The relative position of the scales for x, y and h, i is shown in Fig. 84. The summation in Eq. (2) can be performed in two steps:

$$\begin{aligned} a(k \mid i) &= \sum_{h=0}^{2^n-1} F(h, i)\, \text{wal}\left(k, \frac{h + \frac{1}{2}}{2^n} - \frac{1}{2}\right) \\ a(k, m) &= \sum_{i=0}^{2^n-1} a(k \mid i)\, \text{wal}\left(m, \frac{i + \frac{1}{2}}{2^n} - \frac{1}{2}\right) \end{aligned} \tag{3}$$

---

[1] This simplification results in a somewhat reduced performance.
[2] *See* footnote 1, p. 122.

Consider the case $2^n = 16$ and let the voltages $F(h, i)$ for a fixed value of i and for $h = 0, \ldots, 15$ be fed to the input terminals $0, i, \ldots, 15, i$ of the circuit in Fig. 85. The voltages $-a(0 \mid i), \ldots, -a(15 \mid i)$ are obtained at the output terminals $i, 0, \ldots, i, 15$. For instance, the voltage

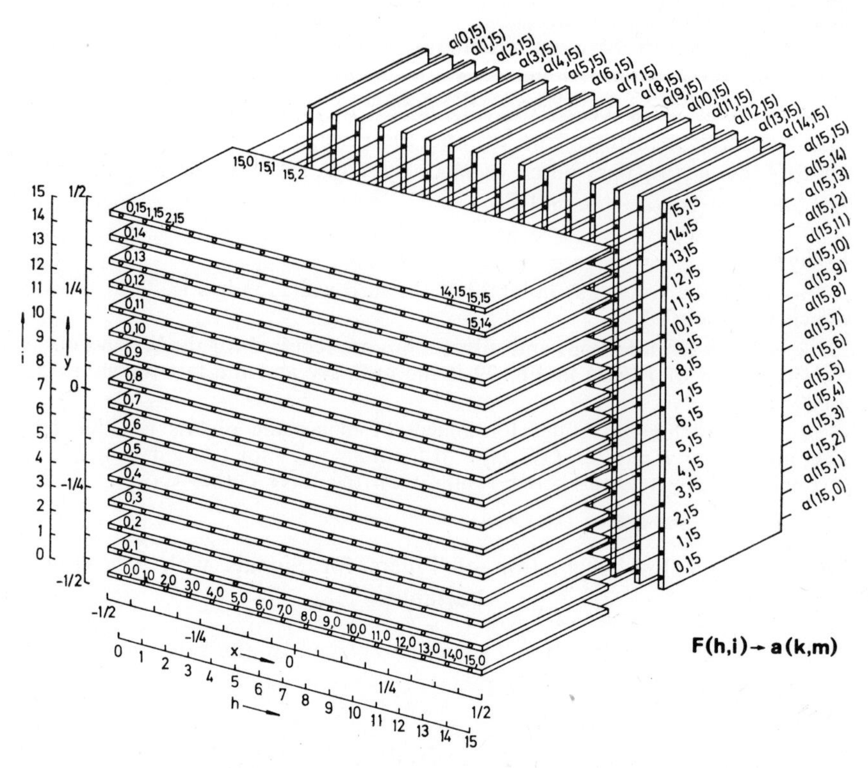

Fig. 84. Practical version of a sequency half filter for functions of two space variables x and y or h and i.

$a(0 \mid i)$ at the output terminal $i, 0$ is the negative sum of all input voltages:

$$a(0 \mid i) = -\sum_{h=0}^{15} F(h, i) = -\sum_{h=0}^{15} F(h, i) \operatorname{wal}\left(0, \frac{h}{16} - \frac{15}{32}\right)$$

The voltage at output terminal $i, 1$ equals:

$$a(1 \mid i) = -\sum_{h=0}^{7} F(h, i) + \sum_{h=8}^{15} F(h, i) = -\sum_{h=0}^{15} F(h, i) \operatorname{wal}\left(1, \frac{h}{16} - \frac{15}{32}\right)$$

The 16 horizontal printed circuit cards in Fig. 84 contain each the circuit of Fig. 85. If the function $F(h, i)$ is fed to the 16^2 input terminals denoted $0, 0$ to $15, 15$ one obtains the voltages $a(0 \mid 15), a(1 \mid 15), \ldots,$

$a(15 \mid 15)$ at the output terminals of the topmost card and generally the voltages $a(0 \mid i), \ldots, a(15 \mid i)$, $i = 0, \ldots, 15$, at any one of the 16 cards.

The second summation transforms $a(k \mid i)$ into $a(k, m)$. This process is the same as before except that the summation is now for $i = 0, \ldots, 15$ rather than for $h = 0, \ldots, 15$. Hence, cards with the circuit of Fig. 85

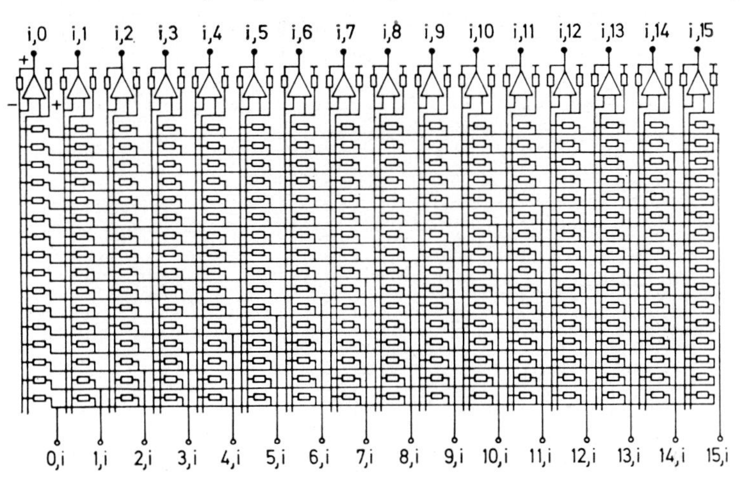

Fig. 85. Circuit diagram for the cards of Fig. 84 using the Walsh-Fourier transform. All resistors have the same value.

can again be used, but they must be positioned vertically as shown in Fig. 84.

The circuit of Fig. 84 is much simpler than the one of Fig. 78 since all cards are equal and the connection between the horizontal and vertical cards can be made by means of a socket rather than by wires that have to be fiddled through the cards. Note that the operation of this filter does not depend on the use of Walsh functions: The circuit of Fig. 85 can be implemented, e.g., for the use of Haar functions; it can also be implemented according to the fast Walsh-Fourier transform rather than the ordinary Walsh-Fourier transform used in Fig. 85; etc. The arrangement of the cards in Fig. 84 is not affected by such a change.

A circuit diagram for the fast Walsh-Fourier transform is shown in Fig. 86. Each block produces the sum and the difference of the two input voltages either directly or with reversed polarity. The signs in the first three columns of blocks are the same as those of the coefficients $s_{k,m}^{j,p}$ in columns 1, 2 and 3 in Table 2. The signs for the fourth column of blocks follow from Eq. (1.2.5-1). The circuit diagrams of the four types of blocks are shown on the right side of Fig. 86.

The fast Walsh-Fourier transform requires 64 operational amplifiers compared with the 16 of the ordinary Walsh-Fourier transform of Fig. 85.

However, only two voltages are summed by any amplifier, which reduces the leakage problem posed by the large resistor array of Fig. 85.

Figure 87 shows a circuit for the functions $\mathrm{ter}(k, x)$ according to Fig. 21 with j and θ replaced by k and x. The fast transform of the ter functions

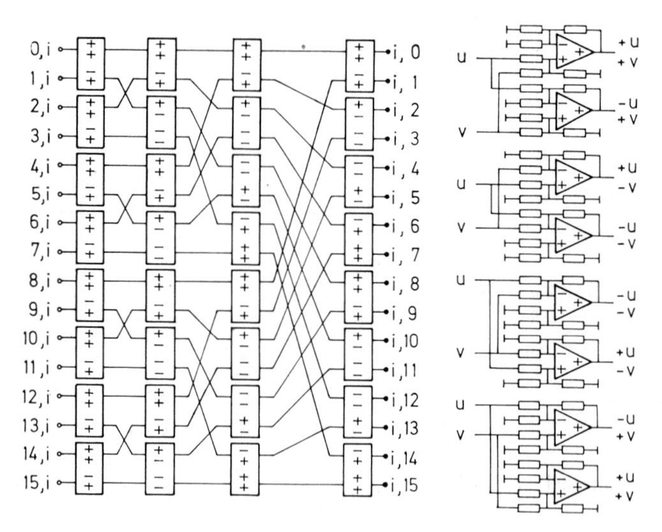

Fig. 86. Circuit diagram for the cards of Fig. 84 using the fast Walsh-Fourier transform. Detailed circuit diagrams of the various blocks are shown on the right.

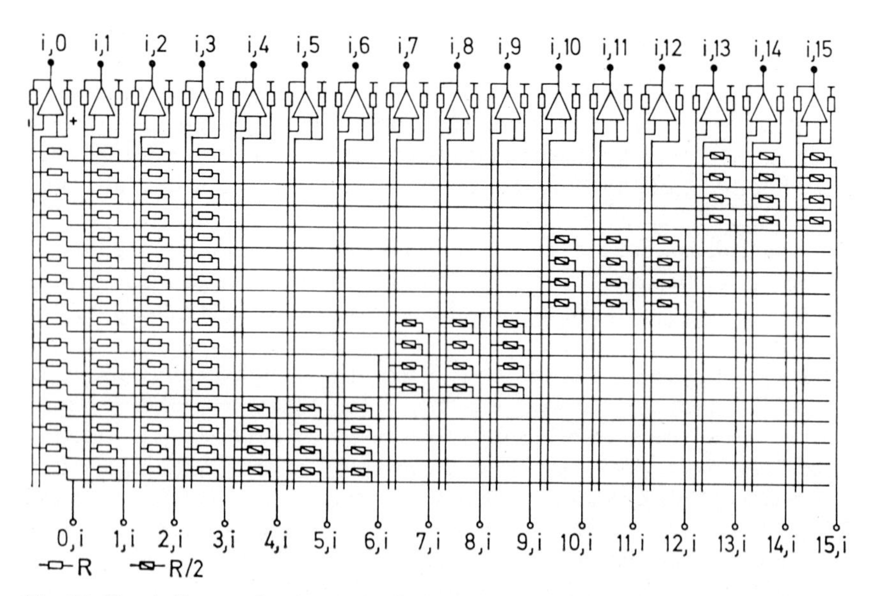

Fig. 87. Circuit diagram for the cards of Fig. 84 using the Ter-Fourier transform according to the ter functions of Fig. 21.

requires fewer additions and subtractions than that of the Walsh functions (Problem 126.2). This shows up in the fewer resistors in Fig. 87 compared with Fig. 85.

Figure 88 shows an example of the state of the art of resolution filters used for television signals. Eight successive signal samples are transformed to yield eight coefficients in the one-dimensional sequency domain. These coefficients are encoded into PCM signals having on the average 2.5 bits per coefficient. The PCM signals are transmitted in a closed loop, reconverted into analog signals and retransformed from the sequency to the

Fig. 88. Original analog TV signal (right) and PCM signal reduced to 2.5 bits/sample by a one-dimensional resolution filter with eight input terminals (left). (Courtesy K. Shibata of Kokusai Denshin Denwa Co., Research Laboratory.)

space domain. The 2.5 bits per coefficient or per signal sample are about one half of what is needed for a comparable picture if the transformation into the sequency domain is omitted. Color television signals require about 3.75 bits per sample if eight successive signal samples are transformed. For details, the reader is referred to the numerous papers by Shibata and co-workers, which will be enjoyed more by those who can read Japanese [2–6]. Similar equipment has been developed by Hatori and Taki of Tokyo University [7].

Problems

233.1 Draw the equivalent of Fig. 85 for Haar functions.

233.2 Draw the equivalent of Fig. 86 for Haar functions.

233.3 Discuss the limitations of the circuit of Fig. 85 due to the finite input impedance of the amplifiers and the non-zero ouput impedance of the voltage sources feeding into the terminals $0, i, \ldots, 15, i$.

233.4 The circuit of Fig. 85 sums 16 voltages at a time and requires 16 operational amplifiers. The circuit of Fig. 86 sums 2 voltages at a time and requires 64 operational amplifiers. Find a circuit that sums 4 voltages at a time and requires 32 operational amplifiers.

2.3.4 Filters for Signals with Three Space Variables

The mathematical basis for filtering signals $F(x, y, z)$ is the three-dimensional Walsh-Fourier transform:

$$a(k, m, n)$$

$$= \int_{-1/2}^{1/2} \int_{-1/2}^{1/2} \int_{-1/2}^{1/2} F(x, y, z) \operatorname{wal}(k, x) \operatorname{wal}(m, y) \operatorname{wal}(n, z) \, dx \, dy \, dz \qquad (1)$$

$$F(x, y, z) = \sum_{k=0}^{\infty} \sum_{m=0}^{\infty} \sum_{n=0}^{\infty} a(k, m, n) \operatorname{wal}(k, x) \operatorname{wal}(m, y) \operatorname{wal}(n, z) \qquad (2)$$

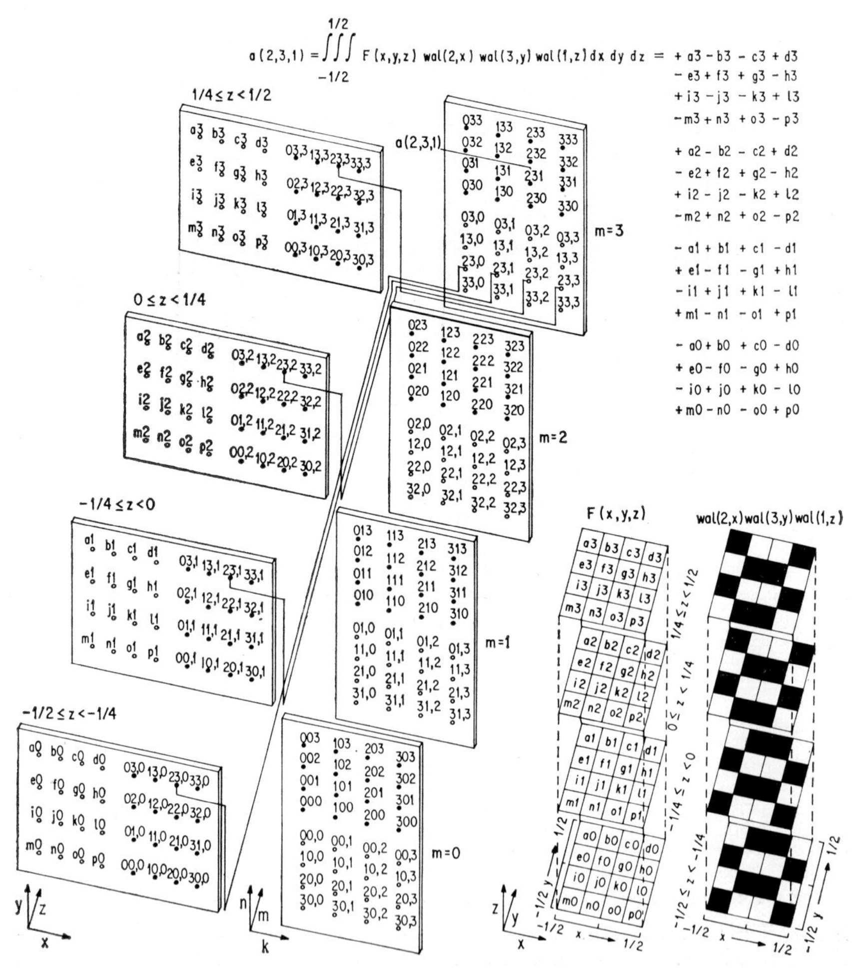

Fig. 89. Principle of a sequency filter for signals with three space variables or two space variables and the time variable.

A filtering effect is obtained by multiplying the transform $a(k, m, n)$ with attenuation coefficients $K(k, m, n)$ or using a variable resolution $R(k, m, n)$ before retransformation.

Figure 89 shows the principle of a circuit that produces $4^3 = 64$ coefficients $a(k, m, n)$ from 4^3 input voltages denoted $a\,r, \ldots, p\,r$, with $r = 0, 1, 2, 3$. These voltages are fed to the input terminals $a\,0, \ldots, p\,3$ of the printed circuit cards on the left side, denoted $-\frac{1}{2} \leqq z < -\frac{1}{4}$ to $\frac{1}{4} \leqq z < \frac{1}{2}$. The transform $a(k, m, n)$ is obtained at the output terminals k, m, n of the cards on the right side, denoted $m = 0, 1, 2, 3$. The circuit diagram of the cards $-\frac{1}{2} \leqq z < -\frac{1}{4}$ to $\frac{1}{4} \leqq z < \frac{1}{2}$ is shown by Fig. 90, while Fig. 91 shows the diagram of the cards $m = 0, 1, 2, 3$.

The terminals $00, 0$ to $33, 3$ of the cards on the left are connected to the terminals of the cards on the right showing the same number. Only the four wires connecting the terminals $23, 0$; $23, 1$; $23, 2$ and $23, 3$ are shown rather than all 64 wires. The need for such a cable connection is inherent for filters for three-dimensional signals. The filters for signals with one variable θ or x in Figs. 62, 65 and 77 require a two-dimensional circuit. Signals with two variables x and y lead to the three-dimensional circuits of Figs. 78 and 84. Signals with three variables require four-dimensional circuits, which can be implemented by means of a cable.

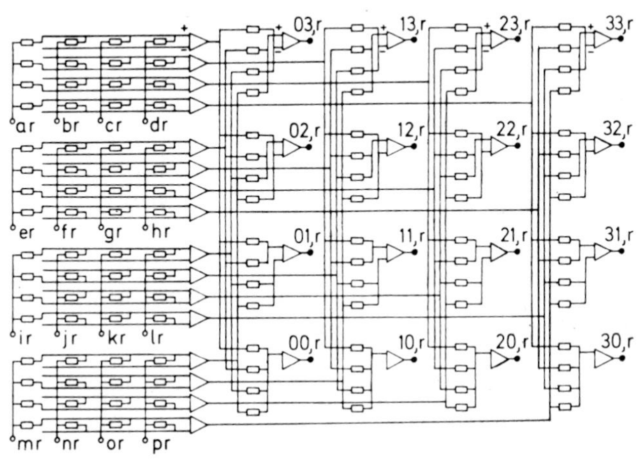

Fig. 90. Circuit diagram for the cards in Fig. 89 denoted by $-1/2 + r/4 \leqq z < -1/4 + r/4$; $r = 0, 1, 2, 3$.

The function $F(x, y, z)$ is shown in the lower right corner of Fig. 89 by four squares in the (x, y)-plane stacked in the direction z. The three-dimensional Walsh function $\mathrm{wal}(2, x)\,\mathrm{wal}(3, y)\,\mathrm{wal}(1, z)$ is shown in the same way; black areas indicate the value $+1$, white areas the value -1. This is the same function as shown in Fig. 29 by black and white cubes.

The transform $a(2, 3, 1)$ is also shown. Note that the signs of $a\,3$ to $p\,0$ correspond to the black (+) and white (−) areas of the function $\mathrm{wal}(2, x)\,\mathrm{wal}(3, y)\,\mathrm{wal}(1, z)$.

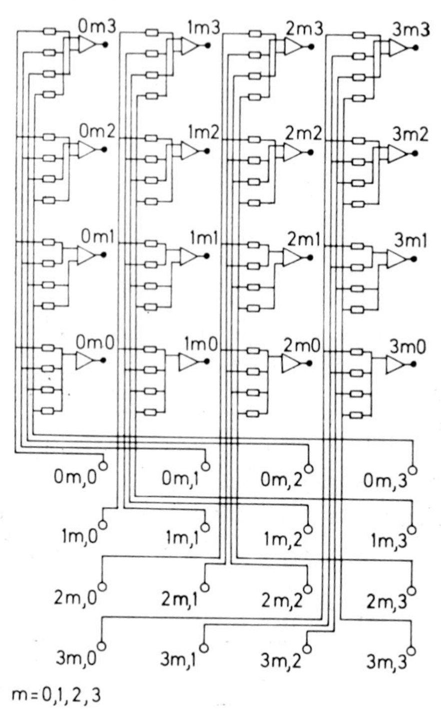

Fig. 91. Circuit diagram for the cards in Fig. 89 denoted by $m = 0, 1, 2, 3$.

A filter for a space-time function $F(x, y, t)$ is obtained from the circuit of Fig. 89 by inserting sample-and-hold circuits between the terminals $a\,0$, $a\,1$, $a\,2$ and $a\,3$; etc. The input signal is then fed to the 4^2 terminals $a\,0, \ldots, p\,0$.

There is no theoretical difficulty in extending the one-, two- and three-dimensional filters of Figs. 77, 78 and 89 to four-dimensional filters for space-time signals $F(x, y, z, t)$. However, the actual representation of a circuit with $4^4 = 256$ input and output terminals is a major drafting project.

Television signals are typical three-dimensional functions $F(x, y, t)$. Sequency filters are at the present the only known means for the implementation of attenuation or resolution filters for them, since the optical methods of Fourier transformation can be applied to two-dimensional spatial signals $F(x, y)$ only. The present state of technology makes two-dimensional sequency filters practical, but three-dimensional filters are very expensive. To make them useful will probably require that the present two-dimensional large scale integrated circuit techniques be advanced to three-dimensional circuits. Furthermore, the only foreseeable economic application of three-dimensional filters for TV signals would be live TV

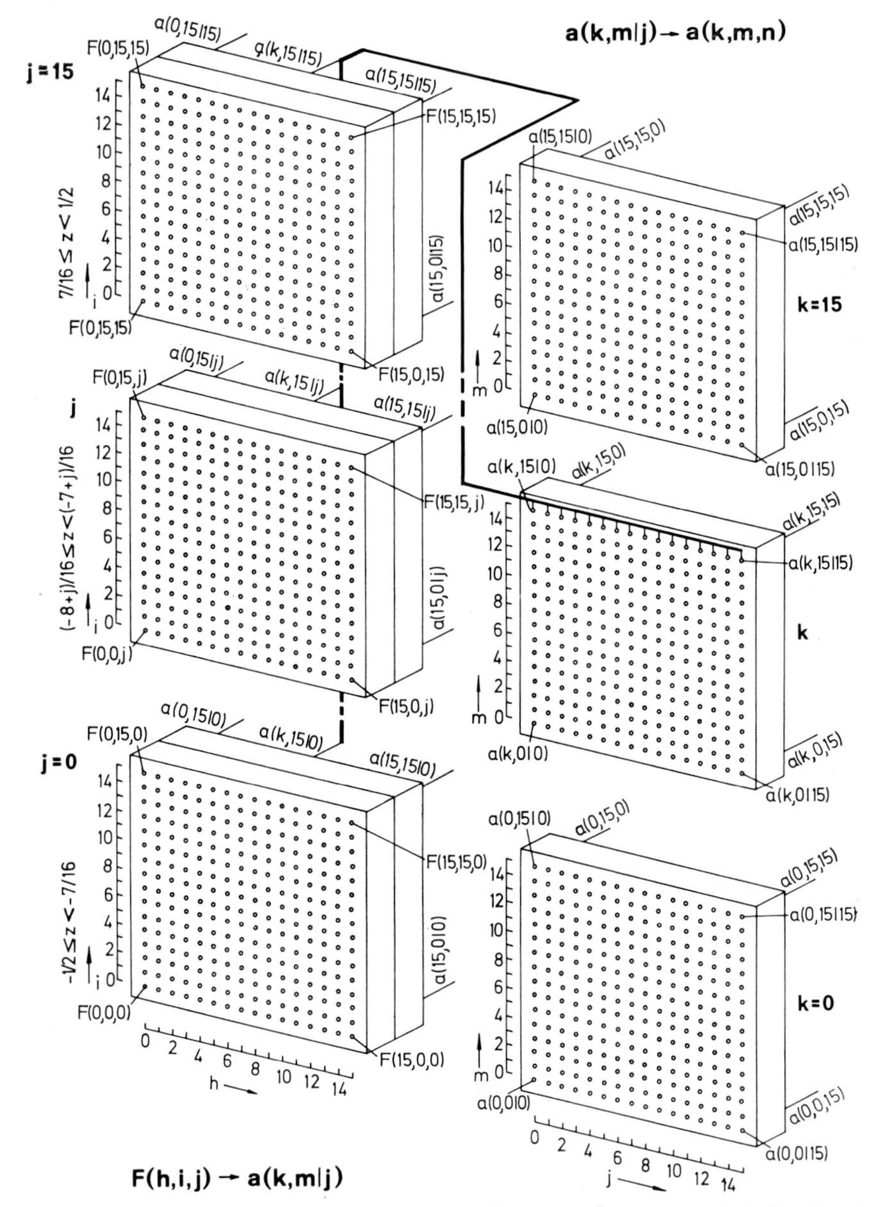

Fig. 92. Practical version of a sequency half filter with 16^3 input terminals for signals with three space variables. The blocks on the left are shown in detail in Fig. 84, the blocks on the right in Fig. 93.

broadcasting from Mars or a similar futuristic project. The reduction of digits in television PCM signals by one or two orders of magnitude by means of a resolution filter would increase the signal-to-noise ratio by 10 to 20 dB.

After these words of caution let us consider methods for simplifying three-dimensional filters. The decisive simplification is brought by sampling filters discussed later on, since the instantaneous filters work 100 to 1000 times faster than required for TV signal filtering and are thus poorly utilized without some time-sharing method. Let us here consider the instantaneous half filter with 16^3 inputs shown in Fig. 92. The space function $F(x, y, z)$ is replaced by the sample function $F(h, i, j)$. The blocks on the left denoted by **j** perform the following transformation ($2^s = 16$ in Fig. 92):

$$a(k, m \mid j) = \sum_{h=0}^{2^s-1} \sum_{i=0}^{2^s-1} F(h, i, j) \operatorname{wal}\left(k, \frac{h+\frac{1}{2}}{2^s} - \frac{1}{2}\right) \operatorname{wal}\left(m, \frac{i+\frac{1}{2}}{2^s} - \frac{1}{2}\right) \tag{3}$$

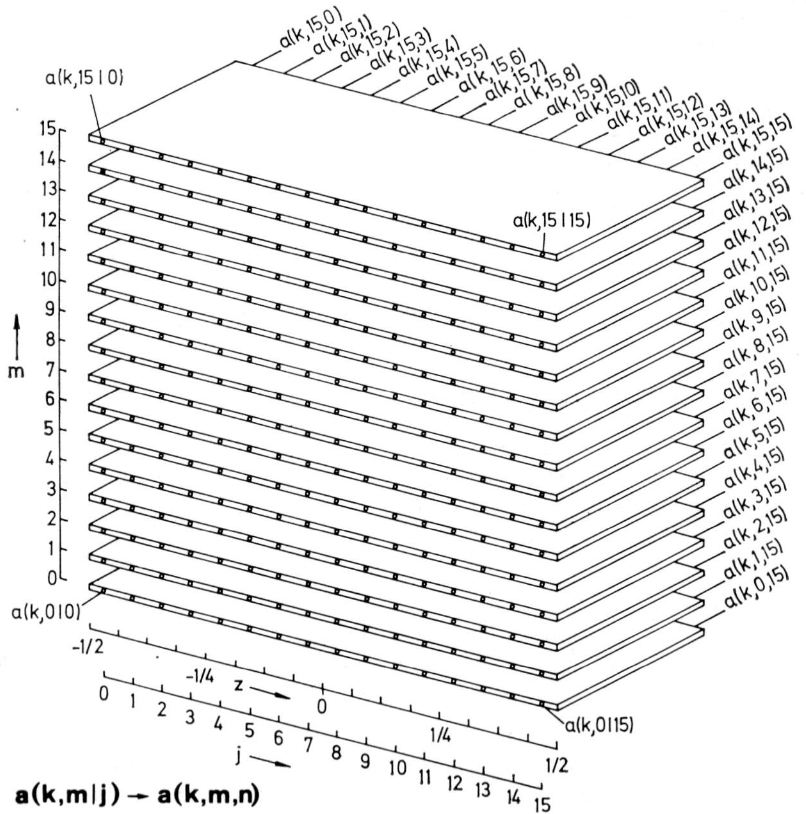

Fig. 93. Half filter for the transformation of the function $a(k, m \mid j)$ into $a(k, m, n)$. Circuit diagrams of the 16 cards are shown in Figs. 85 to 87.

The blocks on the right denoted by **k** complete the three-dimensional transformation:

$$a(k, m, n) = \sum_{j=0}^{2^s-1} a(k, m \mid j) \operatorname{wal}\left(n, \frac{j + \frac{1}{2}}{2^s} - \frac{1}{2}\right) \tag{4}$$

The transformation in Eq. (3) is performed by the circuit of Fig. 84, of which 2^s are required, while Eq. (4) represents 2^{2s} one-dimensional transformations. Figure 93 shows a suitable circuit; each of the 16 cards has the same circuit as the cards in Fig. 84. A total of 16^3 wires is required to connect the output terminals of the blocks on the left side of Fig. 92 with the input terminals of the blocks on the right side. The cable with 16 wires shown should clarify the pattern of the connections.

The number of printed circuit cards according to Figs. 85, 86 or 87 required for Fig. 92 is $2 \times 16^2 + 16^2 = 768$. This number is frightening in terms of individual components technology, since it translates into 221 184 resistors required for the ordinary Walsh transform. The situation is more hopeful in terms of thick or thin film technology since all cards are equal and all resistors have the same value.

Problems

234.1 Draw the circuits of Figs. 90 and 91 for Haar functions.

234.2 Write the output voltage at the point 132 of card $m = 3$ in Fig. 89 as shown for the point 231.

234.3 Draw the connections of the wires leading to the input terminals $a(k,7,0)$, ..., $a(k, 7, 15)$ of the block **k** in Fig. 92.

2.4 Sampling Filters for Space Signals

2.4.1 Generators for Space Variable Walsh Functions

Generators for space variable Walsh functions that automatically produce a set of such functions are required for the implementation of sampling filters. It is worthwhile to contemplate the difficulty of designing a tunable sinusoidal generator, which produces space functions with 20 to 20000 oscillations per meter, with the difficulty of designing one that produces time functions with 20 to 20000 oscillations per second. The differing degree of difficulty carries over to other types of sinusoidal equipment and goes a long way to explain why time signals are so much more important in communications than space signals.

Several generators for space variable Walsh functions will be considered here to show how concepts familiar in the time domain carry over to the space domain. The circuit of Fig. 94 and its pulse diagram in Fig. 95 are the space equivalent of the time function generator of Figs. 53 and 54. There is no problem in interpreting the Walsh functions wal$(1111, x)$ to wal(0001, x) shown on top of Fig. 95, their derivatives wal′(1111, x) to wal′(0001, x), the trigger functions tri(1000, x) to tri(0001, x) obtained from the derivatives, the sum function sum(1101, x) of the triggers tri(1000, x), tri(0100, x) and tri(0001, x), and the resulting Walsh function wal(1101, x). There is also no problem converting such a pulse diagram into a circuit that produces time variable Walsh functions wal(k, θ).

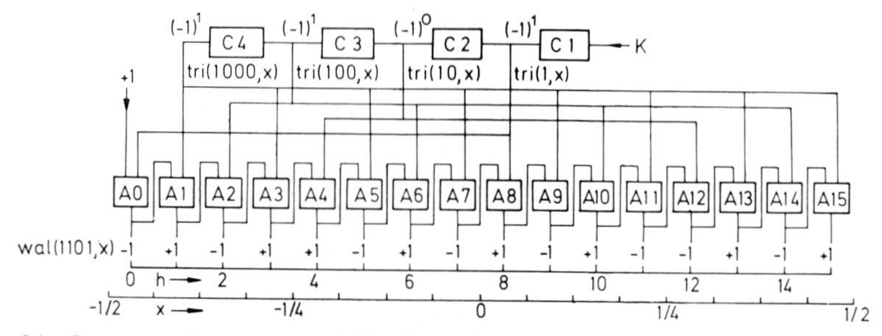

Fig. 94. Generator for space variable Walsh functions wal(k, x) using the derivatives of Rademacher functions. A, half adder; C, stage of a shift register or counter.

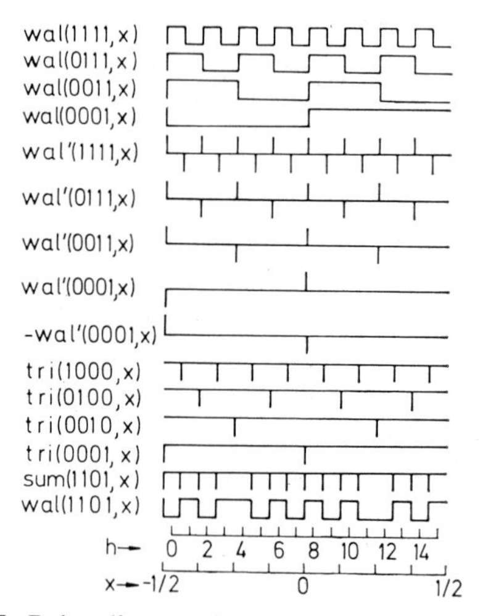

Fig. 95. Pulse diagram for the generator of Fig. 94.

The meaning of trigger pulse in time for a binary circuit is that the polarity of a voltage with value ± 1 is reversed at this point in time and remains unchanged until the next trigger pulse. Translating this into space means that the polarity of a voltage of value ± 1 is reversed at the point in space where the trigger pulse is applied and remains unchanged until the next trigger pulse. For instance, the trigger pulse tri(0001, x) in Fig. 95 indicates that the output voltages in the interval $0 \leqq x < \frac{1}{2}$ should have the opposite polarity from that in the interval $-\frac{1}{2} \leqq x < 0$. A closer study of Fig. 94

shows that the trigger voltage $\mathrm{tri}(0001, x)$ is applied at the left end and in the middle of the interval $-\frac{1}{2} \leqq x < \frac{1}{2}$. One may readily verify that the other trigger voltages in Fig. 94 are also applied at the space points[1] required by Fig. 95. Let the number 1101 be stored in the stages C 4 to C 1 of Fig. 94. The output voltages are then $\mathrm{tri}(1000, x) = -1$, $\mathrm{tri}(100, x) = -1$, $\mathrm{tri}(10, x) = +1$ and $\mathrm{tri}(1, x) = -1$. These voltages are applied to 16 half adders A. Let the voltage $+1$ be applied to the second input terminal of A 0. The output voltages of the 16 half adders represent then the Walsh function $\mathrm{wal}(1101, x)$ by samples taken in the middle of the pulses as shown by the scale (h) in Figs. 94 and 95.

Another Walsh function generator is shown by Fig. 96. It contains four rows of eight half adders that are arranged like the functions $\mathrm{wal}(1111, x)$ to $\mathrm{wal}(0001, x)$ in Fig. 95. Some of the half-adders are numbered, others not, but this may be ignored at the moment. Again, let the

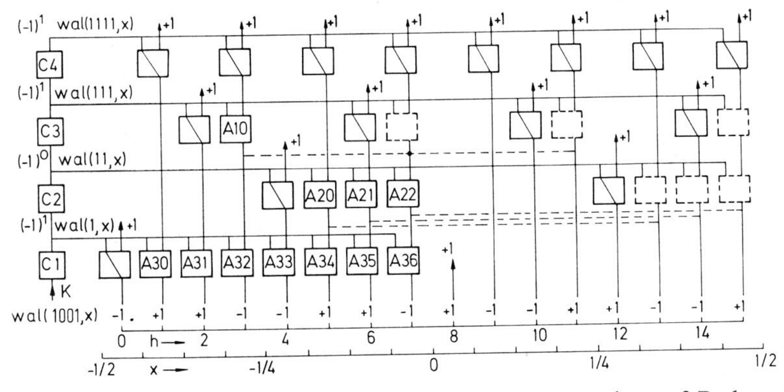

Fig. 96. Generator for space variable Walsh functions using products of Rademacher functions $\mathrm{wal}(2^s - 1, x)$. C1 to C4 are stages of a shift register or counter. All other blocks are half adders. $\mathrm{wal}(1001, x) = \mathrm{wal}(1111, x)\,\mathrm{wal}(111, x)\,\mathrm{wal}(1, x)$.

number 1101 be stored in the stages C 1 to C 4. The following voltages are applied to the four rows of half adders: $\mathrm{wal}(1111, x) = -1$, $\mathrm{wal}(111, x) = -1$, $\mathrm{wal}(11, x) = +1$ and $\mathrm{wal}(1, x) = -1$. One may readily verify that the output voltages represent the function $\mathrm{wal}(1111, x)\,\mathrm{wal}(111, x)\,\mathrm{wal}(1, x) = \mathrm{wal}(1001, x)$. The principle used by this generator is the multiplication of Rademacher functions $\mathrm{wal}(2^s - 1, x)$ in analogy to the time function generator of Fig. 52. The half-adders shown by solid lines but without numbers in Fig. 96 may be omitted, since one input voltage is always $+1$ and the output voltage is thus equal to the second input voltage. Furthermore, the half-adder shown by dashed lines without numbers may be omitted if the four connections shown by dashed lines are made. The resulting circuit is essentially equal to the one of Fig. 52.

[1] There should be no difficulty explaining the space shift of $-1/32$.

Another generator based on the functions wal$(2^s, x)$ is shown in Fig. 97. The half adders A are arranged according to the functions wal$(1, x)$ to wal$(1000, x)$ in Fig. 98. The output voltages represent the function wal(k, x) if the number k is stored in binary form in the shift register or counter stages C 1 to C 4, which is more convenient than in Fig. 96. If the time

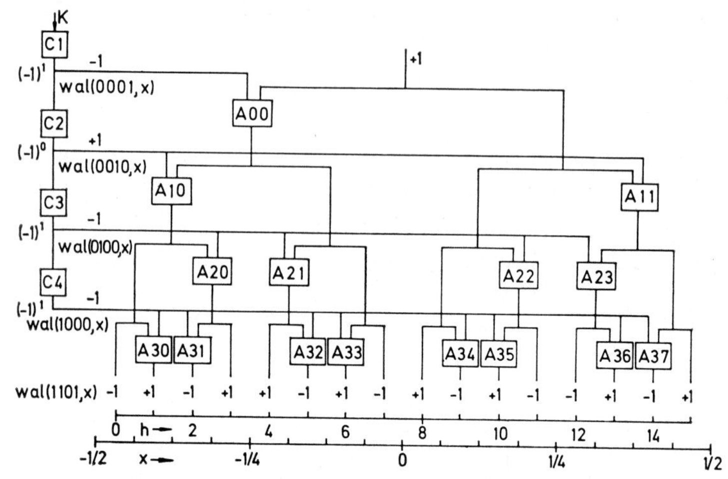

Fig. 97. Generator for space variable Walsh functions using products of the shifted Rademacher functions wal$(2^s, x)$. A, half adder; C, stage of a shift register or counter. wal$(1101, x) = $ wal$(1000, x)$ wal$(100, x)$ wal$(1, x)$.

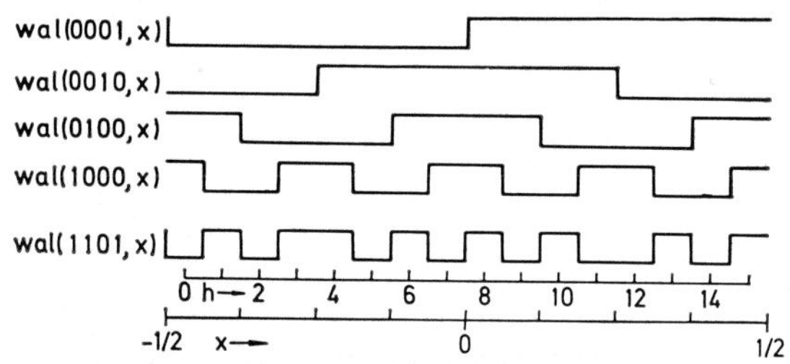

Fig. 98. Pulse diagram for the generator of Fig. 97.

variable function wal(j, θ) instead of the voltage $+1$ is fed to the terminal shown on top of Fig. 96 one obtains the space-time function wal(k, x) wal(j, θ) at the output terminals. Note that the circuit of Fig. 94 can also be used in this fashion, but switching delays accumulate more rapidly.

Problems

241.1 Explain the relationship of the number 1101 stored in the stages C 1 to C 4 of Fig. 96 and the function wal$(1001, x)$ generated, using the concept of Gray code [1].

241.2 If the voltage $+1$ rather than wal(j, θ) is fed to the terminal on top of Fig. 97 one may eliminate three half-adders. Which ones?

241.3 Redesign the circuit of Fig. 96 so that the functions $\text{wal}(j, \theta)\,\text{wal}(k, x)$ can be generated.

241.4 Design a generator for time functions $\text{wal}(j, \theta)$ according to the principle of Fig. 97. Which difficulty occurs that is absent in the space function generator?

2.4.2 Sampling in Two or Three Space Dimensions by Block Pulses and Walsh Functions

Sampling in two space dimensions is usually based on the system of block pulses $\text{blo}(k, x)\,\text{blo}(m, y)$ shown in Fig. 23. There it is pointed out that the black square representing the value $+1$ moves like the illuminated spot on a TV tube that is scanned from left to right and from bottom to top. This type of sampling is a decomposition of a space function $F(x, y)$ into the system of block pulses.

A possible implementation of a Walsh function sampling device will be explained with reference to Fig. 99. The left part of the illustration

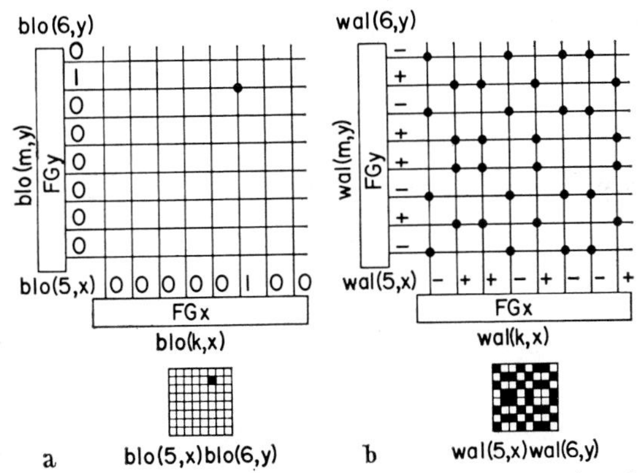

Fig. 99. Sampling by block pulses (a) and Walsh functions (b) using the crossbar principle in two dimensions.

shows a crossbar sampler using block pulses. The function generator FGx produces the function $\text{blo}(5, x)$ which is represented by a voltage 0 at all vertical bars except at the bar 6 to which the voltage 1 is applied. Similarly, the generator FGy produces the function $\text{blo}(6, y)$ which is represented by a voltage 0 at all horizontal bars except et the bar 7, to which the voltage 1 is applied. The crossing of the bars with voltage 1 applied is indicated by a black dot. This dot represents the function $\text{blo}(5, x)\,\text{blo}(6, y)$.

Sampling by the crossbar principle using Walsh functions is shown in Fig. 99b. The function $\text{wal}(5, x)$ represented by positive and negative voltages applied to the vertical bars is supplied by the function generator FGx, while the generator FGy supplies the function $\text{wal}(6, y)$ to the horizontal bars. The crossings of bars with equal applied voltages are indicated by the black dots. The dot pattern corresponds to the black areas of the function wal$(5, x)$ wal$(6, y)$. The white areas correspond to the crossings where a voltage difference exists.

The crossbar principle is presently the most likely one to be used if the flat TV screen ever becomes practical. However, Fig. 99 may also represent information storage in a magnetic core memory. The usual way is to store one bit in one core or one word in one storage location. This corresponds to storage according to block pulses. Storage according to Walsh functions would distribute each bit over all cores or each word over all storage locations. It is interesting to note that the destruction of one storage location destroys the information stored in this location completely and leaves all other information unchanged in the case of block pulse storage. In the case of Walsh function storage, all information would be degraded – resulting, e.g., in smaller output pulses – but no information would be completely destroyed. In theory, such a storage would remain operable as long as more than half the storage locations were operating.

Let us extend the crossbar principle to three space dimensions. The bars have to be replaced by planes and the bar crossings by the intersection points of three perpendicular planes. These planes are represented in Fig. 100 by four parallel rods. Each one of the four inputs in the x, y and z

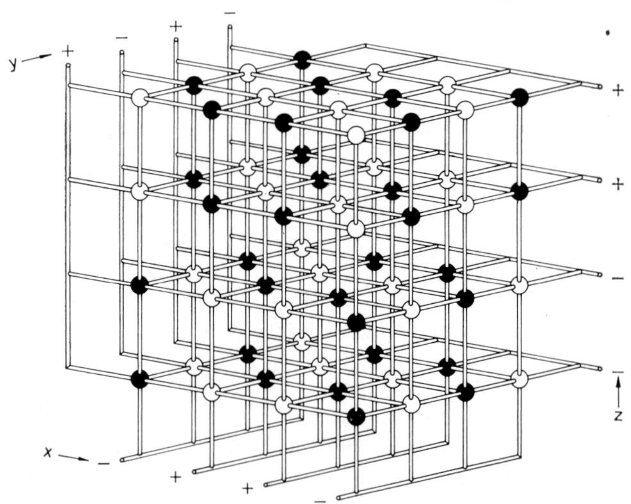

Fig. 100. Principle of the crossbar sampler in three dimensions using Walsh functions. The function shown is wal$(2, x)$ wal$(3, y)$ wal$(1, z)$ with wal$(2, x) = -1+1+1-1$, wal$(3, y) = +1-1+1-1$ and wal$(1, z) = -1-1+1+1$.

direction denoted by + or − is connected to four rods located in a plane. The inputs in the x direction feed to rods in a plane perpendicular to the x axis, the inputs in the y direction to rods in a plane perpendicular to the y axis, and the inputs in the z direction to rods in a plane perpendicular to the z axis. The black spheres in Fig. 100 are located like the black cubes in Fig. 29. They represent the value $+1$ of the function $\mathrm{wal}(k, x)\,\mathrm{wal}(m, y)\,\mathrm{wal}(n, z)$. The white spheres represent the value -1.

A small change is required to turn the principle shown by Fig. 100 into a practical circuit. According to Fig. 101 the inputs in the x and y direction are fed to exclusive OR-gates XOR located at each crossing of the input rods. The outputs of the gates feed into vertical rods. Let us

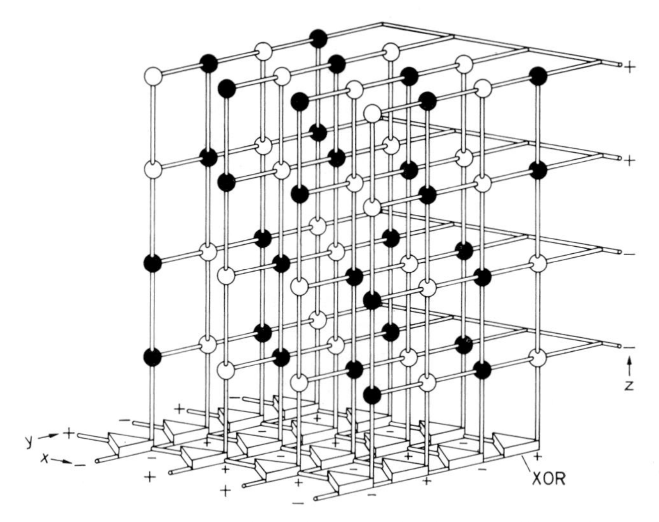

Fig. 101. Principle of an improved crossbar sampler in three dimensions using Walsh functions. XOR, exclusive OR-gate.

assume that the voltage $+10$ V is fed to each x and y input denoted by $+$, and the voltage -10 V to each x and y input denoted by $-$. The voltages $+10$ V and -10 V will then be applied to the vertical rods as shown. Let the voltages $+10$ V and -10 V also be applied to the inputs denoted by $+$ and $-$ in the z direction. The black spheres are then located at the rod crossings with no voltage difference, while the white spheres are located at the crossings with a voltage difference of 20 V.

The spheres in Fig. 101 are arranged like a stack of printed circuit cards that have connectors at the bottom and rear edge. This is easy to implement. The three intersecting sets of rods in Fig. 100 would be much harder to implement.

For the sake of completeness let us briefly consider the extension of the crossbar principle to signals $F(x, y, z, t)$ with three space variables and the time variable. This may be done by time or space sharing. Time sharing means that the three generators producing the functions $\mathrm{wal}(k, x)$, $\mathrm{wal}(m, y)$ and $\mathrm{wal}(n, z)$ are switched so fast that all functions $\mathrm{wal}(k, x)\,\mathrm{wal}(m, y) \times$ $\times\,\mathrm{wal}(n, z)$ are generated before the signal $F(x, y, z, t)$ changes appreciably. In terms of TV type resolution, this requires the generation of 512^3

functions 30 times per second[1]. Space sharing calls for the replacement of the spheres in Fig. 101 by exclusive OR-gates. There would be 4^2 gates in each one of the four x z-planes spanned by the rods in Fig. 101. The outputs of each set of 4^2 gates would feed into the vertical rods of a cubical rod structure exactly like the one in Fig. 101. Four such structures would be required if the time functions $\mathrm{wal}(0, t/T)$ to $\mathrm{wal}(3, t/T)$ are used. The four inputs in the z-direction of Fig. 101 would be connected in parallel and one value + or − of the time function $\mathrm{wal}(j, t/T)$ would be fed to all of them. In this way one could actually construct equipment that represents the intersection of four three-dimensional spaces in one point. The graphic representation of the principle would require four illustrations like Fig. 101 plus a fifth with the spheres replaced by exclusive OR-gates.

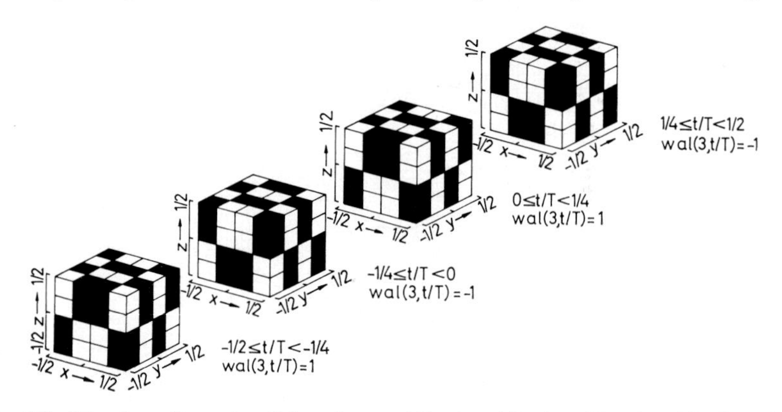

Fig. 102. The four-dimensional function $\mathrm{wal}(2, x)\,\mathrm{wal}(3, y)\,\mathrm{wal}(1, z)\,\mathrm{wal}(3, \theta)$. Black volumes represent +1, white volumes −1.

This is too hard to draw. We show instead the four-dimensional function $\mathrm{wal}(2, x)\,\mathrm{wal}(3, y)\,\mathrm{wal}(1, z)\,\mathrm{wal}(3, t/T)$ in Fig. 102, which may help in understanding this paragraph.

Problems

242.1 Let the 2×8 bars in Fig. 99 be replaced by 2×512 bars. How much time at most is available for the generation of one function $\mathrm{wal}(k, x)\,\mathrm{wal}(m, y)$ if the whole system of functions is to be generated 30 times per second?

242.2 Replace the 3×4 "planes" in Figs. 100 and 101 by 3×512 planes. How many different functions can be generated and how much time at most is available for the generation of one if the whole system of functions is to be generated 30 times per second?

242.3 Draw one of the four cubes in Fig. 102 as a lattice with spheres as shown in Fig. 101 and indicate the voltages applied by the space function $\mathrm{wal}(2, x) \times$ $\times\,\mathrm{wal}(3, y)\,\mathrm{wal}(1, z)$.

[1] In reality the numbers would be much smaller, for a variety of reasons, but still impressive.

242.4 Which four-dimensional functions can be represented in the way shown in Fig. 102: a) two-valued functions as shown in Figs. 17 and 18; b) three-valued functions as shown in Figs. 19 to 21; c) functions consisting of a finite number of samples with a finite number of different possible amplitudes; d) same as *c* but number of possible amplitudes is not finite; e) denumerably many samples with a finite or infinite number of different possible amplitudes; f) continuous functions like $\sin x \cdot \sin y \cdot \sin z \cdot \sin \theta$.

2.4.3 Sampling Filters and Displays for Signals with Two or Three Space Variables

One of several possible ways to implement a two-dimensional sampling filter for Walsh functions is shown in Fig. 103. Two function generators FGx and FGy feed functions $\mathrm{wal}(k, x)$ and $\mathrm{wal}(m, y)$ to the vertical and

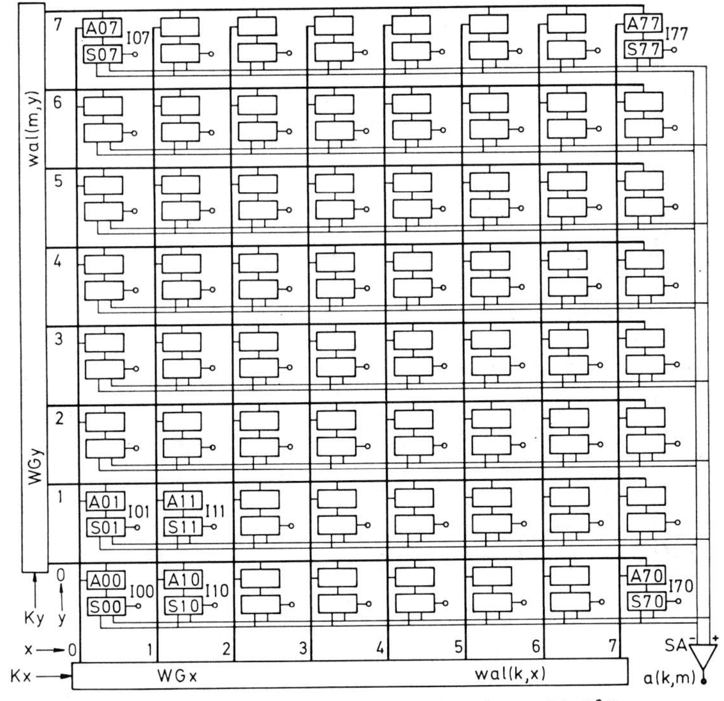

Fig. 103. Crossbar sampler for Walsh functions with 8^2 inputs.

horizontal crossbars. Half adders A 00 to A 77 and single-pole, double-throw switches S 00 to S 77 are located at each bar crossing. If the voltages at two crossing bars are equal, the input terminal of the switch is connected to the adding (+) input of the summing amplifier SA; if they are not equal, the input terminal is connected to the subtracting (−) input of SA. The output voltage of the summing amplifier SA represents the

Walsh-Fourier transform $a(k, m)$ of the signal applied to the input terminals I 00 to I 77 of the sampler.

The filtering process of an attenuation filter requires that each sample $a(k, m)$ is multiplied by an attenuation coefficient $K(k, m)$ and that the inverse transformation is performed. In the simplest case $K(k, m)$ is either 1 or 0. The coefficients $a(k, m)$ for which $K(k, m)$ is zero do not need to be produced at all. Consider as example a low-pass filter according to Fig. 82 that multiplies all coefficients $a(k, m)$ by 1 for $k \leqq 6$, $m \leqq 6$ and by 0 for $k > 6$, $m > 6$. The Walsh function generator WGy would then produce the function $\mathrm{wal}(0, y)$, while the generator WGx would produce successively the functions $\mathrm{wal}(0, x), \mathrm{wal}(1, x), \ldots, \mathrm{wal}(6, x)$. The coefficients $a(0, 0), a(0, 1), \ldots, a(6, 0)$ are obtained at the output of the summing amplifier SA. The generator WGy then produces the function $\mathrm{wal}(1, y)$ and the generator WGx produces again $\mathrm{wal}(0, x), \mathrm{wal}(1, x), \ldots, \mathrm{wal}(6, x)$. The coefficients $a(1, 0), a(1, 1), \ldots, a(6, 1)$ are obtained. This process continues until the function $\mathrm{wal}(6, y)$ is produced by generator WGy and the functions $\mathrm{wal}(0, x), \mathrm{wal}(1, x), \ldots, \mathrm{wal}(6, x)$ by generator WGx. If the Walsh function generators are advanced from one function to the next by trigger pulses via binary counters – as, e.g., in Fig. 94, 96 and 97 – the filtering process is performed by resetting the counters whenever they reach a predetermined value. This may be a fixed value as just discussed, but there is no great problem in resetting according to the conditions $k + m \leqq r$ or $k^2 + m^2 \leqq r^2$ which yields more complicated low-pass filters according to Fig. 82.

Let us consider resolution filters. The voltage $a(k, m)$ in Fig. 103 would be fed to an analog/digital converter. The generators WGx and WGy would produce the functions $\mathrm{wal}(k, x)$ and $\mathrm{wal}(m, y)$ as above. The number of bits into which $a(k, m)$ would be encoded would be made a function of k and m.

The input signal in Fig. 103 could be provided by an array of microphones, e.g., for the production of images by sound waves in water, by photocells in a pattern reader, or by any other transducer that converts an observable quantity into a voltage. Progress in large-scale integration may eventually make it possible to produce such a sampler with 512×512 input terminals instead of the 8×8 shown[1]. By using phototransistors as input devices one would then have an image scanner with a filter built in. Only three external connections beyond the power supply leads would be required: The terminals Kx and Ky for trigger pulses and the output terminal of the summing amplifier SA. The Walsh function generators WGx and WGy for 512 functions would not be much more complex than the ones for 16 functions discussed in section 2.4.1. A very large number

[1] The development of silicon diode arrays with 500000 individual diodes on a single slide of silicon for camera tubes was recently reported [2].

of half-adders A and switches S would be required ($512^2 = 262144$), but they would all be equal. Not all of them would have to work properly. It is known from computer simulation that one may expect a reduction of the number of digits by a resolution filter to about $\frac{1}{8}$ compared with sampling by block pulses. This not only decreases the bandwidth of the transmitted signal but also the necessary switching speed of the sampling device.

A sampling filter for a function with three space variables is obtained by replacing the spheres in Fig. 101 by half adders Ahij and single-pole, double-throw switches Shij according to Fig. 104. The two output terminals of Shij are connected to the adding and subtracting input terminals of a summing amplifier as in Fig. 103. The input terminal Ihij of the switch may be connected to any transducer that provides a voltage as output. Three Walsh function generators are required to feed the x, y and z rods according to Fig. 101. Otherwise, the three-dimensional filter hardly differs from the two-dimensional one. There is no difficulty continuing on to four-dimensional filters for signals $F(x, y, z, t)$, but there is little incentive to do so at present.

The sampling devices discussed so far correspond to the decomposition of images by two-dimensional block pulses using a Nipkow disc[1] or a more modern TV pick-up tube. There still remains the problem of reconverting the filtered transform $a(k, m)$ or $a(k, m, n)$ to the space signals. Figure 105 shows the principle of this reconversion. A crossbar device

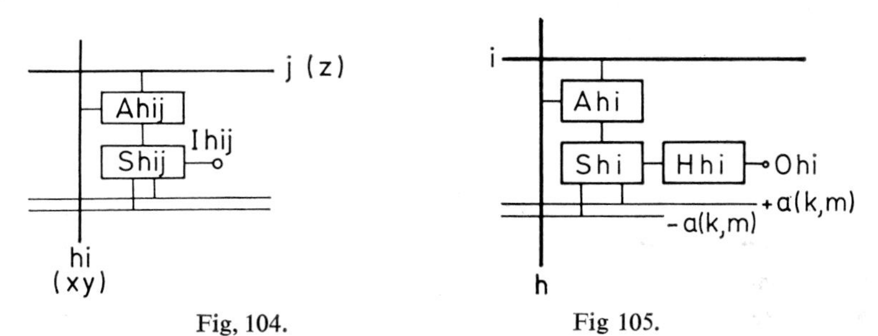

Fig. 104. Circuit to be substituted for the spheres in Fig. 101 if used for sampling. Ahij, exclusive OR-gate; Shij, single-pole, double-throw switch.

Fig. 105. Crossbar display for Walsh functions.

like that of Fig. 103 drives at each crossing a half-adder Ahi which in turn operates a single-pole, double-throw switch Shi. The voltages $+a(k, m)$ and $-a(k, m)$ are fed to all switches. Either $+a(k, m)$ or $-a(k, m)$ passes through the switch to the output terminal Ohi. This completes the recon-

[1] A sampling device analogous to the Nipkow disc could be constructed for Walsh functions as well.

version for those cases where a positive or negative output voltage is wanted.

A problem arises in the reconversion of optical images. A positive voltage is easily transformed into light by means of a glow tube or a similar device. A negative voltage, however, cannot be used to subtract already emitted light. The storage device Hhi is used in Fig. 105 to overcome this problem. A positive or negative charge proportionate to $+a(k,m)$ or $-a(k,m)$ is fed through Shi into Hhi. At the end of a scanning cycle—that is, after the voltages $+a(k,m)$ and $-a(k,m)$ for all values of k and m have been applied to Shi—the charge in Hhi is converted into light. Only charges of one polarity can be in the storage devices Hhi at this moment if all coefficients $a(k,m)$ have been transmitted unchanged. This is not necessarily so if filtering is used, but the problem is overcome by input and output transducers with a logarithmic characteristic [1]. The implementation of the circuit of Fig. 105 in semiconductor technology using, e.g., metastable states for storage and the exponential variation of distribution curves to obtain a logarithmic characteristic, is a challenging task for scientists working in the semiconductor field. The practical incentive for such a development is the expected reduction of the bandwidth of TV signals to about $\frac{1}{8}$, but there is also the prospect of a significant increase of brightness compared with crossbar displays using block pulses.

For a partial explanation of the last statement consider Fig. 106a. The block pulse $\operatorname{blo}(5,x)$ represented by a voltage $+10$ V is applied to the vertical bars, and a block pulse $\operatorname{blo}(6,y)$ represented by a voltage -10 V is applied to the horizontal bars. A voltage difference of 20 V is produced

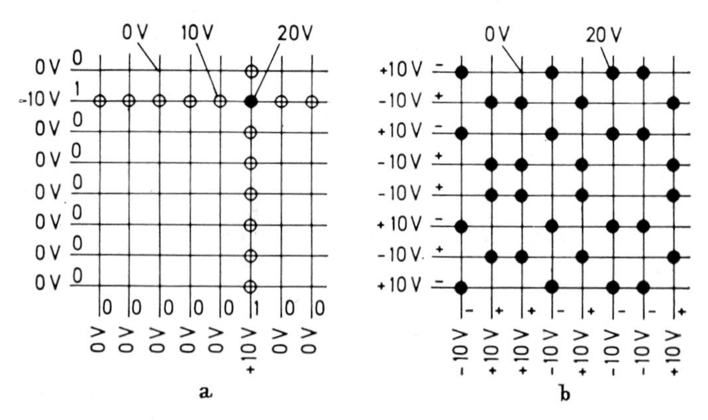

Fig. 106. Voltage differences in a simplified crossbar display using block pulses (a) and Walsh functions (b).

where the function $\operatorname{blo}(5,x)\operatorname{blo}(6,y)$ is $+1$. The voltage difference at most other bar crossings is zero, but there are a few with 10 V difference. Let a voltage-difference-to-light converter be placed at the bar crossings, e.g., a glow tube. The largest permissable voltage difference would be limited to twice the difference that just starts light emission. Furthermore, light can only be emitted from one bar crossing at a time. These two restrictions limit the brightness of a produced image.

Let us now turn to Fig. 106b that shows the function wal$(5, x)$ applied to the vertical and the function wal$(6, y)$ applied to the horizontal bars. Note that the sign of 10 V corresponds to that of the Walsh function at the vertical bars but is reversed at the horizontal bars, just as it was for the block pulse. Only voltage differences of 20 V and 0 V occur now. There is no intermediate voltage difference and hence no restriction on the permissable largest voltage difference. Furthermore, light is emitted from half the bar crossing at a time. The latter effect becomes, of course, even more important in three-dimensional displays. A modulation may be achieved in Fig. 106a and b by varying the time during which the voltages are applied.

Problems

243.1 Draw a detailed circuit diagram of Fig. 104.

243.2 Let a glow tube be connected to Ohi in Fig. 105 and let a capacitor be used for the storage Hhi. Draw the circuit diagram.

243.3 A trigger pulse is used in electronic photo flash units to discharge a capacitor at a certain moment through a glow tube. Draw a corresponding trigger circuit for problem 243.2.

243.4 The flying spot scanner inherently decomposes into the system of block pulses. How can it be made to decompose into the system of Walsh functions?

243.5 Let electrons be emitted by an array of 512^2 emitters in a plane and focused in another plane as done by an electron microscope. Could this principle be used to decompose a signal $F(x, y)$ into Walsh functions? How could the Walsh function pattern be generated in one or both of the two planes?

2.5 Digital Sequency Filters

2.5.1 Filters Based on the Generalized Fourier Transform

One of the most promising applications of sequency filters based on Walsh functions are digital filters. One reason is that numerical Walsh-Fourier transformation of signals requires additions and subtractions only, while numerical Fourier transformation requires multiplications.

For an explanation of a digital sequency filter consider the block diagram of Fig. 107. A signal $F(\theta)$ is passed through a sequency low-pass filter LP. Let the step function $F_1(\theta)$ at its output have steps of duration $T/16$. The amplitudes of the steps are converted by the analog/digital converter AD into binary digital characters, which are delivered at the rate of 16 characters per time unit T to the digital storage ST 1. Let a set of

16 consecutive characters be denoted by $A, B, \ldots, H, \ldots, P$. The fast Walsh-Fourier transform of section 1.25 may be used to obtain from these 16 characters the 16 coefficients $a(0), a_s(1), a_c(1), \ldots, a_s(8)$. Additions and subtractions only have to be performed by the arithmetic unit AU to obtain one, several or all of these coefficients. Assume the coefficient $a_c(5)$ alone is computed as shown in Fig. 107. The 16 coefficients $-a_c(5)$,

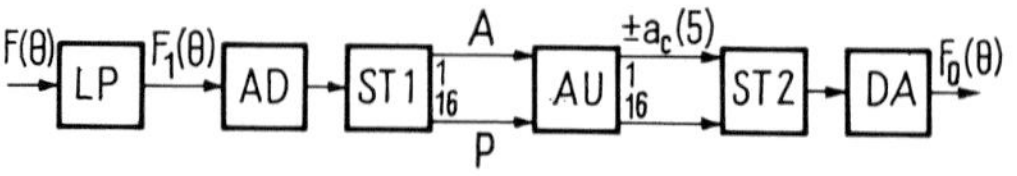

Fig. 107. Block diagram of a digital sequency filter. LP, sequency low-pass filter of Fig. 55; AD, analog/digital converter; ST, digital storage; AU, arithmetic unit performing additions and subtractions; DA, digital/analog converter.

$+a_c(5)$, $+a_c(5)$, $-a_c(5), \ldots$ with the signs corresponding to the signs of $\mathrm{cal}(5, \theta)$ in Fig. 2 are transferred into the digital storage ST 2. Reading these 16 coefficients out serially through a digital/analog converter DA yields the analog output signal $F_0(\theta)$. The connection between input signal $F(\theta)$ and output signal $F_0(\theta)$ follows from Eqs. (2.1.2-3) and (2.1.3-4):

$$a_c(5) = \int_{-1/2}^{1/2} F(\theta) \, \mathrm{cal}(5, \theta) \, d\theta = a_c(\mu) \qquad 5 \leqq \mu < 6$$

$$F_0(\theta) = a_c(5) \, \mathrm{cal}(5, \theta - 1) = \int_5^6 a_c(\mu) \, \mathrm{cal}(\mu, \theta) \, d\theta \qquad (1)$$

Let $F_0(\theta)$ be obtained by feeding $\mathrm{cal}(5, \theta)$ into the sequency filter of Fig. 56. Let $F_0(\theta)$ deviate from its nominal value according to a Gaussian distribution with mean square deviation 10^{-6} referred to a unit voltage. The crosstalk attenuation of the coefficients $a_c(i)$, $i \neq 5$, and $a_s(i)$ is then -60 dB. A much higher crosstalk attenuation can be obtained by a digital filter. Let $F_1(0)$ in Fig. 107 have the mean square deviation 10^{-6} referred to a unit voltage. If 2^n samples of $F_1(\theta)$ are used to compute $a_c(5)$, and if the analog/digital conversion introduces a negligible error, one obtains a mean square deviation of $a_c(5)$ of $10^{-6}/2^n$ referred to a unit voltage. The crosstalk attenuation in dB is thus $10 \log 10^{-6}/2^n = -(60 + 10n \log 2)$. For $2^n = 16$ as used in Fig. 107 one obtains a crosstalk attenuation of $-(60 + 40 \log 2) = -72$ dB.

Problems

251.1 Draw the block diagram of a digital filter corresponding to the analog filter of Fig. 60.

251.2 Draw the block diagram of a digital multiplicity filter with $j \leqq 48$, $q \leqq 3$ according to Fig. 63.

251.3 Draw the block diagram of a digital asynchronous filter according to Figs. 64 and 65.

2.5.2 Filters Based on Difference Equations

It has been pointed out earlier that the difference equations describing LCS resonance filters are instructions for the design of digital filters. A digital resonance filter for the function sal$(4, \theta)$ is shown in Fig. 108. An analog signal $F(\theta)$ is fed through a sequency low-pass filter SL and

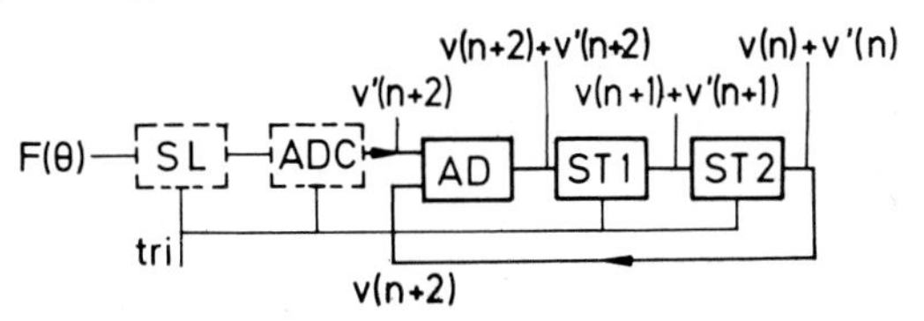

Fig. 108. Recursion filter for the function sal$(4, \theta)$. SL, sequency low-pass filter; ADC, analog/digital converter; AD, adder; ST, storage.

the analog/digital converter ADC. Digital signals $v'(n), v'(n+1), v'(n+2), \ldots$ are applied at the times $n, n+1, n+2, \ldots$ to one input of the adder AD, while a signal $v(n), v(n+1), v(n+2), \ldots$ is applied to the other input. The output of the adder represents $v'(n+2) + v(n+2)$ at the time $n + 2$. The previously added signals $v'(n+1) + v(n+1)$ and $v'(n) + v(n)$ are represented by the outputs[1] of the storage ST 1 and ST 2. The output of ST 2 is fed back to the one input terminal of the adder AD. The following relation must hold for this feedback loop:

$$v(n+2) = v(n) + v'(n) \tag{1}$$

This is the difference equation (2.2.1-5) of the voltage $v_1(n)$ of an LCS resonance filter for the function sal$(4, \theta)$. The filter is somewhat trivial since the Rademacher functions sal$(2^s, \theta)$ do not require time variable circuits. Some fresh points of theory are encountered by investigating a resonance circuit for the function sal$(3, \theta)$ according to Figs. 67 and 68.

In analogy to Eq. (2.2.1-4) one derives the following sets of equations from Fig. 67a and line *8* of Fig. 68:

$$v_1(n+1) = v_2(n) \qquad v_2(n+1) = v_1(n) + v_1'(n) \tag{2}$$

$$v_1(n+2) = v_1(n+1) + v_1'(n+1) \qquad v_2(n+2) = v_2(n+1) \tag{3}$$

$$v_1(n+3) = v_2(n+2) \qquad v_2(n+3) = v_1(n+2) + v_1'(n+2) \tag{4}$$

$$v_1(n+4) = v_2(n+3) \qquad v_2(n+4) = v_1(n+3) + v_1'(n+3) \tag{5}$$

$$n = 0, 4, 8, 12, \ldots, 4i, \ldots$$

The variables v_1 and v_2 in Eq. (3) are separated, and there is no difficulty separating them in Eqs. (4) and (5). The separation in Eq. (2) succeeds by

[1] The storages ST shall contain a buffer storage so that they can be read in and out at the same time.

substituting $n - 4$ for n in Eq. (5):

$$v_1(n+1) = v_1'(n-1) + v_1(n-1) \quad \text{for} \quad n \geqq 4; \quad v_1(1) = v_2(0) \quad \text{for} \quad n = 0 \tag{6}$$

$$v_1(n+2) = v_1'(n+1) + v_1(n+1) \tag{7}$$

$$v_1(n+3) = v_1'(n) \qquad + v_1(n) \tag{8}$$

$$v_1(n+4) = v_1'(n+2) + v_1(n+2) \tag{9}$$

$$n = 4i$$

A similar set of equations is obtained for v_2. The four Eqs. (6) to (9) are equivalent to one difference equation with variable coefficients. Using the Kronecker symbol $\delta_{j,k}$ one may write it as follows:

$$\begin{aligned} &\delta_{4i+1,m}[v_1(m) - v_1'(m-2) - v_1(m-2)] + \\ &+ \delta_{4i+2,m}[v_1(m) - v_1'(m-1) - v_1(m-1)] + \\ &+ \delta_{4i+3,m}[v_1(m) - v_1'(m-3) - v_1(m-3)] + \\ &+ \delta_{4i+4,m}[v_1(m) - v_1'(m-2) + v_1(m-2)] = 0 \end{aligned} \tag{10}$$

$$m = 1, 2, 3, \ldots$$

Figure 109 shows a digital circuit for the difference equations (6) to (9) or (10) with the signals at various points indicated for eight successive times. The circuit consists of the adder AD and two storages ST, just like the circuit of Fig. 108. In addition, two switches S 1 and S 2 are used that provide time-variable feedback links.

Let us assume the signal $v'(n+1)$ is fed from an analog/digital converter to one input of the adder AD and the signal $v(n+1)$ to the other, as shown in Fig. 109a. The signals $v'(n) + v(n)$ and $v'(n-1) + v(n-1)$ are stored in ST 1 and ST 2. The relation $v(n+1) = v'(n-1) + v(n-1)$ must hold if the feedback loop is closed from ST 2 to AD via switch S 2. This represents the solution of Eq. (6).

A shift pulse is then applied that shifts the signal $v'(n+1) + v(n+1)$ into storage ST 1, and $v'(n) + v(n)$ into ST 2 as shown by Fig. 109b. Immediately afterwards the switches S 1 and S 2 are switched from position 1 to position 0, and the next signal $v'(n+2)$ is fed to adder AD (Fig. 109c). The feedback loop from ST 1 to AD is closed via switch S 1. The relation $v(n+2) = v'(n+1) + v(n+1)$ must hold. This is the solution of Eq. (7). Note that the signal $v'(n) + v(n)$ occurs on both input and output terminals of storage ST 2.

A shift pulse is now applied. The signal $v'(n+2) + v(n+2)$ is read into storage ST 1 while the signal $v'(n) + v(n)$ is recycled and remains in storage ST 2, as shown in Fig. 109d. The switches ST 1 and ST 2 are returned from position 0 to position 1 and the next signal $v'(n+3)$ is applied to the adder AD (Fig. 109e). The feedback loop is closed from ST 2 to AD via S 2, and the relation $v(n+3) = v'(n) + v(n)$ must hold. This is the solution of Eq. (8).

Another shift brings $v'(n+3) + v(n+3)$ and $v'(n+2) + v(n+2)$ into the storages ST 1 and ST 2 (Fig. 109f). Feeding the signal $v'(n+4)$ to the adder yields the condition $v(n+4) = v'(n+2) + v(n+2)$ for

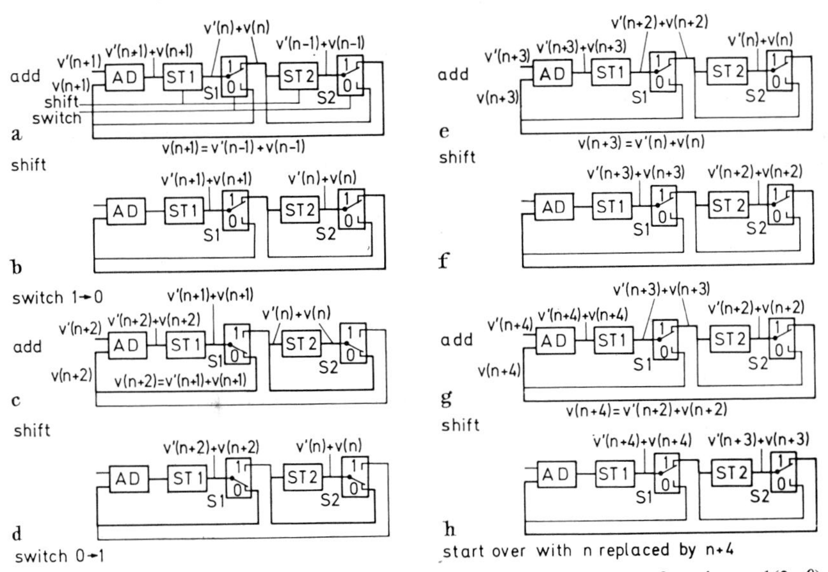

Fig. 109. Recursion filter with time variable feedback loops for the function $\text{sal}(3, \theta)$, showing the eight different states of one recursion cycle. AD, adder; ST, storage; S, switch.

the feedback loop in Fig. 109g. This is the solution of Eq. (9). A shift pulse brings the circuit into the state shown by Fig. 109h. If now the next signal $v'(n+5)$ is applied to AD one obtains again the state shown by Fig. 109a with n replaced by $n+4$.

It is evident that digital filters like those of Figs. 108 and 109 could be used to detect binary sequences of radar returns or Reed-Muller coded signals[1].

Problems

252.1 Discuss the operation of the circuit of Fig. 109 if set to resonate with the function $\text{cal}(3, \theta)$ (use Fig. 71).

252.2 Repeat problem 252.1 for $\text{sal}(1, \theta)$.

252.3 Find the digital equivalent of the circuit of Fig. 72.

---

[1] The connection between Reed-Muller codes and Walsh functions is treated in chapter 7.

3. DIRECT TRANSMISSION OF SIGNALS

3.1 Orthogonal Division as Generalization of Time and Frequency Division

3.1.1 Representation of Signals

Consider a telegraphy alphabet containing a finite number of characters. An example is the teletype alphabet having 32 characters. It is usual to represent them by sets of 5 coefficients with value $+1$ or -1:

character 1: $+1 \quad +1 \quad +1 \quad +1 \quad +1$

character 2: $+1 \quad +1 \quad +1 \quad +1 \quad -1$ etc.

In general, the characters may consist of sets of m coefficients having arbitrary values rather than just the values $+1$ or -1. The following notation is appropriate in this case:

$$\text{character } \chi\text{:} \quad a_\chi(0), a_\chi(1), \ldots, a_\chi(j), \ldots, a_\chi(m-1) \tag{1}$$

The representation of characters by time functions is another important representation. Consider m time functions $f(j, \theta)$. Let the functions be multiplied by the coefficients $a_\chi(j)$ and the products be added. One obtains the representation of the character χ by the time function $F_\chi(\theta)$:

$$F_\chi(\theta) = \sum_{j=0}^{m-1} a_\chi(j) f(j, \theta) \tag{2}$$

The coefficients $a_\chi(j)$ may be regained individually from $F_\chi(\theta)$, if the system of functions $\{f(j, \theta)\}$ is linearly independent. The process is particularly simple if the functions are orthogonal. Let them be orthogonal and normalized in the interval $-\frac{1}{2} \leqq \theta < \frac{1}{2}$. The coefficient $a_\chi(k)$ is obtained by multiplying $F_\chi(\theta)$ with $f(k, \theta)$ and integrating the product; the shorter expression "correlating $F_\chi(\theta)$ with $f(k, \theta)$" is generally used for this process:

$$\int_{-1/2}^{1/2} F_\chi(\theta) f(k, \theta)\, d\theta = \sum_{j=0}^{m-1} a_\chi(j) \int_{-1/2}^{1/2} f(j, \theta) f(k, \theta)\, d\theta = a_\chi(k) \tag{3}$$

Let m equal 5; let $a_\chi(0)$, $a_\chi(2)$, $a_\chi(3)$ equal $+1$ and $a_\chi(1)$, $a_\chi(4)$ equal -1. $F_\chi(\theta)$ has then the shape shown in Fig. 3, if the functions $f(j, \theta)$ are the block pulses of Fig. 3. $F_\chi(\theta)$ represents voltage or current of the usual teletype signals as function of time.

The values of the coefficients $a_\chi(j)$ transmitted by the signal $F_\chi(\theta)$ of Fig. 3 may also be obtained by amplitude sampling at proper times. Hence, the terms transmission by time multiplex or by time division are used.

The block pulses of Fig. 3 may also be interpreted as frequency functions $f(j, \nu)$. The character χ is then represented by the frequency function $F_\chi(\nu)$. If $F_\chi(\nu)$ is applied to 5 suitable frequency band-pass filters, one may recover the coefficients by sampling the output voltages of these filters. Frequency multiplex or frequency division are usual terms for this type of transmission.

Recovery of the transmitted coefficients by sampling in time or frequency domain without further computation is not possible for most systems of orthogonal functions[1]. Recovery by means of their orthogonality is always possible according to Eq. (3). The terms orthogonal division or orthogonal multiplex are appropriate in this case. The advantage of orthogonal division is that the number of useful systems of functions is much larger than for time or frequency division. Hence, there is more freedom of choice for the best system for a particular application.

Theoretical investigations frequently represent characters by vectors in a signal space. How is this vector representation related to the representation by orthogonal functions? Consider m-dimensional, rectangular cartesian coordinates having the unit vectors $\mathbf{e}_j$. The length of these vectors equals the integral of the square of the orthonormal functions $f(j, \theta)$:

$$\int_{-1/2}^{1/2} f^2(j, \theta)\, d\theta = \mathbf{e}_j \cdot \mathbf{e}_j = 1 \tag{4}$$

The scalar product of two vectors $\mathbf{e}_j$ and $\mathbf{e}_k$, $j \neq k$, vanishes since they are perpendicular to each other. The connection between orthogonal function and vector representation may thus be expressed by the orthogonality relation:

$$\int_{-1/2}^{1/2} f(j, \theta) f(k, \theta)\, d\theta = \mathbf{e}_j \cdot \mathbf{e}_k = \delta_{jk} \tag{5}$$

A character χ is represented by the vector $\mathbf{F}_\chi$ in signal space:

$$\mathbf{F}_\chi = \sum_{j=0}^{m-1} a_\chi(j)\, \mathbf{e}_j \tag{6}$$

Instead of m orthogonal vectors one may also use m linearly independent vectors. This representation is obtained, if the functions $f(j, \theta)$ are not orthogonal but linearly independent.

As a practical example consider a teletype character $F_\chi(\theta)$ composed of 5 sine and cosine elements according to Fig. 1:

$$\begin{aligned} F_\chi(\theta) = a_\chi(0) f(0, \theta) &+ a_\chi(1)\sqrt{(2)}\sin 4\pi\theta + a_\chi(2)\sqrt{(2)}\cos 4\pi\theta + \\ &+ a_\chi(3)\sqrt{(2)}\sin 6\pi\theta + a_\chi(4)\sqrt{(2)}\cos 6\pi\theta \\ &-\tfrac{1}{2} \leqq \theta < \tfrac{1}{2};\ \theta = t/T \end{aligned} \tag{7}$$

[1] More than one amplitude sample is then needed to compute the coefficients. Such a process is, however, a method to compute the integral in Eq. (3) and this is not what is generally understood as time or frequency division.

T equals 150 ms if the duration of a teletype character is 150 ms, which is a much-used standard. The coefficients $a_\chi(j)$ are $+1$ or -1 for a balanced system, and $+1$ or 0 for an on-off system. Let $F_\chi(\theta)$ be applied at the receiver simultaneously to 5 multipliers which multiply $F_\chi(\theta)$ with the 5 functions $f(0, \theta)$ to $\sqrt{(2)} \cos 6\pi\, \theta$. The output voltages of the 5 multipliers are integrated during the time interval $-\frac{1}{2} \leqq \theta < \frac{1}{2}$. The output voltages of the five integrators represent the values of the coefficients $a_\chi(j)$ of (7) at the time $\theta = \frac{1}{2}$. Fig. 110 shows oscillograms of the output voltages of the 5 integrators during the interval $-\frac{1}{2} \leqq \theta < \frac{1}{2}$. There are 32 different traces for each of the 5 output voltages due to the 32 characters of

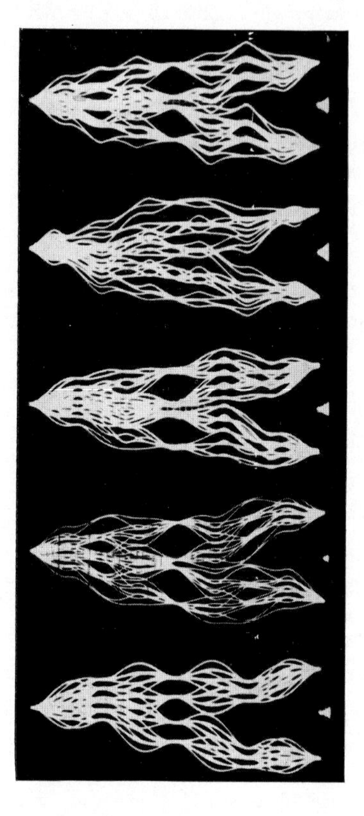

Fig. 110. Detection of the coefficients $+1$ and -1 by cross-correlation of 32 different teletype signals composed of sine and cosine pulses. Duration of the traces $T = 150$ ms (Courtesy H. Wagenlehner of Battelle-Institut Frankfurt.)

the teletype alphabet. 16 traces reach a positive value $(+1)$ for $\theta = \frac{1}{2}$ and 16 a negative value (-1). This indicates a balanced teletype system. In an on-off system, 16 traces would assume the value 0 instead of a negative value at $\theta = \frac{1}{2}$. The apparent lack of symmetry between the traces ending at $+1$ and those ending at -1 is caused by an additional signal $\sqrt{(2)} \sin 2\pi\, \theta$ added to the characters for synchronization. The elements $\sqrt{(2)} \sin 2\pi\, \theta$ and $\sqrt{(2)} \cos 2\pi\, \theta$ do not appear in Eq. (7) for this reason.

All three discussed representations of signals contain the coefficients $a_\chi(j)$. The vectors $\mathbf{e}_j$ permit the representation of m coefficients by one vector $\mathbf{F}_\chi$, the time functions $f(j, \theta)$ the representation by one time function $F_\chi(\theta)$. Some signals, such as the output voltage of a microphone, are usually available as time functions. Their representation by coefficients will be discussed in section 3.1.3.

Problems

311.1 Write the character $+1 +1 -1 -1 +1$ by means of vectors and Walsh functions.

311.2 What is the difference between orthogonal and linearly independent vectors?

311.3 The connection between Hadamard matrices and Walsh functions was discussed in section 1.1.5. Let the row j of a matrix represent the components of one vector in Eq. (5) and the row k the components of the other vector. What does the integral in Eq. (5) mean in terms of matrix algebra?

3.1.2 Examples of Signals

Figure 111a shows two characters $\mathbf{F}_0 = \mathbf{e}_0$ and $\mathbf{F}_1 = -\mathbf{e}_0$ in signal space. The same characters are shown below as time functions for the block pulse $f(0, \theta)$ or the Walsh pulse $\text{sal}(1, \theta)$:

$$F_0(\theta) = +f(0, \theta), \quad F_1(\theta) = -f(0, \theta)$$

or

$$F_0(\theta) = +\text{sal}(1, \theta), \quad F_1(\theta) = -\text{sal}(1, \theta)$$

Figures 111b and c show characters $\mathbf{F}_\chi$ and $F_\chi(\theta)$ constructed from the vector $\mathbf{e}_0$ and the values $+1$, 0, -1 or $+1$, $+\frac{1}{3}$, $-\frac{1}{3}$, -1 for the coefficients $a_\chi(0)$.

Figures 111d, e and f show characters constructed from vectors $\mathbf{e}_0$ and $\mathbf{e}_1$, or from two functions. Written in detail, the characters of Fig. 111d have the following form:

$$\mathbf{F}_0 = \mathbf{e}_0 + \mathbf{e}_1, \quad \mathbf{F}_1 = \mathbf{e}_0 - \mathbf{e}_1, \quad \mathbf{F}_2 = -\mathbf{e}_0 + \mathbf{e}_1, \quad \mathbf{F}_3 = -\mathbf{e}_0 - \mathbf{e}_1$$

or

$$F_0(\theta) = f(0, \theta) + f(1, \theta), \qquad F_1(\theta) = f(0, \theta) - f(1, \theta)$$

$$F_2(\theta) = -f(0, \theta) + f(1, \theta), \qquad F_3(\theta) = -f(0, \theta) - f(1, \theta)$$

or

$$F_0(\theta) = \text{wal}(0, \theta) + \text{sal}(1, \theta), \qquad F_1(\theta) = \text{wal}(0, \theta) - \text{sal}(1, \theta)$$

$$F_2(\theta) = -\text{wal}(0, \theta) + \text{sal}(1, \theta), \qquad F_3(\theta) = -\text{wal}(0, \theta) - \text{sal}(1, \theta)$$

The functions $f(0, \theta)$, $f(1, \theta)$, $\text{wal}(0, \theta)$ and $\text{sal}(1, \theta)$ are shown below Fig. 111d; the characters $F_0(\theta)$ to $F_3(\theta)$ composed of these functions are shown above them.

The terms binary, ternary and quarternary may be applied to the characters of Fig. 111, since the individual vectors or functions are multiplied by coefficients that assume 2, 3 or 4 different values. Figure 112 shows

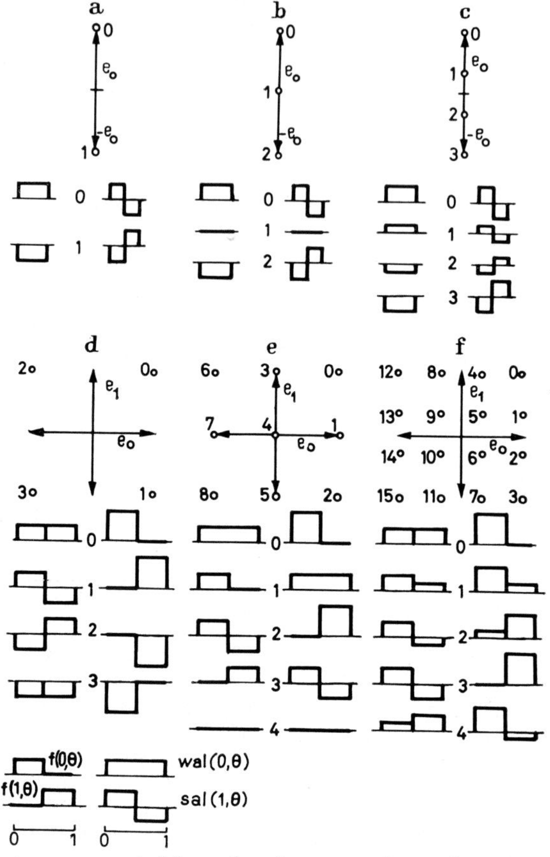

Fig. 111. Characters represented by points in one- and two-dimensional signal spaces and by time functions.

that terms like "binary character" are generally not applicable, if a character consists of more than one vector or function. Figure 112a shows the three characters of a so-called transorthogonal alphabet. The characters read in vector representation as follows:

$$\mathbf{F}_0 = \tfrac{1}{2}\sqrt{(3)}\,\mathbf{e}_0 + \tfrac{1}{2}\mathbf{e}_1, \quad \mathbf{F}_1 = 0\mathbf{e}_0 - 1\,\mathbf{e}_1, \quad \mathbf{F}_2 = -\tfrac{1}{2}\sqrt{(3)}\,\mathbf{e}_0 + \tfrac{1}{2}\mathbf{e}_1$$

$\mathbf{e}_0$ is multiplied by one of the three coefficients $+\frac{1}{2}\sqrt{(3)}, 0$ or $-\frac{1}{2}\sqrt{(3)}$, $\mathbf{e}_1$ by one of the two coefficients $\frac{1}{2}$ or -1. If the vectors $\mathbf{e}_0$ and $\mathbf{e}_1$ are rotated relative to the signal points, representations are obtained that have three different coefficients for each vector, or two different coefficients for $\mathbf{e}_0$ and three for $\mathbf{e}_1$. Signals composed of the functions $f(0, \theta)$ and $f(1, \theta)$ or wal$(0, \theta)$ and sal$(1, \theta)$ are shown below the vector model:

$$F_0(\theta) = \tfrac{1}{2}\sqrt{(3)} f(0, \theta) + \tfrac{1}{2} f(1, \theta), \quad F_1(\theta) = -f(1, \theta),$$

$$F_2(\theta) = -\tfrac{1}{2}\sqrt{(3)} f(0, \theta) + \tfrac{1}{2} f(1, \theta)$$

$$F_0(\theta) = \tfrac{1}{2}\sqrt{(3)}\, \mathrm{wal}(0, \theta) + \tfrac{1}{2}\mathrm{sal}(1, \theta), \quad F_1(\theta) = -\mathrm{sal}(1, \theta)$$

$$F_2(\theta) = -\tfrac{1}{2}\sqrt{(3)}\, \mathrm{wal}(0, \theta) + \tfrac{1}{2}\mathrm{sal}(1, \theta)$$

Figure 112b shows the four characters of a so-called biorthogonal alphabet:

$$\mathbf{F}_0 = \mathbf{e}_0, \quad \mathbf{F}_1 = \mathbf{e}_1, \quad \mathbf{F}_2 = -\mathbf{e}_1, \quad \mathbf{F}_3 = -\mathbf{e}_0$$

These characters look very similar to those of Fig. 111d. The similarity disappears, if the characters are composed of more than two vectors or functions.

The dashed lines in Fig. 112 show distances between certain signal points. All signal points of the transorthogonal alphabet (Fig. 112a) have the same distance from each other. The vectors from signal points 0 to 1,

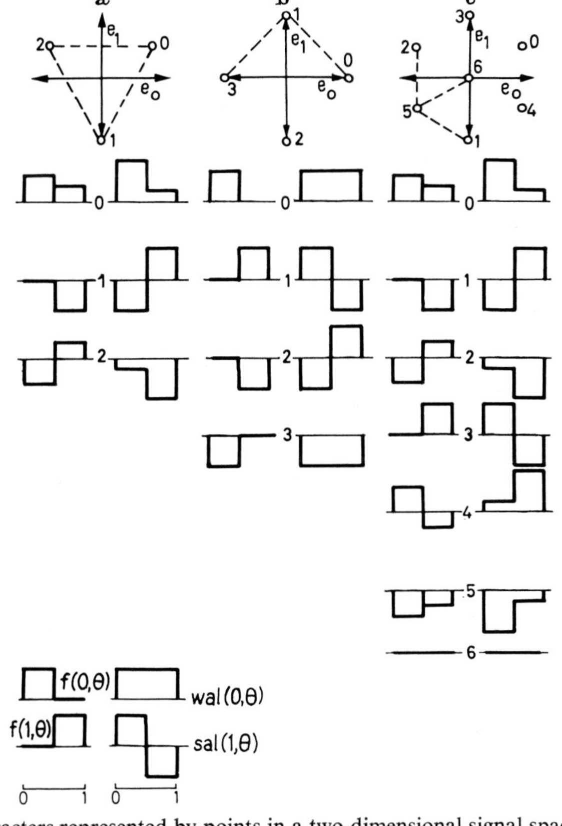

Fig. 112. Characters represented by points in a two-dimensional signal space and by time functions.

1 to 2 and 2 to 0 are $\mathbf{F}_1 - \mathbf{F}_0$, $\mathbf{F}_2 - \mathbf{F}_1$ and $\mathbf{F}_0 - \mathbf{F}_2$. The square of their length equals 3:

$$(\mathbf{F}_1 - \mathbf{F}_0)^2 = (-\tfrac{1}{2}\sqrt{(3)}\,\mathbf{e}_0 - \tfrac{3}{2}\mathbf{e}_1)^2 = \tfrac{3}{4} + \tfrac{9}{4} = 3$$
$$(\mathbf{F}_2 - \mathbf{F}_1)^2 = (\tfrac{1}{2}\sqrt{(3)}\,\mathbf{e}_0 + \tfrac{3}{2}\mathbf{e}_1)^2 = \tfrac{3}{4} + \tfrac{9}{4} = 3$$
$$(\mathbf{F}_0 - \mathbf{F}_2)^2 = (\sqrt{(3)}\,\mathbf{e}_0)^2 = 3$$

If the characters are represented by functions rather than by vectors one must replace scalar products by the integrals of the products of the respective functions as shown by Eq. (3.1.1-5). It follows:

$$\int_{-1/2}^{1/2} [F_1(\theta) - F_0(\theta)]^2\, d\theta = \int_{-1/2}^{1/2} [-\tfrac{1}{2}\sqrt{(3)}\, f(0,\theta) - \tfrac{3}{2} f(1,\theta)]^2\, d\theta = 3$$
$$\int_{-1/2}^{1/2} [F_2(\theta) - F_1(\theta)]^2\, d\theta = \int_{-1/2}^{1/2} [\tfrac{1}{2}\sqrt{(3)}\, f(0,\theta) - \tfrac{3}{2} f(1,\theta)]^2\, d\theta = 3$$
$$\int_{-1/2}^{1/2} [F_0(\theta) - F_2(\theta)]^2\, d\theta = \int_{-1/2}^{1/2} [\sqrt{(3)}\, f(0,\theta)]^2\, d\theta = 3$$

$F_1(\theta) - F_0(\theta)$ is the function that must be added to the character $F_0(\theta)$ in order to obtain the character $F_1(\theta)$. $\int_{-1/2}^{1/2} [F_1(\theta) - F_0(\theta)]^2\, d\theta$ is the energy required to transform character $F_0(\theta)$ into character $F_1(\theta)$, if the integral $\int_{-1/2}^{1/2} F_\chi^2(\theta)\, d\theta$ is the energy of the character $F_\chi(\theta)$. The square of the distance of a signal point from the origin represents the energy of that character.

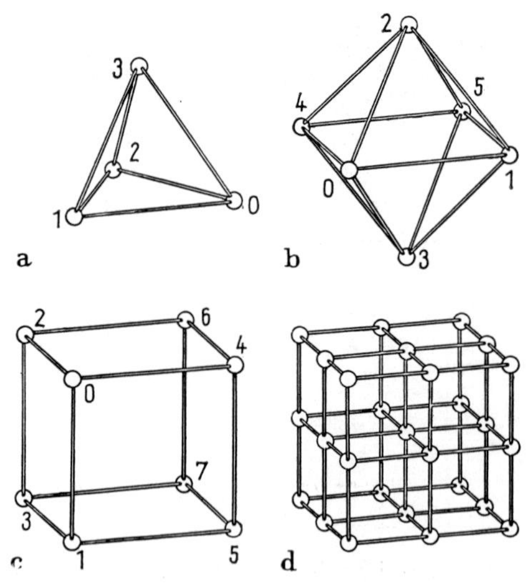

Fig. 113. Characters represented by points in a three-dimensional signal space.

Figure 113 shows characters composed of three vectors. The spheres represent the signal points. The rods between them represent the distances between adjacent points. No unit vectors $\mathbf{e}_0$, $\mathbf{e}_1$ and $\mathbf{e}_2$ are shown. Normalization is different from Figs. 111 and 112. It is chosen so that the distances between adjacent signal points in Figs. 113a, b and c are equal. The values of the coefficients in vector space depend on the orientation of the unit vectors $\mathbf{e}_j$. For instance, the four characters of the transorthogonal alphabet of Fig. 113a may be written as follows:

$$\mathbf{F}_0 = +\frac{1}{2}\mathbf{e}_0 + \frac{1}{6}\sqrt{(3)}\,\mathbf{e}_1 - \frac{1}{12}\sqrt{(2)}\sqrt{(3)}\,\mathbf{e}_2$$

$$\mathbf{F}_1 = 0\,\mathbf{e}_0 - \frac{1}{3}\sqrt{(3)}\,\mathbf{e}_1 - \frac{1}{12}\sqrt{(2)}\sqrt{(3)}\,\mathbf{e}_2$$

$$\mathbf{F}_2 = -\frac{1}{2}\mathbf{e}_0 + \frac{1}{6}\sqrt{(3)}\,\mathbf{e}_1 - \frac{1}{12}\sqrt{(2)}\sqrt{(3)}\,\mathbf{e}_2$$

$$\mathbf{F}_3 = 0\,\mathbf{e}_0 + 0\,\mathbf{e}_1 + \frac{1}{4}\sqrt{(2)}\sqrt{(3)}\,\mathbf{e}_2$$

The energy of all four characters is equal:

$$\mathbf{F}_0^2 = \mathbf{F}_1^2 = \mathbf{F}_2^2 = \mathbf{F}_3^2 = \tfrac{3}{8}$$

The distances between the four signal points are also equal:

$$(\mathbf{F}_0 - \mathbf{F}_1)^2 = (\mathbf{F}_0 - \mathbf{F}_2)^2 = \cdots = (\mathbf{F}_2 - \mathbf{F}_3)^2 = 1$$

Figure 114a shows a representation of these four characters by three block pulses $f(0, \theta)$, $f(1, \theta)$ and $f(2, \theta)$ as well as by three Walsh pulses $\mathrm{wal}(0, \theta)$, $\mathrm{sal}(1, \theta)$ and $\mathrm{cal}(1, \theta)$.

In the case of the biorthogonal alphabet of Fig. 113b it is reasonable to orient the coordinate system so that two opposed signal points are located on each axis. The following simple vector representation results:

$$\mathbf{F}_0 = \mathbf{e}_0, \quad \mathbf{F}_1 = \mathbf{e}_1, \quad \mathbf{F}_2 = \mathbf{e}_2, \quad \mathbf{F}_3 = -\mathbf{e}_2, \quad \mathbf{F}_4 = -\mathbf{e}_1, \quad \mathbf{F}_5 = -\mathbf{e}_0$$

These characters composed of three block pulses or three Walsh functions are shown in Fig. 114b.

The characters of the alphabet of Fig. 113c may be written in a particularly simple form, if the axes of the coordinate system intersect the surfaces of the cube at their centers:

$$\begin{aligned}
\mathbf{F}_0 &= \mathbf{e}_0 + \mathbf{e}_1 + \mathbf{e}_2 & \mathbf{F}_4 &= -\mathbf{e}_0 + \mathbf{e}_1 + \mathbf{e}_2\\
\mathbf{F}_1 &= \mathbf{e}_0 + \mathbf{e}_1 - \mathbf{e}_2 & \mathbf{F}_5 &= -\mathbf{e}_0 + \mathbf{e}_1 - \mathbf{e}_2\\
\mathbf{F}_2 &= \mathbf{e}_0 - \mathbf{e}_1 + \mathbf{e}_2 & \mathbf{F}_6 &= -\mathbf{e}_0 - \mathbf{e}_1 + \mathbf{e}_2\\
\mathbf{F}_3 &= \mathbf{e}_0 - \mathbf{e}_1 - \mathbf{e}_2 & \mathbf{F}_7 &= -\mathbf{e}_0 - \mathbf{e}_1 - \mathbf{e}_2
\end{aligned}$$

Figure 114c shows these characters composed of three block pulses and three Walsh pulses.

The perspicuity of the vector representation is lost, if the characters consist of more than three vectors. The characters of some alphabets may readily be specified for four or more vectors or functions. This is true, e.g.,

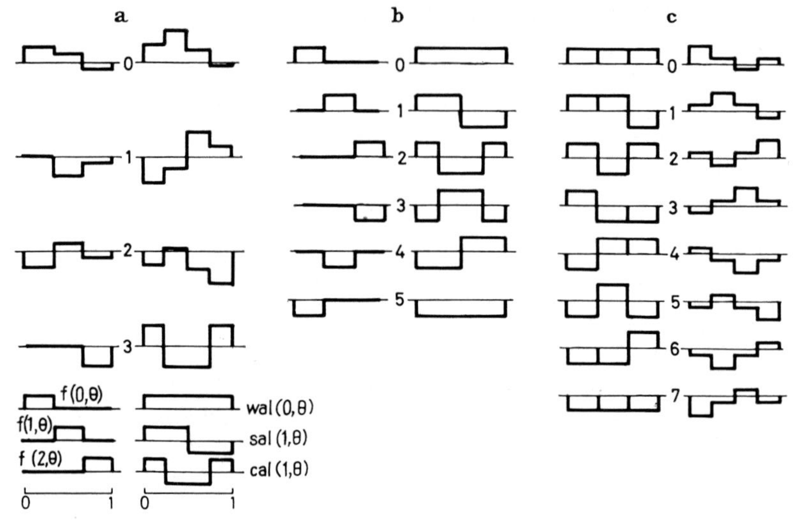

Fig. 114. Characters of Fig. 113 represented by time functions.

for the characters of the biorthogonal and the binary alphabets of Figs. 113b and c. Transorthogonal alphabets already require considerable computation. One may compose $m+1$ characters of a transorthogonal alphabet from m functions. These $m+1$ characters are specified by $m(m+1)$ coefficients $a_\chi(j)$; $j=0,\ldots,m-1$, $\chi=0,\ldots,m$. The following conditions are available for their computation:

a) The energy of all $m+1$ characters is equal. This statement yields m conditions.

b) The distances between the $m+1$ characters are equal. There are $m+(m-1)+(m-2)+\cdots+1=\frac{1}{2}m(m+1)$ distances between $m+1$ characters.

A total of $\frac{1}{2}m(m+3)$ equations are available for the determination of the $m(m+1)$ coefficients. A considerable number of coefficients can be chosen freely or fixed by additional conditions.

Problems

312.1 Compute the energy of the following four signals and the distances between them: $+1+1+1$, $-1-1+1$, $-1+1-1$, $+1-1-1$.

312.2 Do the same for the signals $+1+1+1-1$, $+1+1-1+1$, $+1-1+1+1$, $-1+1+1+1$.

312.3 Plot the signals of 3.121 and 312.2 with block pulses and Walsh pulses as in Fig. 114.

3.1.3 Amplitude Sampling and Orthogonal Decomposition

The sampling theorem of Fourier analysis states that a signal consisting of a superposition of periodic sine and cosine functions $\sin 2\pi f t$ and $\cos 2\pi f t$ with frequencies in the interval $0 \leqq f \leqq \Delta f$ is completely determined by $2\Delta f$ amplitude samples per second if Δf is measured in hertz [1–6]. This sampling theorem has been generalized by Kluvanec for other complete systems of orthogonal functions [7]. In essence, Kluvanec's sampling theorem states that a signal consisting of a superposition of functions $f_c(\phi T, t/T)$ and $f_s(\phi T, t/T)$ with sequencies in the interval $0 \leqq \phi \leqq \Delta\phi$ is completely determined by $2\Delta\phi$ amplitude samples per second if $\Delta\phi$ is measured in zps. It will be shown in this section that amplitude sampling of a frequency limited signal is mathematically equivalent to its decomposition into the incomplete orthogonal system $[\sin\pi(\theta + j)]/\pi(\theta + j); j = 0, \pm 1, \pm 2, \ldots$ The corresponding result for sequency limited signals composed of Walsh functions has been discussed in section 2.1.2. It turned out to be so simple that it is evident without calculation.

A frequency limited signal $F(\theta)$ with no components having a normalized frequency $\nu = fT > \frac{1}{2}$ may be expanded in a series of orthogonal functions that vanish outside the interval $-\frac{1}{2} \leqq \nu \leqq \frac{1}{2}$. Sine-cosine pulses, Walsh pulses, Legendre polynomials, etc. are suitable functions. The following system of sine-cosine pulses will be used, the phase angle $\frac{1}{4}\pi$ being introduced to simplify the result:

$$\left.\begin{aligned}
&g(j,\nu) = \begin{cases} g(0,\nu) = 1 \\ g(2i,\nu) = \sqrt{(2)}\cos(2\pi i \nu + \tfrac{1}{4}\pi) \\ g(2i-1,\nu) = \sqrt{(2)}\sin(2\pi i \nu + \tfrac{1}{4}\pi) \end{cases} \\
&g(j,\nu) = 0 \quad \text{for } \nu > \tfrac{1}{2} \text{ and } \nu < -\tfrac{1}{2} \\
&\qquad j = 0, 2i, 2i-1;\ i = 1, 2, \ldots
\end{aligned}\right\} \tag{1}$$

The Fourier transform $G(\nu)$ of a signal $F(\theta)$ is expanded in a series of these pulses:

$$\left.\begin{aligned}
G(\nu) &= \alpha(0) + \sum_{i=1}^{\infty} [\alpha(2i-1)\sqrt{(2)}\sin(2\pi i \nu + \tfrac{1}{4}\pi) + \\
&\qquad + \alpha(2i)\sqrt{(2)}\cos(2\pi i \nu + \tfrac{1}{4}\pi)] \\
\alpha(0) &= \int_{-1/2}^{1/2} G(\nu)\,d\nu, \quad \alpha(2i) = \int_{-1/2}^{1/2} G(\nu)\sqrt{(2)}\cos(2\pi i \nu + \tfrac{1}{4}\pi)\,d\nu \\
\alpha(2i-1) &= \int_{-1/2}^{1/2} G(\nu)\sqrt{(2)}\sin(2\pi i \nu + \tfrac{1}{4}\pi)\,d\nu
\end{aligned}\right\} \tag{2}$$

The inverse Fourier transform yields $F(\theta)$:

$$F(\theta) = \int_{-\infty}^{\infty} [G(\nu)(\cos 2\pi \nu \theta + \sin 2\pi \nu \theta)\,d\nu$$

The sum (2) is substituted for $G(\nu)$. Keeping in mind that $G(\nu)$ is zero outside the interval $-\frac{1}{2} \leqq \nu \leqq \frac{1}{2}$ one obtains:

$$F(\theta) = \alpha(0)\frac{\sin\pi\,\theta}{\pi\,\theta} + \sum_{i=1}^{\infty}\left[\alpha(2i-1)\,\frac{\sin\pi(\theta-i)}{\pi(\theta-i)} + \right.$$
$$\left. + \,\alpha(2i)\frac{\sin\pi(\theta+i)}{\pi(\theta+i)}\right] \tag{3}$$

A frequency limited signal $F(\theta)$ may thus be represented by a series of the incomplete orthogonal system of $(\sin x)/x$ functions. It follows from section 1.1.3 that these functions are orthogonal. One may prove it directly by evaluating the integral

$$\int_{-\infty}^{\infty} \frac{\sin\pi(\theta+k)}{\pi(\theta+k)}\,\frac{\sin\pi(\theta+j)}{\pi(\theta+j)}\,d\theta = \delta_{jk} \tag{4}$$
$$k, j = 0, \pm 1, \pm 2, \ldots$$

The coefficients $\alpha(0)$, $\alpha(2i-1)$ and $\alpha(2i)$ of Eq. (3) may be obtained by sampling the amplitude of the signal $F(\theta)$ at the times $\theta = t/T = 0, \pm 1, \pm 2, \ldots$ For instance, all functions $[\sin\pi(\theta-i)]/\pi(\theta-i)$ and $[\sin\pi(\theta+i)]/\pi(\theta+i)$ are zero for $\theta = 0$ and $(\sin\pi\,\theta)/\pi\,\theta$ is 1. Hence, it holds $F(0) = \alpha(0)$.

It follows from Eq. (4) that the coefficients $\alpha(0)$, $\alpha(2i)$ and $\alpha(2i-1)$ may also be obtained by orthogonal decomposition of $F(\theta)$ by $(\sin x)/x$ functions.

$$\int_{-\infty}^{\infty} F(\theta)\,\frac{\sin\pi(\theta+j)}{\pi(\theta+j)}\,d\theta = \begin{cases} \alpha(0) = F(0) & \text{for } j = 0 \\ \alpha(2i) = F(-i) & \text{for } j = i \\ \alpha(2i-1) = F(i) & \text{for } j = -i \end{cases} \tag{5}$$

The equivalence of amplitude sampling and orthogonal decomposition is not restricted to frequency limited signals. Let a finite number of discrete oscillations $A_h \sin 2\pi\,\nu_h\,\theta$ and $B_h \cos 2\pi\,\nu_h\,\theta$ with $\nu_h > \frac{1}{2}$ be added to $F(\theta)$. An ideal low-pass filter with cut-off frequency $\nu = \frac{1}{2}$ would suppress these additional oscillations, and amplitude sampling would again yield the coefficients $\alpha(0)$, $\alpha(2i)$ and $\alpha(2i-1)$. Orthogonal decomposition of the new signal $F(\theta) + A_h \sin 2\pi\,\nu_h\,\theta + B_h \cos 2\pi\,\nu_h\,\theta$ also yields $\alpha(0)$, $\alpha(2i)$ and $\alpha(2i-1)$, since the functions $A_h \sin 2\pi\,\nu_h\,\theta$ and $B_h \cos 2\pi\,\nu\,\theta$ yield no contribution:

$$\int_{-\infty}^{\infty} [F(\theta) + A_h \sin 2\pi\,\nu\,\theta + B_h \cos 2\pi\,\nu_h\,\theta]\,\frac{\sin\pi(\theta+j)}{\pi(\theta+j)}\,d\theta$$
$$= \int_{-\infty}^{\infty} F(\theta)\,\frac{\sin\pi(\theta+j)}{\pi(\theta+j)}\,d\theta \tag{6}$$
$$\nu_h = f_h T > \tfrac{1}{2}, f_h > 1/2T;\ \theta = t/T$$

It remains to be shown that continuous bands of oscillations do not yield any contribution either. Let a function $D(\theta)$ be added to $F(\theta)$, which contains no oscillation with frequency $|\nu| < \frac{1}{2}$. The Fourier transform of $D(\theta)$ must then be zero in the interval $-\frac{1}{2} \leqq \nu \leqq \frac{1}{2}$. On the other hand, the Fourier transform of the functions $[\sin\pi(\theta + j)]/\pi(\theta + j)$ is zero outside this interval. The two Fourier transforms are thus orthogonal to each other and the same must hold for the time functions:

$$\int_{-\infty}^{\infty} D(\theta) \frac{\sin\pi(\theta + j)}{\pi(\theta + j)} d\theta = 0 \tag{7}$$

The sampling theorem may readily be extended from the low-pass to the band-pass case. Consider the delta functions $\delta(\theta - i)$, $i = 0, \pm 1, \pm 2, \ldots$, shown in Fig. 115. Their Fourier transform,

$$\int_{-\infty}^{\infty} \delta(\theta - i)(\cos 2\pi \nu \theta + \sin 2\pi \nu \theta) \, d\theta = \sqrt{(2)} \sin(2\pi i \nu + \pi/4) \tag{8}$$

is also shown; use is made of the relation

$$\sin(-2\pi i \nu + \pi/4) = \cos(2\pi i \nu + \pi/4)$$

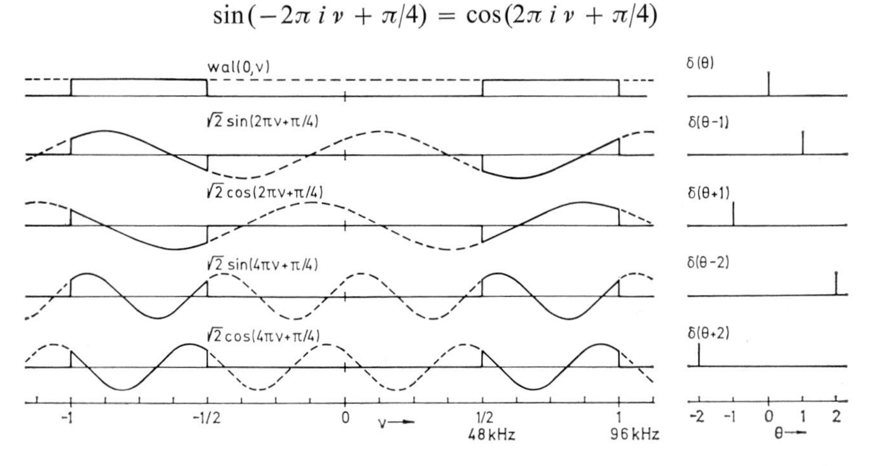

Fig. 115. Delta functions $\delta(\theta - i)$ and their Fourier transforms $\sqrt{(2)} \sin(2\pi i \nu + \pi/4)$, $i = 0, \pm 1, \pm 2, \ldots$ The scale in kHz holds for 12×8000 delta functions per second.

If all oscillations outside the band $0 \leqq \nu \leqq \frac{1}{2}$ are suppressed by a low-pass filter one obtains the inverse transform

$$\int_{-1/2}^{1/2} \sqrt{(2)} \sin(2\pi i \nu + \pi/4)(\cos 2\pi \nu \theta + \sin 2\pi \nu \theta) d\nu = \frac{\sin\pi(\theta - i)}{\pi(\theta - i)} \tag{9}$$

Let us now consider the more general case that all oscillations outside the band $\nu_0 \leqq \nu \leqq \nu_0 + \frac{1}{2}$ are suppressed:

$$\int_{-\nu_0-1/2}^{-\nu_0} \sqrt{(2)} \sin(2\pi\, i\, \nu + \pi/4)(\cos 2\pi\, \nu\, \theta + \sin 2\pi\, \nu\, \theta)\, d\nu +$$

$$+ \int_{\nu_0}^{\nu_0+1/2} \sqrt{(2)} \sin(2\pi\, i\, \nu + \pi/4)(\cos 2\pi\, \nu\, \theta + \sin 2\pi\, \nu\, \theta)\, d\nu = f(\nu_0, \theta - i)$$

$$= \cos 2\pi\, \nu_0 (\theta - i) \frac{\sin \pi(\theta - i)}{\pi(\theta - i)} + \cos \pi(\theta - i) \frac{\sin 2\pi\, \nu_0(\theta - i)}{\pi(\theta - i)} -$$

$$- \frac{\sin 2\pi\, \nu_0(\theta - i)}{\pi(\theta - i)} \tag{10}$$

For $\nu_0 = 0$ one obtains again the result of Eq. (9). Other useful results are obtained for $\nu_0 = \frac{1}{2}, 1, \frac{3}{2}, \ldots$ For instance, $\nu_0 = \frac{1}{2}$ yields:

$$f\left(\frac{1}{2}, \theta - i\right) = \frac{\sin 2\pi(\theta - i)}{\pi(\theta - i)} - \frac{\sin \pi(\theta - i)}{\pi(\theta - i)} \tag{11}$$

This function is shown for $i = 0$ by the solid line in Fig. 116. It has zeros at the points $\theta = \pm 1, \pm 2, \ldots$, just like the function $(\sin \pi\, \theta)/\pi\, \theta$ which

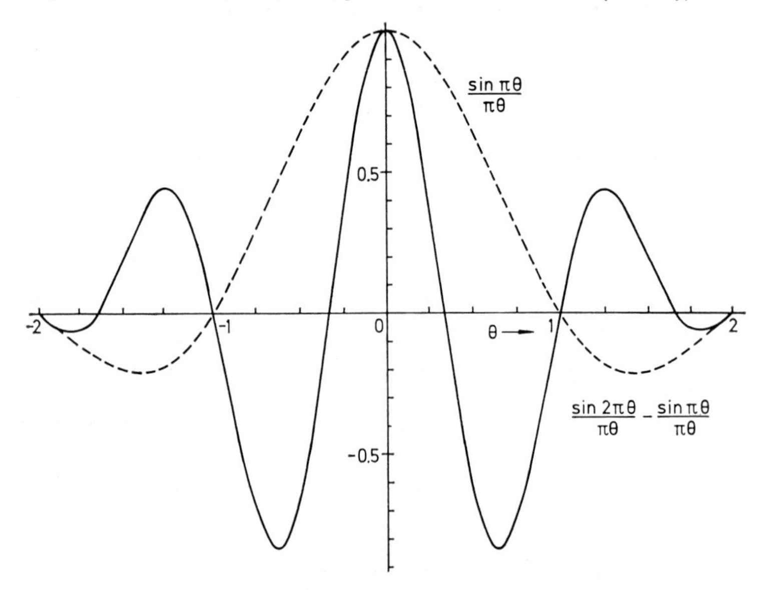

Fig. 116. Delta function $\delta(\theta)$ after passing through frequency filters with pass-band $0 \leqq \nu \leqq \frac{1}{2}$ (dashed) and $\frac{1}{2} \leqq \nu \leqq 1$ (solid).

is shown by the dashed line. Hence, sampling a signal with the amplitudes $A(i)$ at the times $\theta = i$ and passing the samples through a band-pass filter with pass-band $\frac{1}{2} \leqq \nu_0 \leqq 1$ yields the functions $A(i)f(\frac{1}{2}, \theta - i)$, while sampling these functions at the times $\theta = i$ yields in turn the amplitude samples $A(i)$.

Problems

313.1 Compute the function $f(1, \theta - i)$ of Eq. (10), plot it and discuss the zero crossings.

313.2 What is the power density spectrum of Eq. (8)?

313.3 Plot the power density spectrum of the functions $f(0, \theta - i)$, $f(\frac{1}{2}, \theta - i)$ and $f(1, \theta - i)$.

313.4 Combine the first two terms of $f(\nu_0, \theta - i)$ in Eq. (10) into one and show that ν_0 does not have to be $\frac{1}{2}, 1, \ldots$

313.5 A class of functions is defined by amplitude samples at equal intervals. The additional condition that the functions be frequency or sequency band-limited reduces the class to one function. It is common knowledge in communications that a function is "fairly well" defined by the location of its zero crossings (clipped speech is understandable; limiting signals to one positive and one negative amplitude value to simplify the correlator works well in radar signal processing; etc.). The zero crossings define a large class of functions that seems to be reduced to a smaller class by the equipment or the human ear. There must be some additional mathematical condition that reduces the class of functions defined by the location of the zero crossings just like the additional condition of bandwidth limitation reduces the class of functions defined by amplitude samples.

3.2 Practical Problems of Transmission

3.2.1 Circuits for Orthogonal Division[1]

Figure 117 shows a block diagram for the transmission of 5 coefficients $a_\chi(j)$ by orthogonal division. A function generator FG generates 5 functions $f(0, \theta), \ldots, f(4, \theta)$ at the transmitter, which are orthogonal in the interval $-\frac{1}{2}T \leqq t < \frac{1}{2}T$. The five coefficients $a_\chi(0), \ldots, a_\chi(4)$ are represented by voltages, which have a constant value during the interval $-\frac{1}{2}T \leqq t < \frac{1}{2}T$. The functions $f(j, \theta)$ are multiplied by the coefficients $a_\chi(j)$ in the multipliers M. The five products $a_\chi(j)f(j, \theta)$ are added by the resistors R and the operational amplifier TA. The resulting signal is transmitted and enters the receiver through the amplifier RA. It is then applied to 5 multipliers M. The signal is multiplied simultaneously with each one of the 5 functions $f(j, \theta)$ used in the transmitter as carriers for the coefficients. Function generators FG in the transmitter and receiver must be synchronized. The 5 products of the received signal with the functions $f(j, \theta)$ are integrated in the integrators I during the interval $-\frac{1}{2}T \leqq \leqq t < \frac{1}{2}T$. The voltages at the integrator outputs represent the coefficients $a_\chi(0)$ to $a_\chi(4)$ at the time $t = \frac{1}{2}T$.

[1] *See* [1—5] for a more detailed discussion of circuits.

Another set of five coefficients denoted by $a_\chi(0)$ to $a_\chi(4)$ is transmitted during the interval $\frac{1}{2}T \leqq t \leqq \frac{3}{2}T$. The functions $f(0, \theta)$ to $f(4, \theta)$ of the function generator FG in the transmitter and receiver are required again.

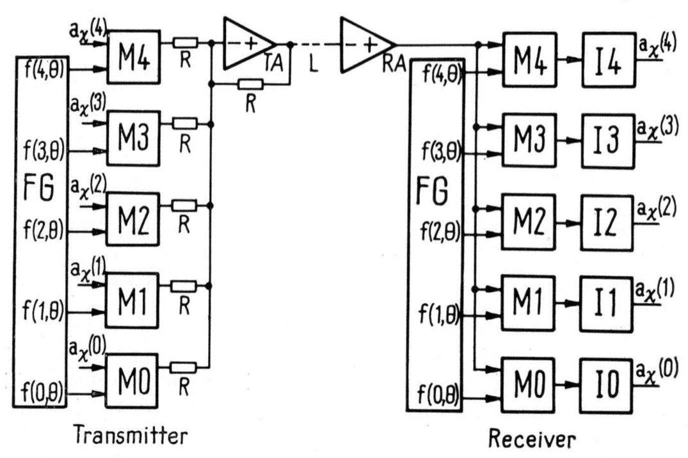

Fig. 117. Block diagram for signal transmission by orthogonal functions $f(j, \theta)$. FG, function generator; M, multiplier; I, integrator; TA, transmitter amplifier; RA, receiver amplifier.

Hence, these functions are periodic with period T. The voltages representing the coefficients $a_\chi(j)$ during the interval $-\frac{1}{2}T \leqq t < \frac{1}{2}T$ in the transmitter are changed suddenly at $t = \frac{1}{2}T$ and represent the coefficients $a_\chi(j)$ during the interval $\frac{1}{2}T \leqq t < \frac{3}{2}T$. The integrators in the receiver are reset at $t = \frac{1}{2}T$ and start integrating the voltages delivered from the multipliers during the interval $\frac{1}{2}T \leqq t < \frac{3}{2}T$.

For practical use the block diagram of Fig. 117 has to be augmented by a synchronization circuit. Furthermore, modems are required to transform the coefficients $a_\chi(j)$ delivered to the transmitter into the proper form and also to transform the coefficients obtained at the receiver at the time $\frac{1}{2}T$ into the desired form.

Several generators for time-variable Walsh functions have been discussed in section 2.1.1. Figure 118 shows a function generator for generation of phase stable sine and cosine oscillations for the pulses of Fig. 1 and 9. The binary counters B 1 and B 2 produce Rademacher functions, from which the filters extract the fundamental sinusoidal functions. The first harmonic has three times the frequency of the fundamental oscillation. In practical applications it is better to leave out the filters and to produce a better approximation of the sine functions by a superposition of Rademacher functions.

There are three basic types of multipliers. The first multiplies two voltages that can assume two values only, say $+1$ V and -1 V. This type of multiplier is implemented by logic circuits and usually referred to as exclusive OR-gate or half-adder. The second type multiplies a voltage V_1 having arbitrary values with a voltage V_2 that can assume a few values

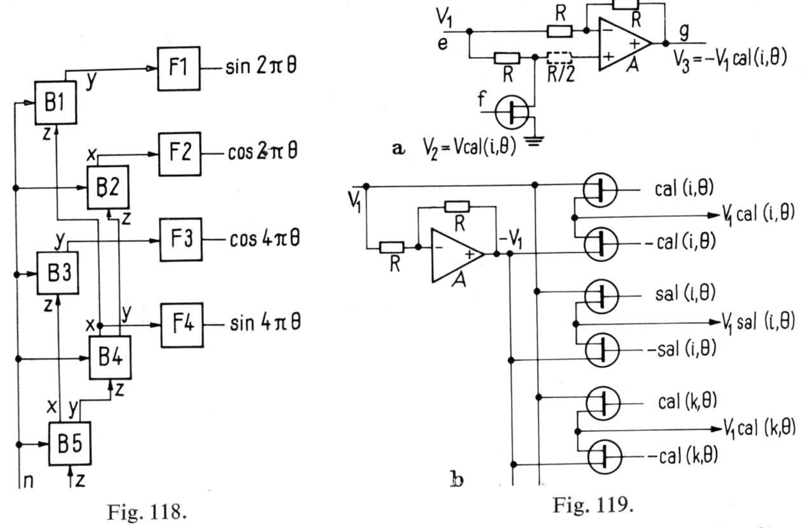

Fig. 118. Fig. 119.

Fig. 118. Generator for phase locked sine and cosine functions. B, binary counter; F, filter; z, input for trigger pulses. x and y are complementary outputs of the counters.

Fig. 119. Multipliers for the multiplication of an arbitrary function by Walsh functions. (a) single multiplication, (b) multiple multiplication (e.g., filter bank).

only. Figure 119a shows an example of this type. Voltage V_2 assumes the values $+1$ or -1 only. The output voltage equals either $+V_1$ or $-V_1$, where V_1 may have any value within the voltage range of the operational amplifier A. The circuit works as follows: The non-inverting input terminal (+) of the amplifier is grounded, if the field effect transistor FET is fully conducting. V_3 must equal $-V_1$ to bring the inverting input terminal (−) also to ground potential. Let FET be non-conducting. The non-inverting terminal is then at V_1 and the inverting terminal must also be at V_1. This requires V_3 to equal V_1. A variation of this multiplier is shown in Fig. 119b.

The third basic type of multiplier multiplies two arbitrary voltages. In principle, this type can be implemented by Hall effect multipliers, field effect transistors and logarithmic elements. These devices are usually unsatisfactory for practical applications due to low impedance, temperature drift, price, etc. Fairly suitable is the diode quad multiplier. Fig. 120 shows a representative circuit. It deviates from the usual one by not using transformers.

A major step in overcoming the perennial problem of analog multipliers was the recent introduction of encapsulated multipliers that can be used like operational amplifiers. One type uses the variable transconductance of semiconductor junctions. It yields errors of about 1% for operating frequencies up to about 1 MHz. Somewhat smaller errors and higher speeds, but at much higher costs, are obtained by resistor-diode networks that produce the square of an input voltage. These networks are used in place of the diodes in a bridge circuit like the one of Fig. 120.

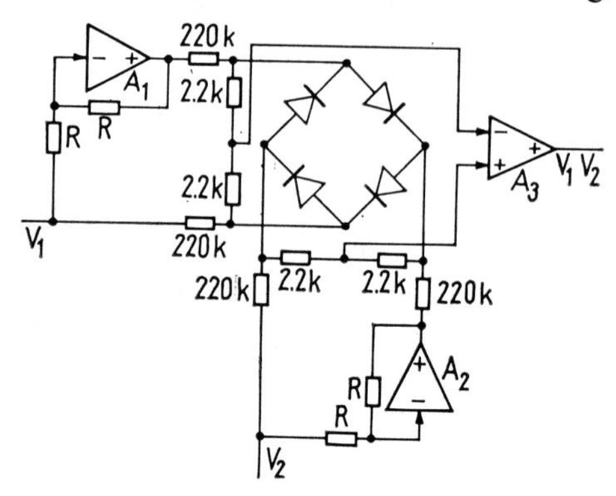

Fig. 120. Multiplier using diode quad.

The voltage V_2 in Fig. 119 assumes the values +1 or −1 only and may be considered to be represented by one binary digit. Four binary digits represent a voltage that can assume 16 values. A corresponding multiplier requires four field effect transistors rather than the one in Fig. 119a and a more complicated resistor network. Such a multiplier is due to P. Schmid. It yields excellent results, but the one voltage must be available in digital form.

Figure 121 shows an integrator. The capacitive feedback of the operational amplifier yields an output voltage that is proportionate to the integral

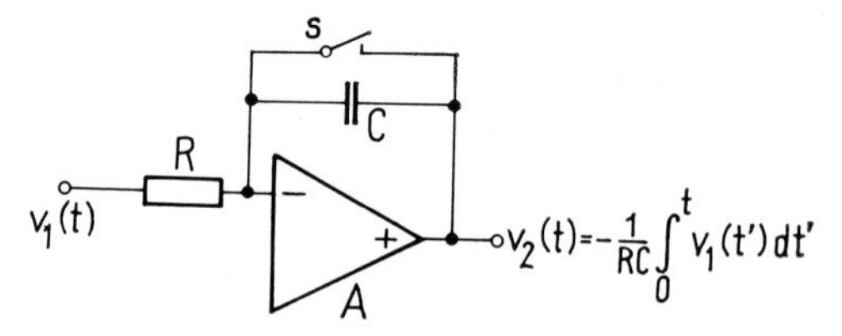

Fig. 121. Integrator.

of the input voltage with great accuracy. The switch s resets the integrator by discharging the capacitor. The practical implementation of this switch is usually by a field effect transistor.

Multiplier, integrator and function generator suffice in principle for the detection of any function. Superior circuits are available for special functions. Figure 122 shows a detector for sine and cosine pulses according to Fig. 1. This circuit makes use of the fact that sine and cosine pulses $\sin 2\pi\, i\, t/T$ and $\cos 2\pi\, i\, t/T$ are eigenfunctions of the following differential equation:

$$y'' + 4\pi^2 i^2 T^{-2} y = 0, \quad y(-\tfrac{1}{2}T) = y(\tfrac{1}{2}T), \quad y'(-\tfrac{1}{2}T) = y'(\tfrac{1}{2}T) \quad (1)$$

The output voltage $v_3(t)$ of amplifier A_1 of Fig. 122 is

$$v_3(t) = -(R_1 C_1)^{-1} \int v_1(t)\, dt - (R_2 C_1)^{-1} \int v_2(t)\, dt \quad (2)$$

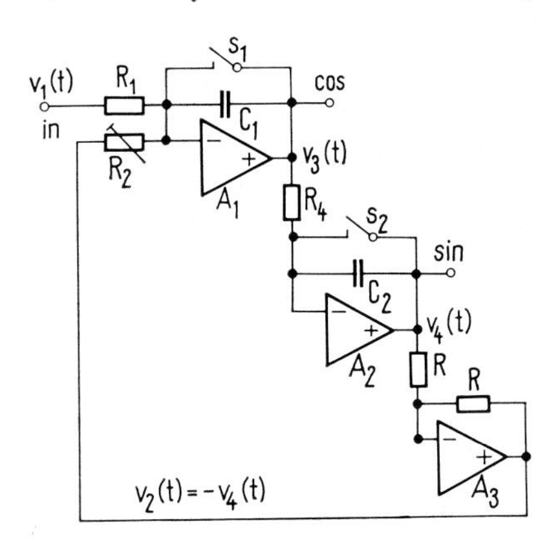

Fig. 122. Detector for sine and cosine pulses $\sin 2\pi\, i\, t/T$ and $\cos 2\pi\, i\, t/T$ according to Fig. 1. $R_2 C_1 = T/2\pi\, i$, $R_4 C_2 = T/2\pi\, i$, $R_1 = \pi i\, R_2$; s_1 and s_2 are closed at $t = \pm T/2$.

The output voltage $v_4(t) = -v_2(t)$ of A_2 equals:

$$\begin{aligned} v_4(t) &= -(R_4 C_2)^{-1} \int v_3(t)\, dt \\ &= (R_1 C_1 R_4 C_2)^{-1} \iint v_1(t')\, dt\, dt' + (R_2 C_1 R_4 C_2)^{-1} \iint v_2(t')\, dt\, dt' \\ &= -v_2(t) \end{aligned} \quad (3)$$

Differentiating twice and reordering the terms yields:

$$v_2''(t) + (R_2 C_1 R_4 C_2)^{-1} v_2(t) = -(R_1 C_1 R_4 C_2)^{-1} v_1(t) \quad (4)$$

Choosing $R_2 C_1 R_4 C_2 = (T/2\pi\, i)^2$ makes the left hand side of Eqs. (1) and (4) identical. The inhomogeneous term $v_1(t)$ is equal to $V_k \cos 2\pi\, k\, t/T$ or $V_k \sin 2\pi\, k\, t/T$ for $-\frac{1}{2}T \leqq t < \frac{1}{2}T$. The shape of $v_1(t)$ outside of this interval is of no interest, since the switches s_1 and s_2 are closed at $t = \pm\frac{1}{2}T$. $v_3(T/2)$ and $v_4(T/2)$ are zero for $i \neq k$ and $R_1 = \pi\, i\, R_2$; $V_i \sin 2\pi\, i\, t/T$ yields $v_3(T/2) = 0$ and $v_4(T/2) = (-1)^i V_i$. Figure 123 shows oscillograms of $v_3(t)$ and $v_4(t)$ for $i = k = 1$. Figure 124 shows oscillograms of $v_3(t)$ for $i = 128$ and input voltages $V_k \cos 2\pi\, k\, t/T$ with k equal to 128, 129 and 130; this means that the circuit is tuned for detection of a cosine pulse of 128 cycles and that cosine pulses with 128, 129 or 130 cycles are fed to its input.

The losses of the circuit of Fig. 122 are comparable to those of mechanical resonators. Q-factors of several thousand at a frequency of 100 Hz are readily obtained without use of regeneration. The frequency range for its application lies between fractions of 1 Hz and about 100 kHz. The lower limit is determined by leakage, the upper by the frequency response of operational amplifiers [6].

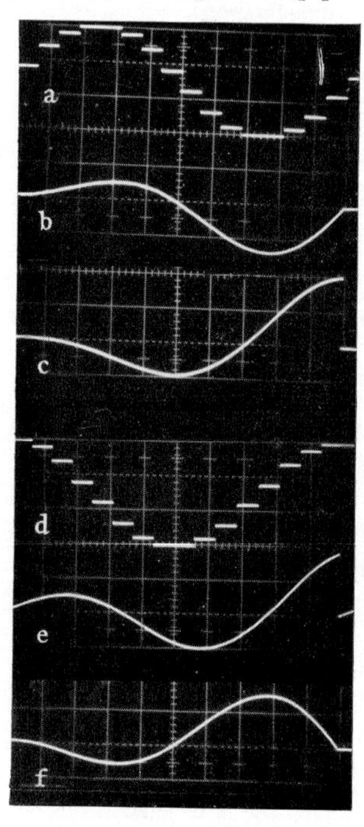

Fig. 123.

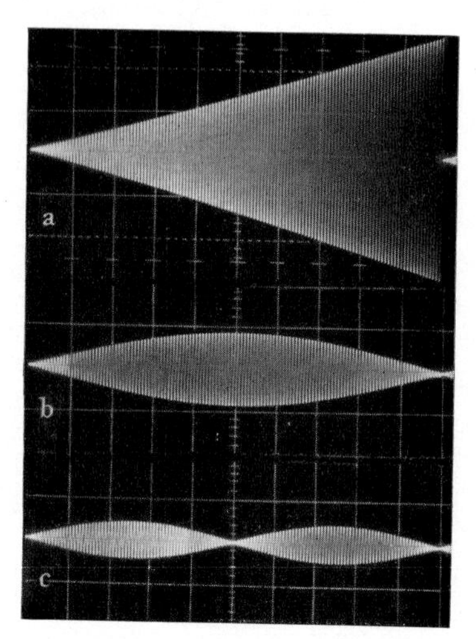

Fig. 124.

Fig. 123. Typical voltages of the circuit of Fig. 122. (*a*) input voltage $v_1(t) = V \sin 2\pi\, t/T$; (*b*) and (*c*): resulting voltages $v_3(t)$ and $v_4(t)$; (*d*): input voltage $v_1(t) = V \cos 2\pi\, t/T$; (*e*) and (*f*): resulting voltages $v_3(t)$ and $v_4(t)$. Horizontal scale: 15 ms/div. (Courtesy P. Schmid, R. Durisch and D. Nowak of Allen-Bradley Co.)

Fig. 124. Typical voltages of the circuit of Fig. 122. Circuit is tuned for the detection of sine and cosine pulses with $i = 128$ cycles. Output voltages $v_3(t)$ shown are caused by input voltages $v_1(t)$ with 128 cycles (*a*), 129 cycles (*b*) and 130 cycles (*c*). Duration of the traces is $T = 78$ ms. (Courtesy P. Schmid, R. Durisch and D. Nowak of Allen-Bradley Co.)

Problems

321.1 The amplifier in Fig. 121 saturates at ± 10 V. The peak input voltage is also ± 10 V. What value must RC have, if the integration interval has the duration T, to avoid saturation and make full use of the peak output voltage?

321.2 A field effect transistor used for the switch s in Fig. 121 will have a resistance of about 100 ohms when conducting and 100 megohms when not conducting. How does this influence the choice of R in the product RC?

321.3 A multiplier by P. Schmid was mentioned as an advancement of the circuit of Fig. 119a. Design one for 16 values $\pm 1/15$, $\pm 3/15, \ldots$ of V_2.

321.4 Generalize the circuit of Fig. 118 so that the functions $\sin 6\pi\,\theta$, $\cos 6\pi\,\theta$ and $\sin 8\pi\,\theta$, $\cos 8\pi\,\theta$ are generated too.

321.5 Replace amplifiers and resistors in Fig. 120 by transformers. What is the usual name of the resulting circuit?

321.6 Design a resistor-diode network that yields approximately the output voltage v^2 if the input voltage is $v \geqq 0$.

321.7 Repeat problem 321.6 for $v \leqq 0$ and combine the two circuits.

3.2.2 Transmission of Digital Signals by Sine and Cosine Pulses[1]

One block pulse of Fig. 3 can be transmitted per second and hertz through an idealized frequency low-pass filter. This is the limit for detection of the block pulses by amplitude sampling without correction of intersymbol interference. The same transmission rate holds for the "raised cosine pulses" in time domain. Some of them are shown in Fig. 125. Those

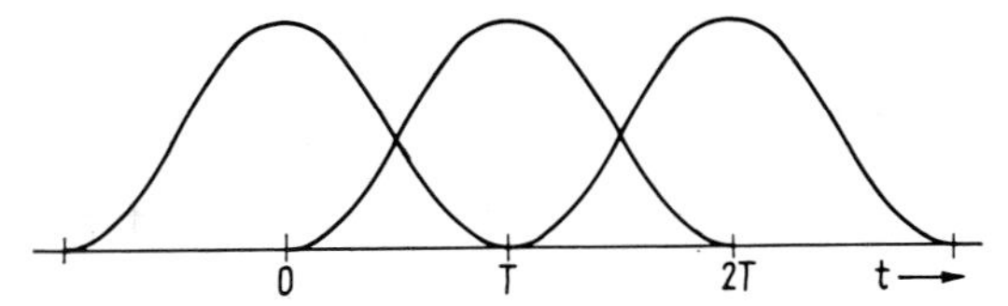

Fig. 125. Raised cosine pulses in time domain: $1 + \cos 2\pi\, t/T$, $1 + \cos 2\pi(t - T)/T$ and $1 + \cos 2\pi(t - 2T)/T$.

pulses are not orthogonal but linearly independent. They may be detected by amplitude sampling. Detection by cross-correlation with sample functions requires circuits to correct the intersymbol interference.

The pulses $[\sin\pi(\theta + j)]/\pi(\theta + j)$ permit one to transmit 2 pulses per second and hertz. However, these pulses cannot be used in practice. Arbitrarily large amplitudes can occur, if a sequence of such pulses is transmitted; any deviation from ideal synchronization may lead to arbitrarily large crosstalk between the pulses. There does not seem to be any way to transmit faster than at half the Nyquist rate if amplitude sampling is used, at least not without paying a power penalty.

Sine and cosine pulses of Figs. 1 or 9 permit transmission rates higher than one pulse per second and hertz [12, 13]. The Nyquist limit of two pulses per second and hertz can be approached arbitrarily close by using more and more complex equipment. This may be seen from Fig. 126 which shows three systems of functions. The first consists of a block pulse of duration T only. Its frequency power spectrum is shown on the right. The

[1] *See* [1—11] for more examples of transmission systems using orthogonal functions.

frequency band required for transmission shall be defined—somewhat arbitrary—as $0 \leqq f \leqq f_g = 1/T$. One block pulse can then be transmitted per second and hertz.

It is reasonable to identify the block pulse as function wal$(0, \theta)$ of Fig. 1 and to transmit a block pulse, a sine pulse and a cosine pulse of duration $3T$ instead of 3 block pulses of duration T each. The power

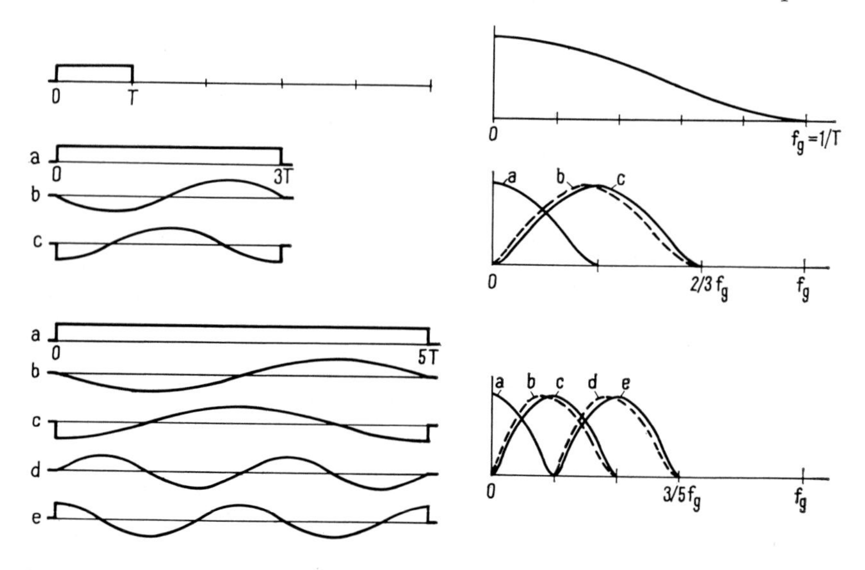

Fig. 126. Comparison of the frequency bandwidth required by various systems of functions.

spectra of the pulses are shown in Fig. 126. The bandwidth required for transmission is reduced to $0 \leqq f \leqq 2f_g/3 = 2/3T$. Hence, 1.5 pulses are transmitted per second and hertz.

Consider a further step. Instead of transmitting a series of 5 block pulses of duration T each, one may transmit simultaneously one block pulse, 2 sine pulses and 2 cosine pulses of duration $5T$. The required frequency band is reduced to $0 \leqq f \leqq 3f_g/5 = 3/5T$ according to Fig. 126. This means that 1.67 pulses are transmitted per second and hertz.

The required frequency band for the simultaneous transmission of one block pulse, i sine and i cosine pulses of duration $(2i + 1)T$ equals $0 \leqq f \leqq (i + 1)/(2i + 1)T$. The transmission rate equals $(2i + 1)/(i + 1)$ pulses per second and hertz. This rate approaches 2 for large values of i.

Table 7 shows values for the number $2i + 1$ of different pulse shapes and for the number $(2i + 1)/(i + 1)$ of pulses transmitted per second and hertz. One may see that the number of different pulse shapes and thus the complexity of the equipment increases rapidly as $(2i + 1)/(i + 1)$ approaches 2.

Table 7. *Number* $2i + 1$ *of different pulse shapes and number* $(2i + 1)/(i + 1)$ *of pulses transmitted per second and hertz for a transmission system using sine and cosine pulses.*

$2i + 1$	$\frac{2i+1}{i+1}$
1	1
3	1.5
5	1.67
7	1.75
9	1.8
11	1.83

Table 8 lists sine and cosine pulses, the frequencies of the periodic waves from which they are gated, and their use in a multichannel teletype system. Transmission is not "start-stop" but synchronous. Teletype signals arriving asynchronously must be fed through a buffer from which they are fed synchronously to the transmitter of Fig. 117. Start and stop

Table 8. *Utilization of a* 120 Hz *wide teletype channel.*
(Transmission rate is 6.67 characters per second; duration of a character is 150 ms. First column lists the pulse, second the frequency of the function from which it is gated, third the subchannel (su) and digit (di) for which the pulse is used. carr. stands for carrier synchronization, sync. for character synchronization.)

pulse	f [Hz]	su	di	pulse	f [Hz]	su	di
$\mathrm{wal}(0, \theta)$	0	carr.		$\sin 18\pi\,\theta$	60	3	5
				$\cos 18\pi\,\theta$	60	4	1
$\sin 2\pi\,\theta$	6.67	sync.		$\sin 20\pi\,\theta$	66.67	4	2
$\cos 2\pi\,\theta$	6.67			$\cos 20\pi\,\theta$	66.67	4	3
$\sin 4\pi\,\theta$	13.33	1	1	$\sin 22\pi\,\theta$	73.33	4	4
$\cos 4\pi\,\theta$	13.33	1	2	$\cos 22\pi\,\theta$	73.33	4	5
$\sin 6\pi\,\theta$	20	1	3	$\sin 24\pi\,\theta$	80	5	1
$\cos 6\pi\,\theta$	20	1	4	$\cos 24\pi\,\theta$	80	5	2
$\sin 8\pi\,\theta$	26.67	1	5	$\sin 26\pi\,\theta$	86.67	5	3
$\cos 8\pi\,\theta$	26.67	2	1	$\cos 26\pi\,\theta$	86.67	5	4
$\sin 10\pi\,\theta$	33.33	2	2	$\sin 28\pi\,\theta$	93.33	5	5
$\cos 10\pi\,\theta$	33.33	2	3	$\cos 28\pi\,\theta$	93.33	6	1
$\sin 12\pi\,\theta$	40	2	4	$\sin 30\pi\,\theta$	100	6	2
$\cos 12\pi\,\theta$	40	2	5	$\cos 30\pi\,\theta$	100	6	3
$\sin 14\pi\,\theta$	46.67	3	1	$\sin 32\pi\,\theta$	106.67	6	4
$\cos 14\pi\,\theta$	46.67	3	2	$\cos 32\pi\,\theta$	106.67	6	5
$\sin 16\pi\,\theta$	53.33	3	3	$\sin 34\pi\,\theta$	113.33		
$\cos 16\pi\,\theta$	53.33	3	4	$\cos 34\pi\,\theta$	113.33		

pulses must be added to the signals at the receiver, so that they may be fed into the usual teletype equipment. The duration of the teletype characters is assumed to be 150 ms. A periodic function $\sin 2\pi\, t/T$ with $T = 150$ ms and $1/T = 6.66$ Hz has a zero crossing with negative slope at beginning and end of the characters and is used as synchronization signal.

A teletype system according to Table 8 can operate some 100 to 200 teletype channels in a telephony channel, depending on the quality of the telephony channel. With such a large number, the question of power loading becomes important. Tests have shown that error rates of 10^{-7} and less can be obtained without exceeding the permissible power loading. This figure holds for transmission between two subscribers, a much more severe condition than transmission between two telephone exchanges. No coding or other error-reducing methods were used. For comparison, the widely used telex system accommodates 24 teletype channels in one telephony channel, but exceeds the permissible power loading by about a factor 3. Exceeding the power loading is quite usual for high speed data transmission systems. There is at least one system that requires the bandwidth of one telephony channel but the power load of eight channels to transmit 2400 bits/s. It should be pointed out that synchronous transmission is very sensitive to phase jumps which occur in switched telephone networks over long distances. Error rates may increase to 10^{-3} and more, depending on how fast lost synchronization can be reestablished.

One reason why sine and cosine pulses yield very reliable transmission is that telephony channels are designed for distortion free transmission of periodic sine and cosine functions. Sine and cosine pulses containing very many cycles come close to the periodic functions and suffer little delay or attenuation distortions. Another reason is that errors in telephone channels are mainly caused by pulse-type interference rather than thermal noise. It will be shown in chapter 6 that thermal noise affects all orthogonal pulse shapes equally. Pulse-type interference, however, affects block pulses more than others, particularly if amplitude sampling is used for detection.

Equipment using sine and cosine pulses has been developed under names like Kineplex [5], OSC, Rectiplex [11], etc. Except for Kineplex, none was ever used to any extent due to the high cost. A digital version of Rectiplex, appropriately named Digiplex[1], has recently been developed and is expected to reduce costs to about one third.

Problems

322.1 One five digit character is transmitted every 100 ms. The bandwidth of the channel is 180 Hz. How many teletype channels can be accommodated theoretically in this bandwidth if 100 ms long sine and cosine pulses are used for the digits and if synchronization is ignored?

322.2 Is the number of channels in problem 322.1 changed significantly if synchronization is not ignored?

322.3 How many different pulse shapes are used in problem 322.1 and how close does one approach the Nyquist limit of 2 pulses transmitted per second and hertz?

---

[1] Digiplex as well as Rectiplex were developed by Kokusai Denshin Denwa Co., which is an operating company for Japan's overseas communication, in cooperation with manufacturing companies.

3.2.3 Conversion of Sequency Limited Signals into Frequency Limited Signals

The problem of high cost of equipment using sine and cosine pulses as discussed in the previous section can be overcome by the substitution of Walsh pulses. For instance, the pulses $\sin 2\pi\, i\, \theta$ and $\cos 2\pi\, i\, \theta$ in Table 8 can be replaced[1] by $\mathrm{sal}(i, \theta)$ and $\mathrm{cal}(i, \theta)$. The resulting signals are step functions like $F(\theta)$ in Fig. 127. The existing communication channels were not designed for the transmission of such sequency limited signals. However, it is easy to convert sequency limited signals into frequency limited ones, which are better matched to existing channels.

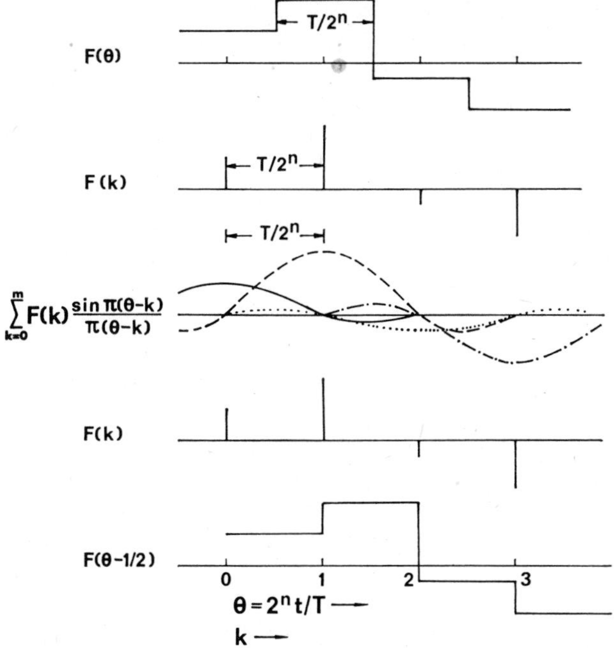

Fig. 127. Conversion of a sequency limited signal $F(\theta)$ into a frequency limited signal $\sum F(k)\,[\sin\pi(\theta-k)]/\pi(\theta-k)$ and reconversion into $F(\theta-\frac{1}{2})$.

Consider the set of $s = 2^n$ Walsh functions $\mathrm{wal}(j, \theta)$ with $j = 0 \ldots 2^n - 1$. Let them be multiplied by 2^n voltages $V(j)$. The sequency limited signal $F(\theta)$ is obtained:

$$F(\theta) = \sum_{j=0}^{s-1} V(j)\,\mathrm{wal}(j,\theta) = V(0)\,\mathrm{wal}(0,\theta) + \sum_{i=1}^{s/2-1} V(2i)\,\mathrm{cal}(i,\theta) + \\ + \sum_{i=1}^{s/2} V(2i-1)\,\mathrm{sal}(i,\theta) \qquad (1)$$

[1] For best results a slightly different allocation of channels is required for Walsh pulses. This is discussed in detail in chapter 4.

The sum has only a finite number of terms. There is no need to consider an infinite sum since this case cannot occur in communications. The highest normalized sequency of any Walsh function contained in this signal is $\frac{1}{2}s = 2^{n-1}$. Hence, the signal is limited to the nonnormalized sequency band $0 \leqq \phi \leqq 2^{n-1}/T$. $F(\theta)$ is a step function as shown in Fig. 127. The steps have the width $T/2^n$ or an integer multiple thereof. Sampling the amplitudes of these steps every $T/2^n$ seconds yields the sampled function $F(k)$. Passing $F(k)$ through an idealized frequency low-pass filter with a sharp cut-off at the frequency $s/2T = 2^{n-1}/T$ produces the sum of $(\sin x)/x$ functions shown. The function $F(k)$ can be regained by sampling this sum of $(\sin x)/x$ functions, and a hold circuit will produce the time-shifted original signal $F(\theta - \frac{1}{2})$.

The idealized frequency low-pass filter with sharp cut-off and the $(\sin x)/x$ functions have a number of undesirable properties. For instance, precise synchronization is required. A low-pass filter with a symmetric roll-off of the attenuation avoids these properties at the price of a larger bandwidth used [1, 2]. If the sample function $F(k)$ in Fig. 127 is fed to an analog/digital converter that produces a stream of block pulses, one has solved the problem of transmitting sequency-limited signals through a digital cable.

Problems

323.1 Do the samples $F(k)$ of $F(\theta)$ in Fig. 127 have to be taken in the middle of the steps? Is any synchronization between $F(\theta)$ and the sampling circuit required if the sampling time and the transient time between the steps are negligibly short compared with $T/2^n$?

323.2 Let a signal consist of a sum of s parabolic cylinder functions of Fig. 11. How many samples will in general determine the signal?

323.3 What is the frequency bandwidth required for the transmission of signals according to Eq. (3.1.1-2), regardless of the functions $f(j, \theta)$ chosen?

3.3 Characterization of Communication Channels

3.3.1 Frequency Response of Attenuation and Phase Shift of a Communication Channel

Communication channels are usually specified by the attenuation and phase shift of harmonic oscillations as function of their frequency. A voltage $V \cos\omega t$ is applied to the input and the steady state voltage at the output $V_c(\omega) \cos\omega [t - t_c(\omega)]$ is measured. The quantities $\log V/V_c(\omega) = a_c(\omega)$ and $\omega\, t_c(\omega) = b_c(\omega)$ are attenuation and phase shift as a function of the frequency ω. The parameter c may be omitted, if attenuation and phase shift of $V \cos\omega t$ and $V \sin\omega t$ are equal. Since it is well known that periodic sine and cosine functions transmit information at rate zero only, it is interesting to investigate why those functions are used for characterization of communication channels.

Let the communication channel be divided into the transmission line and the circuitry at its ends. The line is described by a partial differential equation or a partial difference-differential equation. The terminal circuitry is described by ordinary differential or difference-differential equations, if its dimensions are not too large. In particular, it will be described by a differential equation with constant coefficients, if the circuit components are such time invariant items as coils, capacitors and resistors. A sinusoidal voltage applied to the input of such a circuit appears in the steady state as an attenuated and phase shifted voltage at the output; the sinusoidal shape and the frequency are preserved. Hence, the circuit may be characterized by the frequency response of attenuation and phase shift. A characterization by other functions—for instance Walsh functions—is perfectly possible, but more complicated since the shape of these functions is changed.

Consider a transmission line described by the telegrapher's equation [1, 2].

$$\frac{\partial^2 w}{\partial x^2} - LC\frac{\partial^2 w}{\partial t^2} - (LA + RC)\frac{\partial w}{\partial t} - RAw = 0 \tag{1}$$

L, C, R, and A are the inductivity, capacity, resistivity and conductivity per unit length. The line is distortion-free if LA is equal RC. Its general solution in this case is as follows:

$$w(x, t) = e^{-at}[f(x - ct) + g(x + ct)] \tag{2}$$

$$a = (LA + RC)/2LC, \quad c = 1/\sqrt{(LC)}, \quad LA - RC = 0$$

$f(x - ct)$ and $g(x + ct)$ are arbitrary functions determined by the initial and boundary conditions. The only change suffered by these function during transmission is an attenuation and a delay. This feature also holds for non distortion-free lines, if they are "electrically short". According to K. W. Wagner [1] a line of length x is electrically short, if the following condition holds for x:

$$x < \frac{2}{R}\sqrt{\left(\frac{L}{C}\right)} = \frac{2Z}{R}, \quad Z = \sqrt{\left(\frac{L}{C}\right)} \tag{3}$$

As an example consider an open wire line. The two conductors are copper wires of 3 mm diameter at a distance of 18 cm. The following typical values apply:

$$L = 2.01 \times 10^{-3}\ \text{H/km}, \quad C = 5.9 \times 10^{-9}\ \text{F/km}$$

$$R = 4.95\ \Omega/\text{km}, \quad A = 0$$

One obtains $Z = 540\,\Omega$ and $2Z/R = 225$ km. This line is like a distortion-free line for distances smaller than 225 km. Inserting regenerative amplifiers at shorter distances, one may transmit signals distortion-free over any distance.

As a further example, consider a telephone cable between exchange and subscriber. The conductors are paper-insulated copper wires of 0.8 mm diameter. The following typical values apply:

$$L = 7 \times 10^{-4}\ \text{H/km}, \qquad C = 3.3 \times 10^{-8}\ \text{F/km}$$

$$R = 70\,\Omega/\text{km}, \qquad Z = 145\,\Omega, \qquad 2Z/R = 4\ \text{km}$$

This line will be electrically short for some subscribers but not for all.

The usual coaxial cables have a wave impedance Z between 50 and $100\,\Omega$. Let the resistivity be $10\,\Omega$/km. $2Z/R$ will then be between 10 and 20 km. This is the order of magnitude of the usual distances between amplifiers. However, one must keep in mind that the telegrapher's equation does not allow for the skin effect. An investigation of the skin effect for functions other than sine and cosine seems to be lacking[1].

Despite these results, sine and cosine functions do play a distinguished role in the theory of transmission lines. One important reason is Bernoulli's method for the solution of partial differential equation with time invariant coefficients. Assume that $w(x, t)$ in Eq. (1) may be represented as the product of a space variable $u(x)$ and a time variable $v(t)$:

$$w(x, t) = u(x)\, v(t) \tag{4}$$

Substitution of $u(x)\, v(t)$ into Eq. (1) yields two ordinary differential equations:

$$\frac{d^2 u}{dx^2} + (\lambda - RA)\, u = 0$$

$$LC \frac{d^2 v}{dt^2} + (LA + RC) \frac{dv}{dt} + v = 0 \tag{5}$$

Their eigenfunctions are $\cos\sqrt{(\lambda - RA)}\, x$, $\sin\sqrt{(\lambda - RA)}\, x$ and $e^{\gamma t}$, where γ is defined as follows:

$$\gamma = -\frac{LA + RC}{2LC} \pm \left[\frac{1}{4}\left(\frac{LA + RC}{LC}\right)^2 - \frac{1}{LC}\right]^{1/2} \tag{6}$$

Bernoulli's method is of great importance for finding solutions of the telegrapher's equation and of other partial differential equations that satisfy certain initial and boundary conditions. However, it is the method of solution that favors sine and cosine functions. Time variable lines would

[1] Superconductive cables are almost distortion-free and transmit switching transients in the nanosecond region [4, 5]. Such superconductive cables could have great practical potential, if organic compounds can be developed that are superconducting at room temperature, as some physicists believe to be possible.

in general not permit a separation of the solution into a time and a space dependent factor.

The propagation of electromagnetic radio waves is described by the wave equation. It is obtained for one-dimensional propagation as a special case of the telegrapher's equation with $R = A = 0$. Its general solution is:

$$w(x, t) = f(x - c\,t) + g(x + c\,t) \tag{7}$$

This solution has the same form as Eq. (2), except that the attenuation term e^{-at} is missing. Hence, a radio link behaves like a distortion-free line. Radio waves do not have to be sine waves or be described by sine functions. Different transmitters do not have to operate in different frequency bands; they may instead operate in different sequency bands. There are excellent practical reasons for allocating radio channels according to frequency, but these reasons are mainly the simplicity of implementing receivers and transmitters rather than laws of nature. It will be shown later on that mobile radio communication is indeed theoretically possible with Walsh waves.

Let us consider characterization of communication channels from another angle. It is reasonable to describe the features of a channel by functions that are distinguished by the transmitted signals. There have been so many pulse shapes proposed and used for digital transmission that it would be hard to claim a particular one as the only useful one for characterizing a channel. This does not hold for telephony signals. It is general practice to regard telephony signals as a superposition of sine and cosine functions. Hence, these functions appear preeminent for the characterization of telephony channels. The difficulty is that there is no overwhelming reason why telephony signals should be regarded as superpositions of sine and cosine functions. Consider voice signals represented by the output voltage of a microphone. A long sustained vowel will produce a voltage consisting with good approximation of a sum of a few sinusoidal oscillations. The system producing the vowel is time-invariant and is activated by the vocal cords with a sine function. Such a system is described by a partial differential equation with time independent coefficients and a sinusoidal excitation function. This is not so for voiceless consonants, particularly sounds like *p*, *t* or *k*. The system producing the sounds is described by a partial differential equation with time-variable coefficients and there is no sinusoidal excitation function. There is no particular reason why one should consider such sounds to consist of a superposition of sine and cosine functions and not of functions of some other complete system of orthogonal functions.

Experimental work by Klein, Boesswetter, Tasto, Lüke, Maile, Robinson and others has shown that voice signals may indeed be considered to be a superposition of Walsh functions. Lüke and Maile have built a telephony multiplex system using filters that permit Walsh functions up to a sequency of 4000 zps to pass through rather than sine-cosine functions up to a frequency of 4000 Hz. There is no discernible difference of performance. Boesswetter [3] has built an analyzer and a synthesizer for a vocoder using 16 filters that filter according to the sequency of Walsh functions rather than according to the frequency of sine functions. Klein has shown for a few examples, that voice decomposed by Walsh functions contains "sequency formants" just as voice decomposed by sine-cosine functions contains frequency formants; these investigations are continuing. A theoretical argument explaining these results is given in section 6.1.1. Sandy has used Walsh functions in theoretical work on speech analysis as early as 1962 [6]. Robinson has shown that a digital vocoder can be implemented efficiently on the basis of Walsh functions, a result that was also predicted by Wien [7–10].

Problems

331.1 Why is a superconductive cable almost distortion-free?

331.2 The telegrapher's equation ignores the skin effect which is usually stated as frequency variation of the resistance. Is it necessary or only convenient to describe the skin effect in terms of frequency?

331.3 What causes most of the distortions in a modern communication system, the coaxial cables and microwave links or the filters in the terminals?

331.4 A differential equation with two variables x and y, in the combination $x^2 + y^2$ only, is usually not separated by Bernoulli's product method but the substitution $r^2 = x^2 + y^2$ is made. Which functions and which transform take the place of sine-cosine functions and the Fourier transform? What are the eigenfunctions of the circular diaphragm of a telephone headset or a loudspeaker?

3.3.2 Characterization of a Communication Channel by Crosstalk Parameters

Having shown that communication channels do not have to be characterized by sine-cosine functions raises the question, how else they may be characterized. A consistent theory of communication based on orthogonal functions requires a method of characterization that will apply to all or at least many systems of orthogonal functions. As a side effect this more general characterization will simplify the discussion of channel capacity in section 7.1.

Consider a complete system of orthogonal functions $\{f(j,\theta)\}$. Let them be divided into even functions $f_c(i,\theta)$, odd functions $f_s(i,\theta)$ and the constant $f(0,\theta)$. The voltage $Vf_c(i,\theta)$ instead of $V_c \cos\omega t$ is applied to the input of a channel. For the time being, let the channel be such that the steady state voltage $V_c(i) f_c[i, \theta - \theta_c(i)]$ is obtained at the output. This will hold for a large class of systems of functions $\{f(j,\theta)\}$ in the case of a distortion-free transmission line. $V_c(i)/V = K_c(i)$ or $\lg V/V_c(i) = a_c(i)$ is the generalized attenuation of the communication channel.

$\theta_c(i) = b_c(i)$ is the generalized delay, since the term phase shift is applicable to sine and cosine functions only.

An input voltage $V f_s(i, \theta)$ produces the output voltage $V_s(i) f_s[i, \theta - \theta_s(i)]$. Attenuation and delay are defined by $V_s(i)/V = K_s(i)$, $\log V/V_s(i) = a_s(i)$ and $\theta_s(i) = b_s(i)$. The constant $V f(0, \theta)$ yields $V(0) f[0, \theta - \theta(0)]$, $V(0)/V = K(0)$, $\log V/V(0) = a(0)$ and $\theta(0) = b(0)$.

Let the functions of the system $\{f(0, \theta), f_c(i, \theta), f_s(i, \theta)\}$ be stretched by the substitution $i \to i/\xi = \mu$. Let i and ξ increase beyond all bounds. The system $\{f_c(\mu, \theta), f_s(\mu, \theta)\}$ is obtained according to section 1.2.2. $K_c(i), K_s(i), a_c(i), a_s(i), b_c(i)$ and $b_s(i)$ become $K_c(\mu), \ldots, b_s(\mu)$. In particular, one obtains for the special functions $f_c(\mu, \theta) = \sqrt{(2)} \cos 2\pi \mu \theta = \sqrt{(2)} \cos \omega t$ and $f_s(\mu, \theta) = \sqrt{(2)} \sin 2\pi \mu \theta = \sqrt{(2)} \sin \omega t$ the frequency functions $K_c(\omega), \ldots, b_s(\omega)$. The indices c and s may be omitted, if sine and cosine functions of the same frequency are equally attenuated and delayed; the functions $K(\omega)$, $a(\omega)$ and $b(\omega)$ are obtained. Hence, the characterization of communication channels by the frequency response of attenuation and phase shift is included as a special case.

In general, the functions $V f_c(i, \theta)$ are not only attenuated and delayed but distorted. The new output function $V g_c(i, \theta)$ is obtained instead of $V_c(i) f_c[i, \theta - \theta_c(i)]$. Let $g_c(i, \theta)$ be expanded into a series of the system $\{f[0, \theta - \theta_c(i)], f_c[k, \theta - \theta_c(i)], f_s[k, \theta - \theta_c(i)]\}$. The values of the delay $\theta_c(i)$ will be defined later on. The variable is now k, while i is a constant $(k = 1, 2, \ldots, i, \ldots)$:

$$\left.\begin{aligned}
g_c(i, \theta) &= K(c\,i, 0) f[0, \theta - \theta_c(i)] + \\
&+ \sum_{k=1}^{\infty} \{K(c\,i, c\,k) f_c[k, \theta - \theta_c(i)] + K(c\,i, s\,k) f_s[k, \theta - \theta_c(i)]\} \\
K(c\,i, 0) &= \int_{-\infty}^{\infty} g_c(i, \theta) f[0, \theta - \theta_c(i)]\, d\theta \\
K(c\,i, c\,k) &= \int_{-\infty}^{\infty} g_c(i, \theta) f_c[k, \theta - \theta_c(i)]\, d\theta \\
K(c\,i, s\,k) &= \int_{-\infty}^{\infty} g_c(i, \theta) f_s[k, \theta - \theta_c(i)]\, d\theta
\end{aligned}\right\} \quad (1)$$

Consider the integral $K(c\,i, c\,k)$ for $i = k$. Its value depends on $\theta_c(i)$. Let $\theta_c(i)$ be chosen so that $K(c\,i, c\,i)$ assumes its absolute maximum. The generalized delay $\theta_c(i) = b_c(i)$ and $\theta_s(i) = b_s(i)$ is then defined so that it approaches the value for the distortion-free line with decreasing distortions.

The coefficients $K(c\,i, 0)$, $K(c\,i, c\,k)$ and $K(c\,i, s\,k)$ are generalizations of the attenuation $K_c(i)$ for a distorting communication channel. $K_c(i)$ has the one variable i and may be represented by a vector. $K(c\,i, 0)$, $K(c\,i, c\,k)$ and $K(c\,i, s\,k)$ have the two variables i and k, and may be represented by a matrix $\mathbf{K}(c\,i)$:

$$\mathbf{K}(c\,i) = \begin{bmatrix} K(c\,1,0)\,K(c\,1,c\,1)\,K(c\,1,s\,1)\,K(c\,1,c\,2)\,K(c\,1,s\,2)\ldots \\ K(c\,2,0)\,K(c\,2,c\,1)\,K(c\,2,s\,1)\,K(c\,2,c\,2)\,K(c\,2,s\,2)\ldots \\ K(c\,3,0)\,K(c\,3,c\,1)\,K(c\,3,s\,1)\,K(c\,3,c\,2)\,K(c\,3,s\,2)\ldots \\ \cdot \\ \cdot \end{bmatrix} \quad (2)$$

The output voltages $V g_s(i, \theta)$ are obtained, if $V f_s(i, \theta)$ instead of $V f_c(i, \theta)$ is applied to the input. Coefficients $K(s\,i, 0)$, $K(s\,i, c\,k)$ and $K(s\,i, s\,k)$ are obtained in analogy to Eq. (1). The matrix $\mathbf{K}(s\,i)$ has the form of the matrix (2), but $c\,i$ is replaced by $s\,i$.

Transmission of $V f(0, \theta)$ yields $V g(0, \theta)$ and the coefficients $K(0, 0)$, $K(0, c\,k)$ and $K(0, s\,k)$ which may be written as line matrix:

$$\mathbf{K}(0) = K(0,0)\,K(0,c\,1)\,K(0,s\,1)\,K(0,c\,2)\,K(0,s\,2)\ldots$$

The three matrices $\mathbf{K}(c\,i)$, $\mathbf{K}(s\,i)$ and $\mathbf{K}(0)$ may be combined into one:

$$\mathbf{K} = \begin{bmatrix} K(0,0) & K(0,s\,1) & K(0,c\,1) & K(0,s\,2) & K(0,c\,2) & \ldots \\ K(s\,1,0) & K(s\,1,s\,1) & K(s\,1,c\,1) & K(s\,1,s\,2) & K(s\,1,c\,2) & \ldots \\ K(c\,1,0) & K(c\,1,s\,1) & K(c\,1,c\,1) & K(c\,1,s\,2) & K(c\,1,c\,2) & \ldots \\ K(s\,2,0) & K(s\,2,s\,1) & K(s\,2,c\,1) & K(s\,2,s\,2) & K(s\,2,c\,2) & \ldots \\ K(c\,2,0) & K(c\,2,s\,1) & K(c\,2,c\,1) & K(c\,2,s\,2) & K(c\,2,c\,2) & \ldots \\ \cdot \\ \cdot \end{bmatrix} \quad (3)$$

The terms outside of the main diagonal of $\mathbf{K}$ vanish, if the functions $f(j, \theta)$ are not distorted. The terms in the main diagonal become the one-dimensional set of attenuation coefficients $K(0) = K(0, 0)$, $K_c(i) = K(c\,i, c\,i)$ and $K_s(i) = K(s\,i, s\,i)$.

The delay times $\theta_c(i)$ of Eq. (1) and the corresponding delay times $\theta_s(i)$ and $\theta(0)$ for the transmission of the functions $V f_s(i, \theta)$ and $V f(0, \theta)$ may also be written as matrix:

$$\boldsymbol{\theta} = \begin{bmatrix} \theta(0) & 0 & 0 & 0 & \ldots \\ 0 & \theta_c(1) & 0 & 0 & \ldots \\ 0 & 0 & \theta_s(1) & 0 & \ldots \\ 0 & 0 & 0 & \theta_c(2) & \ldots \\ \cdot & \cdot & \cdot & \cdot & \ldots \\ \cdot & \cdot & \cdot & \cdot & \ldots \end{bmatrix} \quad (4)$$

The two matrices $\mathbf{K}$ and $\boldsymbol{\theta}$ characterize the communication channel for the system of functions $\{f(j, \theta)\}$.

Distortions in a channel cause crosstalk in multiplex transmission. One application of the matrix $\mathbf{K}$ is for the correction of this crosstalk. Hence, *cross-talk matrix* is an appropriate term for $\mathbf{K}$, while $\boldsymbol{\theta}$ may be called the delay matrix.

4. CARRIER TRANSMISSION OF SIGNALS

4.1 Amplitude Modulation (AM)

4.1.1 Modulation and Synchronous Demodulation

The transmission of constants $a_\chi(j)$ by a system of orthogonal functions $\{f(j, \theta)\}$ has been discussed in the previous chapter. The transmission of time functions $F(\theta)$ or $F_\chi(\theta)$ by means of a system of time functions $\{\Phi(k, \theta)\}$ will be discussed now. These functions $\Phi(k, \theta)$ will be called carriers. $F(\theta)$ denotes any time function, e.g., the output voltage of a microphone. The notation $F_\chi(\theta)$ is used to emphasize time functions that contain a finite number of coefficients $a_\chi(j)$, such as teletype signals:

$$F_\chi(\theta) = \sum_{j=0}^{m-1} a_\chi(j) f(j, \theta)$$

The carriers $\Phi(k, \theta)$ are predominantly sine and cosine functions at the present. There is, however, neither a mathematical nor a physical reason why other functions could not be used. This holds for transmission via wire lines, wave guides, radio links, etc. Periodic trains of block pulses are used to some extent as carriers in cables.

Functions that form a group with respect to multiplication are particularly well suited as carriers from the mathematical point of view. Amplitude modulation of such functions yields inherently a single sideband modulation. Multiplex systems using such carriers do not need single sideband filters. The term single sideband modulation is used here with a more general meaning than usual. The exact meaning of this and other terms used in a generalized sense is best explained by an example.

Consider amplitude modulation of a cosine function by a signal $F(\theta)$. Let $F(\theta)$ be expanded into a Fourier *series* in the interval $-\frac{1}{2} \leqq \theta < \frac{1}{2}$:

$$F(\theta) = a(0) + \sqrt{(2)} \sum_{i=1}^{\infty} [a_c(i) \cos 2i\pi\theta + a_s(i) \sin 2i\pi\theta] \qquad (1)$$

Let $F(\theta)$ pass through a low-pass filter that suppresses all terms of the sum with index $i > k$. Such filters can be implemented very much like the sequency filters in section 2.1.2, but implementation is of no importance here. The filtered signal $F\dagger(\theta)$ has $i = k$ rather than $i = \infty$ as upper limit of the sum of Eq. (1). Amplitude modulation of the carrier $\sqrt{(2)}\cos\Omega_0\,\theta$ by $F\dagger(\theta)$ yields:

$$F\dagger(\theta)\sqrt{(2)}\cos\Omega_0\,\theta = \sum_{i=1}^{k}[a_c(i)\cos(\Omega_0 - 2\pi\,i)\,\theta - a_s(i)\sin(\Omega_0 - 2\pi\,i)\,\theta]$$

$$+ a(0)\sqrt{(2)}\cos\Omega_0\,\theta + \sum_{i=1}^{k}[a_c(i)\cos(\Omega_0 + 2\pi\,i)\,\theta + a_s(i)\sin(\Omega_0 + 2\pi\,i)\theta] \tag{2}$$

The first sum represents the lower sideband. It follows the term with the frequency Ω_0 of the carrier, which is produced by the dc component of $F\dagger(\theta)$. The second sum represents the upper sideband.

Let $F(\theta)$ be expanded into a Walsh series:

$$F(\theta) = a(0) + \sum_{i=1}^{\infty}[a_c(i)\,\mathrm{cal}(i,\theta) + a_s(i)\,\mathrm{sal}(i,\theta)]$$

$$= a(0) + \sum_{i=1}^{\infty}[a_c(i)\,\mathrm{wal}(2i,\theta) + a_s(i)\,\mathrm{wal}(2i-1,\theta)]$$
$$-\tfrac{1}{2} \leqq \theta < \tfrac{1}{2} \tag{3}$$

Let $F(\theta)$ pass through a sequency low-pass filter that suppresses all terms with index $i > k$. The filtered signal $F\dagger\dagger(\theta)$ has $i = k$ as upper limit of the sum of Eq. (3). The series expansion of $F\dagger(\theta)$ and $F\dagger\dagger(\theta)$ have thus the same number of terms. Amplitude modulation of a Walsh carrier $\mathrm{wal}(j,\theta)$ by $F\dagger\dagger(\theta)$ yields:

$$F\dagger\dagger(\theta)\,\mathrm{wal}(j,\theta) = a(0)\,\mathrm{wal}(j,\theta) + \sum_{i=1}^{k}\{a_c(i)\,\mathrm{wal}(2i\oplus j,\theta) +$$
$$+ a_s(i)\,\mathrm{wal}[(2i-1)\oplus j,\theta)]\} \tag{4}$$

Comparison of Eqs. (4) and (2) shows that the dc component $a(0)$ is transmitted in both cases by the unchanged carrier. There is, however, one sum only in Eq. (4). Depending on the value of j, this sum may describe an "upper", "lower", or "partly upper, partly lower sideband". Consider, for example, a number $2k$ having η digits in binary notation. Let j be larger than $2k$ and let j have zeros at the η lowest binary places. The following relations hold:

$$\left.\begin{array}{r} 2i\oplus j = j+2, j+4, \ldots, j+2k > j \\ (2i-1)\oplus j = j+1, j+3, \ldots, j+2k-1 > j \\ i = 1, 2, \ldots, 2k < j \end{array}\right\} \tag{5}$$

All indices $2i\oplus j$ and $(2i-1)\oplus j$ of the sum of Eq. (4) are larger than the index j of the carrier $\mathrm{wal}(j,\theta)$ for this choice of j. This corresponds to an upper sideband modulation.

As a further example, let j have ones at the η lowest binary places. One obtains in this case:

$$2i \oplus j = j - 2, j - 4, \ldots, j - 2k < j$$
$$(2i - 1) \oplus j = j - 1, j - 3, \ldots, j - 2k < j \quad (6)$$

Now the indices $2i \oplus j$ and $(2i - 1) \oplus j$ in the sum of Eq. (4) are all smaller than the index of the carrier $\mathrm{wal}(j, \theta)$. This corresponds to a lower sideband modulation.

The numbers $2i \oplus j$ and $(2i - 1) \oplus j$ will be for certain values of i larger than j and for other values smaller than j, if j has neither zeros only nor ones only on the η lowest binary places. This corresponds to a partly upper, partly lower sideband modulation.

Why does amplitude modulation of sine and cosine carriers yield two sidebands, but amplitude modulation of Walsh carriers only one sideband? For the answer consider the multiplication theorems of sine and cosine:

$$\left.\begin{aligned}
2\cos i\,\theta \cos k\,\theta &= +\cos(i - k)\,\theta + \cos(i + k)\,\theta \\
2\sin i\,\theta \cos k\,\theta &= +\sin(i - k)\,\theta + \sin(i + k)\,\theta \\
2\cos i\,\theta \sin k\,\theta &= -\sin(i - k)\,\theta + \sin(i + k)\,\theta \\
2\sin i\,\theta \sin k\,\theta &= +\cos(i - k)\,\theta - \cos(i + k)\,\theta
\end{aligned}\right\} \quad (7)$$

There is a sum of two sine or cosine functions on the right hand sides of these equations. Let $\cos k\,\theta$ or $\sin k\,\theta$ be carriers and $\cos i\,\theta$ or $\sin i\,\theta$ Fourier components of a signal that are amplitude modulated onto those carriers. An upper and a lower side-oscillation is produced. Hence, the double sideband modulation of sine and cosine carriers is a consequence of the multiplication theorems of Eq. (7).

Let us consider once more the multiplication theorems of Walsh functions:

$$\left.\begin{aligned}
\mathrm{cal}(i, \theta)\,\mathrm{cal}(k, \theta) &= \mathrm{cal}(r, \theta) & r &= i \oplus k \\
\mathrm{sal}(i, \theta)\,\mathrm{cal}(k, \theta) &= \mathrm{sal}(r, \theta) & r &= [k \oplus (i - 1)] + 1 \\
\mathrm{cal}(i, \theta)\,\mathrm{sal}(k, \theta) &= \mathrm{sal}(r, \theta) & r &= [i \oplus (k - 1)] + 1 \\
\mathrm{sal}(i, \theta)\,\mathrm{sal}(k, \theta) &= \mathrm{cal}(r, \theta) & r &= (i - 1) \oplus (k - 1)
\end{aligned}\right\} \quad (8)$$

There is only one Walsh function on the right hand side of Eqs. (8). Let $\mathrm{cal}(k, \theta)$ or $\mathrm{sal}(k, \theta)$ be carriers and $\mathrm{cal}(i, \theta)$ or $\mathrm{sal}(i, \theta)$ Walsh components of a signal that are amplitude modulated onto the carriers. There is not one upper and one lower "side-function" but one function only. This it the reason why amplitude modulation of Walsh functions yields a single sideband modulation.

A circuit for amplitude modulation of a Walsh carrier is shown in Fig. 128a.

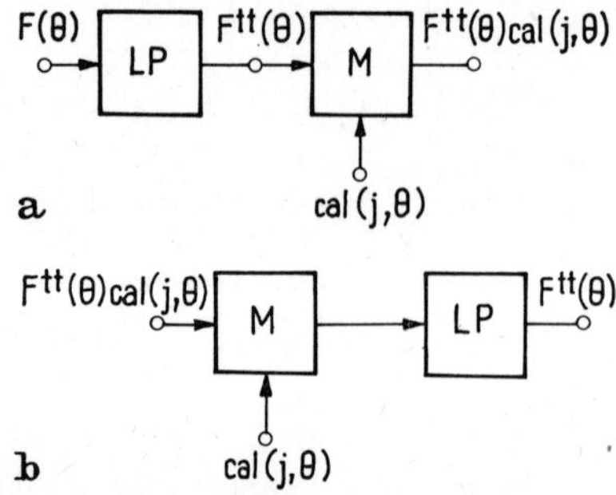

Fig. 128. Amplitude modulator (a) and demodulator (b) for Walsh carriers. LP, sequency low-pass filter; M, multiplier for Walsh functions.

The cosine carrier $\sqrt{(2)}\cos\Omega_0\,\theta$ modulated by $F\dagger(\theta)$ in Eq.(2) may be demodulated by multiplication with $\sqrt{(2)}\cos\Omega_0\,\theta$:

$$[F\dagger(\theta)\sqrt{(2)}\cos\Omega_0\,\theta]\sqrt{(2)}\cos\Omega_0\,\theta = F\dagger(\theta)\,(1+\cos 2\Omega_0\,\theta) \qquad (9)$$

The first term on the right hand side represents the demodulated signal. The second term must be suppressed by a filter. Half the power is lost on the average by this filtering. This power loss is unimportant, if the product $F\dagger(\theta)\sqrt{(2)}\cos\Omega_0\,\theta$ can be amplified before multiplication with $\sqrt{(2)}\cos\Omega_0\,\theta$.

Let a signal $D\dagger(\theta)$ be transmitted by a carrier $\sqrt{(2)}\cos\Omega_1\,\theta$. Synchronous demodulation by $\sqrt{(2)}\cos\Omega_0\,\theta$ yields:

$$[D\dagger(\theta)\sqrt{(2)}\cos\Omega_1\,\theta]\sqrt{(2)}\cos\Omega_0\,\theta = D\dagger(\theta)[\cos(\Omega_0-\Omega_1)\,\theta+\cos(\Omega_0+\Omega_1)\,\theta] \qquad (10)$$

Let the signals $F\dagger(\theta)$ and $D\dagger(\theta)$ contain oscillations with frequencies in the interval $0 \leqq \nu \leqq \nu_g$ only and let the demodulated signals pass through a frequency filter with cut-off frequency ν_g. $F\dagger(\theta)$ will be received without interference from $D\dagger(\theta)$, if the frequencies of the carriers $\sqrt{(2)}\cos\Omega_0\,\theta$ and $\sqrt{(2)}\cos\Omega_1\,\theta$ satisfy the following conditions:

$$|(\Omega_0-\Omega_1)/2\pi| - \nu \geqq \nu_g, \quad 0 \leqq \nu \leqq \nu_g \qquad (11)$$

Let the signal $F\dagger(\theta)\sqrt{(2)}\cos\Omega_0\,\theta$ be first multiplied by an auxiliary carrier $\sqrt{(2)}\cos\Omega_h\,\theta$ and let the product then be modulated by multiplication with $2\cos(\Omega_0-\Omega_h)\,\theta$:

$$\{[F\dagger(\theta)\sqrt{(2)}\cos\Omega_0\,\theta]\sqrt{(2)}\cos\Omega_h\,\theta\}\,2\cos(\Omega_0-\Omega_h)\,\theta$$
$$= F\dagger(\theta)\,[1+\cos 2(\Omega_0-\Omega_h)\,\theta+\cos 2\Omega_h\,\theta+\cos 2\Omega_0\,\theta] \qquad (12)$$

The desired term $F\dagger(\theta)$ is obtained. The three undesired terms on the right hand side must be filtered.

Let a signal $D\dagger(\theta)\sqrt{(2)}\cos(\Omega_0 - 2\Omega_h)\theta$ be received. Direct synchronous demodulation according to Eq. (9) yields:

$$[D\dagger(\theta)\sqrt{(2)}\cos(\Omega_0 - 2\Omega_h)\theta]\sqrt{(2)}\cos\Omega_0\theta = D\dagger(\theta)[\cos 2\Omega_h\theta + \cos 2(\Omega_0 - \Omega_h)\theta] \quad (13)$$

The signal of Eq. (13) may be filtered if the frequency bandwidth of $D\dagger(\theta)$ is sufficiently small. Hence, there is no interference between $F\dagger(\theta)$ and this image signal. This is not so, if the signal $D\dagger(\theta)\sqrt{(2)}\cos(\Omega_0 - 2\Omega_h)\theta$ is first multiplied by an auxiliary carrier $\sqrt{(2)}\cos\Omega_h\theta$ and then demodulated by multiplication with $2\cos(\Omega_0 - \Omega_h)\theta$:

$$\{[D\dagger(\theta)\sqrt{(2)}\cos(\Omega_0 - 2\Omega_h)\theta]\sqrt{(2)}\cos\Omega_h\theta\}2\cos(\Omega_0 - \Omega_h)\theta$$
$$= D\dagger(\theta)[\cos 2\Omega_h\theta + \cos 2(\Omega_0 - 2\Omega_h)\theta + 1 + \cos 2(\Omega_0 - \Omega_h)\theta] \quad (14)$$

The term $D\dagger(\theta)$ appears on the right hand side of Eq. (14). $F\dagger(\theta)$ is affected by the image signal $D\dagger(\theta)$. One may see from Eqs. (7) and (14) that the reception of image signals is a consequence of the multiplication theorems of sine and cosine. There would be no interference by image signals, if there were one term rather than two on the right hand side of Eq. (7).

Let us consider the same processes, if Walsh carriers are used instead of sine-cosine carriers. Let the signal $F\dagger\dagger(\theta)\,\mathrm{wal}(j,\theta)$ of Eq. (4) be multiplied by $\mathrm{wal}(j,\theta)$:

$$[F\dagger\dagger(\theta)\,\mathrm{wal}(j,\theta)]\,\mathrm{wal}(j,\theta) = F\dagger\dagger(\theta)\,\mathrm{wal}(0,\theta) = F\dagger\dagger(\theta)$$
$$j \oplus j = 0 \quad (15)$$

There is no high sequency term to be filtered, contrary to synchronous demodulation of sine-cosine carriers. However, this difference usually means very little, since filtering is required anyway in multichannel systems in order to separate signals from different channels. To show that, let a signal $D\dagger\dagger(\theta)$ be transmitted by a second carrier $\mathrm{wal}(l,\theta)$:

$$D\dagger\dagger(\theta) = b(\theta) + \sum_{i=1}^{k}[b_c(i)\,\mathrm{cal}(i,\theta) + b_s(i)\,\mathrm{sal}(i,\theta)]$$
$$= \sum_{r=0}^{2k} c(r)\,\mathrm{wal}(r,\theta) \quad (16)$$

Synchronous demodulation of $D\dagger\dagger(\theta)\,\mathrm{wal}(l,\theta)$ by $\mathrm{wal}(j,\theta)$ yields:

$$[D\dagger\dagger(\theta)\,\mathrm{wal}(l,\theta)]\,\mathrm{wal}(j,\theta) = D\dagger\dagger(\theta)\,\mathrm{wal}(l \oplus j,\theta)$$
$$= \sum_{r=0}^{2k} c(r)\,\mathrm{wal}(l \oplus j \oplus r,\theta) \quad (17)$$

Let $F\dagger\dagger(\theta)$ and $D\dagger\dagger(\theta)$ contain Walsh functions $\mathrm{wal}(0,\theta)$, $\mathrm{cal}(i,\theta)$ and $\mathrm{sal}(i,\theta)$ with $i \leqq k$ only, or Walsh functions $\mathrm{wal}(r,\theta)$ with $r \leqq 2k$ only. Let further the demodulated signals be filtered by a sequency low-pass filter with cut-off sequency $\mu = k$. No Walsh function $\mathrm{wal}(l \oplus j \oplus r,\theta)$

of Eq. (17) will pass through this filter, if the following condition is satisfied:

$$l \oplus j \oplus r \geqq 2k, \qquad r = 0, 1, \ldots, 2k \tag{18}$$

Only two of the many possible ways to satisfy Eq. (18) will be discussed. Let the number $2k$ have η binary digits. Let j and l be larger than $2k$ and let them have zeros only at the η lowest binary places. One obtains:

$$\begin{aligned} l \oplus r &= l, l+1, l+2, \ldots, l+2k \\ l \oplus j \oplus r &= l \oplus j, l \oplus j + 1, \ldots, l \oplus j + 2k \end{aligned} \tag{19}$$

The condition

$$l \oplus j > 2k \tag{20}$$

must be satisfied, in order for Eq. (18) to hold. Adding j modulo 2 on both sides of Eq. (20) yields:

$$l > 2k \oplus j = 2k + j \tag{21}$$

The last transformation uses the relation $j \oplus j = 0$ and the fact that $2k$ has η binary places only, while j has zeros at its η lowest binary places.

Adding l modulo 2 in Eq. (20) yields a second possibility to satisfy Eq. (18):

$$j > 2k \oplus l = 2k + l \tag{22}$$

Conditions like Eqs. (21) and (22) divide the sequency spectrum into sequency channels just as the frequency spectrum is divided into frequency channels by the requirement of certain frequency bandwidths for the channels. A more general method for allocating sequency channels for Walsh carriers based on group theory will be given later.

It is possible to multiply a signal $F^{\dagger\dagger}(\theta)\,\mathrm{wal}(j, \theta)$ first with an auxiliary carrier $\mathrm{wal}(h, \theta)$ and then demodulate it by multiplication with a carrier $\mathrm{wal}(j \oplus h, \theta)$:

$$\begin{aligned} &\{[F^{\dagger\dagger}(\theta)\,\mathrm{wal}(j, \theta)]\,\mathrm{wal}(h, \theta)\}\,\mathrm{wal}(j \oplus h, \theta) \\ &\qquad = [F^{\dagger\dagger}(\theta)\,\mathrm{wal}(j \oplus h, \theta)]\,\mathrm{wal}(j \oplus h, \theta) = F^{\dagger\dagger}(\theta) \end{aligned} \tag{23}$$

There is no interference by image signals, if Walsh carriers are used. To show this, let a received signal $D^{\dagger\dagger}(\theta)\,\mathrm{wal}(l, \theta)$ be multiplied first by $\mathrm{wal}(h, \theta)$ and then by $\mathrm{wal}(j \oplus h, \theta)$:

$$\begin{aligned} &\{[D^{\dagger\dagger}(\theta)\,\mathrm{wal}(l, \theta)]\,\mathrm{wal}(h, \theta)\}\,\mathrm{wal}(j \oplus h, \theta) \\ &\qquad = D^{\dagger\dagger}(\theta)\,\mathrm{wal}(l \oplus h \oplus j \oplus h, \theta) = D^{\dagger\dagger}(\theta)\,\mathrm{wal}(l \oplus j, \theta) \\ &\qquad = \sum_{r=0}^{2k} c(r)\,\mathrm{wal}(l \oplus j \oplus r, \theta) \end{aligned} \tag{24}$$

$D^{\dagger\dagger}(\theta)\,\mathrm{wal}(l \oplus j, \theta)$ contains no component that could pass through a sequency low-pass filter with cut-off sequency $\mu = k$, as long as the condition in Eq. (18) is satisfied. The absence of image signals can be traced to the occurrence of only one Walsh function on the right hand side of the multiplications theorems in Eq. (8).

Figure 128b shows a block diagram for the synchronous demodulation of Walsh carriers.

Amplitude modulation of functions of other complete, orthogonal systems may be discussed in very much the same way. No other systems have shown practical advantages so far, but this may well be due to an insufficient effort. Most of the better known functions have multiplication theorems that produce not one or two terms as in the case of Walsh or sine-cosine functions, but an infinite series of terms. Carriers of periodic block pulses form an orthogonal system but not a complete one. Their amplitude modulation differs strongly from that of sine-cosine or Walsh functions.

Problems

411.1 Substitute the system of Paley functions of section 1.1.5 for the Walsh functions and investigate modulation and demodulation.

411.2 Amplitude modulated sinusoidal carriers are often demodulated by rectification. Can this process be used for Walsh carriers?

411.3 It is sometimes claimed that amplitude modulation can only be performed by a nonlinear system. Is this correct if the definition of section 1.3.2 is used for a linear system?

4.1.2 Correction of Time Differences in Synchronous Demodulation

Consider a frequency band limited signal $F(\theta)\sqrt{(2)}\cos\Omega_0\theta$. It shall be synchronously demodulated by multiplication with a local carrier $\sqrt{(2)}\cos(\Omega_0\theta+\alpha)$ which has the phase difference α with reference to the received carrier $\sqrt{(2)}\cos\Omega_0\theta$ [1]:

$$F(\theta)\sqrt{(2)}\cos\Omega_0\theta\sqrt{(2)}\cos(\Omega_0\theta+\alpha)=F(\theta)\,[\cos\alpha+\cos(2\Omega_0\theta+\alpha)] \quad (1)$$

Let the signal be frequency-shifted by an auxiliary local carrier $\sqrt{(2)}\cos(\Omega_h\theta+\alpha_h)$ and then be demodulated synchronously by the local carrier $2\cos[(\Omega_0-\Omega_h)\theta+\alpha_s]$:

$$\{[F(\theta)\sqrt{(2)}\cos\Omega_0\theta]\sqrt{(2)}\cos(\Omega_h\theta+\alpha_h)\}\,2\cos[(\Omega_0-\Omega_h)\theta+\alpha_s]$$

$$=F(\theta)\{\cos\alpha+\cos[2(\Omega_0-\Omega_h)\theta-\alpha_0]+\cos(2\Omega_h\theta+\alpha_0)+\cos(2\Omega_0\theta+\alpha)\}$$

$$\alpha=\alpha_h+\alpha_s,\ \alpha_0=\alpha_h-\alpha_s \quad (2)$$

Equations (1) and (2) contain the desired signal $F(\theta)$ multiplied by $\cos\alpha$ and high frequency terms which can be suppressed by filters. There are a number of methods for the removal of $\cos\alpha$. One may derive, e.g., a sine oscillation $\sqrt{(2)}\sin(\Omega_0\theta+\alpha)$ from the local cosine carrier $\sqrt{(2)}\cos(\Omega_0\theta+\alpha)$. Multiplication of the received signal by this sine oscillation yields:

$$F(\theta)\sqrt{(2)}\cos\Omega_0\theta\sqrt{(2)}\sin(\Omega_0\theta+\alpha)=F(\theta)\,[\sin\alpha+\sin(2\Omega_0\theta+\alpha)] \quad (3)$$

Let us assume $F(\theta)$ may be written as sum $F(\theta) = 1 + M F^\dagger(\theta)$ where $F^\dagger(\theta)$ is a signal that contains practically no energy below a certain frequency and M is the modulation index. The right hand side of Eq. (3) assumes the form:

$$\sin\alpha + MF^\dagger(\theta)\sin\alpha + [1 + MF^\dagger(\theta)]\sin(2\Omega_0\theta + \alpha) \tag{4}$$

The second and third term can be suppressed by a frequency low-pass filter. The term $\sin\alpha$ remains. It may be used in a feedback loop to shift the local carrier $\sqrt{(2)}\cos(\Omega_0\theta + \alpha)$ and thus $\sqrt{(2)}\sin(\Omega_0\theta + \alpha)$ in such a way that $\sin\alpha$ vanishes. α then equals zero or an integer multiple of π and $\cos\alpha$ equals $+1$ or -1. Let the feedback loop be stable for $\alpha = 0$, $\pm 2\pi, \pm 4\pi, \ldots$ and unstable for $\alpha = \pm\pi, \pm 3\pi, \ldots$ The values $\cos\alpha = -1$ are then unstable. Figure 129 shows a block diagram of a receiver that corrects the phase difference in this way. α is assumed to be zero except in the feedback loop, where values holding for $\alpha \neq 0$ are shown. A very detailed treatment of synchronous demodulation of sinusoidal carriers is given by Viterbi [2].

Consider the correction of a time difference, if Walsh carriers are used. The signal $F^{\dagger\dagger}(\theta)\,\mathrm{wal}(j,\theta)$ of Eq. (4.1.1-15) shall be demodulated

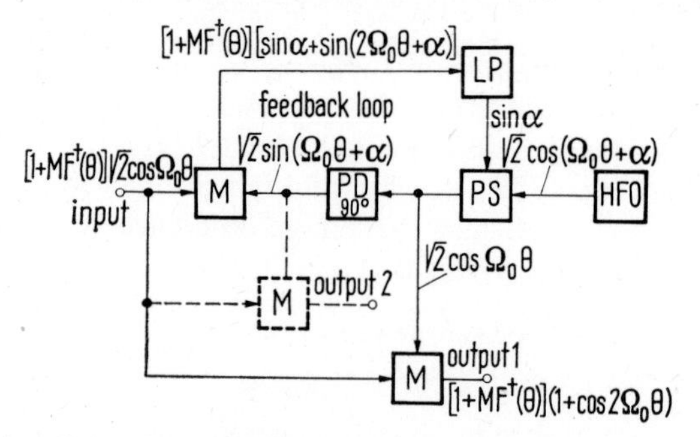

Fig. 129. Correction of the phase difference between received and local carrier $\sqrt{(2)}\cos\Omega_0\theta$ and $\sqrt{(2)}\cos(\Omega_0\theta + \alpha)$. M, multiplier; LP, frequency low-pass filter; HFO, high frequency oscillator; PS, variable phase shifter; PD, fixed phase shifter for 90°. α is put equal 0, except in the feedback loop.

by multiplication with the local carrier $\mathrm{wal}(j, \theta - \theta_v)$. The carriers $\mathrm{wal}(j,\theta)$ and $\mathrm{wal}(j, \theta - \theta_v)$ are periodic functions and do not vanish outside the interval $-\frac{1}{2} \leqq \theta < \frac{1}{2}$. The demodulated signal has the following form:

$$F^{\dagger\dagger}(\theta)\,\mathrm{wal}(j,\theta)\,\mathrm{wal}(j,\theta-\theta_v) \tag{5}$$

The product of $\mathrm{wal}(j,\theta)$ and $\mathrm{wal}(j,\theta)$ is known, but not that of $\mathrm{wal}(j,\theta)$ and $\mathrm{wal}(j,\theta-\theta_v)$. The problem is similar to that of multiplying $\sqrt{(2)}\cos\Omega_0\theta$ with $\sqrt{(2)}\cos(\Omega_0\theta+\alpha)$ in Eqs. (1). This multiplication cannot be performed with the multiplication theorems in Eq. (4.1.1-7) alone, one needs in addition the shift theorems of sine and cosine functions:

$$\cos(\alpha - \beta) = \cos\alpha \cos\beta + \sin\alpha \sin\beta \quad \text{etc.} \tag{6}$$

$\sqrt{(2)} \cos(\Omega_0 \theta + \alpha)$ must be decomposed by this shift theorem, and the multiplication theorems in Eq. (4.1.1-7) may then be applied. Multiplication and shift theorems are essentially the same for sine and cosine functions, since Eqs. (4.1.1-7) are multiplication theorems if read from left to right and shift theorems if read from right to left. Walsh functions have very simple binary shift theorems in Eq. (1.1.4-13),

$$\text{wal}(j, \theta \oplus \theta_v) = \text{wal}(j, \theta)\, \text{wal}(j, \theta_v),$$

but Eq. (5) contains the ordinary subtraction sign and not a modulo 2 addition or subtraction sign.

Certain special cases of the shift theorem of Walsh functions may be derived readily. Figure 2 shows that the periodically continued functions $\text{sal}(1, \theta)$ and $\text{cal}(1, \theta)$ are transformed into each other by a shift of $\pm\frac{1}{4}$ or $\pm\frac{1}{4}T$ in unnormalized notation; the shift equals $\pm\frac{1}{8}$ for $\text{sal}(2, \theta)$ and $\text{cal}(2, \theta)$, $\pm\frac{1}{4}$ for $\text{sal}(3, \theta)$ and $\text{cal}(3, \theta)$, etc. Let i be a power of 2; the following general formula holds:

$$\text{cal}(2^k, \theta - 2^{-k-2}) = \text{sal}(2^k, \theta); \quad k = 0, 1, 2, \ldots \tag{7}$$

Consider the more general case holding for any integer value of i:

$$\text{cal}(i, \theta + \theta_0) = \text{sal}(i, \theta) \tag{8}$$

Table 9 shows values of θ_0 for $i = 1, \ldots, 32$. θ_0 is determined by Eq. (7) for $i = 2^k$. These values are marked by a star in Table 9. One may see that $\theta_0 = \frac{1}{4}$ for $i = 3$ is equal $\theta_0 = -\frac{1}{4}$ for $i = 1$ with the sign reversed. θ_0 for $i = 3$ may thus be called the "image" of θ_0 for $i = 1$ with reference to line $i = 2 = 2^1$. One may readily see that θ_0 for $i = 5, 6, 7$ is the image of θ_0 for $i = 3, 2, 1$ with reference to line $i = 4 = 2^2$. This law of images may be written as follows:

$$\left.\begin{aligned} \text{cal}(2^k + j, \theta + \theta_0) &= \text{sal}(2^k + j, \theta) \\ \text{cal}(2^k - j, \theta - \theta_0) &= \text{sal}(2^k - j, \theta) \\ k = 1, 2, \ldots; \; j &= 1, 2, \ldots, 2^k - 1 \end{aligned}\right\} \tag{9}$$

Equations (7) and (9) are the special shift theorem of the Walsh functions. It corresponds to the relation $\sin x = \cos(x - \frac{1}{2}\pi)$ for sine and cosine functions.

The following relations hold for the sal functions instead of (7) and (9):

$$\text{sal}(2^k, \theta - 2^{-k-2}) = -\text{cal}(2^k, \theta) \tag{10}$$

$$\begin{aligned} \text{sal}(2^k + j, \theta + \theta_0) &= -\text{cal}(2^k + j, \theta) \\ \text{sal}(2^k - j, \theta + \theta_0) &= -\text{cal}(2^k - j, \theta) \\ k = 1, 2, \ldots; \; j &= 1, 2, \ldots, 2^k - 1 \end{aligned} \tag{11}$$

Equations (9) to (11) yield:

$$\begin{aligned} \operatorname{cal}(i, \theta + \theta_0) &= -\operatorname{cal}(i, \theta - \theta_0) \\ \operatorname{sal}(i, \theta + \theta_0) &= -\operatorname{sal}(i, \theta - \theta_0) \end{aligned} \tag{12}$$

or

$$\operatorname{cal}(i, \theta + \theta_1) = -\operatorname{cal}(i, \theta), \quad \operatorname{sal}(i, \theta + \theta_1) = -\operatorname{sal}(i, \theta) \tag{13}$$

$$\theta_1 = -2|\theta_0|$$

Values of θ_1 are shown in Table 9.

It would be cumbersome to obtain θ_0 and θ_1 for large values of i by an extension of Table 9. One can obtain θ_1 much faster by writing i as binary number. θ_1 equals $-\frac{1}{2}$, if the lowest binary digit is 1. An inspection of Table 9 readily shows that θ_1 is $-\frac{1}{2}$ for all odd values of i. θ_1 is $-\frac{1}{4}$, if the lowest binary digit is a 0 and the second lowest is a 1. Generally holds: θ_1 equals -2^{-k-1}, if the k lowest binary digits are zero.

Table 9. *Some values of θ_0 and θ_1 for the special shift theorem of the periodic Walsh functions* cal(i, θ) *and* sal(i, θ)

i				i			
dec.	binary	θ_0	θ_1	dec.	binary	θ_0	θ_1
1	000001	*−1/4	−1/2	17	010001	−1/4	−1/2
2	000010	*−1/8	−1/4	18	010010	−1/8	−1/4
3	000011	+1/4	−1/2	19	010011	+1/4	−1/2
4	000100	*−1/16	−1/8	20	010100	−1/16	−1/8
5	000101	−1/4	−1/2	21	010101	−1/4	−1/2
6	000110	+1/8	−1/4	22	010110	+1/8	−1/4
7	000111	+1/4	−1/2	23	010111	+1/4	−1/2
8	001000	*−1/32	−1/16	24	011000	+1/32	−1/16
9	001001	−1/4	−1/2	25	011001	−1/4	−1/2
10	001010	−1/8	−1/4	26	011010	−1/8	−1/4
11	001011	+1/4	−1/2	27	011011	+1/4	−1/2
12	001100	+1/16	−1/8	28	011100	+1/16	−1/8
13	001101	−1/4	−1/2	29	011101	−1/4	−1/2
14	001110	+1/8	−1/4	30	011110	+1/8	−1/4
15	001111	+1/4	−1/2	31	011111	+1/4	−1/2
16	010000	*−1/64	−1/32	32	100000	*−1/128	−1/64

The absolute value of θ_0 is derived in the same way from the binary representation of i. $|\theta_0|$ equals 2^{-k-2}, if the k lowest binary digits are zero. θ_0 equals $-|\theta_0|$, if the digit $k + 2$ is 0; θ_0 equals $+|\theta_0|$ if the digit $k + 2$ is 1. Consider as example the numbers $i = 20$ and $i = 28$ in Table 9. The two lowest binary digits ($k = 2$) are zero; this yields $|\theta_0| = 2^{-2-2} = 1/16$. The fourth binary digit ($k + 2 = 4$) is 0 for $i = 20$ and θ_0 equals $-1/16$; for $i = 28$ the fourth digit is 1 and θ_0 equals $+1/16$. A proof of the rules for determination of θ_0 and θ_1 was given by Pichler [3].

A circuit for the correction of a time difference between received carrier and local carrier may be based on the special shift theorem of Walsh functions (Fig. 130). Let us assume the signal $[1 + M F^{\dagger\dagger}(\theta)] \operatorname{cal}(i, \theta)$ is received. $F^{\dagger\dagger}(\theta)$ is a signal that has passed through a sequency low-pass filter. A local carrier $\mathrm{cal}(i, \theta - \theta_v)$ is produced in the function generator FG. The local carrier passes a variable delay circuit RV. The carrier $\mathrm{cal}(i, \theta)$ is obtained at the output of RV once the circuit is locked onto the received

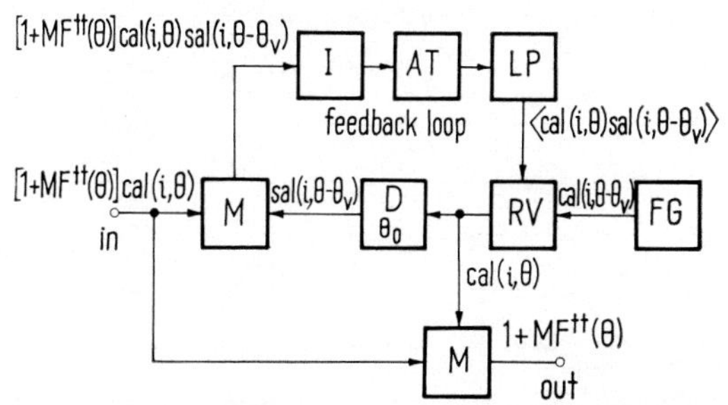

Fig. 130. Correction of a time difference between received and local carrier $\mathrm{cal}(i, \theta)$ and $\mathrm{cal}(i, \theta - \theta_v)$; $i =$ power of 2. M, multiplier; FG, function generator; RV, variable delay circuit; D, fixed delay by θ_0; I, integrator; AT, amplitude sampler; LP, averager. θ_v is put equal to 0, except in the feedback loop.

carrier. A further delay circuit with fixed delay produces the carrier $\mathrm{cal}(i, \theta - \theta_v + \theta_0) = \mathrm{sal}(i, \theta - \theta_v)$. The received signal is multiplied by $\mathrm{sal}(i, \theta - \theta_v)$ and the product is integrated during the orthogonality interval $-\frac{1}{2} + \theta_v \leqq \theta < \frac{1}{2} + \theta_v$ of $\mathrm{sal}(i, \theta - \theta_v)$. The output voltage of the integrator I is sampled at the times $\frac{1}{2} + \theta_v, \frac{3}{2} + \theta_v, \frac{5}{2} + \theta_v, \ldots$ by the sampler AT, and is fed to an averaging circuit TP. This circuit averages over many sampled amplitudes. The following average is obtained at the output of TP due to the fact that the integrator I averages over the intervals $-\frac{1}{2}k + \theta_v \leqq \theta < \frac{1}{2}k + \theta_v$:

$$\langle [1 + MF^{\dagger\dagger}(\theta)]\, \mathrm{cal}(i, \theta)\, \mathrm{sal}(i, \theta - \theta_v)\rangle \tag{14}$$

Let us assume that the average of the second term,

$$\langle MF^{\dagger\dagger}(\theta)\, \mathrm{cal}(i, \theta)\, \mathrm{sal}(i, \theta - \theta_v)\rangle, \tag{15}$$

increases more slowly with increasing averaging time than the average

$$\langle \mathrm{cal}(i, \theta)\, \mathrm{sal}(i, \theta - \theta_v)\rangle \tag{16}$$

of the first term. The term in Eq. (16) dominates then in the output voltage of the averager. It may be used to shift the local carrier $\mathrm{cal}(i, \theta - \theta_v)$ and thus $\mathrm{sal}(i, \theta - \theta_v)$, so that Eqs. (16) and (15) vanish. The values of θ_v for which Eqs. (16) and (15) vanish are obtained from the following integral:

$$\langle \mathrm{cal}(i, \theta)\, \mathrm{sal}(i, \theta - \theta_v)\rangle = \int_{-1/2}^{1/2} \mathrm{cal}(i, \theta)\, \mathrm{sal}(i, \theta - \theta_v)\, d\theta = F_{ci,si}(\theta_v) \tag{17}$$

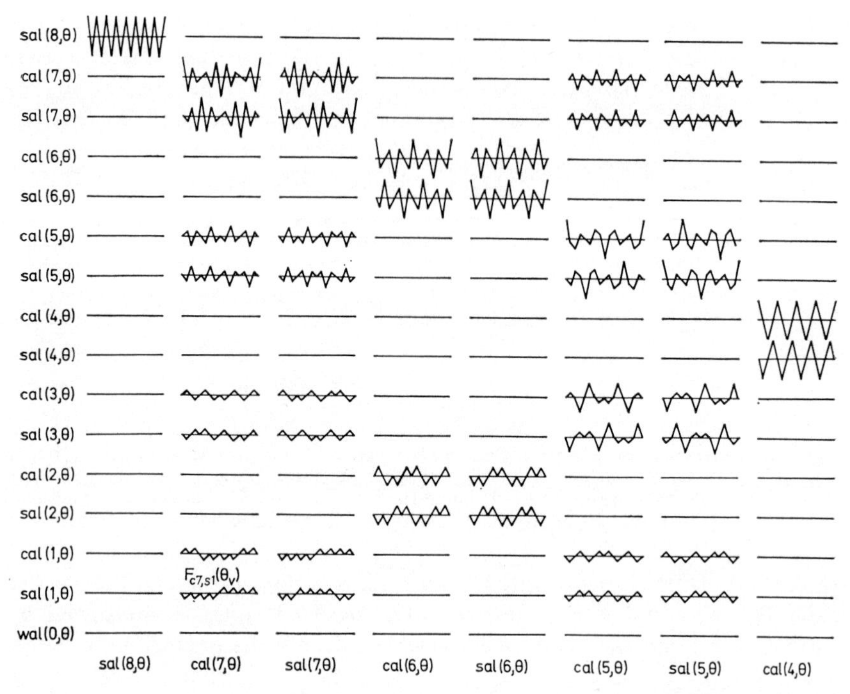

Fig. 131. Correlation functions

Figure 131 shows some functions $F_{ci,ci}(\theta_v)$ and $F_{si,si}(\theta_v)$ in the main diagonal. $F_{ci,si}(\theta_v)$ is shown just below the main diagonal and $F_{si,ci}(\theta_v)$ just above. The interval $0 \leqq \theta_v \leqq 1$ is shown; the functions have to be continued periodically outside this interval. $F_{ci,si}(\theta_v)$ is shown enlarged in Fig. 132. The dashed lines shall give some insight into their structure; a more detailed discussion of the correlation functions of Walsh functions would lead too deeply into abstract mathematics[1]. One may see, however, from Fig. 132 that $F_{ci,si}(\theta_v)$ vanishes for $i = 1, 2, 4, 8, \ldots, 2^k$, if θ_v equals zero or an integer multiple of $\pm 2\theta_0 = \pm 1/2i$. This result may also be obtained from Eqs. (7) and (13). The feedback loop in Fig. 130 may be made stable for $\theta_v = 0, \pm 2/2i, \pm 4/2i, \ldots$ and unstable for $\theta_v = \pm 1/2i, \pm 3/2i, \ldots$.

Consider the Walsh functions of Fig. 2 continued periodically to the left and right. A shift of $\mathrm{sal}(i, \theta)$, $i = 2^k$, by $\theta_v = 0, \pm 2/2i, \pm 4/2i, \ldots$ yields again the periodic function $\mathrm{sal}(i, \theta)$. Things are more complicated if i is not a power of 2. $F_{ci,si}(\theta_v)$ vanishes for certain values $\theta_v = \theta_v'$, but $\mathrm{sal}(i, \theta - \theta_v')$ is in general not identical with $\mathrm{sal}(i, \theta)$. Methods to

---

[1] An extensive investigation of the correlation functions is contained in the thesis of Wetscher of the *Innsbrucker Kreis* [4].

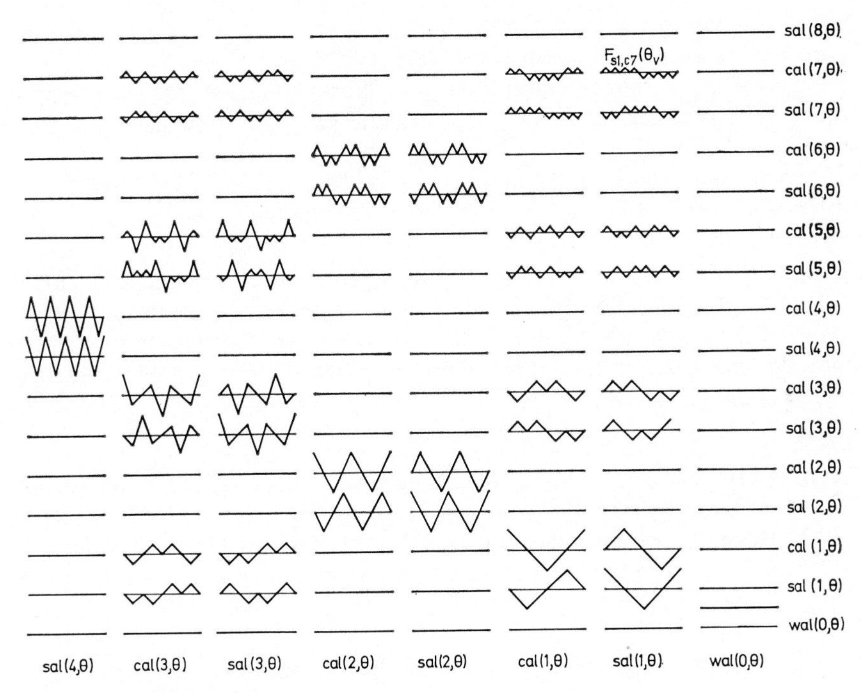

for periodic Walsh functions.

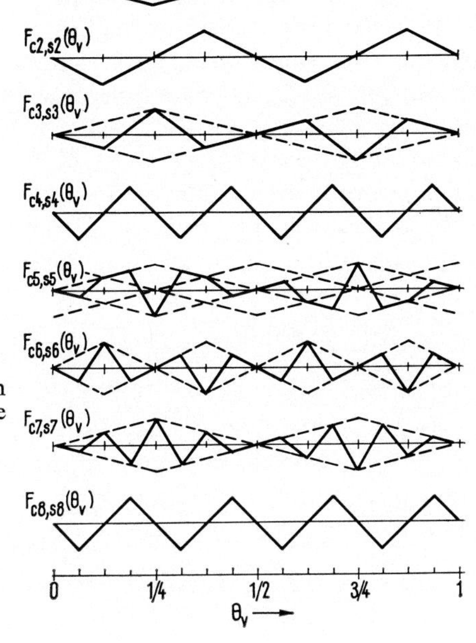

Fig. 132. Cross-correlation functions $F_{ci,si}(\theta_v)$ of some Walsh functions.

cope with this situation will be discussed later on in the section on synchronous reception of Walsh waves in radio transmission.

A Walsh function tracking filter according to Fig. 130 has been developed by Lüke and Maile of AEG-Telefunken for a sequency multiplex system.

There are a number of ways to generalize, improve or simplify the discussed method for the correction of time differences. For instance, the signal $[\mathrm{cal}(r,\theta) + M F^{\dagger\dagger}(\theta)]\,\mathrm{cal}(i,\theta)$ may be transmitted instead of $[1 + M F^{\dagger\dagger}(\theta)]\,\mathrm{cal}(2^k,\theta)$ if $r \oplus i$ equals a power of 2. The three blocks I, AT and TP in Fig. 130 may be combined into one. The feedback voltage $\sin\alpha$ in Fig. 129 and $\langle \mathrm{cal}(i,\theta)\,\mathrm{sal}(i,\theta-\theta_v)\rangle$ in Fig. 130 may be fed into the oscillator or function generator rather than into a phase shifter PS or delay circuit RV.

Problems

412.1 Determine θ_0 and θ_1 for $\mathrm{sal}(17,\theta)$, $\mathrm{sal}(19,\theta)$ and $\mathrm{sal}(29,\theta)$ by extending Table 9 and by means of the rules following Eq. (13).

412.2 Design a variable delay circuit that could be used for RV in Fig. 130.

412.3 Substitute $\sin 2\pi i\theta$ and $\cos 2\pi i\theta$ for $\mathrm{sal}(i,\theta)$ and $\mathrm{cal}(i,\theta)$ in Fig. 131 and plot the correlation functions. How many are not identically zero?

4.1.3 Methods of Single Sideband Modulation

Amplitude modulation of sine or cosine carriers yields a double sideband modulation due to the multiplication theorems of these functions. There are a number of methods for the elimination of one sideband that can be analyzed very well by orthogonal functions.

Consider two transmitters, both radiating sinusoidal functions of frequency Ω_0, but having a phase difference of $\frac{1}{2}\pi$. The carriers, amplitude modulated by time functions $F^\dagger(\theta)$ and $D^\dagger(\theta)$, shall have the form $F^\dagger(\theta)\times$ $\times\sqrt{(2)}\cos\Omega_0\theta$ and $D^\dagger(\theta)\sqrt{(2)}\sin\Omega_0\theta$. It is assumed that the frequency Ω_0 can be reproduced exactly at the receiver, but that there is a phase difference α between the received carriers $\sqrt{(2)}\cos\Omega_0\theta$, $\sqrt{(2)}\sin\Omega_0\theta$ and the locally produced carriers $\sqrt{(2)}\cos(\Omega_0\theta+\alpha)$, $\sqrt{(2)}\sin(\Omega_0\theta+\alpha)$. Multiplication of a received signal $S(\theta)$,

$$S(\theta) = F^\dagger(\theta)\sqrt{(2)}\cos\Omega_0\theta + D^\dagger(\theta)\sqrt{(2)}\sin\Omega_0\theta \tag{1}$$

by $\sqrt{(2)}\cos(\Omega_0\theta+\alpha)$ and $\sqrt{(2)}\sin(\Omega_0\theta+\alpha)$ yields:

$$\begin{aligned} S(\theta)\sqrt{(2)}\cos(\Omega_0\theta+\alpha) &= F^\dagger(\theta)\cos\alpha + D^\dagger(\theta)\sin\alpha + \\ &+ F^\dagger(\theta)\cos(2\Omega_0\theta+\alpha) + D^\dagger(\theta)\sin(\Omega_0\theta+\alpha) \end{aligned} \tag{2}$$

$$\begin{aligned} S(\theta)\sqrt{(2)}\sin(\Omega_0\theta+\alpha) &= -F^\dagger(\theta)\sin\alpha + D^\dagger(\theta)\cos\alpha + \\ &+ F^\dagger(\theta)\sin(2\Omega_0\theta+\alpha) - D^\dagger(\theta)\cos(2\Omega_0\theta+\alpha) \end{aligned} \tag{3}$$

The terms on the right hand sides of Eqs. (2) and (3) multiplied by $\cos(2\Omega_0\theta+\alpha)$ or $\sin(2\Omega_0\theta+\alpha)$ contain very high frequency components only; they shall be suppressed by filters. The right hand sides contain then $F^\dagger(\theta)$ or $D^\dagger(\theta)$ only if the phase difference α vanishes. Hence, two carriers of equal frequency but a phase difference $\frac{1}{2}\pi$ may transmit two independent signals $F^\dagger(\theta)$ and $D^\dagger(\theta)$ without mutual interference. Putting it differently, each frequency channel can be subdivided into two phase channels which will be denoted here as sine and cosine channel. Synchronous demodulation permits utilization of both phase channels.

A certain time function may always be transmitted through the sine channel, but never through the cosine channel in order to make them distinguishable. One may, for instance, replace $F^\dagger(\theta)$ in Eq. (3) by $1 + F^\dagger(\theta)$ and require that $F^\dagger(\theta)$ and $D^\dagger(\theta)$ have practically no energy below a certain frequency. The signal $S(\theta)$ may then be demodulated by the circuit of Fig. 129. The signal $[1 + M\,F^\dagger(\theta)]\,(1 + \cos 2\Omega_0\,\theta)$ is obtained at output 1, and $D(\theta)\,(1 + \cos 2\Omega_0\,\theta)$ is obtained at output 2.

Single sideband modulation is an excellent practical means for transmitting through sine and cosine channels. To show this, let a signal $F(\theta)$ be expanded into a series of sine and cosine pulses according to Fig. 1:

$$F(\theta) = a(0) f(0,\theta) + \sqrt{(2)} \sum_{i=1}^{\infty} [a_c(i)\cos 2\pi i\,\theta + a_s(i)\sin 2\pi i\,\theta] \qquad (4)$$
$$-\tfrac{1}{2} \leqq \theta < \tfrac{1}{2}$$

Fourier transforms $g(0,\nu)$, $g_c(i,\nu)$ and $g_s(i,\nu)$ of these pulses are given by Eq. (1.1.3-11). The first five transforms are shown in Fig. 6.

Let us denote the products $f(0,\theta)\sqrt{(2)}\cos\Omega_0\,\theta$, $f(0,\theta)\sqrt{(2)}\sin\Omega_0\,\theta$, $2\cos 2\pi i\,\theta\cos\Omega_0\,\theta$, $2\cos 2\pi i\,\theta\sin\Omega_0\,\theta$, $2\sin 2\pi i\,\theta\cos\Omega_0\,\theta$ and $2\sin 2\pi i\,\theta \times$ $\times \sin\Omega_0\,\theta$ by $d_{0,c}(\theta)$, $d_{0,s}(\theta)$, $d_{ci,c}(\theta)$, $d_{ci,s}(\theta)$, $d_{si,s}(\theta)$ and $d_{si,s}(\theta)$. The Fourier transforms of these products are denoted by $h_{0,c}(\nu), \ldots, h_{si,s}(\nu)$:

$$\left.\begin{aligned} h_{0,c}(\nu) &= \tfrac{1}{2}\sqrt{(2)}\,[g(0,\nu-\nu_0) \underset{(-)}{+} g(0,\nu+\nu_0)] \\ h_{ci,c}(\nu) &= \tfrac{1}{2}\sqrt{(2)}\,[g_c(i,\nu-\nu_0) \underset{(-)}{+} g_c(i,\nu+\nu_0)] \\ h_{si,c}(\nu) &= \tfrac{1}{2}\sqrt{(2)}\,[(-)g_s(i,\nu-\nu_0) + g_s(i,\nu+\nu_0)] \\ \nu_0 &= \Omega_0/2\pi \end{aligned}\right\} \qquad (5)$$

The signs in parentheses hold for the Fourier transforms $h_{0,s}(\nu)$, $h_{ci,s}(\nu)$ and $h_{si,s}(\nu)$.

The Fourier transforms $G_c(\nu)$ and $G_s(\nu)$ of the functions $F(\theta)\sqrt{(2)} \times$ $\times \cos\Omega_0\,\theta$ and $F(\theta)\sqrt{(2)}\sin\Omega_0\,\theta$ are obtained from Eqs. (4) and (5):

$$\begin{aligned} G_c(\nu) &= a(0)\,h_{0,c}(\nu) + \sum_{i=1}^{\infty} [a_c(i)\,h_{ci,c}(\nu) + a_s(i)\,h_{si,c}(\nu)] \\ G_s(\nu) &= a(0)\,h_{0,s}(\nu) + \sum_{i=1}^{\infty} [a_c(i)\,h_{ci,s}(\nu) + a_s(i)\,h_{si,s}(\nu)] \end{aligned} \qquad (6)$$

Consider the case $a_c(1) = a_s(1) = 1$ and all other coefficients equal zero to get an understanding of the shape of $G_c(\nu)$ and $G_s(\nu)$. The resulting Fourier transforms of the functions

$$\sqrt{(2)}\cos 2\pi\,\theta\,\sqrt{(2)}\cos\Omega_0\,\theta, \quad \sqrt{(2)}\cos 2\pi\,\theta\,\sqrt{(2)}\sin\Omega_0\,\theta$$
$$\sqrt{(2)}\sin 2\pi\,\theta\,\sqrt{(2)}\cos\Omega_0\,\theta, \quad \sqrt{(2)}\sin 2\pi\,\theta\,\sqrt{(2)}\sin\Omega_0\,\theta$$

are shown in the first four lines of Fig. 133. Note that $\sqrt{(2)}\cos\pi 2\,\theta$ and $\sqrt{(2)}\sin 2\pi\,\theta$ are cosine and sine pulses, which are zero outside the interval $-\frac{1}{2} \leqq \theta < \frac{1}{2}$.

The following single sideband signals may be derived from the transforms in the first four lines of Fig. 133:

$$\left.\begin{aligned} f_{cs}(\theta) &= \cos 2\pi\,\theta\,\sin\Omega_0\,\theta + \sin 2\pi\,\theta\cos\Omega_0\,\theta \\ f_{cc}(\theta) &= \cos 2\pi\,\theta\cos\Omega_0\,\theta - \sin 2\pi\,\theta\,\sin\Omega_0\,\theta \\ f_{sc}(\theta) &= \cos 2\pi\,\theta\,\sin\Omega_0\,\theta - \sin 2\pi\,\theta\cos\Omega_0\,\theta \\ f_{ss}(\theta) &= \cos 2\pi\,\theta\cos\Omega_0\,\theta + \sin 2\pi\,\theta\,\sin\Omega_0\,\theta \end{aligned}\right\} \tag{7}$$

The Fourier transforms of these functions are shown in lines *5* to *8* of Fig. 133. The functions $f_{cs}(\theta)$ and $f_{cc}(\theta)$ have almost all of their energy in the upper sideband $\nu > \Omega_0/2\pi$. $f_{sc}(\theta)$ and $f_{ss}(\theta)$ have most of their energy in the lower sideband $\nu < \Omega_0/2$. Both phase channels are used, since all four signals in Eq. (7) contain the sine carrier $\sin\Omega_0\,\theta$ and the cosine carrier $\cos\Omega_0\,\theta$. The practical implementation of single sideband modulation according to Eq. (7) is usually called second method or phase shift method of SSB modulation [2]: A signal $F(\theta)$ is modulated onto the carrier $\sin\Omega_0\,\theta$, and the same signal with all oscillations 90° phase shifted is modulated onto the carrier $\cos\Omega_0\,\theta$; sum or difference of the modulated carriers yields single sideband signals. The first method of SSB modulation obtains the same result by suppressing one sideband by means of a filter.

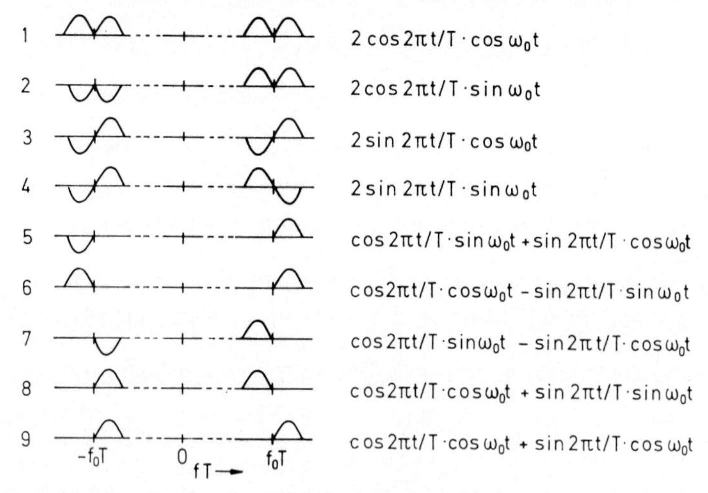

Fig. 133. Fourier transforms of sine and cosine carriers amplitude modulated by sine and cosine pulses.

Line *9* in Fig. 133 shows why negative frequencies cannot be disregarded. This transform looks like the transforms of lines *5* and *6* for positive values of $\nu = fT$; nevertheless, this is not a single sideband signal.

Single sideband and double sideband modulation permit the same number of channels in a certain frequency band, if the two phase channels of each frequency channel are used. The exploitation of double sideband modulation in this way—usually referred to as quadrature modulation—is handicapped by high crosstalk in the case of telephony transmission. Single sideband modulation, on the other hand, causes more distortions in digital signal transmission if SSB filters are used. A double sideband transmitter transmits all energy either through the sine or the cosine channel of a certain frequency band; a single sideband transmitter transmits all energy through the sine as well as the cosine channel of a frequency band half as wide. Thermal noise influences both methods equally, provided of course that phase-sensitive filtering is used for double sideband signals; otherwise one would receive the signal from one phase channel but the noise from both phase channels.

The investigation of amplitude modulation by means of sine and cosine pulses suffers from the fact that these functions are not frequency limited and are cumbersome to plot. The results are simpler to obtain with Walsh functions. Consider the Walsh functions of Fig. 2 as frequency functions $\mathrm{wal}(0, \nu)$, $\mathrm{cal}(i, \nu) = \mathrm{wal}(2i, \nu)$ and $\mathrm{sal}(i, \nu) = \mathrm{wal}(2i - 1, \nu)$ instead of time functions. The following time functions are obtained by a Fourier transformation:

$$\begin{aligned} w(2k, \theta) &= \int_{-\infty}^{\infty} \mathrm{wal}(2k, \nu) \cos 2\pi \nu \theta \, d\nu \\ w(2k+1, \theta) &= \int_{-\infty}^{\infty} \mathrm{wal}(2k+1, \nu) \sin 2\pi \nu \theta \, d\nu, \qquad k = 0, 1, 2, \ldots \end{aligned} \tag{8}$$

The functions $w(j, \theta) \cos \Omega_0 \theta$ and $w(j, \theta) \sin \Omega_0 \theta, j = 2k$ or $2k + 1$, have the following Fourier transforms:

$$\left.\begin{aligned} 2 \int_{-\infty}^{\infty} w(2k, \theta) \cos \Omega_0 \theta \cos 2\pi \nu \theta \, d\theta &= \mathrm{wal}(2k, \nu - \nu_0) + \\ &\quad + \mathrm{wal}(2k, \nu + \nu_0) \\ 2 \int_{-\infty}^{\infty} w(2k, \theta) \sin \Omega_0 \theta \sin 2\pi \nu \theta \, d\theta &= \mathrm{wal}(2k, \nu - \nu_0) - \\ &\quad - \mathrm{wal}(2k, \nu + \nu_0) \\ 2 \int_{-\infty}^{\infty} w(2k+1, \theta) \cos \Omega_0 \theta \sin 2\pi \nu \theta \, d\theta &= \mathrm{wal}(2k+1, \nu - \nu_0) + \\ &\quad + \mathrm{wal}(2k+1, \nu + \nu_0) \\ 2 \int_{-\infty}^{\infty} w(2k+1, \theta) \sin \Omega_0 \theta \cos 2\pi \nu \theta \, d\theta &= -\mathrm{wal}(2k+1, \nu - \nu_0) + \\ &\quad + \mathrm{wal}(2k+1, \nu + \nu_0) \end{aligned}\right\} \tag{9}$$

The following signals having all energy in the upper or lower sidebands only may be derived from the time functions in Eq. (8):

$$\left.\begin{array}{ll} w(0,\theta)\cos\Omega_0\theta - w(1,\theta)\sin\Omega_0\theta, & w(0,\theta)\cos\Omega_0\theta + w(1,\theta)\sin\Omega_0\theta \\ w(0,\theta)\sin\Omega_0\theta + w(1,\theta)\cos\Omega_0\theta, & w(0,\theta)\sin\Omega_0\theta - w(1,\theta)\cos\Omega_0\theta \\ w(2,\theta)\cos\Omega_0\theta - w(3,\theta)\sin\Omega_0\theta, & w(2,\theta)\cos\Omega_0\theta + w(3,\theta)\sin\Omega_0\theta \\ w(2,\theta)\sin\Omega_0\theta + w(3,\theta)\cos\Omega_0\theta, & w(2,\theta)\sin\Omega_0\theta - w(3,\theta)\cos\Omega_0\theta \end{array}\right\} \quad (10)$$

Four Fourier transforms of the functions in Eq. (10) are shown in Fig. 134. The arrows indicate the direction in which the absolute value of

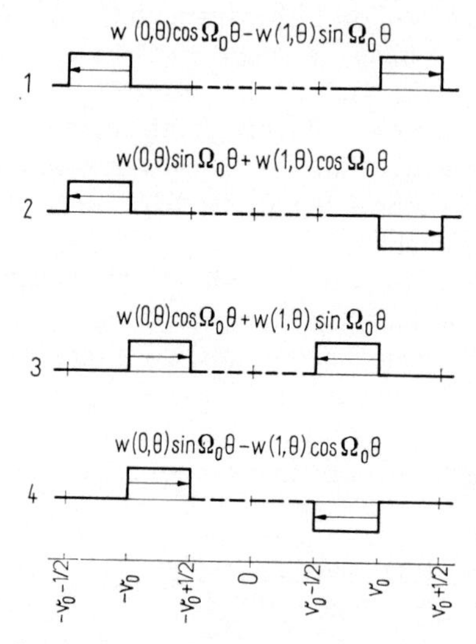

Fig. 134. Fourier transforms of some frequency limited single sideband signals; $\nu_0 = \Omega_0/2\pi$.

the frequency of $\text{wal}(2k, \nu)$ and $\text{wal}(2k+1, \nu)$ increases. The direction of the arrows remains unchanged for the upper sidebands and is reversed for the lower ones.

A block diagram for the second method of single sideband modulation is shown in Fig. 135a. The frequency limited signal $F(\theta)$ is fed through two phase shifting networks. Two signals appear at their outputs, whose oscillation components have a phase difference of 90° but are otherwise equal. The carriers $\cos 2\pi\nu_0\theta$ and $\sin 2\pi\nu_0\theta$ are amplitude modulated. The sum of the products yields an upper sideband signal, the difference a lower sideband signal.

A very similar method is due to Saraga [3]. The carriers $\cos 2\pi\nu\theta$ and $\sin 2\pi\nu\theta$ are added to the phase shifted signals according to Fig. 135b. The two sums are multiplied together. An upper sideband signal is generated; in addition, signals are produced in the baseband and around double the carrier frequency $2\nu_0$. A simple band-pass filter suppresses these undesirable signals.

A further single sideband modulation method is due to Weaver [4]. Figure 136 shows a block diagram for its implementation. A signal $F(\theta)$ with no energy outside the band $0 \leqq f \leqq 1/T$ or $-1 \leqq \nu = fT \leqq +1$ is

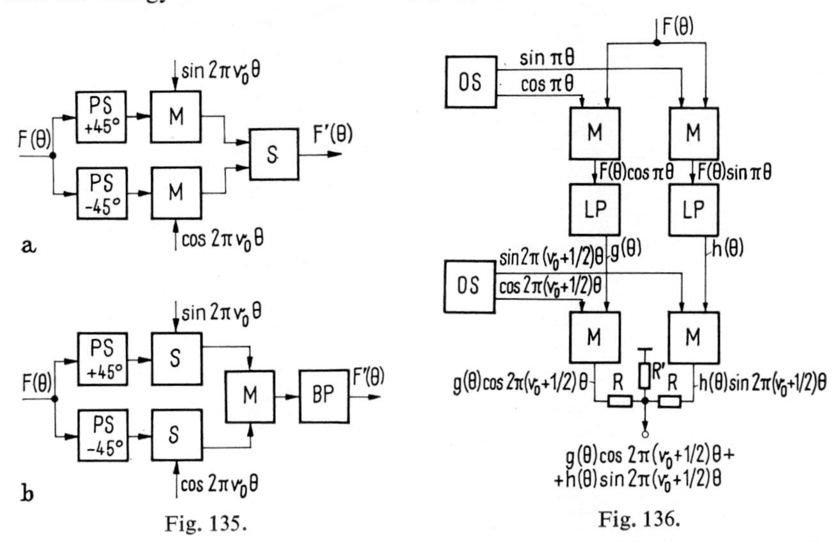

Fig. 135. Fig. 136.

Fig. 135. Outphasing method (a) and Saraga's fourth method of single sideband modulation (b) of a carrier with frequency ν_0 by a frequency limited signal $F(\theta)$. PS, phase shifting network; M, multiplier; S, adder; BP, band-pass filter; $F'(\theta)$ single sideband signal.

Fig. 136. Weaver's third method of single sideband modulation of a carrier with frequency $\nu_0 + \frac{1}{2}$ by a frequency limited signal $F(\theta)$. OS, oscillator; M, multiplier; LP, low-pass filter.

modulated onto the carriers $\sin \pi \theta$ and $\cos \pi \theta$ with frequency $\nu_1 = f_1 T = \frac{1}{2}$ or $f_1 = 1/2T$. The frequency of the carrier is in the middle of the band used. The modulated carriers pass through low-pass filters with cut-off frequencies $\nu_g = f_g T = \frac{1}{2}$. The filtered signals are modulated onto the high frequency carriers $\sin 2\pi(\nu_0 + \frac{1}{2})\,\theta$ and $\cos 2\pi(\nu_0 + \frac{1}{2})\,\theta$. The sum yields an upper sideband signal, the difference a lower sideband signal.

For an explanation of Weaver's method let the frequency limited input signal $F(\theta)$ be expanded into a series of the functions $w(2k, \theta)$ and $w(2k+1, \theta)$ derived by the Fourier transform in Eq. (8) from the Walsh functions:

$$F(\theta) = \sum_{k=0}^{\infty} [a(2k)\, w(2k, \theta) + a(2k+1)\, w(2k+1, \theta)] \tag{11}$$

It suffices to trace one even and one odd function of the series in Eq. (11) through the circuit of Fig. 136 rather than $F(\theta)$. The simplest functions,

$w(0, \theta)$ and $w(1, \theta)$ are used. Their Fourier transforms $\text{wal}(0, \nu)$ and $\text{wal}(1, \nu)$ are shown in Fig. 137, line *1*. The arrows point in the direction of increasing absolute values of ν. Modulation of $\cos\pi\theta$ shifts the Fourier

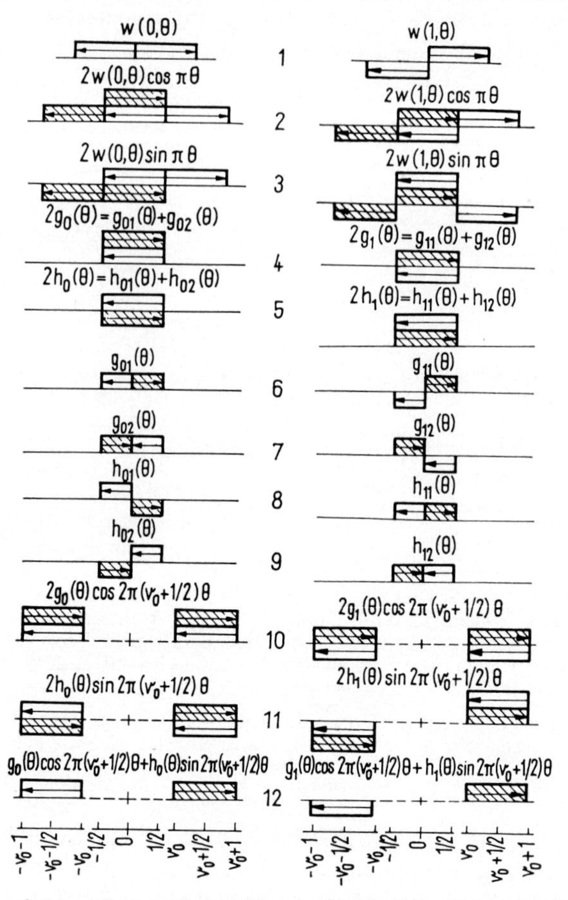

Fig. 137. Fourier transforms of the third method of single sideband modulation.

transforms of line *1* by $\frac{1}{2}$ to the right and to the left (line *2*). The transform shifted to the left is shown hatched for clarity. The two shifted transforms are shown superimposed where they overlap and have equal signs.

Modulation of $\sin\pi\theta$ shifts the transform of $w(0, \theta)$ by $\frac{1}{2}$ to the right and the transform multiplied by -1 by $\frac{1}{2}$ to the left, while the signs are reversed for $w(1, \theta)$ (line *3*). Low-pass filters suppress all components outside the band $-\frac{1}{2} \leqq \nu \leqq \frac{1}{2}$ (lines *4* and *5*). The resulting signals $g_0(\theta)$ and $h_1(\theta)$ have even transforms (lines *4* and *5*): Lines *6* to *9* show the even transforms of $g_{01}(\theta)$, $g_{02}(\theta)$ and $h_{11}(\theta)$, $h_{12}(\theta)$ which may be superimposed to yield $g_0(\theta)$ and $h_1(\theta)$; also shown are the odd transforms of

$h_{01}(\theta)$, $h_{02}(\theta)$ and $g_{11}(\theta)$, $g_{12}(\theta)$ which yield the transforms of $h_0(\theta)$ and $g_1(\theta)$ superimposed.

The transforms of lines *6* to *9* have the shape of wal$(0, \nu)$ and wal$(1, \nu)$. Hence, one obtains the transforms of the following functions with the help of Eq. (9),

$$g_0(\theta)\cos 2\pi(\nu_0 + \tfrac{1}{2})\,\theta = [g_{01}(\theta) + g_{02}(\theta)]\cos 2\pi(\nu_0 + \tfrac{1}{2})\,\theta \tag{12}$$

to

$$h_1(\theta)\sin 2\pi(\nu_0 + \tfrac{1}{2})\,\theta = [h_{11}(\theta) + h_{12}(\theta)]\sin 2\pi(\nu_0 + \tfrac{1}{2})\,\theta \tag{13}$$

as shown in lines *10* and *11*. The transforms of lines *4* and *5* are shifted by $\nu_0 + \frac{1}{2}$ to the right and left; the shifted transforms are multiplied by $+1$ or -1 according to the four possible products of even or odd transforms with sine or cosine carrier as shown in Eq. (9).

The sum of lines *10* and *11* yields the Fourier transforms of an upper sideband signal (line *12*). The difference yields the transform of a lower sideband signal.

The generally used method of single sideband modulation is the suppression of one sideband by a filter. This filter causes distortions which are particularly objectionable for the transmission of digital signals. Figure 138a shows the frequency power spectrum of a signal with practically no energy outside the band $0 \leqq \nu \leqq 2\nu_0$. It is not necessary for the following investigation that the power spectrum actually be rectangular in the band $0 \leqq \nu \leqq 2\nu_0$ as shown. Figures 138b to d show the shift of three such signals into adjacent bands by means of amplitude modulation of carriers with frequencies ν_c, $\nu_c + 2\nu_0$ and $\nu_c + 4\nu_0$. Band-pass filters having transmission functions as shown by the dashed lines, suppress the lower sidebands. Figure 138e shows the sum of the upper sidebands. The oscillations in the hatched frequency areas are partly attenuated and their phase shift does not vary linearly with frequency. This causes signal distortions. At the receiver, the signals are separated by band-pass filters, and additional distortions are introduced (Fig. 138, *f* to *h*). The power spectra of the demodulated signals are shown in Fig. 138, *i* to *k*. The hatched areas indicate where oscillations are improperly attenuated and phase shifted, and thus cause signal distortions.

There are two ways to keep the distortions small. One may shape the signals, so that most of their energy is located in frequency bands where the single sideband filters cause little distortion. Or one may locate the edges of the single sideband filters far away from the frequency bands which contain most of the signal energy. The first method is used in vestigial sideband modulation[1]. This method is particularly useful, if digital signals are to be transmitted by time division through existing telephony channels [6]. A detailed account of this method is given by Bennet and

[1] Vestigial sideband modulation goes back to Nyquist [5].

Davey [7]. The second method is used in transposed sideband modulation [8]. Its principle will be discussed with reference to Fig. 139. The signals have practically all their energy in the frequency band $2\nu_0 \leqq \nu \leqq 4\nu_0$ (Fig. 139). The width of the empty band $0 < \nu < 2\nu_0$ is neither zero nor

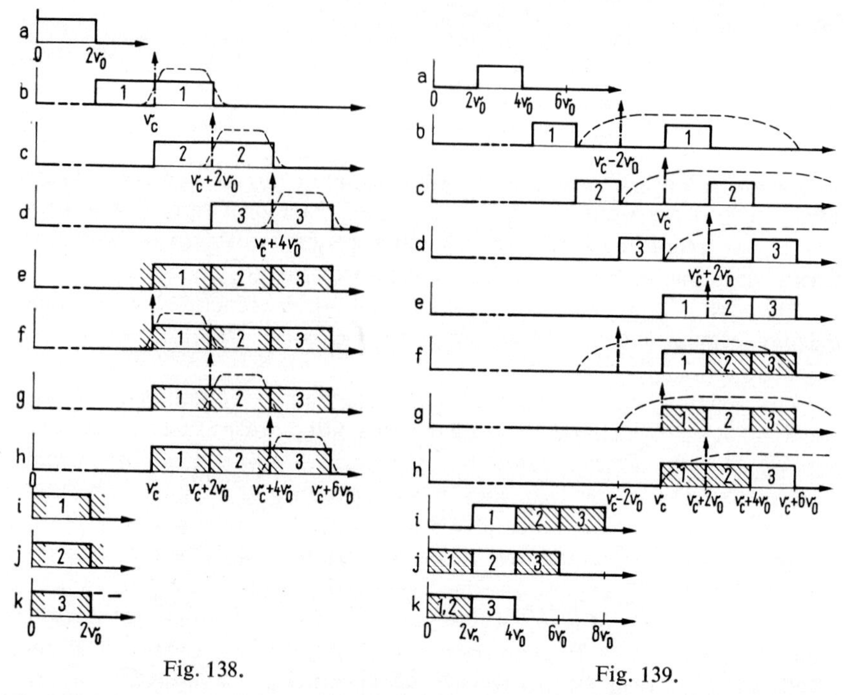

Fig. 138. Fig. 139.

Fig. 138. Power spectra for the modulation and demodulation of three signals by single sideband modulation. Bandwidth of the signals is $2\nu_0$; lowest frequency of the signals is 0.

Fig. 139. Power spectra for the modulation and demodulation of three signals by transposed sideband modulation. Bandwidth of the signals is $2\nu_0$; lowest frequency of the signals is $2\nu_0$.

small compared to the bandwidth $\Delta\nu = 2\nu_0$ of the signal. It is not important that $2\nu_0$ equals $\Delta\nu$. It is only necessary that the empty band $0 < \nu < 2\nu_0$ be wider than the frequency band in which the single sideband filters cause distortions.

Figures 139b to d show the shift of three such signals into adjacent frequency bands by amplitude modulation of carriers with the frequencies $\nu_c - 2\nu_0$, ν_c and $\nu_c + 2\nu_0$. Band-pass filters having the transmission functions shown by the dashed lines suppress the lower sidebands. The sum of the three upper sidebands is shown in Fig. 139e. The signals are not distorted, since there is no energy in the frequency areas where the band-pass filters distort. At the receiver, the signals are separated by band-pass filters. Distortions are introduced in the frequency areas shown hatched (Fig. 139f to h). The power spectra of the demodulated signals are shown in Fig. 139i to k. The non-distorted power spectra of the demodulated signals are again located in the band $2\nu_0 \leqq \nu \leqq 4\nu_0$ as in Fig. 139a. The distorted and folded-over oscillations are located in the unused bands $0 \leqq \nu \leqq 2\nu_0$ and $\nu > 4\nu_0$.

Figure 140a shows a signal $v_1(t)$ that can be transmitted by transposed sideband modulation:

$$v_1(t) = +V \sin 30\pi\, t/T - V \sin 34\pi\, t/T$$

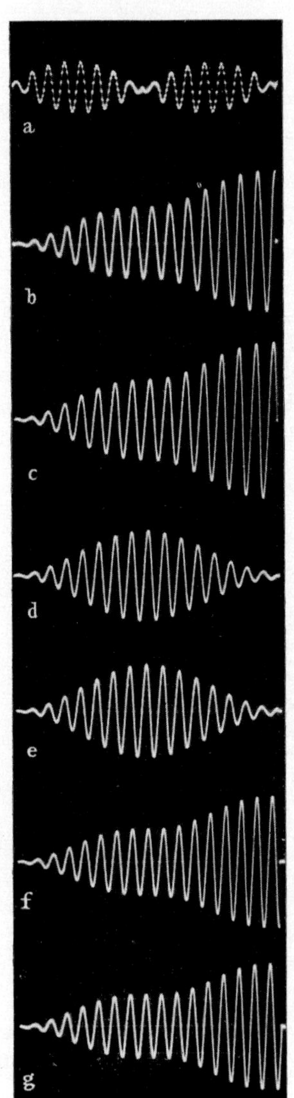

Fig. 140. Detection of digital signals. (a) signal $V(+\sin 30\pi\,\theta - \sin 34\pi\,\theta)$; ouput voltages of the function detectors for $\cos 30\pi\,\theta$ (b), $\sin 30\pi\,\theta$ (c), $\cos 32\pi\,\theta$ (d), $\sin 32\pi\,\theta$ (e), $\cos 34\pi\,\theta$ (f) and $\sin 34\pi\,\theta$ (g). Duration of the traces: $T = 150$ ms. (Courtesy P. Schmid, D. Nowak and R. Durisch of Allen-Bradley Co.)

The binary character $+1 \, -1$ is transmitted by this signal. $\sin 30\pi\, t/T$ has 15 oscillations in an interval of duration T and $\sin 34\pi\, t/T$ has 17 oscillations. It follows from Fig. 141 that the energy of $v_1(t)$ is concentrated in the band $(15-1) \leqq \nu = fT \leqq (17+1)$. The lower frequency limit is equal to $2\nu_0 = 14$; the bandwidth is $\Delta\nu = 4$. The width of the empty

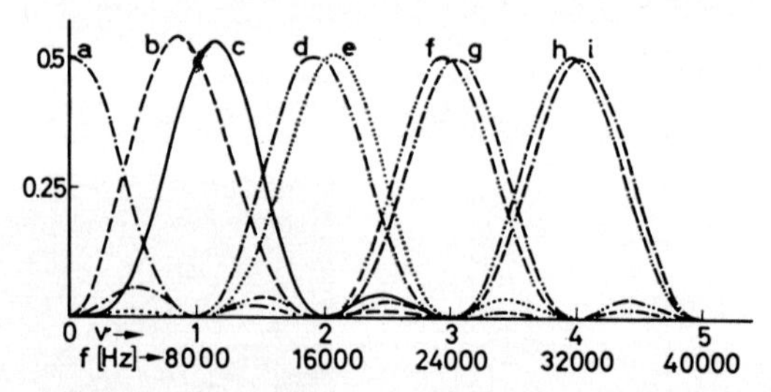

Fig. 141. Frequency power spectra of the following pulses according to Figs. 1 and 126: $\frac{1}{2}\sqrt{(2)}\,\mathrm{wal}(0,\theta)$ (a), $\sin 2\pi\theta$ (b), $\cos 2\pi\theta$ (c), $\sin 4\pi\theta$ (d), $\cos 4\pi\theta$ (e), $\sin 6\pi\theta$ (f), $\cos 6\pi\theta$ (g), $\sin 8\pi\theta$ (h), $\cos 8\pi\theta$ (i). $\theta = t/T$, $\nu = fT$, $-\frac{1}{2} \leqq \theta < \frac{1}{2}$; f is written in hertz for $T = 125\ \mu s$.

band $0 \leqq \nu \leqq 2\nu_0$ is $14\Delta\nu/4$ and is thus much wider than in Fig. 139. The coefficients $+1 \, -1$ can be regained from this signal according to Figs. 140b and f by means of a function detector as shown in Fig. 122. It is quite unimportant what other signals are outside the band $14 \leqq \nu \leqq 18$. Figures 140c, d, e, and g show that a wrong signal produces very little output voltage at the sampling time.

Problems

413.1 Repeat the plots of Fig. 133 with $\cos 4\pi\, t/T$ and $\sin 4\pi\, t/T$ substituted for $\sin 2\pi\, t/T$ and $\cos 2\pi\, t/T$.

413.2 Repeat the plots of Fig. 134 for $w(2,\theta)$ and $w(3,\theta)$.

413.3 Repeat the plots of Fig. 137 for $w(2,\theta)$ and $w(3,\theta)$.

413.4 Twelve telephone channels are multiplexed into a group that occupies the frequency band from 60 kHz to 108 kHz. Repeat the plots of Fig. 139 for this case.

4.2 Multiplexing of Time Variable Signals

4.2.1 Multiplex Systems

One of the most important applications of amplitude modulation is in multiplexing. Consider a frequency multiplex system for telephony. m telephony signals are passed through frequency low-pass filters and amplitude modulated onto m sine or cosine carriers. In principle, the modulated carriers are added and may then be transmitted via a common link. Single sideband filters are inserted after the modulators to suppress one sideband.

There are several methods to separate frequency multiplex signals at the receiver. Synchronous demodulation is one such method and it can be applied to non-sinusoidal carriers as well. The received multiplex signal is multiplied in m modulators by the same m carriers that were used for multiplexing at the transmitter. The carriers in the receiver must be synchronized to those in the transmitter. This means the frequency must be right and the phase difference very small. The demodulated signals pass through m low-pass filters which suppress the contributions from signals of wrong channels. A practical frequency multiplex system differs of course from this principle, since specific features of sine and cosine functions are utilized in practical systems. Here the emphasis is on those features, which sine and cosine functions share with other complete systems of orthogonal functions.

The two methods of quadrature modulation and single sideband modulation are known in frequency multiplexing, and they were discussed in the previous section. Corresponding modulation methods exist for sequency multiplexing [8]. Lacking better terms, one may denote them by quadrature and single sideband modulation too. There are two Walsh functions $\mathrm{cal}(i, \theta)$ and $\mathrm{sal}(i, \theta)$ for each sequency i. Quadrature modulation means that $\mathrm{cal}(i, \theta)$ as well as $\mathrm{sal}(i, \theta)$ are amplitude modulated by two independent signals. Single sideband modulation means that either $\mathrm{cal}(i, \theta)$ or $\mathrm{sal}(i, \theta)$ only are modulated. One sequency sideband is generated in either case, but the carrier sequencies have to be spaced twice as wide apart for quadrature modulation as for single sideband modulation.

For explanation of the principle of sequency multiplexing by means of Walsh carriers refer to Fig. 142. The output voltages of two microphones are applied to points a and a'. They are passing through two sequency low-pass filters LP. Step voltages appear at their outputs b and b'. These are fed to the multipliers M and amplitude modulate two periodic Walsh carriers applied to points c and c'. The modulated carriers d and d' are added in S and the output voltage e is obtained. This voltage is multiplied at the receiver in two multipliers M by the same Walsh functions as used at the transmitter. The two voltages appearing at the outputs g and g' of the multiplier are fed through sequency low-pass filters LP that are equal to those used in the transmitter. The step voltages at the outputs h and h' are equal to those at b and b'. They may be fed directly into a telephone headset. The low-pass filters of the transmitter produce a delay of 125 μs and those of the receiver produce another 125 μs delay. The dashed sections of the time diagram of Fig. 142 indicate these delays.

Let us discuss sequency multiplexing in more detail from the purely academic point of view in order to elucidate its strong and weak points. Practical requirements will be considered in the following two sections.

Figure 143 shows a multiplex system with 1024 channels for transmission in one direction. Multiplexing of analog signals will be discussed.

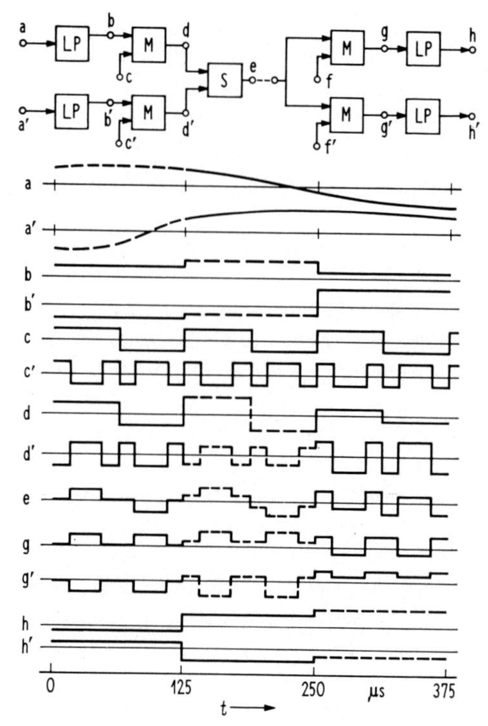

Fig. 142. Principle of a sequency multiplex system. LP, sequency low-pass filter; M, multiplier; S, adder.

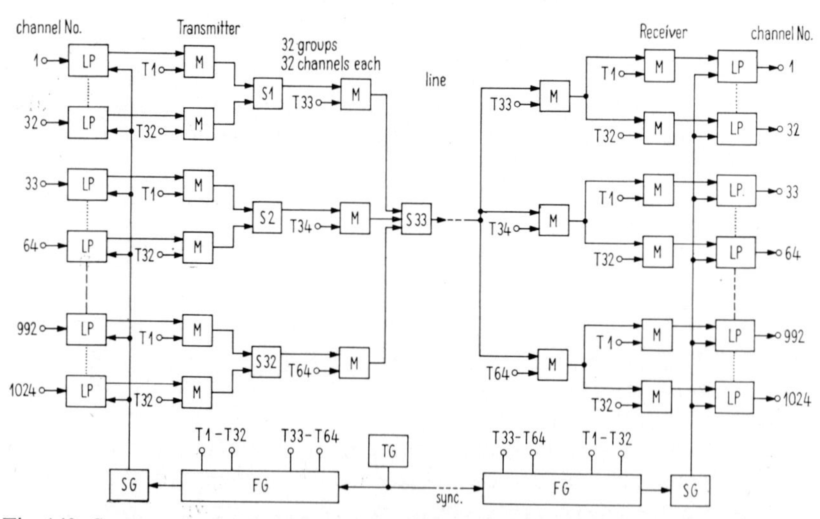

Fig. 143. Sequency multiplex system with 1024 telephony channels for transmission in one direction. LP, sequency low-pass filter; M, multiplier; S, adder; TG, FG, SG, trigger, function and timing generator.

There is no problem in modifying input and output circuits for other types of signals. For instance, the voltages $+V$ and $-V$ only are applied to the inputs of the channels for transmission of binary digital signals. Eight channels are required for the transmission of a usual PCM telephony signal. Such details are omitted from the further discussion, since they are no more important for the principle of sequency multiplexing than for that of time or frequency multiplexing.

The two-wire line coming from a subscriber is split by a hybrid circuit into a transmitting and a receiving branch. A signal on the transmitting branch is applied to one of the 1024 inputs of the transmitter. It passes a sequency low-pass filter LP with cut-off sequency of 4 kzps. The switches of the low-pass filters are driven by pulses of the timing generator SG. The input signal $F(\theta)$ is transformed into a step function $F^{\dagger\dagger}(\theta)$; $F(\theta)$ and $F^{\dagger\dagger}(\theta)$ are shown in Fig. 144 but without the delay of 125 μs between $F(\theta)$ and $F^{\dagger\dagger}(\theta)$.

After filtering, the signal is amplitude modulated onto one of 32 Walsh carriers T 1 to T 32 in one of the multipliers M. The first four carriers $\mathrm{wal}(0, \theta)$ to $\mathrm{wal}(3, \theta)$ are shown in Fig. 144. Duration T and position of their orthogonality interval coincides with the steps of the signal $F^{\dagger\dagger}(\theta)$.

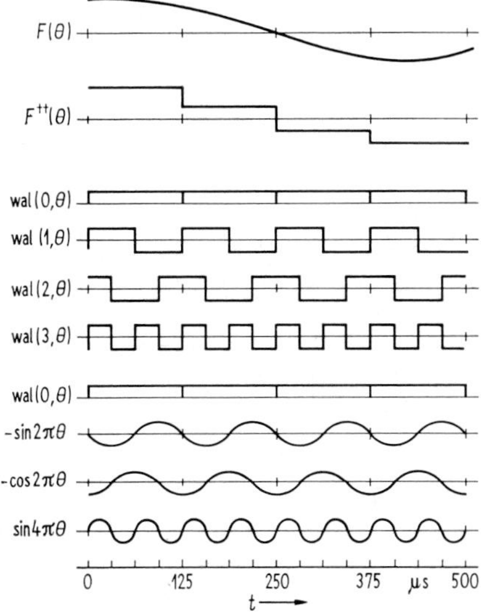

Fig. 144. Time diagram for the multiplex system of Fig. 143.

32 modulated carriers are combined by the adders S 1 to S 32 into one group. As a result, 32 groups with 32 channels each are obtained as shown in Fig. 143. The figure 32 is chosen as example only; principles for judiciously choosing size of groups and supergroups will be discussed later.

The output voltages of the adders are amplitude modulated onto the Walsh carriers T 33 to T 64 in the multipliers M. Adder S 33 adds the resulting 32 voltages. A step voltage is obtained at the output of S 33. The width of the steps is equal to $(32)^{-2} \times 125\,\mu s \cong 122$ ns. The amplitude of this output signal assumes $8000 \times (32)^2 = 8192000$ independent amplitudes per second. The signal occupies the sequency band $0 \leqq \phi \leqq$ $\leqq 4.096$ Mzps. This multiplex signal may be transmitted directly to the receiver or it may be used to modulate a sine or Walsh carrier.

At the receiver the signal is first multiplied in 32 multipliers M with the carriers T 33 to T 64 and then in 32 multipliers with the carriers T 1 to T 32. The demodulated signals pass then through the sequency low-pass filters LP, which are equal to those in the transmitter.

The block diagram of Fig. 143 holds for quadrature modulation as well as for single sideband modulation. The two methods differ only in the carriers used.

Table 10. *Generation of carriers* T 1 *to* T 32 *by the multiplication* $\mathrm{wal}(k,\theta)\,\mathrm{wal}(l,\theta) = \mathrm{wal}(j,\theta)$ *and of the carriers* T 33 *to* T 64 *by the multiplication* $\mathrm{wal}(k,\theta)\,\mathrm{wal}(l,\theta)\,\mathrm{wal}(31,\theta) = \mathrm{wal}(j,\theta)$

carrier	index			carrier	index		
	j	k	l		j	k	l
T 1	0	0	0	T 33	0	0	0
T 2	1	1	0	T 34	32	63	0
T 3	2	3	1	T 35	64	127	32
T 4	3	3	0	T 36	96	127	0
T 5	4	7	3	T 37	128	255	96
T 6	5	7	2	T 38	160	255	64
T 7	6	7	1	T 39	192	255	32
T 8	7	7	0	T 40	224	255	0
T 9	8	15	7	T 41	256	511	224
T 10	9	15	6	T 42	288	511	192
T 11	10	15	5	T 43	320	511	160
T 12	11	15	4	T 44	352	511	128
T 13	12	15	3	T 45	384	511	96
T 14	13	15	2	T 46	416	511	64
T 15	14	15	1	T 47	448	511	32
T 16	15	15	0	T 48	480	511	0
T 17	16	31	15	T 49	512	1023	480
T 18	17	31	14	T 50	544	1023	448
T 19	18	31	13	T 51	576	1023	416
T 20	19	31	12	T 52	608	1023	384
T 21	20	31	11	T 53	640	1023	352
T 22	21	31	10	T 54	672	1023	320
T 23	22	31	9	T 55	704	1023	288
T 24	23	31	8	T 56	736	1023	256
T 25	24	31	7	T 57	768	1023	224
T 26	25	31	6	T 58	800	1023	192
T 27	26	31	5	T 59	832	1023	160
T 28	27	31	4	T 60	864	1023	128
T 29	28	31	3	T 61	896	1023	96
T 30	29	31	2	T 62	928	1023	64
T 31	30	31	1	T 63	960	1023	32
T 32	31	31	0	T 64	992	1023	0

The Walsh functions $\text{wal}(0, \theta)$ to $\text{wal}(31, \theta)$ are best used for the carriers T 1 to T 32. Their generation by means of the multiplication theorem $\text{wal}(k, \theta)\,\text{wal}(l, \theta) = \text{wal}(k \oplus l, \theta)$ from the Rademacher functions $\text{wal}(1, \theta), \text{wal}(3, \theta), \text{wal}(7, \theta), \ldots, \text{wal}(2^n - 1, \theta), \ldots$ is shown by Table 10. Rademacher functions can be generated by binary counters as shown in Figs. 52 and 53.

The carriers T 33 to T 64 must be chosen so that no crosstalk is produced and no sequency bandwidth is wasted. Table 11 shows a possible choice of these carriers. Walsh functions are shown in this table in the notation $\text{wal}(j, \theta)$ as well as $\text{cal}(i, \theta)$, $\text{sal}(i, \theta)$. One may see that the last five digits of the normalized sequency written as a binary number is always zero. The 32 carriers T 1 to T 32 may be fitted between any two of the carriers T 33 to T 64. The available sequency band is completely used, there are no lost sequency bands between adjacent channels.

Concepts of group theory may be used to progress beyond the purely empirical way of choosing the carriers. The Walsh functions $\text{wal}(0, \theta)$ to $\text{wal}(1023, \theta)$ form a group with 2^{10} elements. The functions $\text{wal}(0, \theta)$ to $\text{wal}(31, \theta)$ are a subgroup with 2^5 elements. There are $2^{10}/2^5 = 32$ cosets of this subgroup. This is just the number of carriers T 33 to T 64. The Walsh functions generated by the modulation of carriers T 33 to T 64 by the functions $\text{wal}(0, \theta), \ldots, \text{wal}(31, \theta)$ are the elements of the 32 cosets. All possible functions $\text{wal}(j, \theta)$ usable as carriers T 33 to T 64 are obtained by multiplying each one of the functions $\text{wal}(0, \theta), \text{wal}(32, \theta), \ldots, \text{wal}(992, \theta)$ of Table 11 with any one function $\text{wal}(0, \theta), \text{wal}(1, \theta), \ldots, \text{wal}(31, \theta)$. Such a multiplication means only a reordering of the elements of each coset. One can multiply $\text{wal}(0, \theta)$ with one of the 32 functions $\text{wal}(0, \theta), \text{wal}(1, \theta), \ldots, \text{wal}(31, \theta)$. One may further multiply $\text{wal}(32, \theta)$ with these 32 functions, then $\text{wal}(64, \theta)$, etc. There are a total of $32^{32} = 2^{160}$ such products, which means there are 2^{160} possible choices of carriers T 33 to T 64, none of which would waste sequency bandwidth or produce crosstalk.

Figure 144 shows sine and cosine carriers besides the Walsh carriers $\text{wal}(0\ \theta)$ to $\text{wal}(3, \theta)$. One may use them as carriers T 1 to T 32 in Fig. 143. The multipliers would have to be of a more complicated type in this case. The 32 modulated carriers could be added without having to pass a single sideband filter. This type of quadrature modulation shows the close connection between frequency and sequency multiplexing.

One may readily see from Fig. 144 that any system of functions that is orthogonal in a finite interval can be used for the carriers T 1 to T 32 of Fig. 143 if the input signals $F(\theta)$ are filtered by sequency low-pass filters. This does not hold for the carriers T 33 to T 64. They must have multiplication theorems similar to those of sine-cosine or Walsh functions.

Figure 143 shows an extra synchronization line between transmitter and receiver. Actually, one or more of the 1024 channels can be used for

Table 11. *The* 2×32 *carriers of the sequency multiplex system of Fig. 143*

	Function						Function				
	cal(i, θ), sal(i, θ)			wal(j, θ)			cal(i, θ), sal(i, θ)			wal(j, θ)	
	c, s	i		j			c, s	i		j	
		decimal	binary	decimal	binary			decimal	binary	decimal	binary
T 1	—	0	00000	0	00000	T 33	—	0	000000000	0	0000000000
T 2	s	1	00001	1	00001	T 34	c	16	000010000	32	0000100000
T 3	c	1	00001	2	00010	T 35	c	32	000100000	64	0001000000
T 4	s	2	00010	3	00011	T 36	c	48	000110000	96	0001100000
T 5	c	2	00010	4	00100	T 37	c	64	001000000	128	0010000000
T 6	s	3	00011	5	00101	T 38	c	80	001010000	160	0010100000
T 7	c	3	00011	6	00110	T 39	c	96	001100000	192	0011000000
T 8	s	4	00100	7	00111	T 40	c	112	001110000	224	0011100000
T 9	c	4	00100	8	01000	T 41	c	128	010000000	256	0100000000
T 10	s	5	00101	9	01001	T 42	c	144	010010000	288	0100100000
T 11	c	5	00101	10	01010	T 43	c	160	010100000	320	0101000000
T 12	s	6	00110	11	01011	T 44	c	176	010110000	352	0101100000
T 13	c	6	00110	12	01100	T 45	c	192	011000000	384	0110000000
T 14	s	7	00111	13	01101	T 46	c	208	011010000	416	0110100000
T 15	c	7	00111	14	01110	T 47	c	224	011100000	448	0111000000
T 16	s	8	01000	15	01111	T 48	c	240	011110000	480	0111100000
T 17	c	8	01000	16	10000	T 49	c	256	100000000	512	1000000000
T 18	s	9	01001	17	10001	T 50	c	272	100010000	544	1000100000
T 19	c	9	01001	18	10010	T 51	c	288	100100000	576	1001000000
T 20	s	10	01010	19	10011	T 52	c	304	100110000	608	1001100000
T 21	c	10	01010	20	10100	T 53	c	320	101000000	640	1010000000
T 22	s	11	01011	21	10101	T 54	c	336	101010000	672	1010100000
T 23	c	11	01011	22	10110	T 55	c	352	101100000	704	1011000000
T 24	s	12	01100	23	10111	T 56	c	368	101110000	736	1011100000
T 25	c	12	01100	24	11000	T 57	c	384	110000000	768	1100000000
T 26	s	13	01101	25	11001	T 58	c	400	110010000	800	1100100000
T 27	c	13	01101	26	11010	T 59	c	416	110100000	832	1101000000
T 28	s	14	01110	27	11011	T 60	c	432	110110000	864	1101100000
T 29	c	14	01110	28	11100	T 61	c	448	111000000	896	1110000000
T 30	s	15	01111	29	11101	T 62	c	462	111010000	928	1110100000
T 31	c	15	01111	30	11110	T 63	c	480	111100000	960	1111000000
T 32	s	16	10000	31	11111	T 64	c	496	111110000	992	1111100000

transmission of a synchronization signal. A Walsh function $\mathrm{wal}(2^n - 1, \theta)$, which is a Rademacher function, is transmitted if a constant voltage is applied to the channel 2^n. The orthogonality of the Rademacher functions is invariant to shifts. Tracking filters that lock onto them can be built with relative ease. The modulated Walsh functions form a statistical background and can be suppressed by long averaging times of the tracking filters as pointed out in section 4.1.2.

Requirements for synchronization and rise times may be inferred from Fig. 145. Let the signal $v_e(t)$ represent the output voltage of a telephony multiplex system with 1024 channels. The width of the steps is 122 ns. The information of the signal is completely contained in its amplitudes. If the signal $v_e(t)$ is transmitted, it suffices—in the absence of noise—to sample the amplitudes of the steps, in order to obtain all the information. The sampling may be done anywhere in the 122 ns long intervals, and this is the tolerance interval for the synchronization. Consider

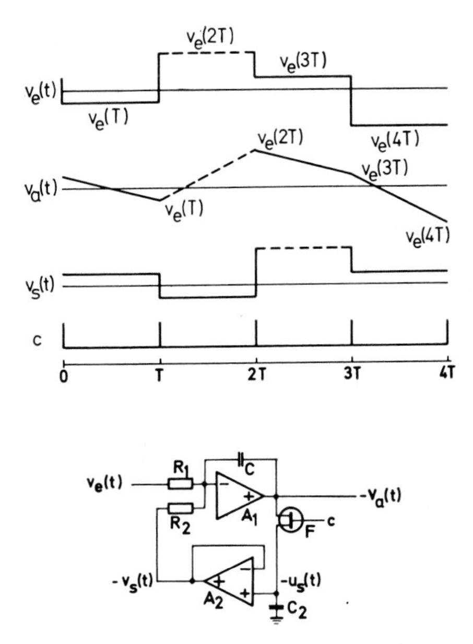

Fig. 145. Finite rise time of a step function and filter for the conversion of $v_e(t)$ into $v_a(t)$. The filter also reconverts $v_a(t)$ into $v_e(t)$.

the rise time. Let the rise time be so slow that it takes 122 ns to change from $v_e(T)$ to $v_e(2T)$, from $v_e(2T)$ to $v_e(3T)$, etc. as shown by $v_a(t)$ in Fig. 145. The original step voltages may be regained by sampling $v_a(t)$ exactly at the points $0, T, 2T, 3T, \ldots$. The maximum rise time is thus 122 ns, if there is no synchronization error. In general, rise time plus synchronization error must be less than 122 ns.

Amplitude sampling is a poor detection method in the presence of noise. However, one may readily see that $v_a(t)$ can be reconverted to $v_e(t)$ by integration, since the integral over $v_a(t)$ taken from, e.g., T to $2T$ is proportionate to $v_e(2T) - v_e(T)$.

The circuit shown in Fig. 145 will transform $v_e(t)$ into $v_a(t)$, theoretically without any ringing. It is a classical problem of frequency theory to approximate a filter which can do this. The simplicity with which this problem can be solved within the wider framework of sequency theory is due to the use of a time variable element—the field effect transistor F—in the filter.

Consider Fig. 146 for a discussion of single sideband modulation of Walsh functions. The original signal $F(\theta)$ and the signal $F^{\dagger\dagger}(\theta)$ filtered by a sequency low-pass filter are shown on top; the time shift between

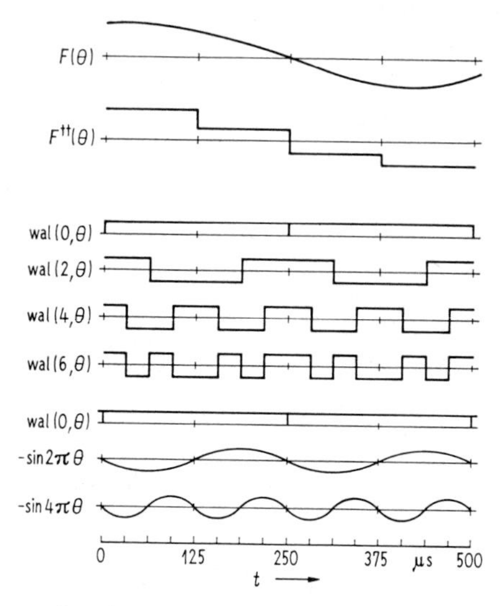

Fig. 146. Time diagram of a single sideband sequency multiplex system. Time base for the low-pass filters is $T = 125$ μs; time base for the carriers is $T = 250$ μs.

$F(\theta)$ and $F^{\dagger\dagger}(\theta)$ is omitted. The Walsh carriers $\mathrm{wal}(0, \theta)$, $\mathrm{wal}(2, \theta)$, $\mathrm{wal}(4, \theta)$, $\mathrm{wal}(6, \theta), \ldots$ are shown. Their time base is 250 μs, which is twice the duration of the steps of $F^{\dagger\dagger}(\theta)$. The filtered signal $F^{\dagger\dagger}(\theta)$ is represented by the following expression in the interval $0 \leqq t \leqq 250$ μs:

$$\begin{gathered} F^{\dagger\dagger}(\theta) = c(0)\,\mathrm{wal}(0, \theta) + c(1)\,\mathrm{wal}(1, \theta) \\ \theta = (t - t_0)/T; \quad t_0 = 125\ \mu\mathrm{s}, \quad T = 250\ \mu\mathrm{s}; \quad -\tfrac{1}{2} \leqq \theta < \tfrac{1}{2} \end{gathered} \tag{1}$$

Amplitude modulation of one of the carriers $\mathrm{wal}(2j, \theta)$ of Fig. 146 by $F^{\dagger\dagger}(\theta)$ yields:

$$F^{\dagger\dagger}(\theta)\,\mathrm{wal}(2j, \theta) = c(0)\,\mathrm{wal}(2j, \theta) + c(1)\,\mathrm{wal}(2j + 1, \theta) \tag{2}$$

Modulation produces just those Walsh carriers $\mathrm{wal}(2j+1,\theta)$ that are left out in Fig. 146. Hence, the modulated Walsh functions occupy the whole available sequency band.

Figure 146 also shows sine functions with a time base of 250 μs. Their amplitude modulation by $F\dagger\dagger(\theta)$ does not yield a (frequency) single sideband modulation. The correspondence between Walsh and sine functions in the case of quadrature modulation is based on the fact that $\mathrm{wal}(0,\theta)$ is the first function of a Walsh as well as a Fourier series. The signal $F\dagger\dagger(\theta)$ of Eq. (1), however, contains the function $\mathrm{wal}(1,\theta)$, which does not belong to a Fourier series.

A characteristic feature of frequency multiplex systems compared with time multiplex systems is the ease with which signals in communication networks or radio signals can be combined and separated. The reason is that frequency multiplex signals are inherently marked by their frequency, which is independent of delay times. Time multiplex signals, on the other hand, having various unknown delay times need some additional marking in order to be separable. Since sequency multiplex signals are also inherently marked by their sequency, one will expect them to lead to communication networks that are very similar to those for frequency multiplex signals.

Figure 147 shows a possible sequency allocation for groups and supergroups in a communication network. This allocation is chosen so that group, supergroup A or supergroup B are cosets of the mathematical subgroup of Walsh functions $\mathrm{wal}(j,\theta)$ with j smaller than 32, 128 or 256.

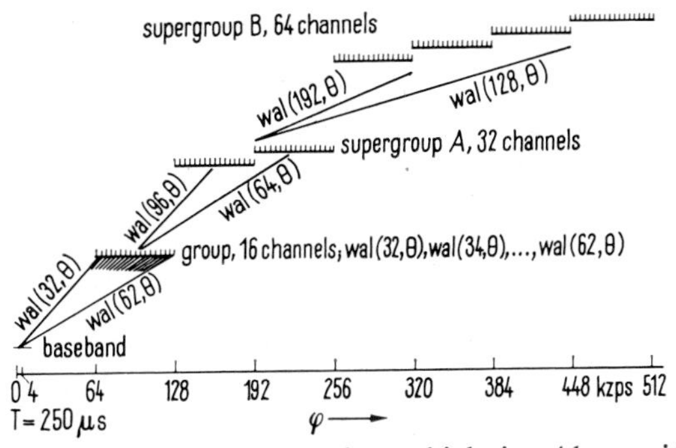

Fig. 147. Occupation of sequency bands by multiplexing 4 kzps wide base bands.

Single sideband modulation and a time base of 250 μs are assumed. The individual channels occupy basebands from 0 to 4 kzps. Sixteen basebands make a group that occupies the sequency band from 64 to 128 kzps. The carriers are $\mathrm{wal}(32+2j,\theta)$; $j = 0, 1, \ldots, 15$.

Amplitude modulation of the carrier $\mathrm{wal}(96,\theta)$ by a group shifts the sequency band into the interval from 128 to 192 kzps; the carrier $\mathrm{wal}(64,\theta)$ shifts a group into the band from 192 to 256 kzps. These 32 channels are marked supergroup A in Fig. 147.

Modulation of a carrier wal$(192, \theta)$ by a supergroup A shifts it into the band from 256 to 384 kzps; the carrier wal$(128, \theta)$ shifts a supergroup A into the band from 384 to 512 kzps. The resulting 64 channels are denoted by supergroup B in Fig. 147. Table 12 shows to which position of the sequency band the channels are shifted. The 16 channels of a group

Table 12. *Transposition of the carriers* wal$(32, \theta)$ . . . wal$(62, \theta)$ *of a group to the carriers* wal$(64, \theta)$ . . . wal$(126, \theta)$ *of a supergroup* A *and the carriers* wal$(128, \theta)$ . . . wal$(256, \theta)$ *of a supergroup* B. *The sequency of the carriers is equal to* $2j$ kzps *for* $T = 250$ µs

Group: 16 channels			Supergroup A: 2 groups			
			carrier wal$(96, \theta)$ 96 = 1100000		carrier wal$(64, \theta)$ 64 = 1000000	
channel	j	j binary	j	j binary	j	j binary
1	32	100000	64	1000000	96	1100000
2	34	100010	66	1000010	98	1100010
3	36	100100	68	1000100	100	1100100
4	38	100110	70	1000110	102	1100110
5	40	101000	72	1001000	104	1101000
6	42	101010	74	1001010	106	1101010
7	44	101100	76	1001100	108	1101100
8	46	101110	78	1001110	110	1101110
9	48	110000	80	1010000	112	1110000
10	50	110010	82	1010010	114	1110010
11	52	110100	84	1010100	116	1110100
12	54	110110	86	1010110	118	1110110
13	56	111000	88	1011000	120	1111000
14	58	111010	90	1011010	122	1111010
15	60	111100	92	1011100	124	1111100
16	62	111110	94	1011110	126	1111110

Supergroup B: 2 supergroups A							
Carrier wal$(192, \theta)$ 192 = 11000000				Carrier wal$(128, \theta)$ 128 = 10000000			
j	j binary	j	j binary	j	j binary	j	j binary
128	10000000	160	10100000	192	11000000	224	11100000
130	10000010	162	10100010	194	11000010	226	11100010
132	10000100	164	10100100	196	11000100	228	11100100
134	10000110	166	10100110	198	11000110	230	11100110
136	10001000	168	10101000	200	11001000	232	11101000
138	10001010	170	10101010	202	11001010	234	11101010
140	10001100	172	10101100	204	11001100	236	11101100
142	10001110	174	10101110	206	11001110	238	11101110
144	10010000	176	10110000	208	11010000	240	11110000
146	10010010	178	10110010	210	11010010	242	11110010
148	10010100	180	10110100	212	11010100	244	11110100
150	10010110	182	10110110	214	11010110	246	11110110
152	10011000	184	10111000	216	11011000	248	11111000
154	10011010	186	10111010	218	11011010	250	11111010
156	10011100	188	10111100	220	11011100	252	11111100
158	10011110	190	10111110	222	11011110	254	11111110

are shifted by the carriers wal(j,θ), $j = 32, 34, \ldots, 62$, from the base-band. For instance, the signal in channel 10 is modulated onto the carrier wal$(50, \theta)$. The sequency of wal$(50, \theta)$ is equal to $2 \times 50 = 100$ kzps and the signal occupies the band from 100 to 104 kzps. The carrier wal$(50, \theta)$ becomes wal$(82, \theta)$ by multiplication with wal$(96, \theta)$, or wal$(114, \theta)$ by multiplication with wal$(64, \theta)$ in supergroup A. Channel 10 occupies the band from $2 \times 82 = 164$ to 168 kzps or the band from $2 \times 114 = 228$ to 232 kzps. Finally, the carrier wal$(50, \theta)$ becomes one of the carriers wal$(146, \theta)$, wal$(178, \theta)$, wal$(210, \theta)$ or wal$(244, \theta)$ in supergroup B. Channel 10 occupies one of the 4 kzps wide bands with lower limit $2 \times 146 = 292$ kzps, $2 \times 178 = 356$ kzps, $2 \times 210 = 420$ kzps or $2 \times 244 = 488$ kzps.

Consider the case of a supergroup B, a supergroup A and a group being transmitted. The signal occupies the sequency band from 64 to 512 kzps according to Fig. 147. One may extract the group by means of a sequency low-pass filter with cut-off sequency of 128 kzps. A low-pass filter with 256 kzps cut-off sequency will extract the group and the supergroup A. This simple kind of filtering is possible only if the cut-off sequency is equal to $2^r \times 4 = 2^{r+2}$ kzps; $r = 0, 1, 2, \ldots$.

Consider as a further example the transmission of a supergroup B. Either the 32 channels in the band from 256 to 384 kzps or the 32 channels in the band from 384 to 512 kzps shall be extracted. Supergroup B is multiplied by the carrier wal$(128, \theta)$. Table 12 shows that the band $256 < \phi < 384$ kzps $(j = 128, \ldots, 190)$ is transposed into the band $0 < \phi < 128$ kzps $(j = 0, \ldots, 62)$; the band $384 < \phi < 512$ kzps $(j = 192, \ldots, 254)$ is transposed into the band $128 < \phi < 256$ kzps $(j = 64, \ldots, 126)$. A sequency low-pass filter having a cut-off sequency of 128 kzps can extract the band $0 < \phi < 128$ kzps. A multiplication by wal$(64, \theta)$ can shift it to the band $128 < \phi < 256$ kzps, which is the band for a supergroup A.

Let supergroup B be multiplied by wal$(192, \theta)$ instead of by wal$(128, \theta)$. The band $256 < \phi < 384$ kzps $(j = 128, \ldots, 190)$ is transposed into the band $128 < \phi < 256$ kzps $(j = 64, \ldots, 126)$, the band $384 < \phi < 512$ kzps $(j = 192, \ldots, 254)$ into the band $0 < \phi < 128$ kzps $(j = 0, \ldots, 62)$. A sequency low-pass filter can extract the band $0 < \phi < 128$ kzps, which now contains the other channels of supergroup B.

Any individual channel in the band $2j < \phi < 2j + 4$ kzps can be extracted by multiplication with wal(j, θ), and filtering by a sequency low-pass filter having 4 kzps cut-off sequency. The filtered signal may then be shifted to any position in the sequency spectrum by multiplying it with the proper Walsh carrier. The extraction of individual channels or groups of channels without need to demodulate and remodulate all channels is very similar to what can be done in time multiplexing. It may be used to route individual channels through a switched communication network [9].

It has been assumed so far that the channels and groups combined into supergroup A or B are synchronized. This assumption holds true if all channels are combined in the same exchange into groups and supergroups. Now consider the case that channels are combined into groups at different exchanges and these groups are combined into supergroups at a higher level exchange. One cannot assume that these groups are synchronized. One may, however, assume that these groups have the same time base $T = 250\ \mu s$.

The combination of unsynchronized groups with equal time base will be discussed with reference to Fig. 148. This figure shows on top the Rademacher functions $\mathrm{wal}(31, \theta)$ and $\mathrm{wal}(127, \theta)$ in the interval $0 \leq \theta < \frac{1}{8}$. The multiplications

$$\mathrm{wal}(127, \theta)\, \mathrm{wal}(63, \theta) = \mathrm{wal}(64, \theta)$$

$$\mathrm{wal}(127, \theta)\, \mathrm{wal}(31, \theta) = \mathrm{wal}(96, \theta)$$

yield the function $\mathrm{wal}(64, \theta)$ and $\mathrm{wal}(96, \theta)$. These are the carriers required for transposition of two groups into one supergroup A according to Fig. 147. They are the reference for synchronization.

Lines *a* of Fig. 148 show symbolically the signals of two nonsynchronous groups. These signals consist of sums of the functions $\mathrm{wal}(32, \theta)$, $\mathrm{wal}(33, \theta), \ldots, \mathrm{wal}(63, \theta)$ according to Figs. 146 and 147; the amplitude of these functions depends on the particular signal transmitted. Figure 2

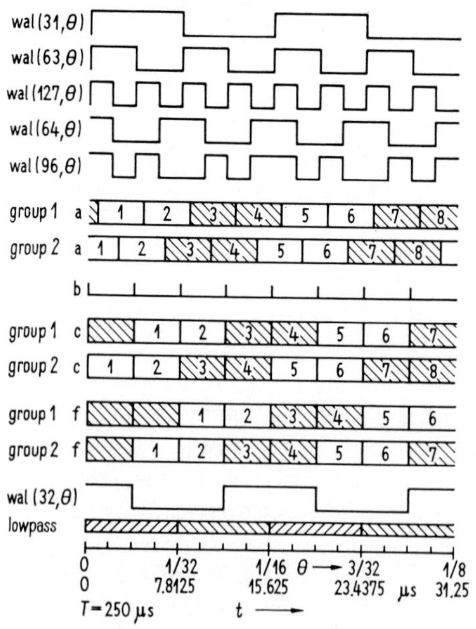

Fig. 148. Principle for the combination of two nonsynchronized groups into a supergroup A according to Fig. 147.

shows that a signal containing the functions $\text{wal}(32, \theta), \text{wal}(33, \theta), \ldots,$ $\text{wal}(63, \theta)$ is a step function with steps $1/64$ wide. The signals in the lines a of Fig. 148 are divided into intervals $1/64$ wide. Their amplitudes are constant in these intervals. The individual intervals are denoted by $1, 2, \ldots$ One may further see from Fig. 2 that a signal containing the functions $\text{wal}(32, \theta), \ldots, \text{wal}(63, \theta)$ and having the amplitude v in the interval $-1/2 \leqq \theta < -1/2 + 1/64$ must have the amplitude $-v$ in the interval $-1/2 + 1/64 \leqq \theta < -1/2 + 2/64$. This result holds generally: the amplitudes have the same absolute value and opposite sign in any two intervals

$$\frac{2k}{64} \leqq \theta < \frac{2k}{64} + \frac{1}{64} \quad \text{and} \quad \frac{2k}{64} + \frac{1}{64} \leqq \theta < \frac{2k}{64} + \frac{2}{64}$$

Lines a of Fig. 148 show such intervals of equal absolute value of the signal alternatively hatched and not hatched. The amplitudes have equal absolute value and opposite sign in the intervals 1 and 2, 3 and 4, 5 and 6, etc.

Sampling the signals of lines a at the time indicated by the trigger pulses of line b and holding the sampled voltages during an interval of duration $1/64$ yields the signals of lines c. They are synchronized with the carriers $\text{wal}(64, \theta)$ and $\text{wal}(96, \theta)$. This synchronization is not yet sufficient. Group 2 in line c begins at time $\theta = 0$ with the intervals 1 and 2 in which the amplitudes have equal absolute value. Group 1 begins with two intervals for which this is not so. Shifting the groups by a sampling and holding circuit yields the signals of lines f. Group 1 is now synchronized correctly but group 2 is not. Group 1 has to be taken from line f and group 2 from line c for modulation of the carriers $\text{wal}(64, \theta)$ and $\text{wal}(96, \theta)$. Note that the problem of synchronization differs from that of time division, since groups 1 and 2 may be shifted with reference to each other by any multiple of $1/32$. Such an arbitrary shift would cause an interchange of channels in time division.

Written symbolically, the signal of supergroup A has the following form:

$$(\text{group } 1)\, \text{wal}(64, \theta) + (\text{group } 2)\, \text{wal}(96, \theta)$$

Demodulation of supergroup A by $\text{wal}(64, \theta)$ or $\text{wal}(96, \theta)$ yields the following two signals:

$$\begin{aligned}
&[(\text{group } 1)\, \text{wal}(64, \theta) + (\text{group } 2)\, \text{wal}(96, \theta)]\, \text{wal}(64, \theta) = \\
&\qquad = (\text{group } 1) + (\text{group } 2)\, \text{wal}(32, \theta) \\
&[(\text{group } 1)\, \text{wal}(64, \theta) + (\text{group } 2)\, \text{wal}(96, \theta)]\, \text{wal}(96, \theta) = \\
&\qquad = (\text{group } 2) + (\text{group } 1)\, \text{wal}(32, \theta)
\end{aligned}$$

One of the terms of each of the right hand sides must be suppressed in order to obtain group 1 or group 2 separately. It is easier to suppress the terms (group 1) or (group 2) than the others. This is of no practical consequence, since the terms (group 1) $\text{wal}(32, \theta)$ and (group 1) $\text{wal}(32, \theta)$

are obtained, which may be demodulated by $\text{wal}(32, \theta)$:

$$[(\text{group 2})\,\text{wal}(32, \theta)]\,\text{wal}(32, \theta) = (\text{group 2})$$
$$[(\text{group 1})\,\text{wal}(32, \theta)\,\text{wal}(32, \theta)] = (\text{group 1})$$

The terms (group 1) and (group 2) can be suppressed by a sequency low-pass filter that integrates over the intervals $0 < \theta < 1/32$, $1/32 < < \theta < 2/32$, etc. These integration intervals are shown in the last line of Fig. 148. Their non-normalized duration is equal to 7.8125 µs for a 1024 channel telephony system. These integration intervals extend over two intervals of group 2 (line c) and over two intervals of group 1 (line f), in which the amplitudes of the signals have equal absolute value and opposite sign. Hence, the integration yields zero and the signals are suppressed.

Multiplication of group 2 in line c and group 1 in line f by $\text{wal}(32, \theta)$ makes the signs in the intervals 1 and 2, 3 and 4, etc., equal. There is no cancellation by integration and the signals (group 2) $\text{wal}(32, \theta)$ or (group 1) $\text{wal}(32, \theta)$ pass through the sequency low-pass filter.

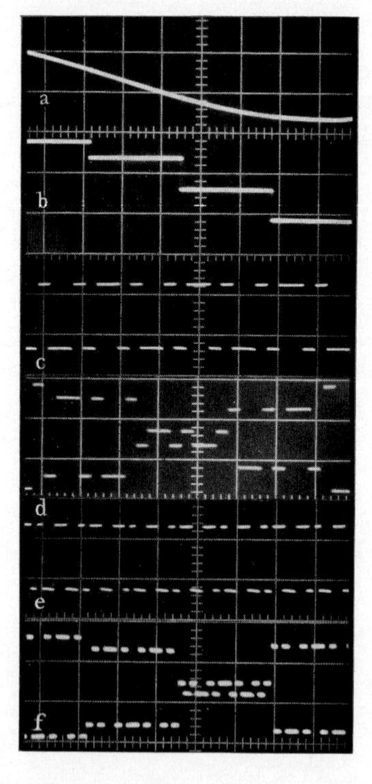

Fig. 149. Oscillograms of sequency multiplexing of a telephone signal. (a) input signal $F(\theta)$; (b) output $F^{\dagger\dagger}(\theta)$ of a sequency low-pass filter; (c) carrier $\text{wal}(5, \theta)$; (d) first modulation $F^{\dagger\dagger}(\theta)\text{wal}(5,\theta)$; (e) carrier $\text{wal}(9,\theta)$; (f) second modulation $F^{\dagger\dagger}(\theta)\text{wal}(5,\theta)\text{wal}(9,\theta) = F^{\dagger\dagger}(\theta)\,\text{wal}(12, \theta)$; horizontal scale 50 µs/div. (Courtesy H. Lüke and R. Maile of AEG-Telefunken AG.)

Figure 149 shows some oscillograms of a sequency multiplex system developed by Lüke and Maile. The carrier $\text{wal}(5, \theta)$ is shown for clarity instead of one of the carriers $\text{wal}(32, \theta), \ldots, \text{wal}(62, \theta)$ in Fig. 147. For the same reason, the carrier $\text{wal}(9, \theta)$ is shown for the second modulation rather than the carrier wal(96, θ) or wal(64, θ) of Fig. 147. The crosstalk attenuation obtained in this equipment was about -55 dB if an extra synchronization line was used, and dropped to about -53 dB if the synchronization signal was transmitted with the telephony signals and extracted by a Walsh function tracking filter. This attenuation would be high enough to meet telephony standards if signal compandors are used. However, sequency multiplexing is mainly of interest for peak power limited digital signals at the present, and -53 dB crosstalk attenuation is more than enough in this case.

Other sequency multiplex systems have been developed by Hübner, Davidson, Bagdasarjanz and Loretan, Roefs and Van Ittersum, Lopez de Zavalia and Moro, Barrett and Gordon [10–15].

Problems

421.1 Which Walsh carriers are used in Fig. 142? Substitute two others and repeat the plot.

421.2 Four signals are to be multiplexed by Walsh functions. Each signal has 8000 independent amplitudes per second. Plot the four signals at the output terminals of the sequency low-pass filters. The four carriers shall have the time base $T = 125\ \mu s$. The first carrier is wal(0, θ). Choose the other three carriers so that neither sequency bandwidth is lost nor overlapping of bands occurs. Plot the modulated carriers. Plot their sum. Plot the sum multiplied with each one of the four carriers. Plot the four products after having passed a sequency low-pass filter with a bandwidth of 4 kzps.

421.3 The transmission line in Fig. 143 shall be a frequency band limited channel. What bandwidth is required? (*See* section 3.2.3.)

4.2.2 Examples of Mixed Sequency-Frequency Multiplex Systems

In order to define areas of application for sequency multiplexing, let us list its characteristic features and compare them with those of time and frequency multiplexing.

a) Equipment can easily be implemented by the present (binary) semiconductor technology, a feature shared with time multiplexing but not with frequency multiplexing.

b) The amplitude distribution of sequency multiplex signals is essentially a Gaussian distribution and very similar to that of frequency multiplex signals; this statement holds for analog as well as for digital signals. Hence, sequency multiplex signals have the amplitude distribution for which the existing frequency multiplex channels have been designed. Time multiplex signals do generally not have a Gaussian amplitude distribution; in particular, the amplitudes of time multiplexed binary digital signals assume two values only.

c) The average power of sequency multiplexed signals is independent of the activity factor[1] if automatic volume control is used. The average power per channel and thus the signal-to-noise ratio increase with a de-

[1] The activity factor gives the fraction of channels actually busy.

creasing activity factor. The same holds true for frequency multiplexing. In the case of time multiplexing, the average power of an active channel is independent of the activity factor and the average power of the multiplex signal declines with the activity factor. In the transmission of peak power limited binary signals in the presence of thermal noise sequency multiplexing yields definitely lower error rates[1] than time multiplexing for activity factors of 50% or less while time multiplexing yields definitely lower error rates for activity factors close to 100%. Since the activity factor of a multiplex system rarely exceeds 25%, means have been developed to increase it for time multiplexing by methods usually referred to as "time assignment". These methods must yield activity factors well above 50%—not counting the time used for the transmission of additional information required by these methods—to make them worthwhile.

d) Errors in digital signal transmission through the switched telephone network are mainly caused by burst-type interference. The error rates of digital time multiplex signals are orders of magnitude above those for frequency multiplex signals under these conditions[2]. Sequency multiplexing should yield about the same error rates as frequency multiplexing.

e) Sequency multiplexing yields presently some 40 dB crosstalk attenuation. This is sufficient for digital signals, most analog telemetry or TV signals, and low grade voice or scrambled voice signals. It is not sufficient for commerical voice channels that should have about 70 dB crosstalk attenuation. Although 70 dB apparent crosstalk attenuation can be attained by compandors, the more prudent approach is to restrict sequency multiplexing for the time being to applications for which 40 dB suffices. The progress of technology will most probably improve the crosstalk attenuation and 70 dB might be a less formidable figure a few years from now.

f) Time multiplexing favors the time-shared use of one complex piece of equipment. Frequency and sequency multiplexing favor the simultaneous use of several but simpler pieces of equipment. Which approach is more economical depends on the state of technology. Time-sharing communication lines and telegraphy equipment were universal before 1920. The introduction of frequency filters emphasized simultaneous use; e.g., several low speed teletypewriters replaced one high speed teleprinter. Time sharing became again attractive after 1950 due to the introduction of the transistor. The present trend to integrated circuits with high tooling-up costs and relatively low costs thereafter may again bring a shift from time sharing to simultaneous use.

g) Time multiplexing as well as sequency[3] multiplexing are less suitable than frequency multiplexing for multiplexing of unsynchronized signals.

---

[1] *See* section 7.2.2.

[2] *See* section 3.2.2.

[3] The succesful solution of problems 128.6 and 128.7 suggests that a general solution of the synchronization problem exists for sequency multiplexing.

Typically, such unsynchronized signals occur in networks where signals are generated at various points and multiplexed at other points.

h) Sequency and time multiplexing require no single sideband filters and thus avoid the costs as well as the distortions caused by them. The distortions are of little importance for voice transmission but are the main reason why the transmission rate of digital signals is generally one third or less of the Nyquist rate.

i) Individual extraction of a channel is done easily in time multiplexing but requires considerable expense in frequency multiplexing. Sequency multiplexing lies between these two extremes.

j) Frequency multiplexing requires individual tuning of filters. Time and sequency multiplexing do not require any individual tuning but use an accurate clock pulse generator instead, that is shared by many circuits.

A possible way to combine the strong points of sequency and frequency multiplexing will be discussed with reference to Fig. 150. It shows frequency multiplexing used above the group level and sequency multiplexing below the group level. The combination of several groups originating

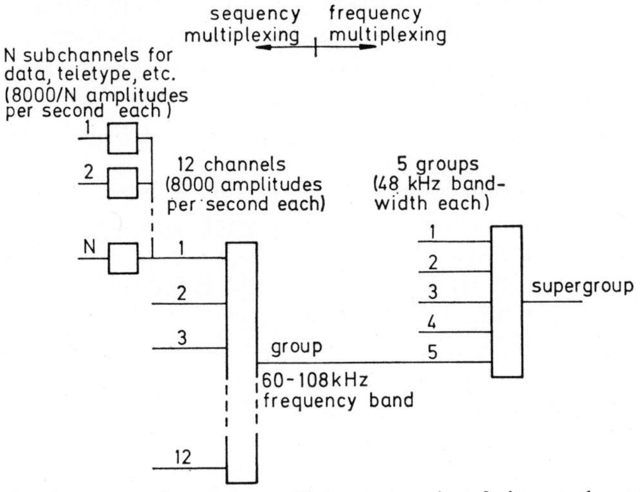

Fig. 150. Mixed sequency-frequency multiplex system interfacing at the group level.

at different locations into a supergroup will thus not require any synchronization between the groups. Equipment costs are less important above the group level than below since they are shared by at least twelve channels. Furthermore, signal distortions caused by frequency filters are less important above the group level since such distortions are caused, at the present, mainly by the single sideband filters of the channels.

Sequency multiplexing below the group level offers the advantage of reduced costs and reduced distortions. Furthermore, the size of the equipment is reduced and the need for individual tuning of filters is eliminated.

Synchronization is required for all signals to be combined into a group, but this is no problem since the multiplexing equipment below the group level will be either on the same rack or at least in the same room. The amplitude distribution of the group signal must be approximately Gaussian to be compatible with existing frequency multiplex equipment above the group level, a requirement that cannot be met by time multiplexing if binary signals are to be transmitted and that would generally require compandors if analog signals are to be transmitted.

According to Fig. 150, the subdivision of a group into twelve channels can be retained even though the number 12 leads to somewhat more complicated circuits than a power of two. The further subdivision of a channel into N subchannels is done again by sequency multiplexing. Typical values for N for teletype transmission are 128, 144 or 192.

Figure 151 shows the block diagram of the multiplex system. It is assumed that 12 voltages V_0 to V_{11}, having a constant value during intervals $0 < t < 125\ \mu s$, $125\ \mu s < t < 250\ \mu s$, etc., are fed to the channel inputs. These voltages may be analog or quantized, in particular they may be quantized to two values $+V$ and $-V$. Modems are required to transform the voltage supplied by the signal source into this form. If the signal source is a microphone, the modem consists of a sequency low-pass filter as shown in Fig. 55. If the signal source is a teletypewriter, the modem consists of sequency multiplexing equipment similar to the one discussed here but working much more slowly.

The twelve Paley functions $\text{pal}(j, \theta)$ of Fig. 18 are used for multiplexing. Figure 152 shows the voltage V_9 on top and the carrier $\text{pal}(9, \theta)$ below. V_9 is constant in the interval $0 < \theta < 1$ or $0 < t < 125\ \mu s$ for $\theta = t/T$ and $T = 125\ \mu s$. This voltage is multiplied with $\text{pal}(9, \theta)$ in the multiplier MT 9 of Fig. 151. The resulting voltage $V_9\,\text{pal}(9, \theta)$ is fed to the summing amplifier SU. There are two wires connecting each multiplier with the summing amplifier SU since the multipliers are actually single-pole, double-throw switches as will be explained later on.

The sum of the output voltages of the 11 multipliers and the voltage V_0 is denoted by $S(\theta)$. A typical voltage $S(\theta)$ is shown in line *4* of Fig. 152. The 12 independent amplitudes of $S(\theta)$ are denoted by $A, B, \ldots, L$. How these amplitudes are derived from the input voltages $V_0, \ldots, V_{11}$ is shown on the right hand side of Fig. 152. The sequency multiplexing process ends here.

The step function $S(\theta)$ is not particularly suited for transmission through a frequency band-limited channel. To make $S(\theta)$ frequency limited one may sample it by the trigger function $\text{tri}(12, \theta)$ shown in Fig. 152. One obtains the very narrow pulses $S(\theta)\,\text{tri}(12, \theta)$. Passing them through a frequency low-pass or band-pass filter produces a frequency band-limited signal that contains the same information as $S(\theta)$. The process of converting a (sequency band-limited) step function into a frequency band-limited function and the reconversion was discussed in sections 3.2.3 and 3.1.3. Let us assume here that the function $S(\theta)\,\mathrm{tri}(12, \theta)$ is produced at the transmitter and is made available at the receiver. By a holding

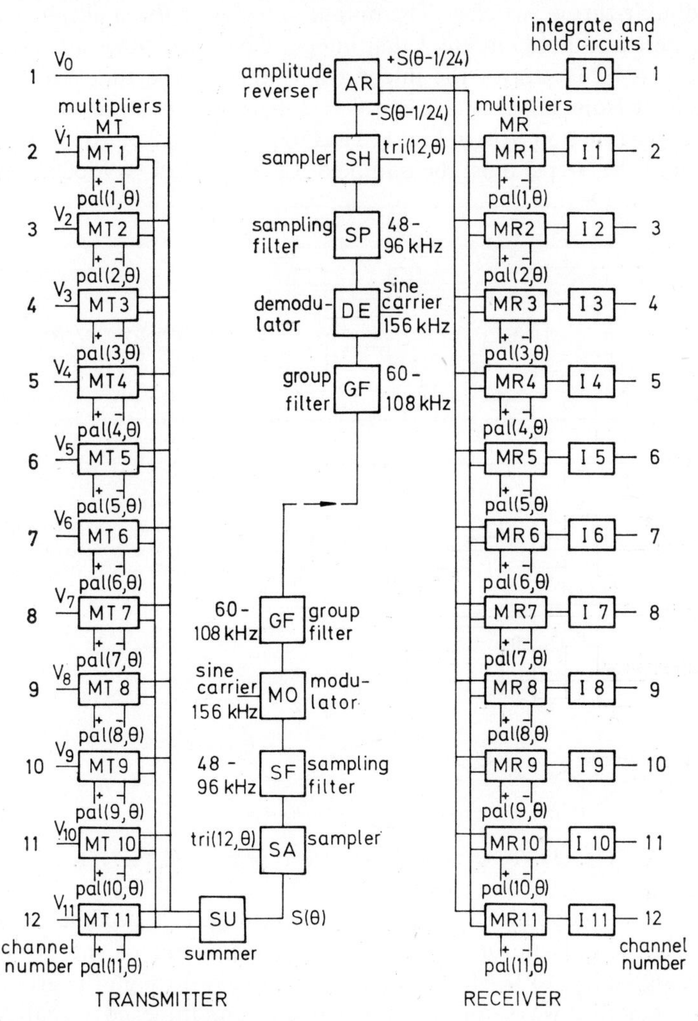

Fig. 151. Block diagram of a 12 channel sequency multiplex system.

circuit one may convert $S(\theta)\,\mathrm{tri}(12, \theta)$ into the delayed function $S(\theta - 1/24)$ as shown in Fig. 152.

Let $(1/12)S(\theta - 1/24)$ be multiplied by $\mathrm{pal}(9, \theta - 1/24)$. The sum of the resulting amplitudes $A, B, \ldots, -L$ yields the voltage V_9 transmitted by the carrier $\mathrm{pal}(9, \theta)$. A practical way to produce this sum is to integrate $(1/12)S(\theta - 1/24)\,\mathrm{pal}(9, \theta - 1/24)$ from $\theta = 1/24$ to $\theta = 1 + 1/24$, since the steps have all the same width. The multiplication is done in Fig. 151 by the amplitude reversing amplifier AR, which produces the voltages $S(\theta - 1/24)$ and $-S(\theta - 1/24)$, and the multipliers MR, which are single-pole, double-throw switches. The output voltages of the multipliers are fed to integrate-and-hold circuits I that integrate over the time interval $1/24 < < \theta < 1 + 1/24$, sample the integrated voltage at the time $\theta = 1 + 1/24$ and hold it from $\theta = 1 + 1/24$ to $\theta = 2 + 1/24$.

The step function $S(\theta)$ in Fig. 152 has steps of $125/12 \times 10^6 = 10.4167\,\mu s$ duration. The steps must be sampled $12 \times 10^6/125 = 96000$ times per

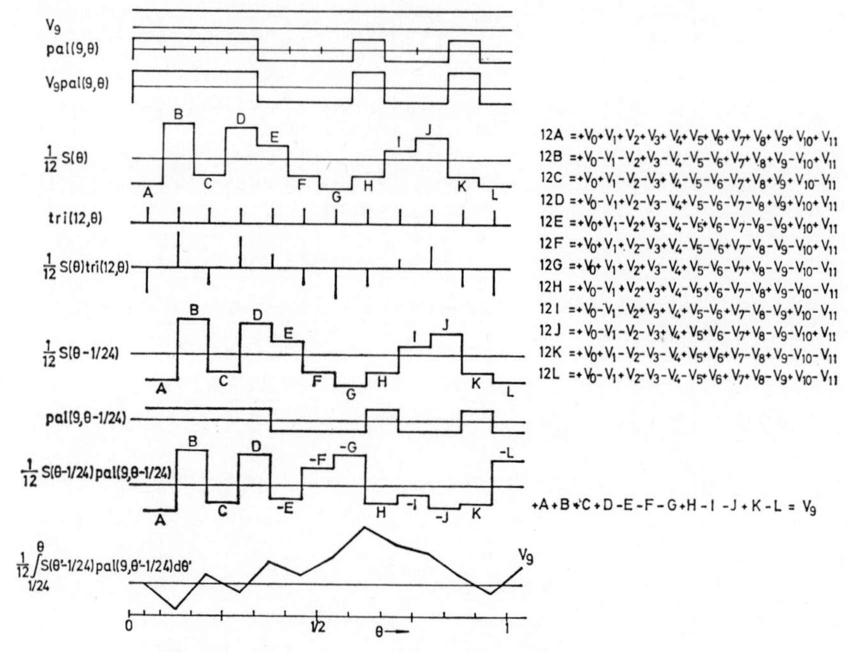

Fig. 152. Time variation of certain voltages in Fig. 151.

second. The frequency band $\frac{1}{2} \leqq \nu \leqq 1$ in Fig. 115 becomes $48 \leqq f \leqq$ $\leqq 96$ kHz. This is unfortunately not the usual band of the group filter and the signal has to be shifted by 16 kHz to pass through the group filter.

Two possible ways to shift the sequency multiplexed signal $S(\theta)$ of Fig. 152 into the group band 60 kHz $\leqq f \leqq$ 108 kHz will be discussed with reference to Fig. 153. Line a shows the frequency power spectrum of a delta pulse used for sampling. Passing the sampled pulses through a low-pass filter with cut-off frequency $f_0 = 48$ kHz produces the power spectrum b. Double sideband modulation of a carrier with frequency $f_c = 60$ kHz produces the power spectrum c. One may suppress the signal in the band 12 kHz $\leqq f \leqq$ 60 kHz as shown in line d. This method requires a low-pass filter with carefully controlled Nyquist [1] roll-off at

$f_0 = 48$ kHz and a high-pass or band-pass filter with carefully controlled roll-off at $f_c = 60$ kHz.

Using a band-pass filter with pass-band 48 kHz $\leqq f \leqq$ 96 kHz one obtains the power spectrum e from the spectrum a. The roll-off of this filter at $f_0 = 48$ kHz and $2f_0 = 96$ kHz must be carefully controlled.

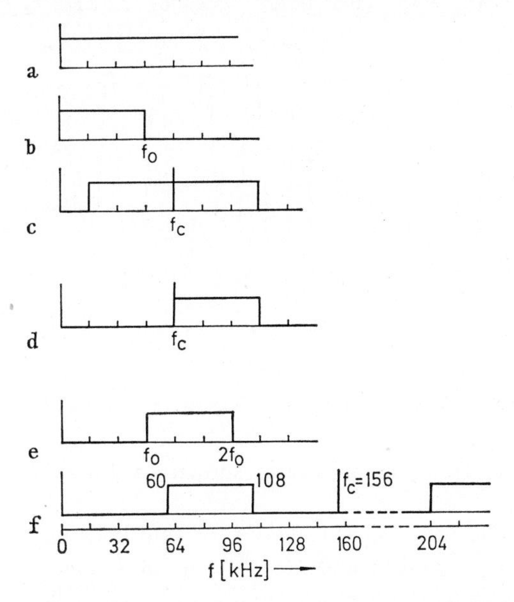

Fig. 153. Frequency shifting by sampling and by amplitude modulation of a sinusoidal carrier.

Shifting the signal into the band 60 kHz $\leqq f \leqq$ 108 kHz seems to be done best by modulating a carrier with frequency $f_c = 156$ kHz and using the lower sideband as shown by the power spectrum f. The upper sideband starting at 204 kHz is so far away that it will be suppressed by the group filter.

The second method requires one high quality filter per group. If several groups are generated in one exchange it will be more economical not to use the easy to produce sampling functions $\mathrm{tri}(12, \theta)$ of Fig. 152 and several expensive sampling filters SF according to Fig. 151, but the hard to produce function $(\sin 2\pi\,\theta)/\pi\,\theta - (\sin \pi\,\theta)/\pi\,\theta$ of Fig. 116 and inexpensive filters SF. The cost of producing the more complicated functions is then shared by several groups. The details of this process will not be discussed here since they have to be dealt with in the following example of a multiplex system, which is compatible at the telephone channel level rather than the group level but still avoids high quality frequency filters for each channel or group.

Substituting $\nu_0 = 4$ into Eq. (3.1.3-10) yields the function

$$f(4, \theta - i) = \frac{\sin 9\pi(\theta - i)}{\pi(\theta - i)} - \frac{\sin 8\pi(\theta - i)}{\pi(\theta - i)} \quad (1)$$

which has a constant power density spectrum in the band $4 \leqq \nu_0 \leqq 4.5$ and no power outside. This band becomes $32\text{ kHz} \leqq f \leqq 36\text{ kHz}$ for

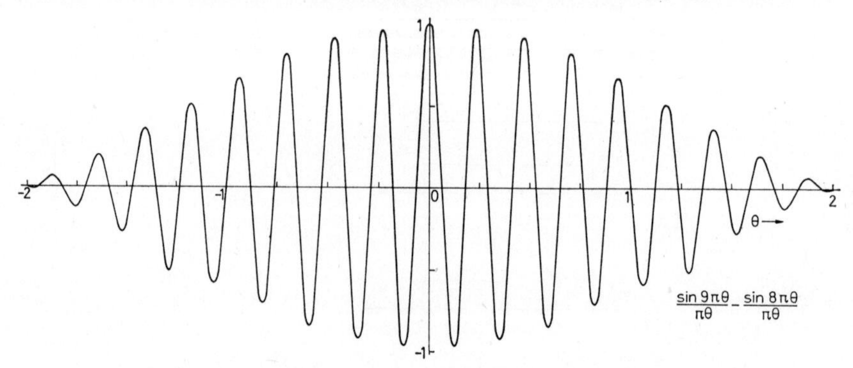

Fig. 154. The function $f(4, \theta)$ according to Eq. (3.1.3-10).

$T = 125\ \mu$s. The function $f(4, \theta)$ is shown in Fig. 154 in the interval $-2 \leqq \theta \leqq 2$.

The use of the function $f(4, \theta - i)$ for multiplexing will be discussed with reference to Figs. 155 and 156. The input signal at the point a 1 is

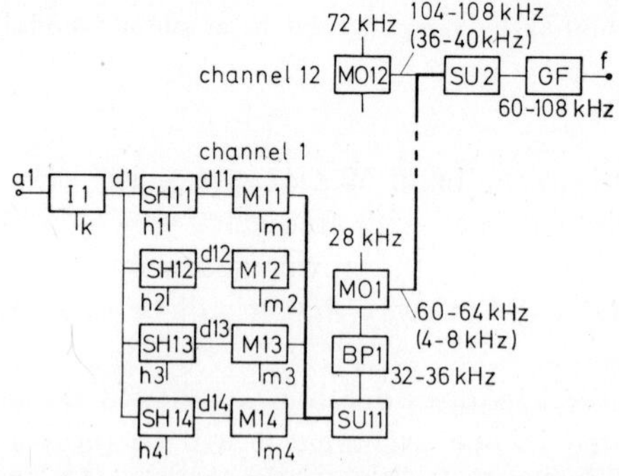

Fig. 155. Block diagram for frequency multiplexing using the function $f(4, \theta)$. I, integrator; SH, sample-and-hold circuit; M, multiplier; SU, summing amplifier; BP, band-pass filter; MO, modulator; GF, group filter.

averaged over periods of duration $T = 125\ \mu$s by the integrator I 1. The output voltage of the integrator at the end of an averaging period is sampled and stored in one of the four sample-and-hold circuits SH 11, . . ., SH 14. Since four holding circuits are used the voltages can be stored $4 \times 125\ \mu$s $= 500\ \mu$s. The four functions $f(4, \theta)$, $f(4, \theta - 1)$, $f(4, \theta - 2)$ and

$f(4, \theta - 3)$ are multiplied in the multipliers M 11, ..., M 14 with the stored voltages and the products are added in the summing amplifier SU 11. The output signal of SU 11 has most of its energy in the frequency

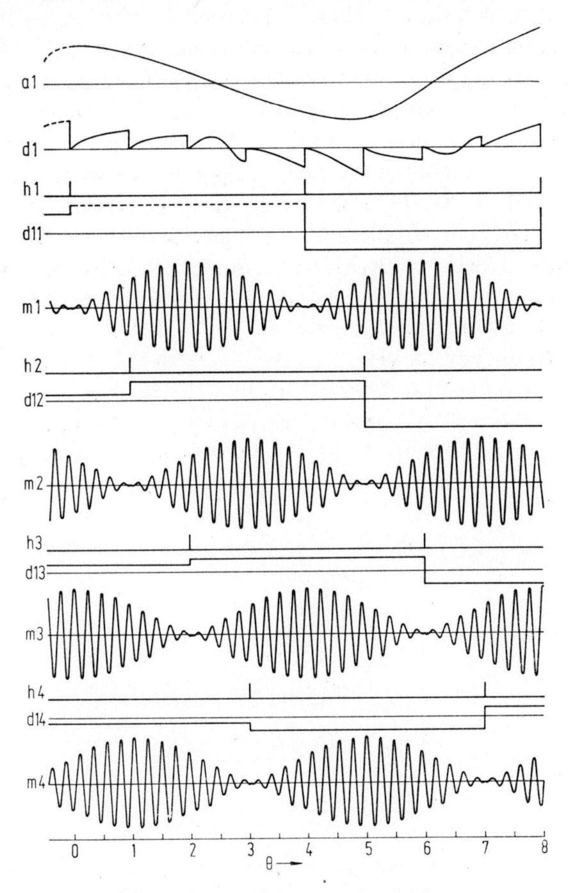

Fig. 156. Time diagram for Fig. 155.

band 32 kHz $\leqq f \leqq$ 36 kHz. It is not exactly band limited since the functions $f(4, \theta - i)$ in Fig. 156 are time limited in the interval $-2 \leqq \theta - i \leqq 2$ and the circuits have practical limitations. A simple band-pass filter BP 1 with about 30 dB attenuation outside the pass-band 32 kHz $\leqq f \leqq$ 36 kHz will increase the total attenuation to about 70 dB. A modulator MO 1 fed with a 28 kHz carrier shifts the signal to the upper sideband 60 kHz $\leqq$ $\leqq f \leqq$ 64 kHz and to the lower sideband 4 kHz $\leqq f \leqq$ 8 kHz.

The eleven carriers with frequencies 32 kHz, 36 kHz, ..., 72 kHz are used for the other eleven channels. Summation of all twelve channels yields the upper sideband signal in the band 60 kHz $\leqq f \leqq$ 108 kHz while the lower sideband signal is located in the band $0 \leqq f \leqq 44$ kHz. A group filter GF suppresses the lower sideband.

The circuit of Fig. 155 looks rather complex but it is well suited for integrated circuit techniques. There is no need to discuss the integrator I 1 and the sample-and-hold circuit. The multipliers M 11, ..., M 14 should be of the Schmid type mentioned in section 3.2.1. The functions $f(4, \theta), \ldots, f(4, \theta - 3)$ are fed as streams of binary numbers to the multipliers. The generation of these numbers by a read-only storage is costly, but the number generator can be shared by many telephone channels. The circuit in Fig. 155 from I 1 to SU 11 contains no coils and needs no tuning. The band-pass filter BP 1 is simple and only one kind of band-pass filter with pass band from 32 kHz to 36 kHz is required for all twelve channels.

For the demultiplexing process consider Figs. 157 and 158. The group signal is fed to twelve modulators MO 1, ..., MO 12 and multiplied by carriers with frequency 28 kHz, 32 kHz, ..., 72 kHz. The resulting signals are fed through band-pass filters BP 1 like the ones in Fig. 155, with an attenuation of about 30 dB outside the pass-band. The output of the filter is fed to the four multipliers M 11, ..., M 14, which are equal to those in

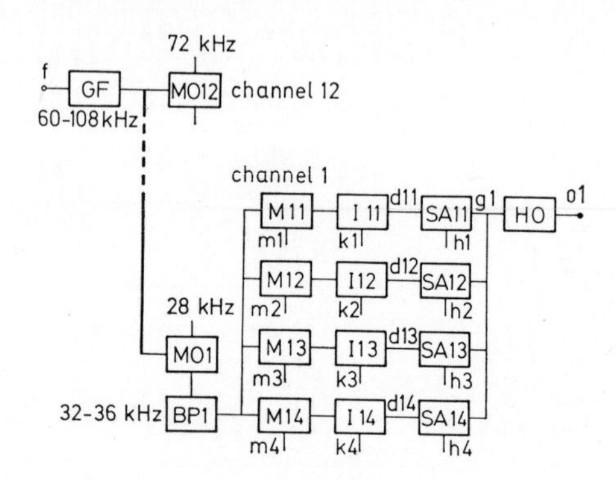

Fig. 157. Demultiplexing using the function $f(4, \theta)$. GF, group filter; MO, modulator; BP, band-pass filter; M, multiplier; I, integrator; SA, sampling switch; HO, holding circuit.

Fig. 155, and multiplied with the functions $f(4, \theta), \ldots, f(4, \theta - 3)$. The products are integrated over periods of 500 µs duration, the integration intervals being staggered by 125 µs as shown by the lines $d11$ to $d14$ in Fig. 158. The output voltages of the integrators are sampled just before resetting by the sampling pulses $h1$ to $h4$. The sampled voltages $g1$ are stored in the holding circuit HO till the next sample is taken. Note that the samples $g1$ represent the voltages in line $d1$ of Fig. 156 at the end of the integration intervals. The holding circuit HO in Fig. 157 transforms the Dirac samples $g1$ intoa step function $o1$ which can be fed directly to a headset.

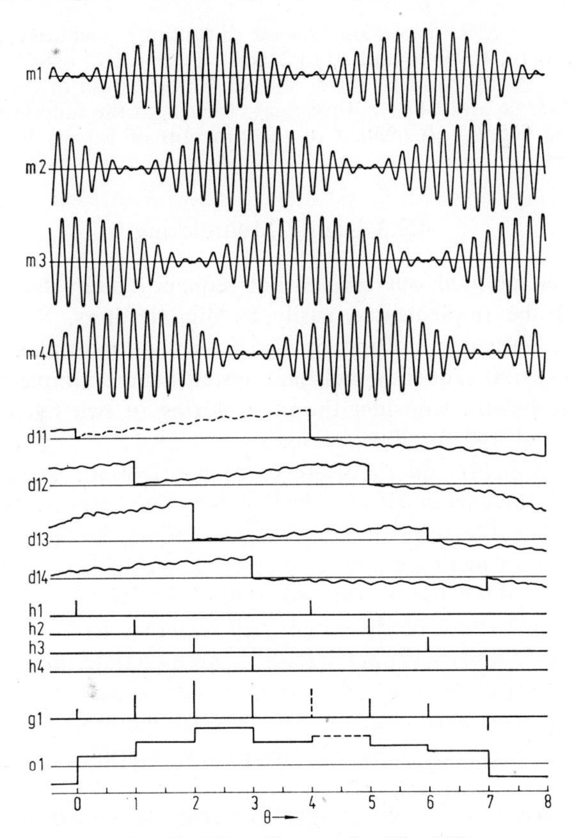

Fig. 158. Time diagram for Fig. 157.

Problems

422.1 Three PCM voice channels are operated at a transmission rate of 192 kilobits per second via a satellite link. Design a sequency multiplex system that will pass them through the group band 60—108 kHz of the telephone network. Compare the amplitude distribution with that of a time multiplex system using 8 amplitude levels.

422.2 Substitute the function $f(7/2, \theta - i)$ for $f(4, \theta - i)$ in Eq. (1) and discuss the resulting changes.

422.3 Design a read-only storage that produces the functions $f(4, \theta) \ldots f(4, \theta - 3)$ in digital form and repeats them periodically.

422.4 Compute the fraction of power outside the band $4 \leqq \nu \leqq 4.5$ if the function $f(4, \theta)$ is time limited in the interval $-2 \leqq \theta \leqq +2$. What fraction of the power is in the bands $3.5 \leqq \nu \leqq 4$ and $4.5 \leqq \nu \leqq 5$? Express this power in dB referred to the power in the band $4 \leqq \nu \leqq 4.5$. How much additional attenuation by a filter is required to obtain 70 dB crosstalk attenuation to adjacent bands?

422.5 Use the formula $\sin 9x = \sin 8x \cos x + \cos 8x \sin x$ to show that $f(4, \theta - i)$ in Eq. (1) represents the single sideband modulated signal $[\sin \pi(\theta - i)]/\pi(\theta - i)$.

422.6 Use the result of problem 313.4 to show that the sampling filters SF and SP as well as modulator MO and demodulator DE in Fig. 151 are not needed.

422.7 A much used multiplexing scheme shifts three telephone channels into the bands 12 to 16, 16 to 20 and 20 to 24 kHz. Four such "pregroups" are then shifted into the group band by lower sideband modulation of carriers with frequency 120, 108, 96 and 84 kHz. Investigate the use of the functions $f(\frac{3}{2}, \theta - i)$, $f(2, \theta - i)$ and $f(\frac{5}{2}, \theta - i)$ instead of the function of Eq. (1) for multiplexing into the pregroup.

4.2.3 Digital Multiplexing

It has been pointed out before that sequency filters based on Walsh functions can be implemented easily as digital filters. Since band-pass filters require sequency shifting of signals just as multiplex systems do, one will expect that sequency multiplex systems can be implemented easily by digital equipment. Consider the multiplexing of two telephony signals according to Fig. 142 for illustration.

Two signals $F_1(\theta)$ and $F_5(\theta)$ are to be multiplexed. These signals are represented by the curves a and a' in Fig. 142. The amplitudes in a particular interval, say the interval 125 μs $< t <$ 250 μs, are transformed into digital form by an analog/digital converter. Table 13 lists the digital representation +1101100 for $F_1^{\dagger\dagger}(\theta)$ and −0110101 for $F_5^{\dagger\dagger}(\theta)$. Multiplexing of these two values will be discussed with reference to Table 13. It is assumed that $F_1(\theta)$ and $F_5(\theta)$ are signals of an 8-channel multiplex system. The 8 Walsh functions wal$(0, \theta)$ to wal$(7, \theta)$ are used as carriers. Only two of the channels carry signals. This corresponds to an activity factor of 0.25, which is representative for telephony multiplex channels during peak traffic.

The two carriers wal$(1, \theta)$ and wal$(5, \theta)$ can be represented by 8 digits +1 and −1 as shown in the columns c and c' of Table 13. The carriers wal$(1, \theta)$ and wal$(5, \theta)$ amplitude modulated by $F_1^{\dagger\dagger}(\theta)$ and $F_5^{\dagger\dagger}(\theta)$ yield the numbers +1101100 and −0110101 multiplied by +1 or −1 as shown in the columns d and d'. The multiplex signal $F(\theta)$ of column e is obtained by adding the two numbers of the same line in column d and d'. The multiplex signal is represented by numbers having one more digit than $F_1^{\dagger\dagger}(\theta)$ or $F_5^{\dagger\dagger}(\theta)$.

The 9 × 8 digits – including the sign – of the signal $F(\theta)$ may be transmitted in many ways. For instance, one could use 72 block pulses with amplitudes +1 or −1. In this case, sequency division would be used for multiplexing and time division for transmission. The 72 pulses would be the same number as for time multiplexing of 8 channels, if one parity check digit were added to the 8 digits of each channel. Such a check digit would permit single error detection but no error correction.

Demodulation of $F(\theta)$ is done by multiplying $F(\theta)$ with wal$(1, \theta)$ and wal$(5, \theta)$. The resulting binary numbers are shown in columns g and g'. Integration of $F(\theta)$ wal$(1, \theta)$ and $F(\theta)$ wal$(5, \theta)$ means adding the 8 numbers in columns g and g', which yields +1101100000 and −0110101000.

Table 13. *Digital sequency multiplexing of two signals* $F_1^{\dagger\dagger}(\theta)$ *and* $F_5^{\dagger\dagger}(\theta)$ *according to Fig. 142.* $c, c', \ldots, g'$ *refer to the respective lines in Fig. 142.* $F(\theta)$ *stands for the sum* $F_1^{\dagger\dagger}(\theta)\,\mathrm{wal}(1,\theta) + F_5^{\dagger\dagger}(\theta)\,\mathrm{wal}(5,\theta)$

c	c'	d	d'	e	g	g'
		$F_1^{\dagger\dagger}(\theta)\times$ $\times\,\mathrm{wal}(1,\theta)$	$F_5^{\dagger\dagger}(\theta)\times$ $\times\,\mathrm{wal}(5,\theta)$	$F(\theta)$	$F(\theta)\,\mathrm{wal}(1,\theta)$	$F(\theta)\,\mathrm{wal}(5,\theta)$
+1	+1	+1101100	−0110101	+00110111	+00110111	+00110111
+1	−1	+1101100	+0110101	+10100001	+10100001	−10100001
+1	−1	+1101100	+0110101	+10100001	+10100001	−10100001
+1	+1	+1101100	−0110101	+00110111	+00110111	+00110111
−1	−1	−1101100	+0110101	−00110111	+00110111	+00110111
−1	+1	−1101100	−0110101	−10100001	+10100001	−10100001
−1	+1	−1101100	−0110101	−10100001	+10100001	−10100001
−1	−1	−1101100	+0110101	−00110111	+00110111	+00110111
					+1101100000	−0110101000
					+1101100	−0110101

Division by 8 yields the original values of $F_1^{\dagger\dagger}(\theta)$ and $F_5^{\dagger\dagger}(\theta)$. The practical way to obtain these numbers would of course be to apply the fast Walsh-Fourier transformation of section 1.2.5 to the multiplex signal $F(\theta)$. Inclusion of this less time-consuming method would obscure the explanation of the principle.

The signal $F(\theta)$ contains only numbers with absolute value 00110111 and 10100001; two of each have negative signs and two of each positive signs. This is typical for two active channels. Hence, if one number is changed due to interference it can be corrected by comparison with the three unchanged numbers. In most cases, it is also possible to correct two errors, and in many cases more than two errors can be corrected. There is thus a definite advantage over time division. The underlying reason is that no useful information is transmitted three fourths of the time if the activity factor is 0.25 and time division is used.

Searle pointed out that the multiplex signal $F(\theta)$ in Table 13 does not need to have more digits than the signals $F_1(\theta)$ and $F_5(\theta)$ from which it is obtained [1]. In order to accomplish this one must replace addition and subtraction by operations that leave the number of digits unchanged. The obvious replacement for addition is modulo 2 addition. In order to find a replacement for subtraction consider Table 14. It shows the truth table for the modulo 2 addition of p and q. Since $0 \oplus 0$ and $1 \oplus 1$ yield the same result one needs a second operation that yields two different results when two zeros or two ones are combined. The Searle operation $p \ominus q$ yields $0 \ominus 0 = 1$ and $1 \ominus 1 = 0$; hence it may be used. A second Searle operation $q \ominus p$, also defined by Table 14, could be used too, but the difference between $p \ominus q$ and $q \ominus p$ is trivial.

Consider now the multiplexing of the two voltages represented by the numbers $P = 11101100$ and $Q = 00110101$ (the signs + and − are replaced by 1 and 0). One obtains with the help of Table 14:

$$P \oplus Q = 11101100 \oplus 00110101 = 11011001 \tag{1}$$

$$P \mathbin{\textcircled{\scriptsize I}} Q = 11101100 \mathbin{\textcircled{\scriptsize I}} 00110101 = 11101100 \tag{2}$$

Using the inverse operations defined in Table 14 one may regain P and Q from $P \oplus Q$ and $P \mathbin{\textcircled{\scriptsize I}} Q$.

Table 14. *Truth tables for the addition modulo 2, $p \oplus q$, the Searle operations $p \mathbin{\textcircled{\scriptsize I}} q$ and $q \mathbin{\textcircled{\scriptsize I}} p$, and the inverse operations yielding again p and q*

$p \oplus q$			$p \mathbin{\textcircled{\scriptsize I}} q$			$q \mathbin{\textcircled{\scriptsize I}} p$			p			q		
	q			q			q			$p \mathbin{\textcircled{\scriptsize I}} q$			$p \mathbin{\textcircled{\scriptsize I}} q$	
	0	1		0	1		0	1		0	1		0	1
p 0	0	1	p 0	1	0	p 0	1	1	$p \oplus q$ 0	1	0	$p \oplus q$ 0	1	0
p 1	1	0	p 1	1	0	p 1	0	0	$p \oplus q$ 1	0	1	$p \oplus q$ 1	1	0

It is evident from Eq. (2) that the operation $P \mathbin{\textcircled{\scriptsize I}} Q$ yields the number P regardless of the value of Q. This is not satisfactory since it means the numbers P and Q are mixed or multiplexed in $P \oplus Q$ only, but not in $P \mathbin{\textcircled{\scriptsize I}} Q$. One way to overcome this difficulty is by operating on pairs of binary digits rather than on single digits. Table 15 shows the truth table for the addition modulo 2 of number pairs p and q, which may of course be infered from Table 14. New in Table 15 is the Searle operation $p \mathbin{\textcircled{\scriptsize I}} q$ on number pairs p and q. The entries in this truth table depend on p *and* q, while the entries for $p \mathbin{\textcircled{\scriptsize I}} q$ and $q \mathbin{\textcircled{\scriptsize I}} p$ in Table 14 depended on either q *or* p only[1].

Table 16 shows in column 0 eight binary characters $A, \ldots, H$ that are to be multiplexed. Column 1 shows eight numbers $A', \ldots, H'$ obtained by modulo 2 addition and Searle operation from the characters $A, \ldots, H$. The same processes yield columns 2 and 3. A comparison with Table 2 in section 1.2.5 shows that Table 16 defines a fast logical Walsh-Fourier transform[2]. The slow way for obtaining it is defined by the operations on the characters $A, \ldots, H$, rather than $A'', \ldots, H''$, in column 3.

[1] Using ordinary addition and subtraction one has the relations $(p+q)+(p-q) = 2p$ and $(p+q)-(p-q) = 2q$. Using modulo 2 addition and Searle operation one obtains in analogy $(p \oplus q) \oplus_i (p \mathbin{\textcircled{\scriptsize I}} q) = p$ and $(p \oplus q) \mathbin{\textcircled{\scriptsize I}}_i (p \mathbin{\textcircled{\scriptsize I}} q) = q$. Table 15 defines the operations $\oplus$, $\mathbin{\textcircled{\scriptsize I}}$ and the inverse operations $\oplus_i$, $\mathbin{\textcircled{\scriptsize I}}_i$.

[2] The logical Walsh transform is introduced here for a very specialized use. It is, however, evident that it could be used instead of the ordinary Walsh transform in applications like radar signal processing (1.2.8), digital filtering (2.5) and generally whenever the Walsh transformation is performed by digital computers or digital rather than analog circuits. The resulting reduction in processing time and equipment complexity is of great interest but little is known about the usefulness of the logical transforms. From the theoretical point of view the logical transform discards the topology of the real numbers for the amplitude y of a function $y(\theta)$ just like this topology was discarded for the variable θ in section 1.2.8.

Table 15. *Truth tables for the addition modulo 2, the Searle operation $p \varobar q$ and the inverse operations for number pairs p and q*

$p \oplus q$		q: 00	01	10	11
p	00	00	01	10	11
	01	01	00	11	10
	10	10	11	00	01
	11	11	10	01	00

$p \varobar q$		q: 00	01	10	11
p	00	00	10	11	01
	01	11	01	00	10
	10	01	11	10	00
	11	10	00	01	11

p		$p \varobar q$: 00	01	10	11
$p \oplus q$	00	00	01	10	11
	01	10	11	00	01
	10	11	10	01	00
	11	01	00	11	10

q		$p \varobar q$: 00	01	10	11
$p \oplus q$	00	00	01	10	11
	01	11	10	01	00
	10	01	00	11	10
	11	10	11	00	01

Table 16. *Digital multiplexing of eight characters $A \ldots H$ by means of the fast logical Walsh-Fourier transformation* The carriers $\mathrm{wal}(j, \theta)$ are ordered according to the Gray code of the normalized sequency.

Carrier	0	1	2	3
$\mathrm{wal}(0, \theta)$	A	$A \oplus B = A'$	$A' \oplus C' = A''$	$A'' \oplus E'' = [(A \oplus B) \oplus (C \oplus D)] \oplus [(E \oplus F) \oplus (G \oplus H)]$
$\mathrm{wal}(1, \theta)$	B	$A \varobar B = B'$	$A' \varobar C' = B''$	$A'' \varobar E'' = [(A \oplus B) \oplus (C \oplus D)] \varobar [(E \oplus F) \oplus (G \oplus H)]$
$\mathrm{wal}(3, \theta)$	C	$C \oplus D = C'$	$B' \oplus D' = C''$	$B'' \oplus F'' = [(A \oplus B) \varobar (C \oplus D)] \oplus [(E \oplus F) \varobar (G \oplus H)]$
$\mathrm{wal}(2, \theta)$	D	$C \varobar D = D'$	$B' \varobar D' = D''$	$B'' \varobar F'' = [(A \oplus B) \varobar (C \oplus D)] \varobar [(E \oplus F) \varobar (G \oplus H)]$
$\mathrm{wal}(7, \theta)$	E	$E \oplus F = E'$	$E' \oplus G' = E''$	$C'' \oplus G'' = [(A \varobar B) \oplus (C \varobar D)] \oplus [(E \varobar F) \oplus (G \varobar H)]$
$\mathrm{wal}(6, \theta)$	F	$E \varobar F = F'$	$E' \varobar G' = F''$	$C'' \varobar G'' = [(A \varobar B) \oplus (C \varobar D)] \varobar [(E \varobar F) \oplus (G \varobar H)]$
$\mathrm{wal}(4, \theta)$	G	$G \oplus H = G'$	$F' \oplus H' = G''$	$D'' \oplus H'' = [(A \varobar B) \varobar (C \varobar D)] \oplus [(E \varobar F) \varobar (G \varobar H)]$
$\mathrm{wal}(5, \theta)$	H	$G \varobar H = H'$	$F' \varobar H' = H''$	$D'' \varobar H'' = [(A \varobar B) \varobar (C \varobar D)] \varobar [(E \varobar F) \varobar (G \varobar H)]$

A numerical example is given in Table 17. The two characters $+1101100$ and -0110101 are to be transmitted by the carriers $\mathrm{wal}(3, \theta)$ and $\mathrm{wal}(6, \theta)$, while the six other carriers transmit zeros only. The binary digits are grouped in pairs to facilitate use of Table 15. The operations shown in Table 16 with the definitions for $p \oplus q$ and $p \obar q$ in Table 15 yield the numbers in columns 1, 2 and 3.

The numbers in column 3 of Table 17 contain a total of 8×8 digits instead of the 9×8 digits of column e in Table 13, while the error correcting and error detecting features are essentially unchanged.

Table 17. *Digital multiplexing of two signals by means of the fast logical Walsh-Fourier transformation (+ is replaced by* 1, *— by* 0)

Carrier	0	1	2	3
$\mathrm{wal}(0, \theta)$	0	0	11 10 11 00	11 01 10 01
$\mathrm{wal}(1, \theta)$	0	0	01 11 01 00	10 00 00 10
$\mathrm{wal}(3, \theta)$	11 10 11 00	11 10 11 00	10 01 10 00	01 01 10 11
$\mathrm{wal}(2, \theta)$	0	10 01 10 00	11 10 11 00	11 01 10 01
$\mathrm{wal}(7, \theta)$	0	00 11 01 01	00 11 01 01	10 00 00 10
$\mathrm{wal}(6, \theta)$	00 11 01 01	00 01 10 10	00 10 11 11	01 01 10 11
$\mathrm{wal}(4, \theta)$	0	0	00 01 10 10	11 01 10 01
$\mathrm{wal}(5, \theta)$	0	0	00 11 01 01	10 00 00 10

The demultiplexing in Table 17 is done by going from column 3 back to column 0, using the inverse operations of Table 15 that yield p and q for any pair of numbers $p \oplus q$ and $p \obar q$.

Problems

423.1 Write Table 13 with the carriers $\mathrm{wal}(3, \theta)$ and $\mathrm{wal}(6, \theta)$ substituted for $\mathrm{wal}(1, \theta)$ and $\mathrm{wal}(5, \theta)$.

423.2 Obtain the numbers in column 3 of Table 17 by the slow transformation defined in column 3 of Table 16.

423.3 Why can a binary digital computer perform the fast logical Walsh-Fourier transformation faster than the fast Walsh-Fourier transformation of section 1.2.5?

423.4 Can you devise a fast logical Haar-Fourier transform? Start with a truth table for modulo 3 addition and find a Searle process that uses ternary digits. Would such a ternary logical Haar-Fourier transform bring a reduction in computing time of a binary computer compared with the transform of section 1.2.6?

423.5 Show that a circuit yielding $p \oplus q$, $p \obar q$ according to Table 15 and the complements of $p \oplus q$, $p \obar q$ can be implemented by eight 2-input AND gates, four 2-input NOR gates and four NAND gates used as inverters, if p, q as well as their complements are available.

423.6 Show that the reconversion of $p \oplus q$, $p \obar q$ and their complements into p, q and their complements according to Table 15 can be done by ten 2-input AND gates, five 2-input NOR gates and five NAND gates used as inverters.

423.7 How is Table 15 changed if the definitions $(p \oplus q) \oplus_i (p \obar q) = 2p$ and $(p \oplus q) \obar_i (p \obar q) = 2q$ are used for the inverse operations to emphasize the similarity to ordinary addition and subtraction?

423.8 The extension of modulo 2 addition to negative numbers is given by Eq. (1.1.4-14). Explain why the following definitions of the Searle operation for negative numbers are useful: $(-a) \textcircled{1} (-b) = -(a \textcircled{1} b)$, $(-a) \textcircled{1} b = -(a \textcircled{1} b)$, $a \textcircled{1} (-b) = a \textcircled{1} b$.

423.9 Derive the rules for the inverse operations $\oplus_i$ and $\textcircled{1}_i$ applied to negative numbers, e.g. $(-a) \oplus_i b = a \oplus_i b$, $(-a) \textcircled{1}_i (-b) = a \textcircled{1}_i b$.

423.10 Show that an Exclusive-OR gate produces the sign of $p \oplus q$ and $p \textcircled{1}_i q$ while no gate is required to produce the sign of $p \textcircled{1} q$ and $p \oplus_i q$, where p and q are positive or negative numbers, if $+$ is represented by 0 and $-$ by 1.

4.3 Time Base, Time Position and Code Modulation

4.3.1 Time Base Modulation (TBM)

Any carrier can be amplitude modulated if it can be written as time function $V\Phi(k, \theta + \theta_0) = V\Phi(k, t/T + t_0/T)$. One will expect that three more individual modulation methods can be defined, since this carrier contains the normalized sequency k, the time base T and the delay t_0 besides the amplitude V. Modulation of T is called a time base modulation.

The basic idea is to replace θ by a function $g(\theta)$. There are several ways to do this. Let $F(\theta)$ be the modulating signal and M a modulation index. One may use the definition:

$$\Phi(k, \theta) = \Phi[k, g(\theta)] \tag{1}$$

$$g(\theta) = \int [1 + MF(\theta)]\, d\theta = \frac{1}{T} \int [1 + MF(t/T)]\, dt \tag{2}$$

This is the approach taken in frequency modulation of sinusoidal carriers. However, the advantages of this approach are strongly connected to the fact that frequency and time are combined as product,

$$\Phi(k, \theta) = \sin k\,\theta \tag{3}$$

for sinusoidal functions. The comma between k and θ in the general case makes the following definition of $g(\theta)$ more advantageous:

$$g(\theta) = \theta[1 + MF(\theta)] = t\frac{1 + MF(\theta)}{T} \tag{4}$$

$$|MF(\theta)| < 1$$

The modulated parameter is now clearly the time base T. Figure 159 shows how a sine function and a Walsh function are changed if the time base T is changed into $3T/4$ and $T/2$.

The modulation index M may be positive or negative. Larger values of $F(\theta)$ reduce the time base for $M > 0$ and increase the time base for $M < 0$. This is in close analogy to frequency modulation, where an increased voltage of the modulating signal may increase or decrease the frequency of the carrier.

Figure 159 shows that the required bandwidth increases with the modulation index M. The shortest time base shown is half as wide as the longest. The frequency or sequency bandwidth occupied by the short functions $\sin 2\pi\,\theta''$ or $\mathrm{sal}(3, \theta'')$ is twice as large as that occupied by the long functions $\sin 2\pi\,\theta$ or $\mathrm{sal}(3, \theta)$. A detailed analysis of energy distribution as function of sequency for various values of M and signals $F(\theta)$ is still lacking.

A possible circuit for time base modulation of Walsh functions is shown in Fig. 160. Let the signal have the shape shown by the first line of the pulse diagram. It is sampled at time $\theta = 0$ by the amplitude sampler AT(a); the sampled voltage is stored in the holding circuit SP(b). An integrator I

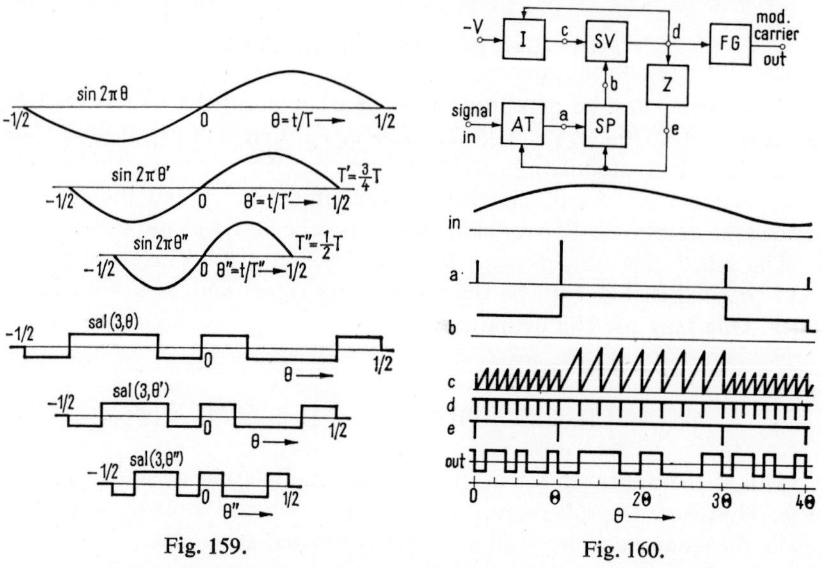

Fig. 159. Fig. 160.

Fig. 159. Time base modulation of a sine and a Walsh function.

Fig. 160. Block and time diagram for time base modulation of Walsh carriers. AT, amplitude sampler; I, integrator; SV, voltage comparator; SP, storage; Z, counter; FG, function generator.

produces a ramp voltage. A voltage comparator SV compares this ramp voltage with the one held in SP and resets integrator I when both voltages become equal. A sawtooth voltage (c) results. The amplitude and duration of the sawteeth are proportionate to the voltage stored in SP.

The pulses (d) from the comparator SV which reset integrator I are also fed into the counter Z. A pulse (e) is generated by Z if a certain number of pulses has been received from SV; this number is 8 in Fig. 160. The pulse (e) clears SP and stores a new amplitude sample of the signal via sampler AT. Note that the distance between sampling points depends on the sampled amplitude. Counter Z is reset, when the pulse (e) is generated. This resetting happens at the times 0, Θ, 3Θ, and 4Θ. The amplitude of the signal at time Θ is twice as large as at time 0. Hence, the saw-teeth are twice as long as before. The 8 pulses (*d*) generated by the voltage comparator SV in the time interval $0 \leqq \theta \leqq \Theta$ have twice the distance as in the interval $0 \leqq \theta \leqq \Theta$. Feeding these pulses into a function generator for Walsh functions generates time base modulated Walsh functions at its output; Fig. 160 shows the modulated carrier $\mathrm{sal}(3, \theta)$.

4.3.2 Time Position Modulation (TPM)

The variable θ of the carrier $V\Phi(k, \theta + \theta_0)$ was replaced by a function $g(\theta)$ in the case of time base modulation. The parameter θ_0 is replaced by a function $h(\theta)$ in the case of time position modulation. Let $F(\theta)$ again denote the modulating signal and M a modulation index. The following definitions are introduced:

$$\begin{aligned} \Phi(k, \theta + \theta_0) &= \Phi[k, \theta + h(\theta)] \\ h(\theta) &= \theta_0 + MF(\theta) \end{aligned} \tag{1}$$

The modulation index M may be positive or negative. $\Phi[k, \theta + \theta_0 + M F(\theta)]$ will be shifted toward larger values of θ for larger values of $F(\theta)$ if M is negative; the opposite holds if M is positive. This corresponds to phase modulation, where the phase of the carrier may be advanced or retarded by a larger amplitude of the signal. Figure 161 shows a sinusoidal carrier $\Phi(1, \theta) = \sin 2\pi\, \theta$ for the three shifts $MF(\theta) = 0, -\frac{1}{4}$ and $-\frac{1}{2}$. Below is shown the Walsh carrier $\Phi(3, \theta) = \mathrm{sal}(3, \theta)$ for the same

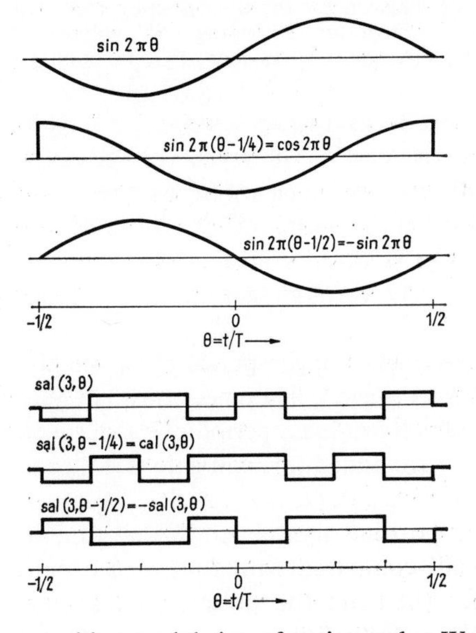

Fig. 161. Time position modulation of a sine and a Walsh function.

three shifts. Note that the section of a function which projects beyond the limits $+\frac{1}{2}$ or $-\frac{1}{2}$ due to a shift is added at the other end of the function.

Figure 162 shows a block diagram and a pulse diagram for time position modulation of Walsh carriers. The amplitude sampler AT samples periodically the amplitude of the input signal at the times $0, \Theta, 2\Theta, \ldots$ and the resulting voltages are held for a certain time (b) in a holding circuit

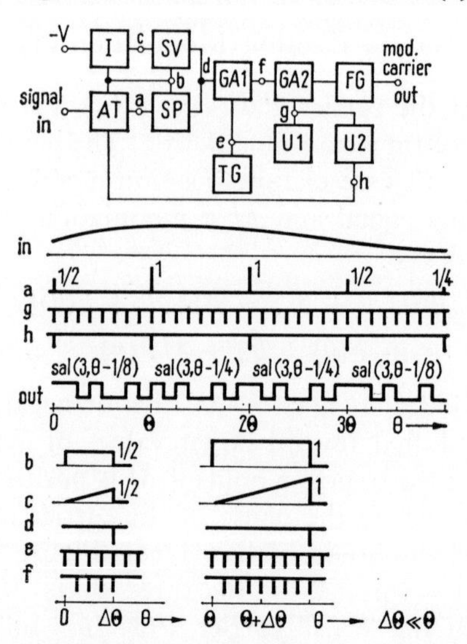

Fig. 162. Block and time diagram for the time position modulation of Walsh carriers. AT, amplitude sampler; I, integrator; SP, storage; SV, voltage comparator; TG, trigger generator; GA, gate; U, divider; FG, function generator.

SP. An integrator I produces a ramp voltage (c). A voltage comparator SV generates a pulse (d) as soon as the ramp voltage reaches the value of the voltage stored in SP. This pulse clears SP and resets integrator I. Positive pulses (b) are obtained at the output of SP, whose duration is proportionate to the amplitude of the sampled voltage. Trigger pulses (e) may pass from the trigger generator TG through gate GA 1 as long as the pulse (b) is present (f).

A divider U 1 produces trigger pulses (g) from the trigger pulses (e), that have a much larger period. They pass through gate GA 2 to the function generator FG, which produces periodic Walsh functions, e.g., $\mathrm{sal}(3, \theta)$. The trigger pulses (f) are added through gate GA 2 to the trigger pulses (g) immediately after the times $0, \Theta, 2\Theta, \ldots$ The output of the function generator FG is a time position modulated Walsh function, if the period of the pulses (e) is small compared with that of the pulses (g). The divider U 2 produces pulses (h) from the pulses (g) that turn on the sampling circuit AT at the times $0, \Theta, 2\Theta, \ldots$

Demodulation circuits for time base and time position modulated Walsh carriers have been devised. They are based on the same principles used for the modulating circuits, but depend strongly on the transmission link envisaged.

4.3.3 Code Modulation (CM)

Modulation of the normalized sequency k of a carrier $V\Phi(k, \theta + \theta_0)$ is called code modulation for the following reason: k distinguishes the functions of a system, which is evident if a particular system of functions is substituted for $\Phi(k, \theta)$, e.g., $\mathrm{wal}(k, \theta)$ with $k = 0, 1, 2, \ldots$ The 128 signals that may be constructed from 7 binary block pulses form such a system with $k = 0, 1, \ldots, 127$. These signals are used for transmission of telephony signals by means of pulse code modulation. This suggests a consideration of modulation of the normalized sequency as a generalization of pulse code modulation.

Code modulation of functions such as $\mathrm{sal}(i, \theta)$ and $\mathrm{cal}(i, \theta)$ means a discontinuous change of the functions, since i can assume integer values only. This is in contrast to amplitude, time base and time position modulation, which permit continuous changes. However, the functions $\mathrm{sal}(\mu, \theta)$ and $\mathrm{cal}(\mu, \theta)$ are defined for all real values of μ with the exception of $\mathrm{sal}(0, \theta)$. Hence, code modulation may be continuous, at least in theory[1].

There is no essential difference between code modulation and time base modulation for sinusoidal functions, since i and θ are connected as product and not separated by a comma as for Walsh and other functions. It holds:

$$\sin i\,\theta = \sin \frac{i}{T}\,t \tag{1}$$

A modulation of i may be interpreted as a modulation of $1/T$ and vice versa.

There are many possible modulators and demodulators for code modulation. Using integer values of the normalized sequency i only, one may produce all functions $\Phi(i, \theta)$ and connect the proper one through a switch to a common line. The demodulator may be based on cross-correlation of the received functions with all possible ones. A more ingenious demodulator for Walsh functions may use the fast Walsh-Fourier transform of section 1.2.5 as was done by Green and collaborators. An even simpler demodulator can be based on the fast logical Walsh-Fourier transform of section 4.2.3.

[1] In reality the situation is reversed since an infinitesimal change of amplitude, time base or time position can neither be produced nor detected.

5. NONSINUSOIDAL ELECTROMAGNETIC WAVES

5.1 *Dipole Radiation of Walsh Waves*

5.1.1 Hertzian Dipole Solution of Maxwell's Equations

Maxwell's equations written in the MKSA system of Giorgi have the following form:

$$\operatorname{curl}\mathbf{H} = \frac{\partial \mathbf{D}}{\partial t} + \mathbf{g}, \qquad \operatorname{curl}\mathbf{E} = -\frac{\partial \mathbf{B}}{\partial t} \tag{1}$$

$$\operatorname{div}\mathbf{D} = \varrho, \qquad \operatorname{div}\mathbf{B} = 0 \tag{2}$$

In empty space the electric and magnetic field strengths **E** and **H** are connected with the electric displacement **D** and the magnetic induction **B** by two additional equations:

$$\mathbf{D} = \varepsilon_0 \mathbf{E}, \qquad \mathbf{B} = \mu_0 \mathbf{H} \tag{3}$$

ε_0 is the dielectric constant and μ_0 the magnetic permeability of empty space. For engineering purposes they are mainly of interest in the following two combinations:

$$Z_0 = \sqrt{\left(\frac{\mu_0}{\varepsilon_0}\right)} \cong 377\,\Omega, \qquad c = \frac{1}{\sqrt{(\varepsilon_0\,\mu_0)}} \cong 3 \cdot 10^8 \text{ m/s} \tag{4}$$

Current density **g** and charge density ϱ are connected by the continuity equation:

$$\operatorname{div}\mathbf{g} + \frac{\partial \varrho}{\partial t} = 0 \tag{5}$$

The computation of electromagnetic field strengths in empty space is simplified by the introduction of a vector potential **A** and a scalar potential ϕ:

$$\mathbf{H} = \frac{1}{\mu_0}\operatorname{curl}\mathbf{A}, \qquad \mathbf{E} = -\frac{\partial \mathbf{A}}{\partial t} - \operatorname{grad}\phi \tag{6}$$

The vector potential is not completely specified and an additional condition may be introduced, which is called Lorentz convention:

$$\operatorname{div}\mathbf{A} + \varepsilon_0\,\mu_0\,\frac{\partial\phi}{\partial t} = 0 \tag{7}$$

Introduction of Eq. (6) into Eq. (1) yields with the help of Eq. (7) equations for the potentials:

$$\nabla^2\mathbf{A} - \varepsilon_0\,\mu_0\,\frac{\partial^2\mathbf{A}}{\partial t^2} = -\mu_0\,\mathbf{g}, \quad \nabla^2\phi - \varepsilon_0\,\mu_0\,\frac{\partial^2\phi}{\partial t^2} = -\frac{\varrho}{\varepsilon_0} \tag{8}$$

The following integrals taken over the whole space are particular solutions of these partial differential equations:

$$\mathbf{A}(x, y, z, t) = \frac{\mu_0}{4\pi}\iiint\frac{\mathbf{g}(\xi, \eta, \zeta, t - r/c)}{r}\,d\xi\,d\eta\,d\zeta \tag{9}$$

$$\phi(x, y, z, t) = \frac{1}{4\pi\,\varepsilon_0}\iiint\frac{\varrho(\xi, \eta, \zeta, t - r/c)}{r}\,d\xi\,d\eta\,d\zeta \tag{10}$$

r is the distance between the points with coordinates x, y, z and ξ, η, ζ:

$$r = [(x-\xi)^2 + (y-\eta)^2 + (z-\zeta)^2]^{1/2} \tag{11}$$

Let two time variable charges $+q(t)$ and $-q(t)$ be located at the ends of a (very short) Hertzian dipole represented by the dipole vector $\mathbf{s}$. A dipole moment and its derivative are defined:

$$\mathbf{p} = q(t)\,\mathbf{s}, \quad \dot{\mathbf{p}} = \frac{d\mathbf{p}}{dt} = i(t)\,\mathbf{s} \tag{12}$$

$i(t)$ is the current flowing in the dipole. It is assumed that s is so short that $q(t)$ and $i(t)$ do not vary over the length of the dipole.

It can be shown[1] that the dipole moment yields the following vector potential:

$$\mathbf{A} = \mu_0\,\frac{\dot{\mathbf{p}}\,(t - r/c)}{4\pi\,r} \tag{13}$$

The Lorentz convention in Eq. (7) yields the scalar potential from Eq. (13):

$$\phi = \frac{1}{4\pi\,\varepsilon_0}\left(\frac{\dot{\mathbf{p}}(t - r/c)\,\mathbf{r}}{c\,r^2} + \frac{\mathbf{p}(t - r/c)\,\mathbf{r}}{r^3}\right) \tag{14}$$

---

[1] The derivation of the electric and magnetic field strengths in Eqs. (15) and (16) from Maxwell's equations is extremely tedious. Most textbooks use the Gaussian cgs system rather than the MKSA system which the engineer wants. Furthermore, a separation of variables by Bernoulli's product method is usually performed, which leads to solutions favoring sine-cosine functions as pointed out in section 3.3.1. The presentation used here follows closely the one in the 18th edition of the renowned book by Abraham, Becker and Sauter [5]. The development of the theory will be much more understandable from Eqs. (15) and (16) on.

To simplify writing the argument $t - r/c$ of the dipole moment will not be written from here on.

Introduction of Eqs. (13) and (14) into Eq. (6) yields the field strengths produced by the Hertzian dipole:

$$\mathbf{E} = \frac{Z_0}{4\pi}\left[\frac{1}{c\,r}\left(\frac{(\ddot{\mathbf{p}}\,\mathbf{r})\,\mathbf{r}}{r^2} - \ddot{\mathbf{p}}\right) + \frac{1}{r^2}\left(\frac{3(\dot{\mathbf{p}}\,\mathbf{r})\,\mathbf{r}}{r^2} - \dot{\mathbf{p}}\right) + \right.$$
$$\left. + \frac{c}{r^3}\left(\frac{3(\mathbf{p}\,\mathbf{r})\,\mathbf{r}}{r^2} - \mathbf{p}\right)\right] \tag{15}$$

$$\mathbf{H} = \frac{1}{4\pi}\left(\frac{1}{c\,r}\,\frac{\ddot{\mathbf{p}} \times \mathbf{r}}{r} + \frac{1}{r^2}\,\frac{\dot{\mathbf{p}} \times \mathbf{r}}{r}\right) \tag{16}$$

The first terms in Eqs. (15) and (16) decline proportionate to $1/r$. They are called wave zone components and denoted by $\mathbf{E}(1/r, t)$ and $\mathbf{H}(1/r, t)$:

$$\mathbf{E}(1/r, t) = \frac{Z_0}{4\pi\,c\,r}\left(\frac{(\ddot{\mathbf{p}}\,\mathbf{r})\,\mathbf{r}}{r^2} - \ddot{\mathbf{p}}\right) = \frac{Z_0\,s}{4\pi\,c\,r}\,\frac{di(t - r/c)}{dt}\,\frac{\mathbf{r} \times (\mathbf{r} \times \mathbf{s})}{s\,r^2} \tag{17}$$

$$\mathbf{r} \times (\mathbf{r} \times \mathbf{s}) = (\mathbf{s}\,\mathbf{r})\,\mathbf{r} - \mathbf{s}(\mathbf{r}\,\mathbf{r})$$

$$\mathbf{H}(1/r, t) = \frac{1}{4\pi\,c\,r}\,\frac{\ddot{\mathbf{p}} \times \mathbf{r}}{r} = \frac{s}{4\pi\,c\,r}\,\frac{di(t - r/c)}{dt}\,\frac{\mathbf{s} \times \mathbf{r}}{s\,r} \tag{18}$$

The wave zone components transport energy to infinity. The power $P(r, t)$ flowing at a certain time t through the surface of a large sphere with radius r is given by the surface integral of Poynting's vector:

$$\mathbf{P} = \mathbf{E}(1/r, t) \times \mathbf{H}(1/r, t) = Z_0\left(\frac{s}{4\pi\,c\,r}\right)^2\left(\frac{di(t - r/c)}{dt}\right)^2\frac{\mathbf{r} \times (\mathbf{r} \times \mathbf{s})}{s\,r^2} \times \frac{\mathbf{s} \times \mathbf{r}}{s\,r}$$
$$= Z_0\left(\frac{s\sin\alpha}{4\pi\,c\,r}\right)^2\left(\frac{di(t - r/c)}{dt}\right)^2\frac{\mathbf{r}}{r} \tag{19}$$

$$\frac{\mathbf{s}\,\mathbf{r}}{s\,r} = \cos\alpha, \qquad \frac{|\mathbf{s} \times \mathbf{r}|}{s\,r} = \sin\alpha$$

$$P(r, t) = \oiint \mathbf{P}\,d\mathbf{O} = Z_0\left(\frac{s}{4\pi\,c\,r}\right)^2\left(\frac{di(t - r/c)}{dt}\right)^2\int_{\alpha=0}^{\pi}\int_{\beta=0}^{2\pi}\sin^2\alpha \cdot r^2\sin\alpha\,d\alpha\,d\beta$$
$$= Z_0\,\frac{s^2}{6\pi\,c^2}\left(\frac{di(t - r/c)}{dt}\right)^2 \tag{20}$$

Introduction of the rms current,

$$I_{\mathrm{rms}}^2 = \langle i^2(t')\rangle, \qquad t' = t - r/c \tag{21}$$

yields the radiation resistance R_s from the average radiated power P:

$$P = \langle P(r, t)\rangle = \langle P(r, t')\rangle = \frac{Z_0\, s^2}{6\pi\, c^2}\left\langle\left(\frac{di(t')}{dt'}\right)^2\right\rangle \tag{22}$$

$$R_s = P/I_{\mathrm{rms}}^2 = \frac{Z_0 s^2}{6\pi\, c^2}\,\frac{\left\langle\left(\frac{di(t')}{dt'}\right)^2\right\rangle}{\langle i^2(t')\rangle} \tag{23}$$

The terms in Eqs. (15) and (16) declining proportionate to $1/r^2$ are denoted $\mathbf{E}(1/r^2, t)$ and $\mathbf{H}(1/r^2, t)$:

$$\mathbf{E}(1/r^2, t) = \frac{Z_0}{4\pi\, r^2}\left(\frac{3(\dot{\mathbf{p}}\,\mathbf{r})\,\mathbf{r}}{r^2} - \dot{\mathbf{p}}\right) = \frac{Z_0\, s\, i(t - r/c)}{4\pi\, r^2}\left(\frac{3(\mathbf{s}\,\mathbf{r})\,\mathbf{r}}{s\, r^2} - \frac{\mathbf{s}}{s}\right) \tag{24}$$

$$\mathbf{H}(1/r^2, t) = \frac{1}{4\pi\, r^2}\,\frac{\dot{\mathbf{p}} \times \mathbf{r}}{r} = \frac{s\, i(t - r/c)}{4\pi\, r^2}\,\frac{\mathbf{s} \times \mathbf{r}}{s\, r} \tag{25}$$

The term $\mathbf{H}(1/r^2, t)$ is called the near zone component of the magnetic field strength since it is the dominating term in Eq. (16) for small values of r. In the case of the electric field strength the term in Eq. (15) declining proportionate to $1/r^3$ is usually called the near zone component:

$$\mathbf{E}(1/r^3, t) = \frac{Z_0\, c}{4\pi\, r^3}\left(\frac{3(\mathbf{p}\,\mathbf{r})\,\mathbf{r}}{r^2} - \mathbf{p}\right)$$

$$= \frac{Z_0\, s\, c \int i(t - r/c)\, dt}{4\pi\, r^3}\left(\frac{3(\mathbf{s}\,\mathbf{r})\,\mathbf{r}}{s\, r^2} - \frac{\mathbf{s}}{s}\right) \tag{26}$$

Let the sinusoidal current $i(t) = I\cos 2\pi\, k\, t/T$ be fed into the Hertzian dipole. Its derivative and its integral will be sinusoidal too. Hence, the three components of the electric field strength in Eqs. (17), (24) and (26) will vary with time like $-\sin 2\pi\, k\, t/T$, $\cos 2\pi\, k\, t/T$ and $+\sin 2\pi\, k\, t/T$. In other words, they will vary like the antenna current except for phase shifts.

Consider now a Walsh current $i(t) = I\,\mathrm{cal}(k, t/T)$ fed into the Hertzian dipole. Its derivative consists of Dirac pulses and its integral of sections of ramp functions. The time variation of the three components of the electric field strength and the two components of the magnetic field strength

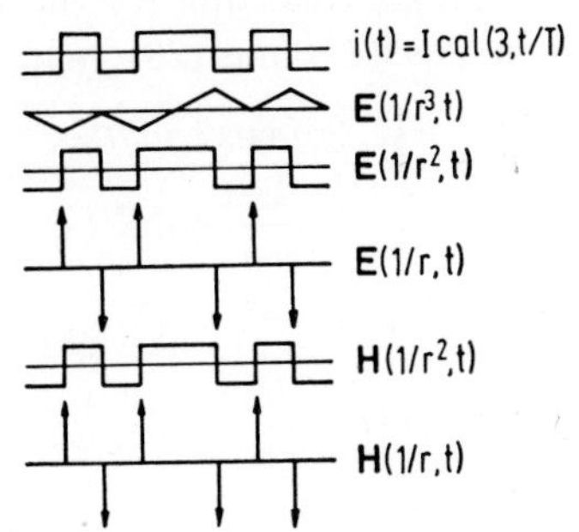

Fig. 163. Walsh shaped antenna current $i(t)$ and the resulting time variations of the elctric and magnetic field strengths produced by a Hertzian electric dipole.

is shown in Fig. 163. Note that the spatial variation of the field strength does not depend on the antenna current.

The wave zone of the magnetic field strength is defined as the zone where $\mathbf{H}(1/r, t)$ is much larger than $\mathbf{H}(1/r^2, t)$. One obtains from Eqs. (18) and (25) the condition:

$$r \gg c \left| \frac{i(t)}{di(t)/dt} \right| \quad \text{wave zone for } \mathbf{H} \tag{27}$$

For the electric field strength the wave zone is defined by the requirement that $\mathbf{E}(1/r, t)$ is much larger than $\mathbf{E}(1/r^2, t)$ and $\mathbf{E}(1/r^3, t)$, which leads to two conditions:

$$r \gg c \left| \frac{i(t)}{di(t)/dt} \right| \frac{|3(\mathbf{s}\,\mathbf{r})\,\mathbf{r} - r^2\,\mathbf{s}|}{|\mathbf{r} \times (\mathbf{s} \times \mathbf{r})|} \quad \text{wave zone for } \mathbf{E} \tag{28}$$

$$r^2 \gg c^2 \left| \frac{\int i(t)\,dt}{di(t)/dt} \right| \frac{|3(\mathbf{s}\mathbf{r})\mathbf{r} - r^2\,\mathbf{s}|}{|\mathbf{r} \times (\mathbf{s} \times \mathbf{r})|} \tag{29}$$

The condition of Eq. (27) defining the wave zone of the magnetic field strength contains no angular term in contrast to the conditions of Eqs. (28) and (29) for the electric field strength. However, the angular term in Eqs. (28) and (29) can be neglected unless the angle between $\mathbf{s}$ and $\mathbf{r}$ is close to 0 or 180° as may be seen by rewriting the angular term as follows:

$$\frac{|3(\mathbf{s}\,\mathbf{r})\,\mathbf{r} - r^2\,\mathbf{s}|}{|\mathbf{r} \times (\mathbf{s} \times \mathbf{r})|} = \frac{[1 + 9\cos^2(\mathbf{r}\,\mathbf{s})]^{1/2}}{\sin(\mathbf{r}\,\mathbf{s})} \tag{30}$$

Let the sinusoidal current $i(t) = I\cos 2\pi\, k\, t/T = I\cos 2\pi\, c\, t/\lambda$ be fed into the Hertzian dipole. The condition in Eq. (29) assumes the following form if the angular term is disregarded:

$$r^2 \gg c^2 \left| \frac{I\dfrac{\lambda}{2\pi\, c}\sin 2\pi\, c\, t/\lambda}{-I\dfrac{2\pi\, c}{\lambda}\sin 2\pi\, c\, t/\lambda} \right| = (\lambda/2\pi)^2 \tag{31}$$

This is the familiar condition for the wave zone of sinusoidal waves. The conditions in Eqs. (27) and (28) yield:

$$r \gg c \left| \frac{I\cos 2\pi\, c\, t/\lambda}{-I\dfrac{2\pi\, c}{\lambda}\sin 2\pi\, c\, t/\lambda} \right| = \frac{\lambda}{2\pi} \left| \frac{\cos 2\pi\, c\, t/\lambda}{\sin 2\pi\, c\, t/\lambda} \right| \tag{32}$$

At the times $t = n\,\lambda/2c$ the right hand side of Eq. (32) is infinite and there is no wave zone. The reason for this is that the wave zone component vanishes at these times and the components declining proportionate to $1/r^2$ are all that is available.

Let us turn to the Walsh functions. At the times when the antenna current $i(t)$ in Fig. 163 is switched from $+I$ to $-I$ or vice versa the wave zone components consist of infinitely large Dirac pulses and the wave zone ranges from zero to infinity. At all other times the wave zone component vanishes and only the components declining proportionate to $1/r^2$ and $1/r^3$ are left.

The highly idealized antenna current of Fig. 163 is replaced in Fig. 164 by a more realistic one with finite switching times ΔT. Its derivative consists

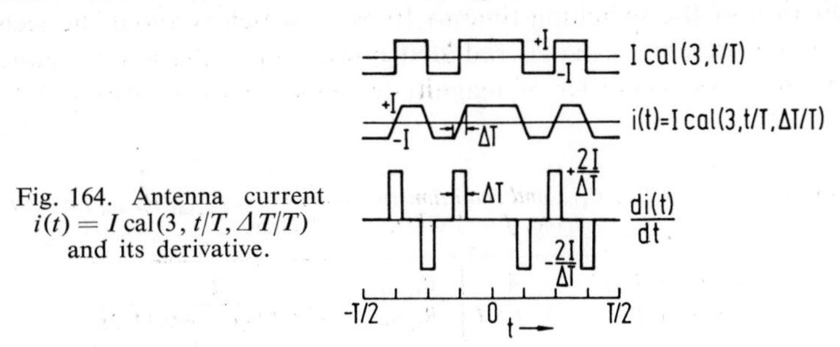

Fig. 164. Antenna current $i(t) = I\,\mathrm{cal}(3, t/T, \Delta T/T)$ and its derivative.

of block pulses of duration ΔT and magnitude $2I/\Delta T$. Consider the general case of a Walsh current $I\,\mathrm{cal}(k, t/T)$ or $I\,\mathrm{sal}(k, t/T)$; the same approximation as for $i(t)$ shall be used. One obtains the following averages $\langle (di/dt)^2\rangle$ and $\langle i^2(t)\rangle$:

$$\left\langle\left(\frac{di}{dt}\right)^2\right\rangle = \left(\frac{2I}{\Delta T}\right)^2 \frac{2k\,\Delta T}{T} = 8I^2\frac{\phi}{\Delta T}, \quad \phi = k/T \tag{33}$$

$$\langle i^2(t)\rangle = I^2\left(1 - \frac{2k\,\Delta T}{T}\right) + \frac{4k}{T}\int_0^{\Delta T/2}\left(\frac{2It}{\Delta T}\right)^2 dt = I^2\left(1 - \frac{4\phi\,\Delta T}{3}\right) \tag{34}$$

Radiated power and radiation resistance follow from Eqs. (22) and (23):

$$P = 4I^2\frac{k}{T\,\Delta T}\frac{Z_0\,s^2}{3\pi\,c^2} = 4I^2\frac{\phi}{\Delta T}\frac{Z_0\,s^2}{3\pi\,c^2} \tag{35}$$

$$R_s = 4\frac{kT}{\Delta T}\left(1 - \frac{4k\,\Delta T}{3T}\right)^{-1}\frac{Z_0\,s^2}{3\pi c^2 T^2} = 4\frac{\phi}{\Delta T}\left(1 - \frac{4\phi\,\Delta T}{3}\right)^{-1}\frac{Z_0\,s^2}{3\pi\,c^2} \tag{36}$$

The sinusoidal current $I\cos 2\pi\,k\,t/T$ or $I\sin 2\pi\,k\,t/T$ yields the following values for radiated power and radiation resistance:

$$P = \pi^2 I^2\frac{k^2}{T^2}\frac{Z_0\,s^2}{3\pi\,c^2} = \pi^2 I^2 f^2\frac{Z_0\,s^2}{3\pi\,c^2}, \quad f = k/T \tag{37}$$

$$R_s = 2\pi^2\frac{k^2}{T^2}\frac{Z_0\,s^2}{3\pi\,c^2} = 2\pi^2 f^2\frac{Z_0\,s^2}{3\pi\,c^2} \tag{38}$$

The relations of Eqs. (37) and (38) for the sine current depend on its frequency f alone, while the relations of Eqs. (35) and (36) for the Walsh current depend on sequency ϕ and switching time ΔT. Theoretically, P and R_s may be made arbitrarily large for a given sequency and antenna by decreasing ΔT. Table 18 shows the quotient of Eqs. (35) and (37) denoted by P_{sal}/P_{sin} and the quotient of Eqs. (36) and (38) denoted by $R_{s,sal}/R_{s,sin}$ for a frequency $f = 1$ GHz and a sequency $\phi = 1$ Gzps. Radiated power and radiation resistance are about equal for a switching time $\Delta T = 100$ ps. A reduction of the switching time to 10 ps – which is about the technical limit at the present – makes radiated power and radiation resistance for Walsh functions one order of magnitude higher than for sinusoidal functions.

Table 18. *Power ratio P_{sal}/P_{sin} and radiation resistance ratio $R_{s,sal}/R_{s,sin}$ for a Hertzian dipole. $f = 1$ GHz, $\phi = 1$ Gzps*

ΔT [ps]	$\frac{P_{sal}}{P_{sin}} = \frac{4}{\pi^2 f \Delta T}$	$\frac{R_{s,sal}}{R_{s,sin}} = \frac{2}{\pi^2 f \Delta T(1 - 4\phi\Delta T/3)}$
100	4	2
10	40	20
1	400	200

Problems

511.1 Plot the time variation of the electric and magnetic field strength in the wave zone and the near zone due to a current $I \sin 10\pi\,\theta$ flowing into a Hertzian dipole. Compare the phases of the four sinusoidal functions obtained.

511.2 Use the current $I\,\mathrm{sal}(5, \theta)$ instead of $I \sin 10\pi\,\theta$ in problem 511.1.

511.3 Plot the power $P(r, t)$ according to Eq. (20) as function of $t - r/c$ for an antenna current $I\,\mathrm{sal}(5, t/T)$ having a switching time ΔT according to Fig. 164.

511.4 What is the shortest switching time that can theoretically be achieved if the peak power of the transmitter is P_{max} and $P(r, t)$ in Eq. (20) can thus not exceed P_{max}?

511.5 Substitute the shortest switching time obtained in problem 511.4 for ΔT in Eq. (35) to obtain the average power P that can be radiated if the peak power of the transmitter is P_{max}.

511.6 A class B amplifier for sinusoidal functions has an efficiency of no more than 50% for maximum power output at a certain supply voltage. This means that an amplifier which can dissipate 1 W can also deliver 1 W of usable power. How much power can this amplifier deliver theoretically if it amplifies Walsh functions?

5.1.2 Near Zone – Wave Zone Effect

The different time variation of the near and wave zone components of Walsh waves makes it possible to design receivers that distinguish between them. The average power of the wave zone component of the magnetic field strength decreases proportionate to $1/r^2$ while that of the near zone component decreases proportionate to $1/r^4$. A comparison of the two powers enables one to determine the distance of the transmitter by measurements at the receiver. A similar remark holds for the electric field strength, Eq. (5.1.1-15), but it contains three terms, the average powers of which decline proportionate to $1/r^2$, $1/r^4$ and $1/r^6$, and the discussion is thus somewhat more complicated.

Consider an antenna current $i(t) = I\,\mathrm{cal}(k, t/T, \Delta T/T)$ with the transient time ΔT according to Fig. 164. Let ΔT be so short that the received energy during the transient time is practically due to the wave zone component only, while during the rest of the time the energy is due to the near zone component only. The average output power of the receiver caused by the magnetic wave zone component is denoted by P_{HW}, while P_{HN} denotes the average output power caused by the near zone component:

$$P_{\mathrm{HW}} = C_1 \left\langle \left(\frac{s}{4\pi\, c\, r} \frac{di(t - r/c)}{dt} \right)^2 \right\rangle \frac{\mathbf{s} \times \mathbf{r}}{s\, r} = C_1 \left(\frac{I s}{\pi\, c \Delta T r} \right)^2 \frac{k \Delta T}{2T} \frac{\mathbf{s} \times \mathbf{r}}{s\, r} \tag{1}$$

$$P_{\mathrm{HN}} = C_2 \left\langle \left(\frac{s\, i(t - r/c)}{4\pi\, r^2} \right)^2 \right\rangle \frac{\mathbf{s} \times \mathbf{r}}{s\, r} = C_2 \left(\frac{I s}{4\pi\, r^2} \right)^2 \left(1 - \frac{2k \Delta T}{T} \right) \frac{\mathbf{s} \times \mathbf{r}}{s\, r} \tag{2}$$

The coefficients C_1 and C_2 depend on the gain of the receiving antenna and the receiver. The ratio $P_{\mathrm{HW}}/P_{\mathrm{HN}}$ is obtained from Eqs. (1) and (2):

$$\frac{P_{\mathrm{HW}}}{P_{\mathrm{HN}}} = 2C_3 \left(\frac{2r}{c\Delta T} \right)^2 \frac{k\Delta T}{T} \left(1 - \frac{2k\Delta T}{T} \right)^{-1} \tag{3}$$

The angular component $(\mathbf{s} \times \mathbf{r})/s\, r$ does not occur in this equation any more. The coefficient C_3 equals 1 if antenna and receiver gain is the same for wave and near zone component. For $C_3 = 1$ and $k\Delta T/T \ll 1$ one obtains from Eq. (3) a formula for the distance r between transmitter and receiver:

$$r = \frac{c\Delta T}{2} \sqrt{\left(\frac{T}{2k\Delta T} \right)} \sqrt{\left(\frac{P_{\mathrm{HN}}}{P_{\mathrm{HW}}} \right)} \tag{4}$$

The distance r_e where P_{HN} and P_{HW} are equal is of interest:

$$r_e = \tfrac{1}{2} c \sqrt{(T \Delta T/2k)} = \tfrac{1}{2} c \sqrt{(\Delta T/2\phi)} \tag{5}$$

The values $\Delta T = 0.04T$ and $k = 3$ hold for the function of Fig. 164. One obtains in this case $r_e = 12000T$ (r_e in kilometers, T in seconds). For $T = 1$ ms one obtains $r_e = 12$ km. The sequency $\phi = k/T$ of this wave is 3000 zps and its average wavelength is $\lambda = c/\phi = 100$ km. This wavelength would be unacceptable for sinusoidal waves in an application like aircraft collision warning, due to the size of the antenna. However, a long average wavelength does not imply a long antenna in the case of Walsh functions.

Let us consider a sinusoidal antenna current. A separation of the wave zone component in Eq. (5.1.1-18) from the near zone component in Eq. (5.1.1-25) is only possible if a synchronization signal is available. Such a synchronization signal can be derived in theory from the electric field strength. The first term in Eq. (5.1.1-16) is essentially the time derivative of the second term. In Eq. (5.1.1-15), on the other hand, the second term is the derivative of the third term, but the first is not exactly the derivative of the second. This results in a phase shift between electric and magnetic field strength which depends on the distance from the transmitter. Hence, the near zone—wave zone effect exists for sinusoidal waves too, but it is not as pronounced as for Walsh waves[1].

Problems

512.1 Distance measurements by electromagnetic waves are usually based on the radar principle. How does power and energy of the reflected radar signal decrease with the distance of the target?

512.2 Compare the possible accuracy of distance measurement by the radar principle and by the near zone — wave zone effect.

512.3 The usefulness of the near zone — wave zone effect could be greatly increased if a component could be found whose power declines proportionate to $1/r^3$ rather than proportionate to $1/r^4$. Is such a component contained in Eqs. (5.1.1-15) and (5.1.1-16)? Multiply these two equations according to Eq. (5.1.1-19). Why is it difficult to use this component? Is there a way to overcome the difficulty with a time variable receiving antenna?

512.4 It is often assumed that the energy carried by components of the electric and magnetic field strength decreasing faster than proportionate to $1/r$ returns to the transmitter. Consider such energy at a certain time in the space between two spheres of radius R and $R + \Delta R$. Let the transmitter be disassembled in a time interval shorter than $2R/c$. Where is the energy going to return to? Where is this energy stored and what will happen to it if there is no perfect vacuum? Do these considerations apply generally to radiators of electromagnetic and other waves?

5.1.3 Magnetic Dipole Radiation

Let the substitutions,

$$\mathbf{E} \rightarrow \sqrt{\left(\frac{\mu_0}{\varepsilon_0}\right)}\,\mathbf{H}, \quad \mathbf{H} \rightarrow -\sqrt{\left(\frac{\varepsilon_0}{\mu_0}\right)}\,\mathbf{E} \tag{1}$$

be made in Eqs. (5.1.1-1) and (5.1.1-2). The charge and current densities shall be zero. The substitutions in Eq. (1) then yield the original set of Maxwell's equations and they represent a new solution for those sections

[1] It is interesting to observe that in this case synchronization is required for sine waves but not for Walsh waves. This result together with the results on sidelobe suppression in the range-Doppler domain (1.2.8) and filters for space variable signals (2.3, 2.4) supports the contention that sine functions appear to be so good for practical applications because only those applications for which they are good have been developed during the last 70 years while the other applications have been neglected. *See* also the comment on Walsh wave radar on page 281.

of space where charge and current density are zero. Poynting's vector in Eq. (5.1.1-19) remains unchanged and the same power is radiated at any moment through the surface of a large sphere. Equations (5.1.1-15) and (5.1.1-16) assume the following form:

$$\mathbf{H} = \frac{1}{4\pi}\left[\frac{1}{c\,r}\left(\frac{(\ddot{\mathbf{p}}\,\mathbf{r})\,\mathbf{r}}{r^2} - \ddot{\mathbf{p}}\right) + \frac{1}{r^2}\left(\frac{3(\dot{\mathbf{p}}\,\mathbf{r})\,\mathbf{r}}{r^2} - \dot{\mathbf{p}}\right) + \frac{c}{r^3}\left(\frac{3(\mathbf{p}\,\mathbf{r})\,\mathbf{r}}{r^2} - \mathbf{p}\right)\right] \tag{2}$$

$$\mathbf{E} = -\frac{Z_0}{4\pi}\left(\frac{1}{c\,r}\,\frac{\ddot{\mathbf{p}} \times \mathbf{r}}{r} + \frac{1}{r^2}\,\frac{\dot{\mathbf{p}} \times \mathbf{r}}{r}\right) \tag{3}$$

It has to be shown what currents and charges produce the magnetic and electric field strengths in Eqs. (2) and (3). $\dot{\mathbf{p}}$ may be rewritten as follows:

$$\dot{\mathbf{p}}(t) = i(t)\,\mathbf{s} = \frac{\mu_0\, i(t)\, s\, \mathbf{s}}{\mu_0\, s} = \frac{\mu_0\, i(t)\,\mathbf{a}}{\mu_0\sqrt{(a)}} = \frac{\mathbf{m}(t)}{\mu_0\sqrt{(a)}} \tag{4}$$

$$\mathbf{m}(t) = \mu_0\, i(t)\,\mathbf{a}$$

$$\mathbf{a} = a\,\frac{\mathbf{s}}{s} \tag{5}$$

$\mathbf{m}$ is the magnetic dipole moment produced by a current $i(t)$ that flows in the sense of a right hand screw around a loop of area a in a plane. The orientation of the plane is defined by the vector $\mathbf{a}$ which is perpendicular to the plane and points in the same direction as $\mathbf{s}$. Equations (2) and (3) may be rewritten with the magnetic dipole moment:

$$\mathbf{H} = \frac{1}{4\pi Z_0}\left[\frac{1}{\sqrt{(a)}\,r}\left(\frac{(\dot{\mathbf{m}}\,\mathbf{r})\,\mathbf{r}}{r^2} - \dot{\mathbf{m}}\right) + \right.$$

$$\left. + \frac{c}{\sqrt{(a)}\,r^2}\left(\frac{3(\mathbf{m}\,\mathbf{r})\,\mathbf{r}}{r^2} - \mathbf{m}\right) + \frac{c^2}{\sqrt{(a)}\,r^3}\left(\frac{3\left(\int \mathbf{m}\,dt\,\mathbf{r}\right)\mathbf{r}}{r^2} - \int \mathbf{m}\,dt\right)\right] \tag{6}$$

$$\mathbf{E} = -\frac{1}{4\pi}\left(\frac{1}{\sqrt{(a)}\,r}\,\frac{\dot{\mathbf{m}} \times \mathbf{r}}{r} + \frac{c}{\sqrt{(a)}\,r^2}\,\frac{\mathbf{m} \times \mathbf{r}}{r}\right) \tag{7}$$

$$\mathbf{m} = \mathbf{m}(t - r/c)$$

The components of the field strengths declining proportionate to $1/r$ become in analogy to Eqs. (5.1.1-17) and (5.1.1-18):

$$\mathbf{H}(1/r, t) = \frac{1}{4\pi Z_0\sqrt{(a)}\,r}\left(\frac{(\dot{\mathbf{m}}\,\mathbf{r})\,\mathbf{r}}{r^2} - \dot{\mathbf{m}}\right) = \frac{\sqrt{(a)}}{4\pi\, c\, r}\,\frac{d i(t - r/c)}{dt}\,\frac{\mathbf{r} \times (\mathbf{r} \times \mathbf{a})}{a\, r^2} \tag{8}$$

$$\mathbf{E}(1/r, t) = -\frac{Z_0\sqrt{(a)}}{4\pi\, c\, r}\,\frac{d i(t - r/c)}{dt}\,\frac{\mathbf{a} \times \mathbf{r}}{a\, r} \tag{9}$$

Power $P(r, t)$, average power P and radiation resistance R_s have the same form as in Eqs. (5.1.1-20), (5.1.1-22) and (5.1.1-23) except that s^2 is replaced by a:

$$P(r, t) = Z_0 \frac{a}{6\pi c^2} \left(\frac{di(t - r/c)}{dt} \right)^2 \tag{10}$$

$$P = \frac{Z_0 a}{6\pi c^2} \left\langle \left(\frac{di(t')}{dt'} \right)^2 \right\rangle \tag{11}$$

$$R_s = \frac{Z_0 a}{6\pi c^2} \frac{\left\langle \left(\frac{di(t')}{dt'} \right)^2 \right\rangle}{\langle i^2(t') \rangle} \tag{12}$$

The components of the field strengths declining proportionate to $1/r^2$ and $1/r^3$ may be rewritten as follows:

$$\begin{aligned} \mathbf{H}(1/r^2, t) &= \frac{c}{4\pi Z_0 \sqrt{(a)}\, r^2} \left(\frac{3(\mathbf{m}\,\mathbf{r})\,\mathbf{r}}{r^2} - \mathbf{m} \right) \\ &= \frac{\sqrt{(a)}\, i(t - r/c)}{4\pi r^2} \left(\frac{3(\mathbf{a}\,\mathbf{r})\,\mathbf{r}}{a\, r^2} - \frac{\mathbf{a}}{a} \right) \end{aligned} \tag{13}$$

$$\mathbf{E}(1/r^2, t) = -\frac{c}{4\pi \sqrt{(a)}\, r^2} \frac{\mathbf{m} \times \mathbf{r}}{r} = -\frac{Z_0 \sqrt{(a)}\, i(t - r/c)}{4\pi r^2} \frac{\mathbf{a} \times \mathbf{r}}{a\, r} \tag{14}$$

$$\begin{aligned} \mathbf{H}(1/r^3, t) &= \frac{1}{4\pi Z_0 \sqrt{(a)}\, r^3} \left(\frac{3\left(\int \mathbf{m}\, dt\, \mathbf{r}\right)\mathbf{r}}{r^2} - \int \mathbf{m}\, dt \right) \\ &= \frac{\sqrt{(a)}\, c \int i(t - r/c)\, dt}{4\pi r^3} \left(\frac{3(\mathbf{a}\,\mathbf{r})\,\mathbf{r}}{a\, r^2} - \frac{\mathbf{a}}{a} \right) \end{aligned} \tag{15}$$

Equation (15) represents the static magnetic field of a magnetic dipole or of an equivalent current flowing in a loop, while Eq. (14) represents the

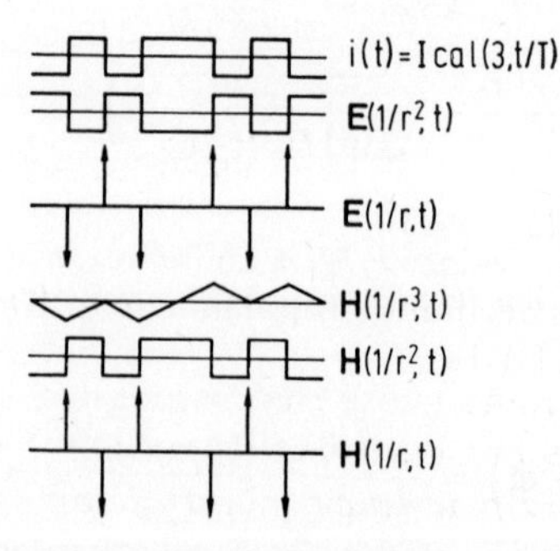

Fig. 165. Walsh shaped antenna current $i(t)$ and the resulting time variations of the electric and magnetic field strengths produced by a Hertzian magnetic dipole.

static electric field of this current. Figure 165 shows the time variation of the various components of the electric and magnetic field strengths.

Problems

513.1 The magnetic dipole moment in Eq. (4) is produced by a current flowing in a loop. What other means exists to produce a magnetic dipole moment that is the more obvious equivalent of the electric dipole?

513.2 Why is there no equivalent to the generation of a magnetic dipole moment by a current flowing in a loop for the electric dipole?

5.1.4 Implementation of Radiators

Antennas in use at the present are designed for the radiation of sinusoidal waves. There are so-called frequency independent antennas that radiate sinusoidal waves of a wide frequency range [1]. Typical representatives are the biconical antenna and the antennas with logarithmic periodicity. These antennas are quite capable of radiating Walsh waves. However, here it is of more interest to find antennas that are specifically designed for the radiation of Walsh waves.

The Hertzian electric and magnetic dipoles can radiate any wave but the power is vanishingly small. Many such dipoles may be used simultaneously to radiate more power. Figure 166a shows several Hertzian dipoles, each fed by one generator. Using several generators would be very expensive for sinusoidal waves. The resonance dipole of Fig. 166b may be

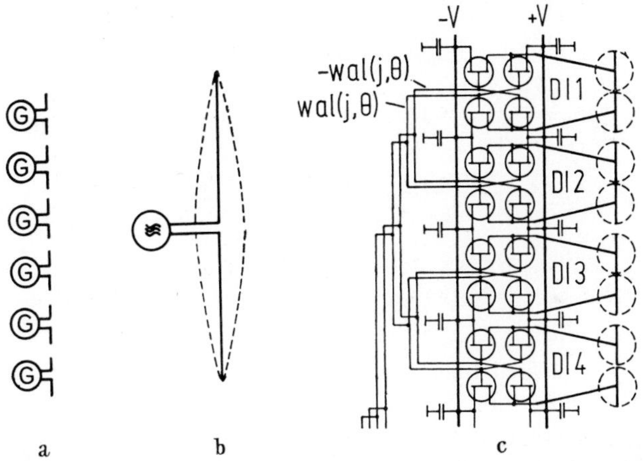

Fig. 166. Row of Hertzian electric dipoles (a), resonance dipole for sinusoidal waves (b) and principle of a Hertzian electric dipole array for Walsh waves (c).

thought of as an ingenious way of feeding many Hertzian dipoles by one generator. A wave fed into the dipole is reflected at its ends. A standing wave is produced if the length of the dipole is an integer multiple of one half of the wave length. Each point of the resonance dipole then acts like a Hertzian dipole, but the amplitude of oscillation of charge and current depends on the location of the point.

Generators for Walsh shaped currents are very inexpensive since they consist of switches that feed positive or negative currents into the dipole. Figure 166c shows an array of four dipoles fed by field effect transistors. Note that semiconductor switches automatically provide a current feed even though the sources for the voltages $+V$ and $-V$ are voltage sources. The wires feeding the driving voltages $\mathrm{wal}(j, \theta)$ and $-\mathrm{wal}(j, \theta)$ to the transistors are arranged so that the delay times are theoretically equal for all transistors.

The Hertzian dipole DI in Fig. 166c consists of circular areas of conducting material. They do not have to be arranged along a line as shown but they can be arranged in two dimensions. Such a two-dimensional arrangement cannot be used for a resonance dipole.

The Hertzian magnetic dipole appears at the present more promising than the electric dipole. Figure 167 shows a linear array of four such dipoles, each one consisting of two circular loops, one shown by a solid line

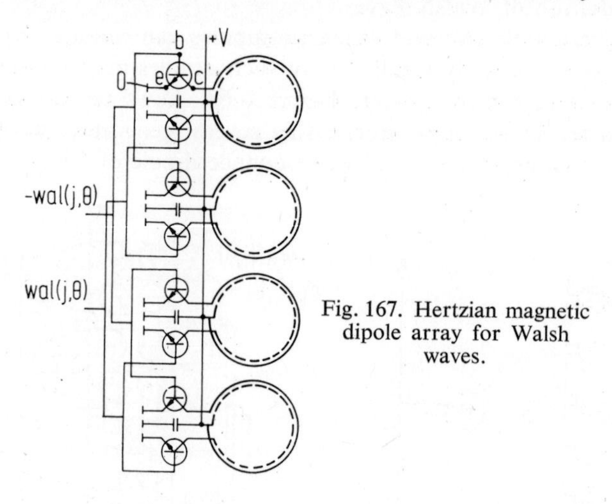

Fig. 167. Hertzian magnetic dipole array for Walsh waves.

the other by a dashed line. A current from the bus $+V$ flows clockwise through the dashed loop and the transistor to ground if $\mathrm{wal}(j, \theta)$ is positive. A negative value of $\mathrm{wal}(j, \theta)$ makes a current flow counter clockwise through the solid loop. This illustration makes it more apparent why one cannot feed the currents $\mathrm{wal}(j, \theta)$ and $-\mathrm{wal}(j, \theta)$ directly to the loops in Fig. 167 or to the dipoles in Fig. 166c. The wires leading to the bases of the transistors radiate in a way that is hard to predict. Hence, the power flowing through them must be kept small. The high power is only flowing through the loops and the transistors to ground. If the transistors are small compared with the area of the loops one can approximate the theoretical assumptions fairly well. The current flowing in the bus $+V$ is constant for transistors with ideally matched switching characteristic. The capacitors along the bus are only required because such an ideal matching is practically impossible.

Figure 168 shows a further step toward a practical radiator. A two-dimensional array of 16 magnetic dipoles, consisting of two circular loops each, can be obtained by etching from a card (1), copper-clad on both sides as used for printed circuits. The loops shown dashed are on the far side of the card, the loops shown solid on the near side. Transistor chips can be soldered to the points e, b and c (emitter, base, collector).

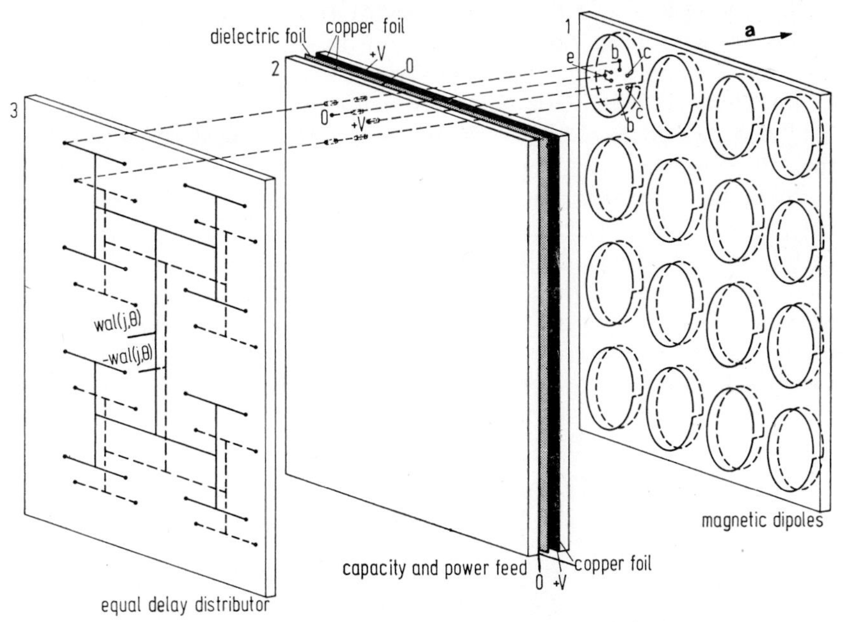

Fig. 168. Two-dimensional array of Hertzian magnetic dipoles.

Conductors from the points e, b and $+V$ – shown for one magnetic dipole only – lead to two cards (2), each copper-clad on one side and separated by a thin dielectric foil. The conductor from e is connected to one copper foil (solid dot) but insulated from the other, to which the conductor from $+V$ is connected. This implements the capacitors of Fig. 167. The conductor from b is insulated from both copper foils. The copper foils must have slots radiating from the point where the conductors from $+V$ and e are connected in order to prevent eddy currents.

A third printed circuit card (3) etched from a sheet copper-clad on both sides provides the connections between the 2×16 conductors from the bases b of the transistors to the two input wires carrying the voltages $\text{wal}(j, \theta)$ and $-\text{wal}(j, \theta)$. The conducting strips on this card are arranged like the streets of Dyadicville in Fig. 49. The delay times are thus equalized. But there are other compensations too: For any charge moving to the right or up there is an equal charge moving to the left or down. For any current flowing in a clockwise loop there is a current flowing in a counter clockwise loop. As a result, these charges and currents cannot produce dipole radiation but only high order multipole radiation.

n magnetic dipoles arranged as in Fig. 168 produce n times the electric and magnetic field strengths in (5.1.3-8) and (5.1.3-9). Poynting's vector and the average power (5.1.3-11) are thus increased by a factor n^2, which means that the average power radiated per dipole and the radiation resistance of each dipole increase by a factor n. This is a well known effect. It is used, e.g., in acoustics when several loudspeakers are located close together and driven in parallel. The higher radiation resistance of the resonance dipole in Fig. 166b compared with that of the Hertzian dipole may also be explained by this superposition of the field strengths rather than the powers.

Since the resonance antenna is so important for sinusoidal waves one may well ask for its equivalent for Walsh waves. It was shown in section 2.2.1 that resonance filters for Walsh functions are obtained from resonance filters for sine-cosine functions by the introduction of a time variable element. Hence, one may expect that the time invariant rod resonating for sine waves can be made to resonate for Walsh waves if it is made time variable. Figure 169 shows the principle. A conductor of length L is short

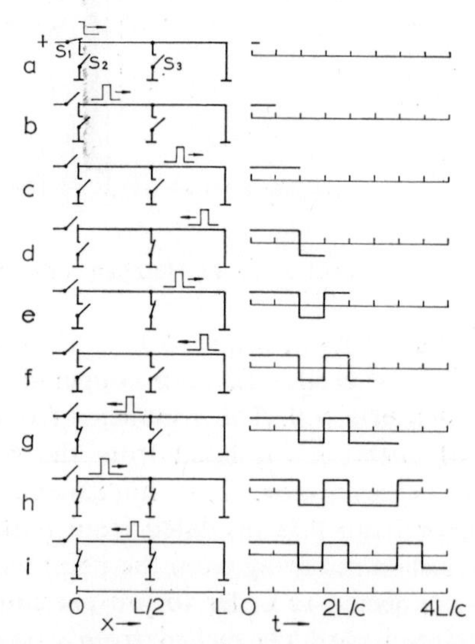

Fig. 169. Principle of a resonance antenna for Walsh waves.

circuited at its end and may in addition be short circuited at $x = 0$ and $x = L/2$ by the switches s_2 and s_3. A short current pulse is applied to the conductor by switch s_1, as shown by Fig. 169*a*. This pulse travels along the conductor to the right (b and c). At the end, $x = L$, of the conductor it is reflected and travels toward $x = L/2$ as shown in d. The time diagrams on the right hand side show the value $+1$ when the current pulse travels to the right and -1 when it travels to the left. The switch s_3 is closed when the pulse arrives at the time $t = 3L/2c$, as shown in d and e. The pulse is thus reflected and travels to the right again. At $x = L$ it is reflected at the time $t = 2L/c$. Switch s_3 is now open so that the pulse can travel to $x = 0$, as shown by f and g, where it arrives at the time $t = 3L/c$. The closed switch s_2 causes a reflection, the pulse travels to the right (h), and is reflected by the closed switch s_3 at the time $7L/2c$, as shown by i. A comparison of the time diagram of Fig. 169 with the Walsh functions of Fig. 2 shows that the function $\mathrm{cal}(3, c\,t/4L + 3/8)$ has been generated.

Problems

514.1 Investigate the radiation from a biconical antenna due to a Walsh shaped input current.

514.2 What is the order of magnitude of the diameter of the loops in Figs. 167 and 168 for a switching time of 1 ns?

514.3 In which direction radiates the antenna of Fig. 168 the maximum power and in which direction is no power radiated?

514.4 Try to design an antenna like the one of Fig. 168 with the magnetic dipoles arranged in a three-dimensional pattern. Would such a radiator be useful in connection with a parabolic mirror?

5.2 Multipole Radiation of Walsh Waves

5.2.1 Radiation of an One-Dimensional Quadrupole

Let us turn from dipole radiation to more general modes. Figure 170 shows that there are two orthogonal modes of dipole radiation: The vertically polarized radiation denoted "dipole 21" and the horizontally polarized radiation denoted "dipole 22." In principle, both modes permit the transmission of independent signals, but crosstalk caused by a rotation of the polarization vector usually prevents the practical use of both dipole modes.

Two dipoles may be combined according to Fig. 170 to produce three orthogonal modes of quadrupole radiation denoted "quadrupole 41, 42, 43." It is also shown in Fig. 170 how these and the higher modes of multipole radiation may be investigated conveniently by means of two-dimensional Walsh functions $\mathrm{wal}(k, x)\,\mathrm{wal}(m, y)$. The dipole 21 is represented by the function $\mathrm{wal}(0, x)\,\mathrm{wal}(1, y)$, the quadrupole 42 by the function $\mathrm{wal}(1, x)\,\mathrm{wal}(1, y)$, etc., if the signs $+$ and $-$ of the poles are identified with the values $+1$ and -1 of the Walsh functions. The unipole radiation corresponding to the function wal(0, x) wal(0, y) does not occur in the radiation of electromagnetic waves due to the preservation of charge, but it is the most important mode of radiation for acoustic waves.

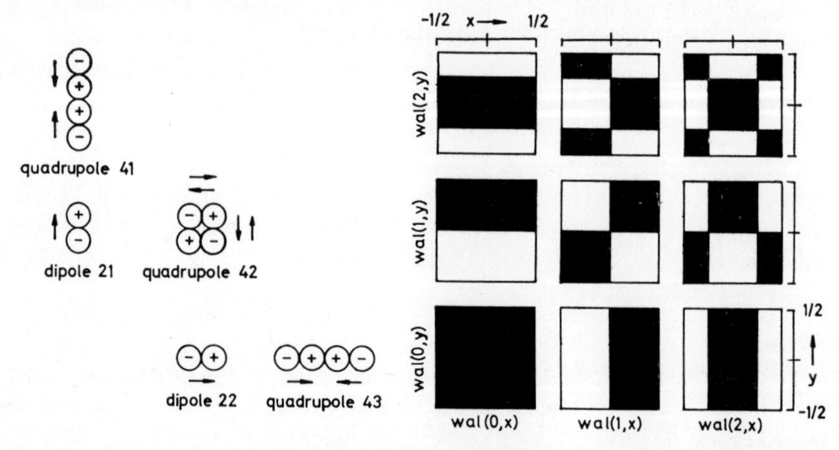

Fig. 170. Dipoles, quadrupoles and their representation by two-dimensional Walsh functions wal(k, x) wal(m, y). The quadrupole 41 can be implemented by two dipoles 21, the quadrupole 43 by two dipoles 22 and the quadrupole 42 by two dipoles 22 or two dipoles 21.

The radiation from a quadrupole 41 or 43 may be calculated with the help of Fig. 171 and the Eqs. (5.1.1-15) to (5.1.1-26) holding for dipole

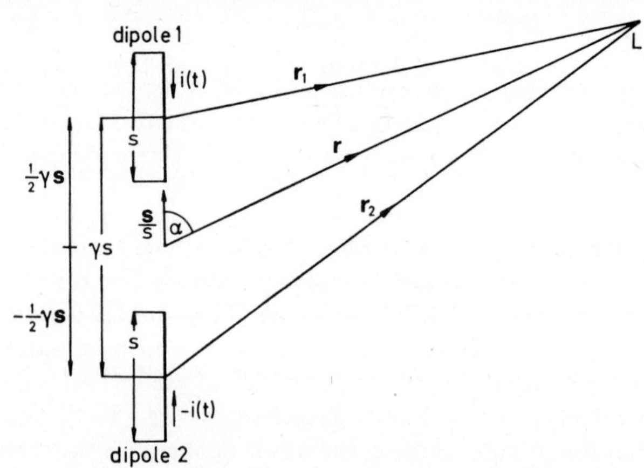

Fig. 171. Radiation from a quadrupole 41 or 43.

radiation. For instance, the electric field strength $\mathbf{E}_{41}(1/r, t)$ in the wave zone is given by the terms declining proportionate to $1/r$ in the equation

$$\mathbf{E}_{41}(1/r, t) = \mathbf{E}(1/r_1, t) - \mathbf{E}(1/r_2, t) \tag{1}$$

where $\mathbf{E}(1/r_1, t)$ and $\mathbf{E}(1/r_2, t)$ are obtained by the substitution of $\mathbf{r}_1$ or $\mathbf{r}_2$ into Eq. (5.1.1-17). These two vectors are defined in Fig. 171:

$$\mathbf{r}_1 = \mathbf{r} - \tfrac{1}{2}\gamma\,\mathbf{s}, \quad \mathbf{r}_2 = \mathbf{r} + \tfrac{1}{2}\gamma\,\mathbf{s} \tag{2}$$

For $r \gg \gamma\, s$ the following relations are obtained:

$$\left.\begin{aligned}
r_1 &= (\mathbf{r}_1\,\mathbf{r}_1)^{1/2} = r\left(1 - \frac{1}{2}\gamma\frac{s}{r}\cos\alpha\right)\\
r_2 &= r\left(1 + \frac{1}{2}\gamma\frac{s}{r}\cos\alpha\right)\\
i(t - r_1/c) &= i(t - r/c) + \frac{1}{2}\gamma\frac{s}{c}\frac{di(t - r/c)}{dt}\cos\alpha\\
i(t - r_2/c) &= i(t - r/c) - \frac{1}{2}\gamma\frac{s}{c}\frac{di(t - r/c)}{dt}\cos\alpha
\end{aligned}\right\} \tag{3}$$

Furthermore, using the relation

$$\mathbf{r}_1 \times \mathbf{s} = \mathbf{r}_2 \times \mathbf{s} = \mathbf{r} \times \mathbf{s} \tag{4}$$

one obtains:

$$\mathbf{E}_{41}(1/r, t) = \frac{Z_0\,\gamma\, s^2 \cos\alpha}{4\pi\, c^2\, r}\,\frac{d^2 i(t - r/c)}{dt^2}\,\frac{\mathbf{r} \times (\mathbf{r} \times \mathbf{s})}{s\, r^2} \tag{5}$$

The magnetic field strength in the wave zone is obtained in a corresponding way:

$$\begin{aligned}
&\mathbf{H}_{41}(1/r, t)\\
&= \mathbf{H}(1/r_1, t) - \mathbf{H}(1/r_2, t) = \frac{\gamma\, s^2 \cos\alpha}{4\pi\, c^2\, r}\,\frac{d^2 i(t - r/c)}{dt^2}\,\frac{\mathbf{s} \times \mathbf{r}}{s\, r}
\end{aligned} \tag{6}$$

The essential difference between Eqs. (5.1.1-17), (5.1.1-18) and (5), (6) is the replacement of the first derivative of the antenna current $i(t)$ by the second derivative.

Poynting's vector is derived from Eqs. (5) and (6):

$$\begin{aligned}
\mathbf{P}_{41} &= \mathbf{E}_{41}(1/r, t) \times \mathbf{H}_{41}(1/r, t)\\
&= Z_0\left(\frac{\gamma\, s^2 \sin\alpha \cos\alpha}{4\pi\, c^2\, r}\right)^2 \left(\frac{d^2 i(t - r/c)}{dt^2}\right)^2 \frac{\mathbf{r}}{r}
\end{aligned} \tag{7}$$

$P_{41}(r, t)$ denotes the power radiated through the surface of a large sphere of radius r:

$$\begin{aligned}
P_{41}(r, t) &= \oiint \mathbf{P}_{41}\, d\mathbf{O}\\
&= Z_0\left(\frac{\gamma\, s^2}{4\pi\, c^2\, r}\right)^2 \left(\frac{d^2 i(t - r/c)}{dt^2}\right)^2 \int_{\alpha=0}^{\pi}\int_{\beta=0}^{2\pi} r^2 \sin^3\alpha \cos^2\alpha\, d\alpha\, d\beta\\
&= Z_0\,\frac{\gamma^2\, s^4}{30\,\pi\, c^4}\left(\frac{d^2 i(t - r/c)}{dt^2}\right)^2
\end{aligned} \tag{8}$$

The average power radiated through the surface of the sphere is denoted by P_{41}:

$$P_{41} = \langle P_{41}(r,t)\rangle = Z_0 \frac{\gamma^2 s^4}{30\pi c^4} \left\langle \left(\frac{d^2 i(t')}{dt'^2} \right)^2 \right\rangle \qquad t' = t - r/c \tag{9}$$

A radiation resistance $R_{s,41}$ may be defined with the help of the rms current I_{rms}:

$$R_{s,41} = P_{41}/I_{\mathrm{rms}}^2 \tag{10}$$

The meaning of rms current is unambiguous in the case of a dipole. A quadrupole, however, has the current $i_1(t) = i(t)$ flowing into dipole 1 and the current $i_2(t) = -i(t)$ flowing into dipole 2 of Fig. 171. The rms current will be defined as follows:

$$I_{\mathrm{rms}}^2 = \langle i_1^2(t)\rangle + \langle i_2^2(t)\rangle = 2\langle i^2(t)\rangle \tag{11}$$

Introduction of Eq. (11) into Eq. (10) yields:

$$R_{s,41} = Z_0 \frac{\gamma^2 s^2}{60\pi c^4} \frac{\left\langle \left(\frac{d^2 i(t')}{dt'^2} \right)^2 \right\rangle}{\langle i^2(t')\rangle} \tag{12}$$

Consider the sinusoidal currents $i_1(t) = I\cos 2\pi k\, t/T = i(t)$ and $i_2(t) = -i(t)$ flowing into the two dipoles of Fig. 171. One obtains:

$$\langle i^2(t')\rangle = I^2 \frac{k}{T} \int_0^{T/k} \cos^2 2\pi k\, t/T \cdot dt = \frac{1}{2} I^2$$

$$\left\langle \left(\frac{d^2 i(t')}{dt'^2} \right)^2 \right\rangle = \frac{1}{2} I^2 \left(\frac{2\pi k}{T} \right)^4$$

Average radiated power and radiation resistance follow from Eqs. (9) and (12):

$$P_{41} = I^2 Z_0 \frac{4\pi^3 \gamma^2 s^4 k^4}{15 c^4 T^4}, \quad R_{s,41} = Z_0 \frac{4\pi^3 \gamma^2 s^4 k^4}{15 c^4 T^4} \tag{13}$$

Average radiated power P and radiation resistance R_s for a Hertzian dipole being fed with a current $i(t) = I\cos 2\pi k\, t/T$ are given by Eqs. (5.1.1-37) and (5.1.1-38). The ratios P_{41}/P and $R_{s,41}/R_s$ are of interest:

$$\frac{P_{41}}{P} = \frac{4}{5}\pi^2 \left(\frac{\gamma s k}{cT} \right)^2 = \frac{4}{5}\pi^2 \left(\frac{\gamma s}{\lambda} \right)^2,$$

$$\frac{R_{s,41}}{R_s} = \frac{2}{5}\pi^2 \left(\frac{\gamma s}{\lambda} \right)^2, \quad \lambda = \frac{cT}{k} \tag{14}$$

γs is essentially the length of the quadrupole according to Fig. 171. Much less power is radiated by a quadrupole wave than by a dipole wave if the wavelength λ is much larger than the length of the quadrupole antenna, which is a basic assumption for the Hertzian dipole. It is well known that this is the reason why light emitted from an atom is predominantly dipole radiation.

Consider now Walsh shaped currents $+I\,\mathrm{cal}(k, t/T)$ and $-I\,\mathrm{cal}(k, t/T)$ fed into the dipoles of Fig. 171. Since the second derivative of the current is required it is not sufficient to use the approximation $I\,\mathrm{cal}(k, t/T, \Delta T/T)$ to the ideal Walsh shaped current as shown in line *2* of Fig. 164. The more sophisticated approximation $i_1(t) = -i_2(t) = I\,\mathrm{cal}(k, t/T, \Delta T/T, \varepsilon)$ as shown in line *2* of Fig. 172 is required. ΔT denotes the time needed for

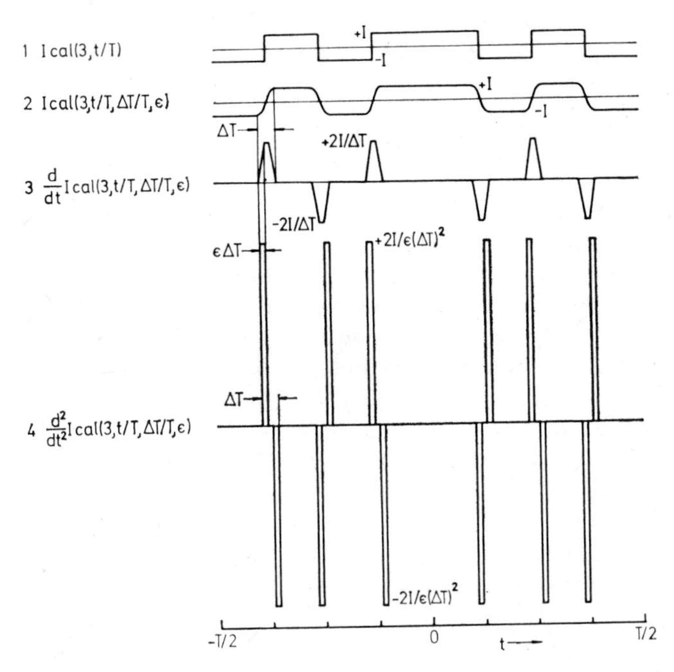

Fig. 172. Time variation of the electric and magnetic field strength at a point in the wave zone due to radiation from a quadrupole. $\Delta T = (3/10)\,(T/8)$, $\varepsilon = 1/3$. The retardation r/c between antenna current and field strengths is disregarded.

the current $i(t)$ to change from $-I$ to $+I$ or vice versa. The derivative $di(t)/dt$ requires the time $\varepsilon\,\Delta T$ to change from 0 to its positive or negative peak value $\pm 2I/\Delta T$ or vice versa.

For small values of ΔT one obtains from Fig. 172 the following values for $\langle i^2(t)\rangle$ and $\langle d^2 i(t)/dt^2\rangle$:

$$i(t) = I\,\mathrm{cal}(k, t/T, \Delta T/T, \varepsilon); \quad \varepsilon \leqq \tfrac{1}{2}, \quad \Delta T \ll T/2^h, \quad 2^{h-1} < 2k \leqq 2^h$$

$$\langle i^2(t')\rangle = I^2, \quad \left\langle \left(\frac{d^2 i(t')}{dt'^2}\right)^2 \right\rangle = \frac{16k\,I^2}{\varepsilon\,T(\Delta T)^3} \tag{15}$$

Radiated power and radiation resistance follow from Eqs. (9) and (12):

$$P_{41} = I^2 Z_0 \frac{8\gamma^2 s^4 k}{15\pi c^4 \varepsilon T(\Delta T)^3}, \quad R_{s,41} = Z_0 \frac{4\gamma^2 s^4 k}{15\pi c^4 \varepsilon T(\Delta T)^3} \tag{16}$$

Average radiated power P and radiation resistance R_s for a Hertzian dipole fed with a current $I\,\mathrm{cal}(k, t/T, \Delta T/T)$ according to Fig. 164 are given by Eqs. (5.1.1-35) and (5.1.1-36). One obtains the following ratios P_{41}/P and $R_{s,41}/R_s$:

$$\frac{P_{41}}{P} = \frac{2}{5}\frac{1}{\varepsilon}\left(\frac{\gamma s}{c\,\Delta T}\right)^2 = \frac{2}{5}\frac{1}{\varepsilon}\left(\frac{T}{k\,\Delta T}\right)^2\left(\frac{\gamma s}{\lambda}\right)^2,$$

$$\frac{R_{s,41}}{R_s} = \frac{1}{5}\frac{1}{\varepsilon}\left(\frac{T}{k\,\Delta T}\right)^2\left(\frac{\gamma s}{\lambda}\right)^2 \tag{17}$$

$\lambda = c\,T/k =$ average wavelength.

The comparison of Eqs. (17) and (14) shows that the ratio P_{41}/P can be increased by a reduction of ΔT and ε in the case of Walsh functions while this possibility does not exist in the case of sinusoidal functions. For a given ratio $\gamma s/\lambda < 1$ the power radiated by a quadrupole Walsh wave may thus exceed the power radiated by a dipole Walsh wave.

One may infer from Fig. 172 that the dipole wave and the quadrupole wave generated by equal antenna currents are orthogonal at a certain space point in the wave zone. The dipole wave generated by the current $i = I\,\mathrm{cal}(3, t/T, \Delta T/T, \varepsilon)$ is proportionate to di/dt, the quadrupole wave is proportionate to $d^2 i/dt^2$. The integral of the product $(di/dt)(d^2 i/dt^2)$ is zero; the integration interval may be $-T/2 < t < +T/2$ or an interval of duration ΔT located symmetrically around the jumps of $I\,\mathrm{cal}(3, t/T, \Delta T/T, \varepsilon)$.

Problems

521.1 Compute the electric and magnetic field strengths of quadrupole 41 radiation that decrease proportionate to $1/r^2$.

521.2 How must the Hertzian dipoles of Figs. 166 to 168 be fed to yield quadrupole 41 radiation?

521.3 What is the minimum switching time for a certain value of ε and a fixed peak power of the transmitter?

5.2.2 Radiation of a Two-Dimensional Quadrupole

Figure 173 shows a quadrupole 42 and defines certain quantities used below. The relations

$$\mathbf{r}_2 = \mathbf{r} - \tfrac{1}{2}\gamma\,\mathbf{u}, \quad \mathbf{r}_1 = \mathbf{r} + \tfrac{1}{2}\gamma\,\mathbf{u} \tag{1}$$

yield the electric and magnetic field strengths in the wave zone analogous to Eqs. (5.2.1-5) and (5.2.1-6) for $r \gg \gamma u$:

$$\begin{aligned}\mathbf{E}_{42}(1/r, t) &= \mathbf{E}(1/r_1, t) - \mathbf{E}(1/r_2, t)\\ &= Z_0 \frac{\gamma u s \cos\beta}{4\pi c^2 r} \frac{d^2 i(t - r/c)}{dt^2} \frac{\mathbf{r} \times (\mathbf{r} \times \mathbf{s})}{s r^2}\end{aligned} \tag{2}$$

$$\begin{aligned}\mathbf{H}_{42}(1/r, t) &= \mathbf{H}(1/r_1, t) - \mathbf{H}(1/r_2, t)\\ &= \frac{\gamma u s \cos\beta}{4\pi c^2 r} \frac{d^2 i(t - r/c)}{dt^2} \frac{\mathbf{s} \times \mathbf{r}}{s r}\end{aligned} \tag{3}$$

Equations (5.2.1-7) to (5.2.1-9) holding for a quadrupole 41 are replaced by the following equations:

$$\mathbf{P}_{42} = Z_0 \left(\frac{\gamma u s \sin\alpha \cos\beta}{4\pi c^2 r}\right)^2 \left(\frac{d^2 i(t - r/c)}{dt^2}\right)^2 \frac{\mathbf{r}}{r} \tag{4}$$

$$P_{42}(r, t) = Z_0 \frac{\gamma^2 u^2 s^2}{30\pi c^4} \left(\frac{d^2 i(t - r/c)}{dt^2}\right)^2 \tag{5}$$

$$P_{42} = Z_0 \frac{\gamma^2 u^2 s^2}{30\pi c^4} \left\langle \left(\frac{d^2 i(t')}{dt'^2}\right)^2 \right\rangle \tag{6}$$

$$R_{s,42} = Z_0 \frac{\gamma^2 u^2 s^2}{60\pi c^4} \frac{\left\langle \left(\frac{d^2 i(t')}{dt'^2}\right)^2 \right\rangle}{\langle i^2(t') \rangle} \tag{7}$$

The equations for quadrupole radiation hold for distances r between transmitter and receiver that satisfy the conditions of Eqs. (5.1.1-27) to

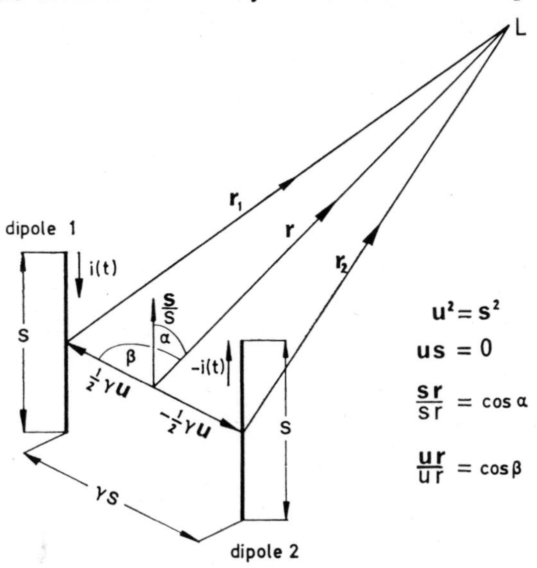

Fig. 173. Radiation from a quadrupole 42.

(5.1.1-29). In addition the relations

$$r \gg \gamma s \quad \text{or} \quad r \gg \gamma u \tag{8}$$

must be satisfied for the quadrupole radiation. For smaller values of r, one will receive the sum of dipole and quadrupole radiation. The field strengths of a dipole radiation vary with time proportionate to di/dt while the field strengths of a quadrupole radiation vary proportionate to $d^2 i/dt^2$. According to Fig. 172 these field strengths will be orthogonal to each other at any point in space for a Walsh shaped antenna current. They may thus be received independently and their average power may be compared. This is a second effect that indicates the distance from the transmitter. A sinusoidal current will produce this effect too, but the sum of the first and second derivative of a sine function is a phase shifted sine function and its two components cannot be separated without synchronization.

Problems

522.1 How must the antenna of Fig. 168 be fed to yield quadrupole 42 radiation?
522.2 In which direction is no power radiated?

5.2.3 Multipole Radiation

Figure 172 shows that $I\,\mathrm{cal}(k, t/T, \Delta T/T, \varepsilon) = i(t)$ is orthogonal to di/dt, and di/dt is orthogonal to $d^2 i/dt^2$. However, $i(t)$ is not orthogonal to $d^2 i/dt^2$. Consider now the next higher approximation $i(t) = I\,\mathrm{cal} \times$ $\times\,(k, t/T, T/T, \varepsilon, \varepsilon')$ as shown in Fig. 174. The second derivative $d^2 i/dt^2$ consists of pairs of positive and negative pulses of trapezoidal shape rather than rectangular shape as in Fig. 172. The first derivative di/dt has rounded corners in contrast to Fig. 172. The third derivative $d^3 i/dt^3$ consists of four rectangular pulses at the jumps of $i(t)$, which would degenerate into Dirac pulses in Fig. 172.

One may infer from Fig. 174 the following orthogonal sets of functions: $\{i(t), di/dt, d^3 i/dt^3\}$, $\{di/dt, d^2 i/dt^2\}$, $\{d^2 i/dt^2, d^3 i/dt^3\}$. The third and higher derivatives of the dipole current occur when one advances from quadrupole radiation to higher order multipole radiation. The field strengths of dipole and quadrupole, or dipole and octupole, or quadrupole and octupole radiation will be orthogonal to each other in the wave zone for a current according to Fig. 174. A sinusoidal current produces orthogonal field strengths for dipole and quadrupole ($d \sin\omega t/dt = \cos\omega t$, $d^2 \sin\omega t/dt^2 = -\omega^2 \sin\omega t$), but not for octupole radiation ($d^3 \sin\omega t/dt^3$ $= -\omega^3 \cos\omega t$). If the current $i(t)$ has the shape of a parabolic cylinder function as shown in Fig. 11 all its derivatives are orthogonal to $i(t)$ and to each other.

Problems

523.1 Derive the electric and magnetic field strength in the wave zone for an octupol.

523.2 Derive the power radiated.

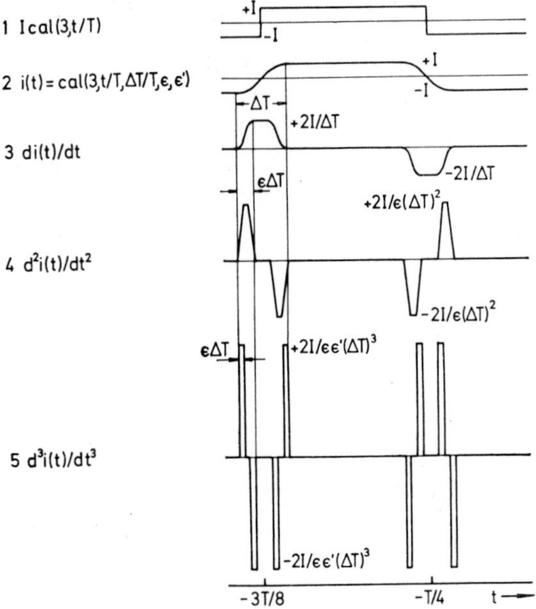

Fig. 174. Third order approximation of the transients of cal$(3, t/T)$ by the function cal$(3, t/T, \Delta T/T, \varepsilon, \varepsilon')$ and resulting time variation of dipole, quadrupole and octupole radiation.

5.3 Interference Effects, Doppler Effect

5.3.1 Radiation Diagram of a Row of Spherical Radiators

Consider a row of radiators located on the line $-D/2 \leqq x \leqq +D/2$ as shown in Fig. 175. The total current fed to all radiators is denoted by $i_a(t)$. A section Δx of the row $-D/2 \leqq x \leqq +D/2$ is fed by the current $i_a(t)\,\Delta x/D$, if all the radiators receive an equal fraction of the total current $i_a(t)$; otherwise the current $i_a(x, t)\,\Delta x/D$ is fed to the radiators within an interval $\pm\Delta x/2$ around the point x. Let there be one Hertzian dipole per interval of length Δx. One obtains from Eq. (5.1.1-17) the electric field strength generated by the dipole located at $x = 0$:

$$\mathbf{E}(1/r, t) = E(t)\frac{1}{r}\,\frac{\mathbf{r} \times (\mathbf{r} \times \mathbf{s})}{s\,r^2} \tag{1}$$

$$E(t) = \frac{Z_0\,s}{4\pi\,c\,D}\,\frac{d i_a(x, t - r/c)}{dt}\,\Delta x \tag{2}$$

The current $i_a(x, t)\,\Delta x/D$ fed to the dipole located at point x produces the voltage $E(t - x\,c^{-1}\sin\phi)$ in the direction $\mathbf{r}$ as shown by Fig. 175:

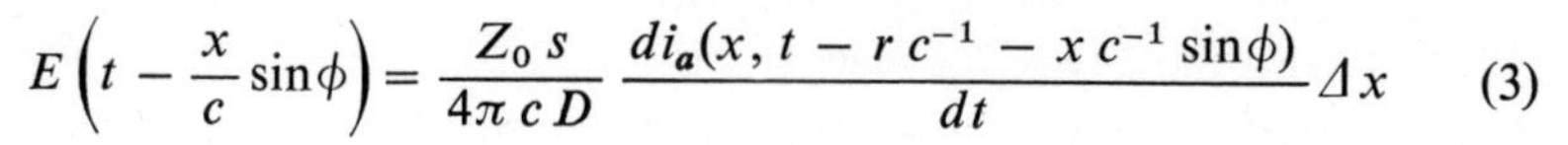

$$E\left(t - \frac{x}{c}\sin\phi\right) = \frac{Z_0\,s}{4\pi\,c\,D}\,\frac{di_a(x, t - r\,c^{-1} - x\,c^{-1}\sin\phi)}{dt}\,\Delta x \tag{3}$$

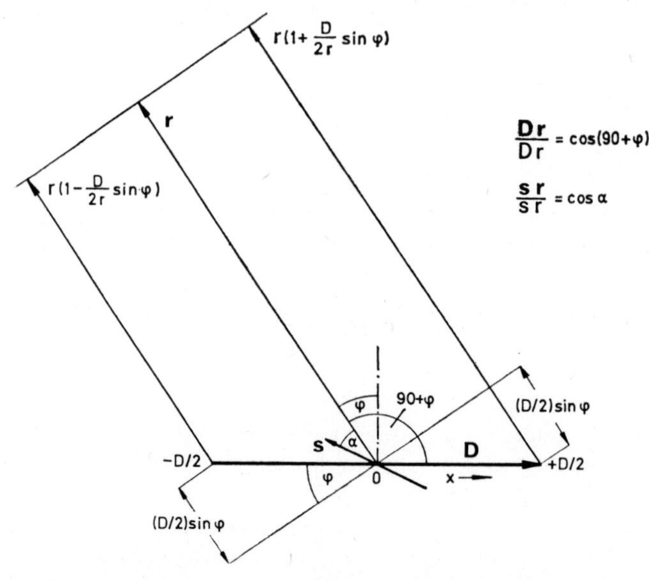

Fig. 175. Radiation from a row **D** of dipoles. **s**, dipole vector (all dipoles parallel); **D**, antenna vector; **r**, vector from center of antenna to receiver at infinity.

Let the line $-D/2 \leqq x \leqq +D/2$ be divided into $2m + 1$ intervals Δx. The field strength $\mathbf{E}_D(1/r, t, \phi)$ produced in the direction $\mathbf{r}$ by $2m + 1$ Hertzian dipoles is the sum of $2m + 1$ field strengths $\mathbf{E}(1/r, t' - n\,\Delta x \times c^{-1}\sin\phi)$, where n runs from $-m$ to $+m$:

$$\begin{aligned}\mathbf{E}_D(1/r, t, \phi) &= \sum_{n=-m}^{+m} E(t' - n\,\Delta x\,c^{-1}\sin\phi)\,\frac{1}{r}\,\frac{\mathbf{r}\times(\mathbf{r}\times\mathbf{s})}{s\,r^2}\\ &= E_D(\phi, t)\,\frac{1}{r}\,\frac{\mathbf{r}\times(\mathbf{r}\times\mathbf{s})}{s\,r^2}\end{aligned} \tag{4}$$

$$\begin{aligned}E_D(\phi, t) &= \frac{Z_0\,s}{4\pi\,c\,D}\sum_{n=-m}^{+m}\frac{d}{dt'}\,i_a(x, t' - n\,\Delta x\,c^{-1}\sin\phi)\\ t' &= t - r/c\end{aligned} \tag{5}$$

The sum in Eq. (5) gives the area of $2m + 1$ strips of width Δx and height $\dfrac{d}{dt'}\,i_a(x, t' - n\,\Delta x\,c^{-1}\sin\phi)$. For simplified computation one may replace it by an integral:

$$E_D(\phi, t) = \frac{Z_0 s}{4\pi c D} \int_{-D/2}^{D/2} \frac{d}{dt'} i_a(x, t' - x c^{-1} \sin\phi)\, dx \tag{6}$$

A comparison of Eqs. (5.1.1-17) and (5.1.1-18) shows that the magnetic field strength produced by the dipole row may be obtained from Eq. (4) by multiplication with $1/Z_0$ and substitution of $[\mathbf{s} \times \mathbf{r}]/s\,r$ for $[\mathbf{r} \times (\mathbf{r} \times \mathbf{s})]/s\,r^2$:

$$\mathbf{H}_D(1/r, t, \phi) = \frac{1}{Z_0} E_D(\phi, t) \frac{1}{r} \frac{\mathbf{s} \times \mathbf{r}}{s\,r} \tag{7}$$

Poynting's vector for the dipole row becomes:

$$\mathbf{P}_D = \mathbf{E}_D(1/r, t, \phi) \times \mathbf{H}_D(1/r, t, \phi) = Z_0^{-1} E_D^2(\phi, t) \sin^2\alpha \frac{\mathbf{r}}{r} \tag{8}$$

where α is the angle between $\mathbf{r}$ and $\mathbf{s}$.

The time average,

$$\langle E_D^2(\phi, t)\rangle / \langle E_D^2(0, t)\rangle = E_D^2(\phi)/E_D^2(0) \tag{9}$$

gives the relative average power radiated in direction $\mathbf{r}$ by a row of spherical radiators. Multiplication of $E_D^2(\phi)/E_D^2(0)$ by $\sin^2\alpha$ yields the relative average power radiated by a row of Hertzian dipoles in direction $\mathbf{r}$.

The extension of these results to quadrupole radiation offers no major problem. The second derivative of $i_a(x, t' - n\Delta x\, c^{-1} \sin\phi)$ rather than the first is used and Poynting's vector contains an additional factor $\gamma\, s \cos\alpha$ in Eq. (5.2.1-7) or $\gamma\, u \cos\beta$ in Eq. (5.2.2-4) that does not occur in Eq. (5.1.1-19).

As an example consider the current $i_a(x, t) = I\,\mathrm{cal}(k, t/T, \Delta T/T)$ as shown for $k = 3$ in Fig. 164. The derivative of the current consists of block pulses of width ΔT and amplitude $2I/\Delta T$; there are $2k/T$ such pulses per second. For $\phi = 0$ one obtains from Eq. (6) during the duration ΔT of a block pulse the following value:

$$E_D(0, t) = \frac{Z_0 s}{4\pi c D} \int_{-D/2}^{D/2} \frac{2I}{\Delta T} dx = \frac{Z_0 I s}{2\pi c \Delta T} \tag{10}$$

$$E_D^2(0) = \langle E_D^2(0, t)\rangle = \left(\frac{Z_0 I s}{2\pi c}\right)^2 \frac{2k}{T \Delta T} \tag{11}$$

A typical block pulse $E_D(0, t)$ and its integral are shown in line *1* of Fig. 176.

Let ϕ be larger than zero. The derivative of $i_a(x, t)$ is zero for $t < 0$, it is $2I/\Delta T$ for $0 \leqq t \leqq \Delta T$ and it is again zero for $t > \Delta T$. One obtains for the derivative of the current $i_a(x, t - r/c - x c^{-1} \sin\phi)$ in Eq. (6):

$$\left.\begin{aligned} \frac{di_a}{dt} &= 0 && \text{for} \quad (c\,t - r)/\sin\phi < x \\ &= 2I/\Delta T && \text{for} \quad (ct - r - c\Delta T)/\sin\phi \leqq x \leqq (ct - r)/\sin\phi \\ &= 0 && \text{for} \quad x < (ct - r - c\Delta T)/\sin\phi \end{aligned}\right\} \tag{12}$$

The integral in Eq. (6) may now be rewritten:

$$E_D(\phi, t) = \frac{Z_0 I s}{2\pi c D \Delta T} \int_L^U dx = E_D(0, t) \frac{U - L}{D} \tag{13}$$

The upper and lower limits U and L are defined as follows:

$$\left.\begin{aligned}
L &= -D/2 && \text{for} \quad (ct - r - c\Delta T)/\sin\phi < -D/2 < \\
& && \qquad < (ct - r)/\sin\phi \\
L &= (ct - r - c\Delta T)\sin\phi && \text{for} \quad -D/2 < (ct - r - c\Delta T)/\sin\phi < \\
& && \qquad < +D/2 \\
U &= (ct - r)/\sin\phi && \text{for} \quad -D/2 < (ct - r)/\sin\phi < +D/2 \\
U &= +D/2 && \text{for} \quad (ct - r - c\Delta T)/\sin\phi < +D/2 < \\
& && \qquad < (ct - r)/\sin\phi
\end{aligned}\right\} \tag{14}$$

One obtains two solutions of Eq. (13) which may be conveniently distinguished by the conditions $\sin\phi \leqq c\Delta T/D$ in Eq. (15) and $\sin\phi \geqq c\Delta T/D$ in Eq. (16):

$$\left.\begin{aligned}
E_D(\phi, t) &= 0 \quad \text{for} \quad t - r/c \leqq -\frac{D\sin\phi}{2c} \\
&= E_D(0, t)\left(\frac{ct - r}{D\sin\phi} + \frac{1}{2}\right) \\
&\qquad \text{for} \quad -\frac{D\sin\phi}{2c} \leqq t - r/c \leqq \frac{D\sin\phi}{2c} \\
&= E_D(0, t) \\
&\qquad \text{for} \quad \frac{D\sin\phi}{2c} \leqq t - r/c \leqq -\frac{D\sin\phi}{2c} + \Delta T \\
&= E_D(0, t)\left(\frac{1}{2} - \frac{ct - r - c\Delta T}{D\sin\phi}\right) \\
&\qquad \text{for} \quad -\frac{D\sin\phi}{2c} + \Delta T \leqq t - r/c \leqq \frac{D\sin\phi}{2c} + \Delta T \\
&= 0 \\
&\qquad \text{for} \quad \frac{D\sin\phi}{2c} + \Delta T \leqq t - r/c
\end{aligned}\right\} \tag{15}$$

$$\left.\begin{aligned}
E_D(\phi, t) &= 0 \quad \text{for} \quad t - r/c \leqq -\frac{D\sin\phi}{2c} \\
&= E_D(0, t)\left(\frac{c\,t - r}{D\sin\phi} + \frac{1}{2}\right) \\
&\qquad \text{for} \quad -\frac{D\sin\phi}{2c} \leqq t - r/c \leqq -\frac{D\sin\phi}{2c} + \varDelta T \\
&= E_D(0, t)\frac{c\,\varDelta T}{D\sin\phi} \\
&\qquad \text{for} \quad -\frac{D\sin\phi}{2c} + \varDelta T \leqq t - r/c \leqq \frac{D\sin\phi}{2c} \\
&= E_D(0, t)\left(\frac{1}{2} - \frac{c\,t - r - c\,\varDelta T}{D\sin\phi}\right) \\
&\qquad \text{for} \quad \frac{D\sin\phi}{2c} \leqq t - r/c \leqq \frac{D\sin\phi}{2c} + \varDelta T \\
&= 0 \quad \text{for} \quad \frac{D\sin\phi}{2c} + \varDelta T \leqq t - r/c
\end{aligned}\right\} \tag{16}$$

Figure 176 shows $E_D(\phi, t)$ and its integral for various values of $D/c\,\varDelta T = \varrho$ and ϕ. In the direction $\phi = 0$ one obtains a block pulse of

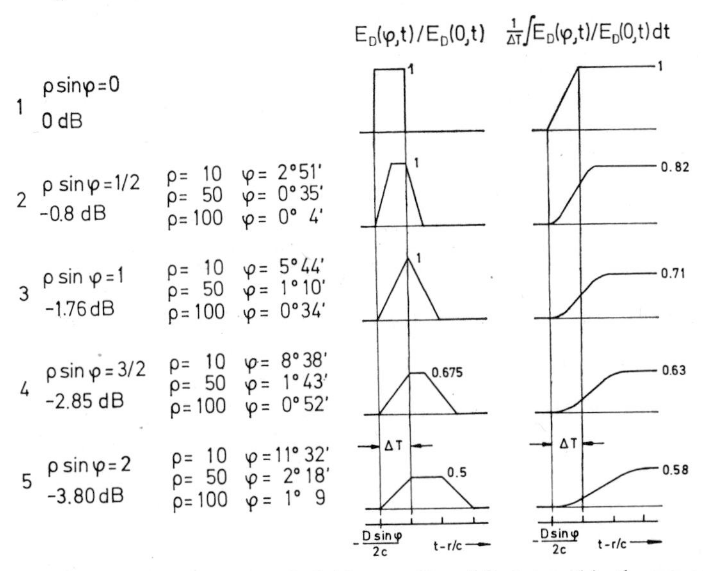

Fig. 176. Time variation of the electric field strength and its integral in the wave zone as function of the angle ϕ and the ratio $D/c\,\varDelta T = \varrho$ caused by a transient from $-I$ to $+I$ of an antenna current $I\,\text{cal}(k, t/T, \varDelta T/T)$.

duration ΔT, since the delay time is the same for any wave $\mathbf{E}(1/r, t)$, and the wave $\mathbf{E}_D(1/r, t, \phi)$ of Eq. (4) differs from $\mathbf{E}(1/r, t)$ by a factor only. For increasing values of ϕ the block pulse of line *1* in Fig. 176 changes to a trapezoidal shape (line *2*) and to a triangular shape (line *3*). For still larger values of ϕ one obtains again trapezoidal pulses but they do not reach the peak value $E_D(\phi, t)/E_D(0, t) = 1$ any more (lines *4* and *5*). The pulses for $\varrho \sin\phi \leqq 1$ are defined by Eq. (15), the ones for $\varrho \sin\phi \geqq 1$ by Eq. (16).

The integrated functions $E_D(\phi, t)$ on the right side of Fig. 176 show the distortions of a transmitted Walsh function for $\phi \neq 0$. The linear transistion from 0 to 1 during the time ΔT in line *1* corresponds to the linear transistion of the antenna current from $-I$ to $+I$ during the time ΔT in the second line of Fig. 164. This linear transition becomes rounded and the transition time increases as $\varrho \sin\phi$ grows. The largest transition time occurs for $\phi = 90°$ and it equals $\Delta T + D/c = \Delta T(1 + \varrho)$.

The time average $E_D^2(\phi)$ follows from Eqs. (9), (11), (14), (15) and (16):

$$E_D^2(\phi) = \frac{2k}{T} \int_L^U E_D^2(\phi, t)\, dt \tag{17}$$

$E_D^2(\phi)/E_D^2(0)$ is plotted for various values of ϱ in Fig. 177. Note that this antenna radiation diagram applies when the positive and negative pulses

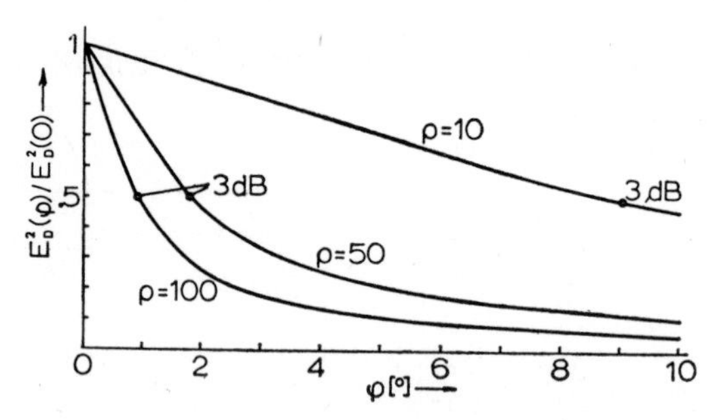

Fig. 177. Radiated average power $E_D^2(\phi)$ normalized by $E_D^2(0)$ as function of the angle ϕ.

of $\mathbf{E}(1/r, t)$ and $\mathbf{H}(1/r, t)$ represented by $di(t)/dt$ in Fig. 164 are so far apart that there is no interference between two adjacent pulses. This condition may be expressed mathematically for a wave generated by a current $I\,\mathrm{cal}(k, t/T, \Delta T/T)$ as follows: The time interval between adjacent pulses equals $T/2^{h-1} - \Delta T$ or $T/2^h - \Delta T$ for $2^{h-1} < 2k \leqq 2^h$. A wave travels $c(T/2^h - \Delta T)$ during the shorter time. There can be no interference between adjacent pulses if the length D of the dipole row in Fig. 175 is shorter than $c(T/2^h - \Delta T)$. More precisely, the diagrams of Fig. 177

hold for angles ϕ for which the condition

$$D \sin\phi \leqq c(T/2^h - \Delta T) \tag{18}$$

is satisfied.

The results obtained for dipoles may be extended qualitatively to quadrupole and higher order multipole radiation as follows: The pulse duration ΔT of the dipole radiation has to be replaced by the pulse duration $\varepsilon \Delta T$ or $\varepsilon' \Delta T$ for quadrupole and octupole radiation according to Figs. 172 and 174. Since $\varepsilon \Delta T$ is smaller than ΔT, and $\varepsilon' \Delta T$ is smaller than $\varepsilon \Delta T$ one obtains higher values for ϱ in Fig. 177 for a fixed length D of a row of quadrupoles or octupoles. Hence, the radiation diagram becomes narrower. Furthermore, since quadrupole radiation consists of pairs of pulses with opposite sign and octupole radiation consists of quadruplets of pulses with opposite signs, there will be interference between these pulses that make the radiation diagram narrower.

Pearlman has computed radiation diagrams according to Fig. 177 for cases for which the condition of Eq. (18) is not satisfied [1]. Figure 178

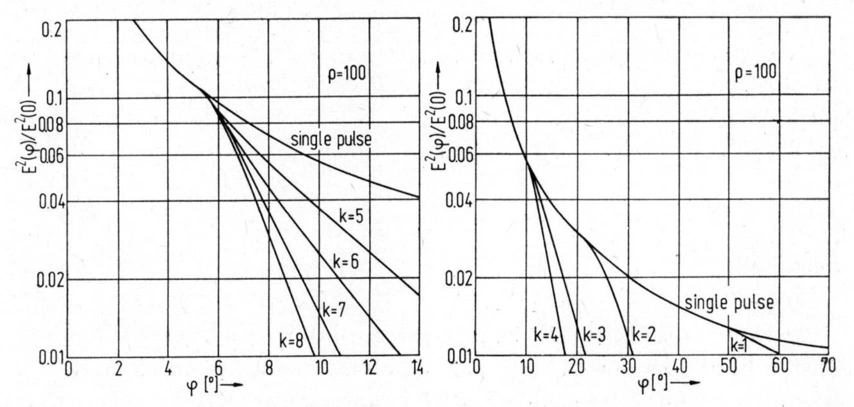

Fig. 178. Radiated average power $E_D^2(\phi)$ normalized by $E_D^2(0)$ as function of the angle ϕ for antenna currents $\text{cal}(k, t/T, \Delta T/T)$ with $k = 1 \ldots 8$. The curve denoted "single pulse" is the same as the one for $\varrho = 100$ in Fig. 177.

gives some of his results for an antenna current $I \,\text{cal}(k, t/T, \Delta T/T)$ with $k = 1, \ldots, 8$. The curve denoted "single pulse" is the same as the curve for $\varrho = 100$ in Fig. 177. The deviations from this curve for smaller values of ϱ are insignificant.

Problems

531.1 Derive plots equivalent to those of Fig. 176 for quadrupole radiation; the antenna current is defined by Fig. 172.

531.2 Using the results of problem 531.1 plot the diagram corresponding to Fig. 177 and compare the plots for dipole and quadrupole radiation for small angles of φ.

5.3.2 Doppler Effect

A sinusoidal electromagnetic wave $E \sin 2\pi f(t - x/c)$ is transformed by the Doppler effect into the wave $E \sin 2\pi f'(t' - x'/c)$. The shifted frequency has the value

$$f' = f \frac{1 - v/c}{\sqrt{(1 - v^2/c^2)}} \tag{1}$$

where v is the relative velocity of transmitter and receiver.

A Walsh wave

$$\mathrm{E}(x, t) = \mathrm{E}\,\mathrm{sal}(\phi T, t/T - x/c\,T) \tag{2}$$

is transformed by the transformation equations of relativistic mechanics

$$t = \frac{t' + v\,x'/c^2}{\sqrt{(1 - v^2/c^2)}}, \quad x = \frac{x' + v\,t'}{\sqrt{(1 - v^2/c^2)}} \tag{3}$$

into the following form:

$$\mathrm{E}(x', t') = \mathrm{E}\,\mathrm{sal}\left(\phi T, \frac{t' - x'/c}{T \dfrac{\sqrt{(1 - v^2/c^2)}}{1 - v/c}}\right) \tag{4}$$

In order to bring Eq. (4) into the form of Eq. (2) one must define the transformed sequency ϕ' and time base T' as follows:

$$\phi' = \phi \frac{1 - v/c}{\sqrt{(1 - v^2/c^2)}} \tag{5}$$

$$T' = T \frac{\sqrt{(1 - v^2/c^2)}}{1 - v/c} \tag{6}$$

It follows:

$$\mathrm{E}(x', t') = \mathrm{E}\,\mathrm{sal}(\phi' T', t'/T' + x'/c\,T') \tag{7}$$

Equations (5) and (1) show that sequency ϕ and frequency f are changed equally by the Doppler effect. The additional change of the time base T according to Eq. (6) generates an invariant of the Doppler effect or of the Lorentz transformation:

$$T' \phi' = T \phi \tag{8}$$

A sine wave with frequency f radiated by a transmitter with relative velocity v cannot be distinguished from one with frequency f' radiated by a transmitter with relative velocity 0. This is generally not so for Walsh functions[1]. One may readily see from Fig. 2 that a reduction of the sequency $\phi = 8/T$ of $\mathrm{sal}(8, \theta)$ to $\phi' = \frac{3}{4}\phi$ yields $\phi' = 6/T$, but the resulting Walsh wave would differ from $\mathrm{sal}(6, \theta)$.

[1] There have been attempts to detect intelligent signals transmitted from other planets. These attempts were based on the assumption that such signals would be sine waves with a frequency in a location of the spectrum where the background noise is particularly low, e.g., at 1.42 GHz. The Lorentz invariance of φT raises the question of whether Walsh waves are not more suitable. For high power transmitters Walsh waves have a considerable advantage over sine waves from the standpoint of transmitter dissipation and efficiency of power conversion. The main advantage of sine waves is at

Problems

532.1 Write the series expansion of Eqs. (1), (5) and (6) up to the first power in v/c. These are the formulas generally used in communications.

532.2 Compute the extreme values of f'/f, φ'/φ and T'/T for a wave radiated from an airplane flying at a speed of Mach 1 and received at the ground.

532.3 Repeat problem 532.2 for a radar signal reflected by the airplane.

5.3.3 Circular Polarization, Interferometry, Shape Recognition

The Walsh functions of Fig. 2 may be considered to represent linearly polarized Walsh waves. The first 5 are shown again in the first column of Fig. 179. The circularly polarized waves of the second column are obtained from them by holding fast the "left ends" of the functions and giving their "right ends" a twist of 360° in the sense of a right hand screw.

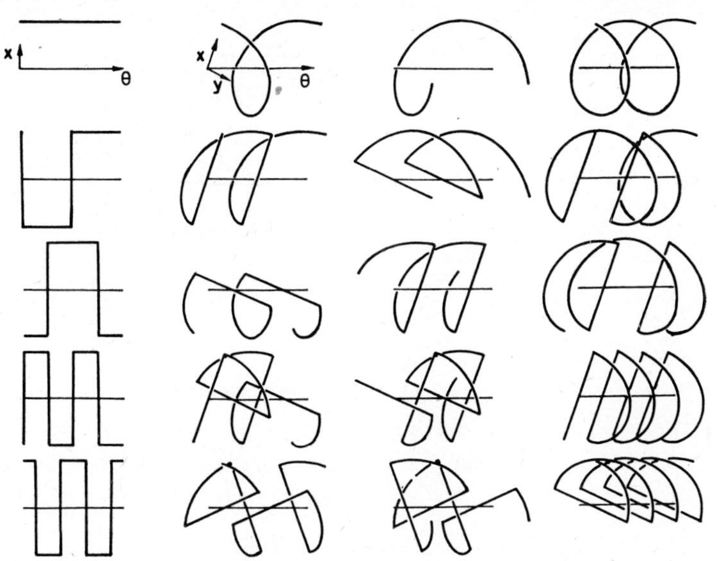

Fig. 179. Circularly polarized Walsh waves.

The third column is obtained by turning the functions of the second column 90° to the right. The fourth column is obtained by twisting the functions of the first column 2 × 360° in the sense of a right hand screw.

The functions of the first line of Fig. 179 are the usual right polarized waves, having the normalized frequencies (turns per unit time) 0, 1, 1 and 2.

---

the present the simplicity of the receiving equipment. Simplicity is, however, not a convincing argument if one reflects on the changes of electronic technology during the last 25 years and the fact that a detectable signal would have to come from a civilization considerably more advanced than ours. The proper approach would be to investigate first which system of functions is most likely to be used and then decide which function of the system is favored by low background noise.

These waves appear here as the special case of right polarized Walsh waves with sequency 0. The functions in the second line have all the normalized sequency 1 and frequencies 0, 1, 1 and 2; the same holds for the functions in line *3*. The sequency of the functions in lines four and five equals 2, the frequencies are again 0, 1, 1 and 2.

Figure 180c shows the principle of interferometric angle measurement. Two receivers at the points *A* and *B* receive waves from a far away transmitter which travel practically parallel along the rays *a* and *b*. A measurement of the propagation time difference $\Delta T = \overline{AC}/c$ yields the angle $\beta = \sin^{-1} c\Delta T/\overline{AB}$. The smallest measurable time difference $\Delta T_{\min}$ depends—for sine as well as for Walsh functions—on the gradient of

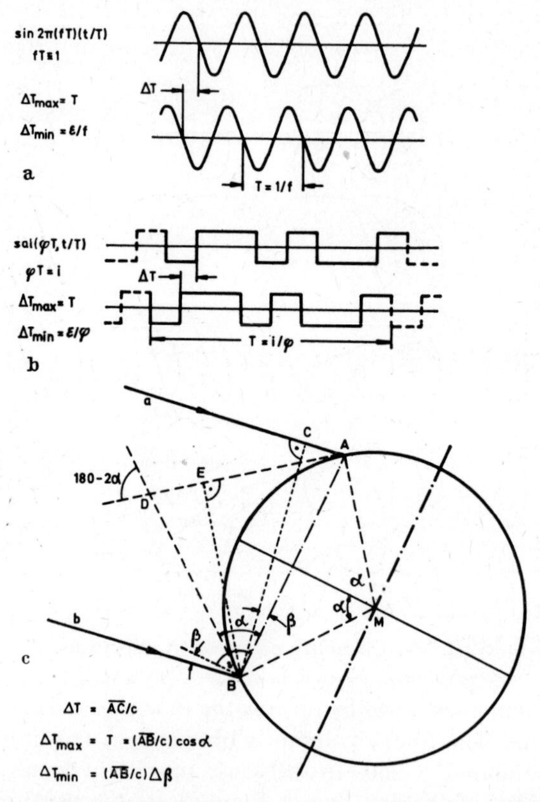

Fig. 180. Interferometric measurement of angles. (a) resolution and resolution range of sine waves; (b) resolution and resolution range of Walsh waves; (c) geometric relations for two receivers A and B positioned on the same meridian.

their zero crossings. Hence, $\Delta T_{\min}$ is proportionate to $1/f$ for sine functions and proportionate to $1/\phi$ for Walsh functions; the proportionality factor is denoted by ε in Fig. 180a and b. The resolution, that is the smallest measurable time ΔT_{min} or the smallest measurable angle $\Delta\beta \cong c\,\Delta T_{min}/\overline{AB}$, is approximately equal for sine and Walsh functions. However, the resolution range is completely different. The largest permissible value of ΔT must lie between $-T/2$ and $+T/2$, if T is the period of the wave, since a wave delayed by a multiple of T is equal to the undelayed wave. Hence, ΔT_{max} equals T. Since T equals $1/f$ in the case of sine functions, ΔT_{max} equals $\Delta T_{min}/\varepsilon$. Certain Walsh functions $\mathrm{sal}(i, \theta)$ have a shortest period $T = i/\phi$ and ΔT_{max} equals then $i\,\Delta T_{min}/\varepsilon$. Such a Walsh function is shown for $i = 3$ in Fig. 180b. Other usable values are $i = 2^k - 1$ according to Table 9 in section 4.1.2. A large value of i increases the resolution range ΔT_{max} without increasing the smallest measurable time difference ΔT_{min}.

The following representative values may be computed from Fig. 180c. Let A and B be two points at $\alpha = 52°$ northern and southern latitude and assume ΔT_{min} to be 10^{-9} s. The distance $\overline{AB}$ is about 10000 km and the usable observation angle $180° - 2\alpha = 76°$. The resolution equals $\Delta\beta = 3 \times 10^{-8}$ or $\Delta\beta = 0.05''$. The value of i is obtained from the following relation:

$$i \geqq \varepsilon\,\Delta T_{max}/\Delta T_{min} = \varepsilon \cos\alpha/\Delta\beta = 2.5 \cdot 10^{8}\varepsilon = 10^{6} \qquad (1)$$

An angle of $0.05''$ corresponds to a distance of about 11 m on the surface of the moon and of about 3 km on the surface of Mars when Mars is close to Earth. For comparison, the smallest resolvable angle of $0.05''$ is about one order of magnitude smaller than the best that can be done with astronomical telescopes.

This method of angle measurement appears attractive for space probe tracking. An accurate knowledge of the distance $\overline{AB}$ would frequently not be required. Such a case is guidance of a space probe to the vicinity of another or to a beacon transmitter. A considerable amount of data processing equipment is required. The previously assumed value $i = 10^6$ means that two Walsh functions consisting of a periodic sequence of 10^6 block pulses have to be compared. A minimum storage capacity of 2×10^6 bits would be required for the comparison. Additional storage capacity would be needed to improve the signal-to-noise ratio by averaging over many multiples of 10^6 pulses. Averaging over, e.g., 1000 multiples would require a total of some 12×10^6 bits storage capacity.

Figure 181 shows a radar R and two point-like targets B 1 and B 2 that are close together. Lines a and b show sine waves reflected from B 1 and B 2. Line c shows the sum of these two sine waves which is received by the radar. A periodic sine wave would look the same whether reflected by two targets or by a single, more reflecting target. The pulsed sine wave of line c shows deviations at beginning and end compared with lines a and b. There are 1000 cycles and only two of them distorted, if the pulse duration is 1 μs and the carrier frequency is 1 GHz. Hence, the energy indicating two targets is in the order of 0.1% of the total energy of the pulse and is insignificant.

Let us consider the reflection of Walsh waves. Lines *d* and *e* show the waves reflected from B 1 and B 2, and line *f* shows their sum. The difference between waves reflected from one or two targets is no longer restricted to beginning and end of a pulse. A periodic Walsh wave would still tell how many targets there are and what the differences of their distances are, although the absolute distance could not be inferred from the shape of the reflected signal.

Since lines *d* to *f* in Fig. 181 show that the sum of several Walsh waves of equal shape but various time shifts may be a differently shaped wave, one must investigate the reflection on the radar dish. The proper approach would be to solve the wave equation for the particular boundary and initial conditions. This has not been done yet. Wave optics has been dominated by sine and cosine functions as much as communications. There is no theory for Walsh waves or complete systems of orthogonal waves. It would be wrong to treat Walsh waves as a superposition of sine and cosine waves and apply the known results of wave optics to these sine and cosine waves. Sequency filters, sequency multiplexing and the results for Walsh wave

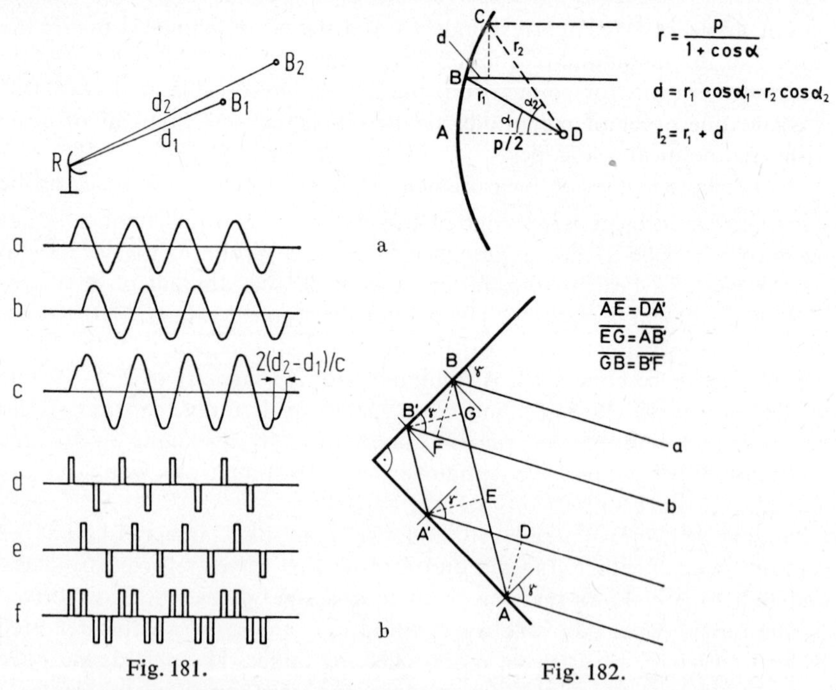

Fig. 181.

Fig. 182.

Fig. 181. Reflection of sine and Walsh waves by two point-like targets.

Fig. 182. Reflection of waves according to geometric optics by a parabolic mirror (a) and two perpendicular mirrors (b).

antennas would never have been found, if the Walsh functions had been treated as a superposition of sine and cosine functions.

Lacking a wave theory, one may use geometrical optics as a first approximation. Figure 182a shows a cut of a parabolic dish. The distances r_2 and $r_1 + d$ are equal. Hence, a Walsh wave radiated from the focal point D will be delayed equally whether reflected at B or C and will add without time shift. Vice versa, a signal reflected by a parabolic dish to the focal point D will not be distorted even though it is not sinusoidal.

Another example of a distortion-free reflector is the perpendicular mirror shown in Fig. 182b. It follows from the geometric relations shown in that figure that the two propagation paths a and b are equally long. Hence, there will be no time shift between Walsh waves reflected from various points of the reflector, and the Walsh wave will be reflected without change of shape. It can be shown that this result also holds for a three-dimensional perpendicular reflector.

In general, a Walsh wave reflected by a target of finite dimension will no longer be a Walsh wave. The shape of the reflected wave will yield information about the geometric size and shape of the target. Consider the reflection of a step, like the one of $\mathrm{sal}(1, \theta)$ at $\theta = 0$ in Fig. 2, from a sphere as shown in Fig. 183a. A correct treatment would again require a solution of the wave equation. A first approximation may, however, be obtained by assuming that a spherical wave is radiated from each point on the surface of the sphere that is illuminated by the incident wave. The reflected step computed under this assumption is shown on the right of Fig. 182a. Initially the wave is reflected by the points on the surface of the sphere close to the plane S only. After the time $t = 2R(1 - \sin\beta)/c$ has elapsed, the wave will be reflected by all points on the sphere having a distance smaller than $\frac{1}{2}c\,t$ from plane S and much more power will be reflected. At the time $t = 2R/c$ all points on the illuminated half of the sphere reflect, and there will be no increase of reflected power for larger values of t. Since all Walsh functions may be considered to be superpositions of step functions with positive or negative amplitude, one may construct the shape of reflected Walsh functions from Fig. 183a.

The computation of the shape of the reflected step wave is as follows: The amplitude due to the reflection from an annular area with distance $\frac{1}{2}c\,t$ from plane S in Fig. 183a is proportionate to its area $2\pi R \cos\beta\, R\, d\beta$, but only the fraction $\sin\beta$ of this area reflects back into the direction of incidence. The voltage u displayed on an oscilloscope as a function of the angle β is thus given by

$$u(\beta) = 2\pi R^2 K \int_0^{\pi/2-\beta} \cos\beta' \sin\beta' \, d\beta' = K\pi R^2 (1 - \sin^2\beta) \qquad (2)$$

$$0 \leqq \beta \leqq \tfrac{1}{2}\pi\,.$$

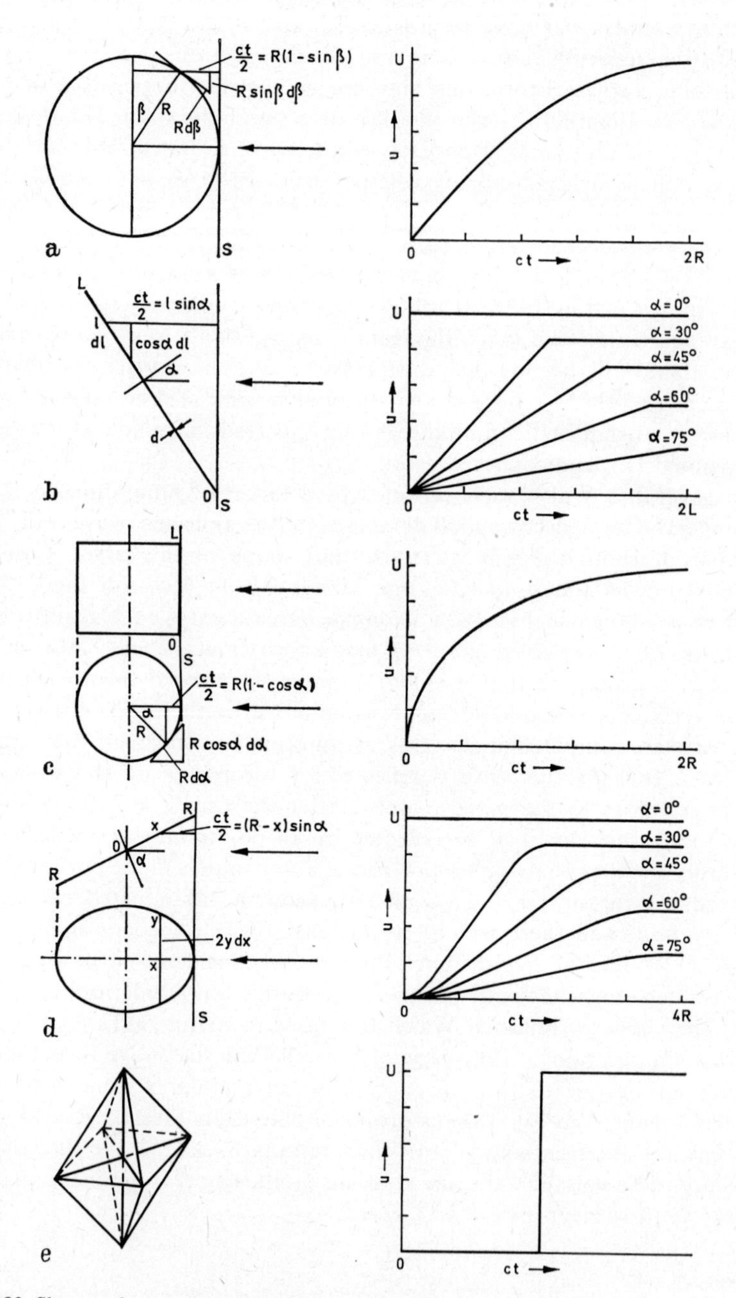

Fig. 183. Shapes of step waves reflected by perfect scatterers of various shapes. (a) sphere; (b) rod of length L and diameter $d \ll L$; (c) cylinder; (d) circular disc of diameter $2R$; (e) radar reflector (3 perpendicular mirrors).

where K is a factor that corrects the dimension and allows for attenuation, amplification, reflectivity, etc.

Since u is displayed on an oscilloscope as a function of time and not of β one may substitute

$$c\,t = 2R(1 - \sin\beta)$$

from Fig. 183 into Eq. (2):

$$u(c\,t) = K\pi R^2\left[1 - \left(1 - \frac{c\,t}{2R}\right)^2\right], \quad c\,t \leqq 2R: \; u(c\,t) = K\pi R^2, \quad c\,t > 2R$$

$u(c\,t)$ is the curve plotted in Fig. 183a.

Figure 183b shows the shape of a step wave reflected by a rod of length L and diameter $d \ll L$ for various angles α of incidence. Figure 183c shows the reflection by a cylinder, if the incidence is perpendicular to the axis. Figure 183d shows the reflection by a circular disc of diameter $2R$ for various angles α of incidence, and Fig. 183e that of a radar reflector consisting of three perpendicular mirrors. The obvious implication of Fig. 183 is that a Walsh wave radar can distinguish small and light-weight radar reflectors directly from the real thing while a sine wave radar requires secondary effects, e.g., different velocity caused by air drag.

The question arises of which Walsh functions would be best for shape recognition. Consider $\text{sal}(8, \theta)$ in Fig. 2. This function is easy to filter and process. However, it has a shortest period of $\theta = \frac{1}{2}$ or $t = \frac{1}{2}T$ and this causes ambiguities, if a target has a larger dimension than $\frac{1}{2}c\,T$, or if there are several targets with distances larger than $\frac{1}{2}c\,T$. The function $\text{sal}(7, \theta)$ is harder to filter and process than $\text{sal}(8, \theta)$, but its shortest period is $\theta = 1$ or $t = T$ and ambiguities will occur for targets with dimension larger than $c\,T$.

The function $\text{sal}(1, \theta)$ is equally simple to filter and process as $\text{sal}(8, \theta)$ and also has the shortest period T. However, $\text{sal}(1, \theta)$ has 2 steps only, while $\text{sal}(7, \theta)$ has 14 steps, and it is the steps that provide information about the shape of the target, not the constant sections of the functions. Hence, the more complicated Walsh functions are better from the theoretical point of view.

Interferometric tracking of space probes and radar target analysis can be and are done by sinusoidal waves too. The point here is, that Walsh waves behave inherently very different from sine waves and thus offer a promising alternative for a more detailed study of resolution and useful signal-to-noise ratios that might be obtained.

The discussion has been restricted to Walsh functions even though the electromagnetic waves in the wave zone are differentiated Walsh functions for dipole radiation and higher derivatives for multipole radiation. A discussion of shape recognition by these waves would require the introduction of the parameters ΔT, $\varepsilon\Delta T$, $\varepsilon'\Delta T$, etc., and it appears premature to carry the theoretical development any further before much more experimental work on electromagnetic Walsh waves has been done.

Problems

533.1 Plot the projection of the curves in Fig. 179 onto the x and y planes.

533.2 Using the results of problem 533.1 suggest a way to generate circularly polarized Walsh waves.

533.3 Using problem 533.2 and the results of section 5.1.1 plot the time variation of the electric and magnetic field strength in the wave zone projected onto the x and y plane as well as in axonometric representation as in Fig. 179.

533.4 How can one increase the resolution range of the sine wave in Fig. 180?

533.5 For a switching time of 1 nanosecond for the antenna current how large must $d_2 - d_1$ be in Fig. 181 to avoid overlapping of the pulses of lines e and f?

5.4 Signal Selection and Synchronization

5.4.1 Separation of Signals in Mobile Communication

One of the most important advantages of sinusoidal waves is the invariance of their orthogonality to time shifts. For explanation consider a sine carrier $\sqrt{(2)}\sin 2\pi n\theta$ amplitude modulated by a signal $F_n(\theta)$. The signal $F_n(\theta)$ is practically constant during any period of n cycles of the carrier $\sqrt{(2)}\sin 2\pi n\theta$. Synchronous demodulation of the modulated carrier may be represented by the following integral:

$$\int_{\theta'-1/2}^{\theta'+1/2} F_n(\theta)\sqrt{(2)}\sin 2\pi n\theta\sqrt{(2)}\sin 2\pi m\theta\, d\theta =$$

$$= F_n(\theta')\int_{\theta'-1/2}^{\theta'+1/2}\sqrt{(2)}\sin 2\pi n\theta\sqrt{(2)}\sin 2\pi m\theta\, d\theta = F_n(\theta')\,\delta_{nm} \tag{1}$$

In the case of mobile radio communication a sum of many modulated carriers with various time shifts is received. Hence, $F_n(\theta)\sqrt{(2)}\sin 2\pi n\theta$ is replaced by

$$\sum_{n=1}^{j} F_n(\theta)\sqrt{(2)}\sin 2\pi n(\theta-\theta_n)$$

and Eq. (1) assumes the following form:

$$\int_{\theta'-1/2}^{\theta'+1/2}\left[\sum_{n=1}^{j} F_n(\theta)\sqrt{(2)}\sin 2\pi n(\theta-\theta_n)\right]\sqrt{(2)}\sin 2\pi m\theta\, d\theta =$$

$$= F_n(\theta')\cos 2\pi n\theta_n\,\delta_{nm} \tag{2}$$

The time shifts θ_n introduce attenuation but not crosstalk. The orthogonality of sine and cosine functions of the same frequency is destroyed by the time shifts but the orthogonality to functions of different frequency is preserved. The subsets of functions $\{\sqrt{(2)}\sin k(\theta - \theta_k)\}$ or $\{\sqrt{(2)}\cos k(\theta - \theta_k)\}$ are orthogonal for any values of θ_k. The underlying reasons for this are the shift theorems of sine and cosine functions:

$$\begin{aligned} \sin k(\theta + \theta_k) &= \sin k\,\theta \cos k\,\theta_k + \cos k\,\theta \sin k\,\theta_k \\ \cos k(\theta + \theta_k) &= \cos k\,\theta \cos k\,\theta_k - \sin k\,\theta \sin k\,\theta_k \end{aligned} \tag{3}$$

Consider now the one-dimensional wave equation,

$$\frac{\partial^2 u}{\partial t^2} = c^2 \frac{\partial^2 u}{\partial x^2} \tag{4}$$

and its general solution

$$u(x, t) = g(t - x/c) + h(t + x/c) \tag{5}$$

The ordinary addition and subtraction signs occur in the arguments $t + x/c$ and $t - x/c$, just as in the shift theorems (3) of sine and cosine functions. It is often believed that sine-cosine functions are therefore uniquely advantaged over all other functions in wave propagation. For purposes of communications it is stated that from the complete system of sine-cosine functions one half—either all the sine or all the cosine functions—can be used as carriers in mobile communication, but less than one half of the functions of other complete systems can be used.

A simple example will show that Walsh functions can do exactly as well as sine-cosine functions: From the complete system of Walsh functions one half—either all the sal or all the cal functions—can be used as carriers. The example will also show that there are infinitely many other systems of functions that are just as good for mobile communication, at least on paper.

The electric or magnetic field strength in the wave zone produced by the antenna current $\mathrm{cal}(j, t/T)$ consists of Dirac pulses at the jumps of $\mathrm{cal}(j, t/T - r/cT)$. Trains of such Dirac pulses produced by the antenna currents $\mathrm{cal}(5, t/T)$ and $\mathrm{cal}(3, t/T)$ are shown in lines *1* and *2* of Fig. 184. It is evident that these pulse trains are orthogonal regardless of time shifts since the one function is zero whenever the other has a Dirac pulse. An exception could only occur if the pulses of lines *1* and *2* would coincide. However, the probability of this happening accidentally is zero since the Dirac pulses are infinitely short while the time intervals between them are finite. The assumption of infinitely short Dirac pulses plays for Walsh functions a similar role as the assumption of infinitely long sinusoidal waves for sine-cosine functions.

Line *3* in Fig. 184 shows the sum of lines *1* and *2*. This is the time variation of the electric and magnetic field strength at the receiver input. Let a switch be operated according to the gating function $g(t)$ of line *4*. This function is synchronized to one of the pulses with period T of $F(t)$; this pulse is denoted by a. The switch is closed periodically 16 times during a time interval of duration T and the gated function $F(t)$ is fed to an integrator. The output voltage of the integrator is the function $\mathrm{cal}(5, t/T)$ with a time shift as shown in line 5. A dc voltage is superimposed on $\mathrm{cal}(5, t/D)$, representing the undetermined constant of the indefinite integral which is of no interest. Regardless of the time shift of $\mathrm{cal}(5, t/T)$ there will be 10 jumps—or 10 zero crossings if the superimposed dc voltage is eliminated—per time interval of duration T. A counter may thus be used to recognize $\mathrm{cal}(5, t/T)$ by its sequency.

Let us assume that the function $\mathrm{cal}(3, t/T)$ rather than $\mathrm{cal}(5, t/T)$ is to be received. The gating function synchronized to the pulse a of $F(t)$ yielded $\mathrm{cal}(5, t/T)$. As soon as the counter has identified this carrier as having the wrong sequency, the gating function $g(t)$ is shifted relative to

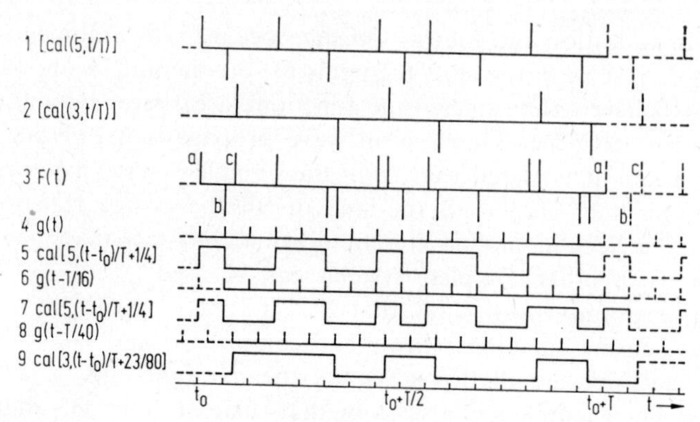

Fig. 184. Selection of the carrier $\mathrm{cal}(3, t/T)$ from the input signal $F(t)$.

$F(t)$ and synchronized to the first pulse following pulse a, which is pulse b. The shifted gating function $g(t - T/16)$ is shown in line 6 while line 7 shows the integral of the Dirac pulses passed through the switch operated according to $g(t - T/16)$. This is again the function $\mathrm{cal}(5, t/T)$. The counter will identify it as having the wrong sequency and the gating function $g(t)$ is shifted again. The first pulse following pulse b is pulse c. The gating function synchronized to pulse c is shown in line 8 and the integral of the passed Dirac pulses in line 9. This function has 6 jumps per time interval of duration T and the counter will identify it as having the desired sequency. The gating function remains synchronized to pulse c. The circuit can separate Walsh carriers with different sequency but it cannot distinguish between a sal and a cal carrier with the same sequency since they differ by a time shift only. The analogy to sine and cosine carriers is perfect.

Assume the current $\mathrm{cal}(3, t/T)$ had been amplitude modulated before it was fed to the transmission antenna. The received function $\mathrm{cal}(3, t/T)$ is then amplitude modulated too. Demodulation can be accomplished by rectification, just as in the case of an amplitude modulated sinusoidal carrier. This process is called asynchronous demodulation. A process which is more complicated but superior in the presence of noise is synchronous demodulation. In the case of a sinusoidal carrier it requires multiplication of the received carrier by a sinusoidal function of equal frequency and phase generated in the receiver. In the case of Walsh carriers multiplication by a Walsh function with equal sequency and "time position" is required. In other words, the carrier of line *9* in Fig. 184 must be multiplied by the synchronized function $\mathrm{cal}[3, (t - t_0)/T + 23/80]$ generated in the receiver. A method to obtain this synchronization will be discussed in the following section; the synchronization of the gating function $g(t)$ to one of the pulses a, b or c of line *3* of Fig. 184 will be dealt with at the same time.

Figure 185 shows the functions $\mathrm{cal}(3, \theta)$ and $\mathrm{cal}(5, \theta)$ together with their first two derivatives. It is evident that the orthogonality of the second derivatives is independent of time shifts just like that of the first derivatives.

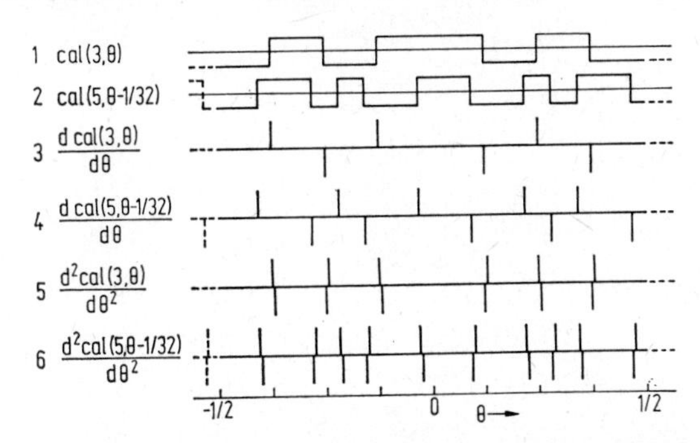

Fig. 185. Invariance to time shifts of the orthogonality of higher derivatives of the Walsh functions.

The same holds true for the orthogonality of the first derivative of $\mathrm{cal}(3, \theta)$ and the second derivative of $\mathrm{cal}(5, \theta)$. A great number of systems of functions suitable in theory for mobile communication may thus be constructed. The selection process at the receiver is essentially the same as discussed above. The only difference is that n integrations are required rather than one to convert the n-th derivative into the original Walsh function.

Problems

541.1 For $T = 100$ microseconds and a switching time $\Delta T = 1$ nanosecond, how many sequency channels can be provided with a probability of 1% for interference between two channels?

541.2 Can the number of channels be increased by assigning additional channels to transmitters using quadrupole radiation?

541.3 Can one use octupole and still higher multipole radiation to provide still more channels without changing the previously introduced transmitters and receivers for dipole radiation?

5.4.2 Synchronous Reception of Walsh Waves

The function $\mathrm{cal}(3, t/T - t_v/T)$ of line *9* in Fig. 184 is shown again in Fig. 186 in the line denoted $t_v = 5T/8$ together with the same function having time shifts from $t_v = 0$ to $t_v = 7T/8$. Figure 186 also shows the function $\mathrm{cal}(3, t/T - t_u/T)$ which is the local carrier produced in the re-

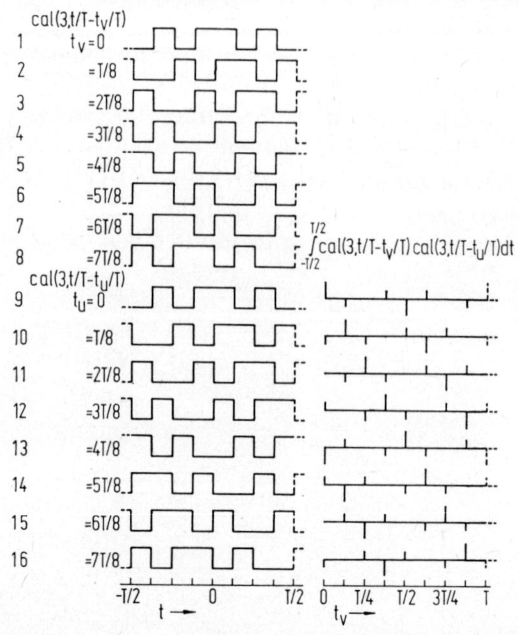

Fig. 186. Synchronization of the selected carrier by means of a cross-correlation function.

ceiver. This carrier is plotted for $t_u = 0$ to $t_u = 7T/8$. The cross-correlation functions of the received and locally produced carrier are shown on the right for $t_v = 0, T/8, \ldots, 7T/8$ and $t_u = 0, T/8, \ldots, 7T/8$. These cross-correlation functions assume the values $+1$, $+\frac{1}{2}$, 0, $-\frac{1}{2}$ and -1. The value $+1$ indicates that t_v equals t_u which means that the two carriers are synchronized. There are two ways to generate the cross-correlation functions: a) one may multiply the received carrier $\mathrm{cal}(3, t/T - t_v/T)$ with the eight local carriers having time shifts $t_u = 0$ to $t_u = 7T/8$; b) one may multiply the locally produced carrier $\mathrm{cal}(3, t/T - t_u/T)$ with the eight received carriers having time shifts $t_v = 0$ to $t_v = 7T/8$. Both methods yield the same result since an interchange of t_u and t_v yields the same cross-correlation functions in Fig. 186.

In the presence of noise it is inconvenient to have to distinguish the value $+1$ from the values $+\frac{1}{2}$ of the cross-correlation function and this problem becomes even more severe for carriers with higher sequency. It is, however, evident that the cross-correlation functions in Fig. 186 show a certain pattern that makes them recognizable even if the peak value $+1$

is blanked out. This suggests using methods of pattern recognition to derive a more useful representation of the cross-correlation function. Let us write the 8 × 8 values of the 8 cross-correlation functions of Fig. 186 as a matrix:

$$\mathbf{C} = \tfrac{1}{2}\begin{bmatrix} +2 & -1 & 0 & +1 & -2 & +1 & 0 & -1 \\ -1 & +2 & -1 & 0 & +1 & -2 & +1 & 0 \\ 0 & -1 & +2 & -1 & 0 & +1 & -2 & +1 \\ +1 & 0 & -1 & +2 & -1 & 0 & +1 & -2 \\ -2 & +1 & 0 & -1 & +2 & -1 & 0 & +1 \\ +1 & -2 & +1 & 0 & -1 & +2 & -1 & 0 \\ 0 & +1 & -2 & +1 & 0 & -1 & +2 & -1 \\ -1 & 0 & +1 & -2 & +1 & 0 & -1 & +2 \end{bmatrix} \qquad (1)$$

If there existed an inverse matrix $\mathbf{C}^{-1}$ one could multiply $\mathbf{C}$ by $\mathbf{C}^{-1}$ and obtain the unit matrix with ones along the main diagonal and zeros everywhere else. Let $\mathbf{C}$ be factored into a Kronecker product of two matrices of lower rank:

$$\mathbf{C} = \tfrac{1}{2}\begin{bmatrix} +2 & -1 & 0 & +1 \\ -1 & +2 & -1 & 0 \\ 0 & -1 & +2 & -1 \\ +1 & 0 & -1 & +2 \end{bmatrix} \times \begin{bmatrix} +1 & -1 \\ -1 & +1 \end{bmatrix} = \tfrac{1}{2}\mathbf{S} \times \mathbf{K} \qquad (2)$$

The matrix $\mathbf{K}$ has no inverse, but the matrix $\mathbf{S}$ has one:

$$\mathbf{S}^{-1} = \tfrac{1}{2}\begin{bmatrix} +2 & +1 & 0 & -1 \\ +1 & +2 & +1 & 0 \\ 0 & +1 & +2 & +1 \\ -1 & 0 & +1 & +2 \end{bmatrix} \qquad (3)$$

$$\mathbf{S}\mathbf{S}^{-1} = \tfrac{1}{2}\begin{bmatrix} +2 & -1 & 0 & +1 \\ -1 & +2 & -1 & 0 \\ 0 & -1 & +2 & -1 \\ +1 & 0 & -1 & +2 \end{bmatrix}\begin{bmatrix} +2 & +1 & 0 & -1 \\ +1 & +2 & +1 & 0 \\ 0 & +1 & +2 & +1 \\ -1 & 0 & +1 & +2 \end{bmatrix} = \begin{bmatrix} 1 & 0 & 0 & 0 \\ 0 & 1 & 0 & 0 \\ 0 & 0 & 1 & 0 \\ 0 & 0 & 0 & 1 \end{bmatrix}$$

It follows:

$$\mathbf{S}^{-1} \times \mathbf{K} = \tfrac{1}{2}\begin{bmatrix} +2 & +1 & 0 & -1 & -2 & -1 & 0 & +1 \\ +1 & +2 & +1 & 0 & -1 & -2 & -1 & 0 \\ 0 & +1 & +2 & +1 & 0 & -1 & -2 & -1 \\ -1 & 0 & +1 & +2 & +1 & 0 & -1 & -2 \\ -2 & -1 & 0 & +1 & +2 & +1 & 0 & -1 \\ -1 & -2 & -1 & 0 & +1 & +2 & +1 & 0 \\ 0 & -1 & -2 & -1 & 0 & +1 & +2 & +1 \\ +1 & 0 & -1 & -2 & -1 & 0 & +1 & +2 \end{bmatrix} = \mathbf{Q} \qquad (4)$$

$$\mathbf{Q}\,\mathbf{C} = \begin{bmatrix} +1 & 0 & 0 & 0 & -1 & 0 & 0 & 0 \\ 0 & +1 & 0 & 0 & 0 & -1 & 0 & 0 \\ 0 & 0 & +1 & 0 & 0 & 0 & -1 & 0 \\ 0 & 0 & 0 & +1 & 0 & 0 & 0 & -1 \\ -1 & 0 & 0 & 0 & +1 & 0 & 0 & 0 \\ 0 & -1 & 0 & 0 & 0 & +1 & 0 & 0 \\ 0 & 0 & -1 & 0 & 0 & 0 & +1 & 0 \\ 0 & 0 & 0 & -1 & 0 & 0 & 0 & +1 \end{bmatrix} \qquad (5)$$

The matrix **Q** is the closest to an inverse matrix of **C**. It is evident from Eq. (5) why an inverse does not exist: The entries -1 indicate that the function $\mathrm{cal}(3, t/T - \frac{1}{2})$ is identical to $-\mathrm{cal}(3, t/T)$, and such a relation holds true for all Walsh functions according to Table 9 in section 4.1.2. This is, of course, hardly a drawback since $+1$ is easier to distinguish from -1 than from 0.

Figure 187 shows on the left the cross-correlation functions from Fig. 186 denoted by V_0 to V_7 and on the right the diagonalized cross-correlation functions according to the matrix (5) denoted U_0 to U_7. Figure 188 shows a circuit that transforms eight voltages V_0 to V_7 that are

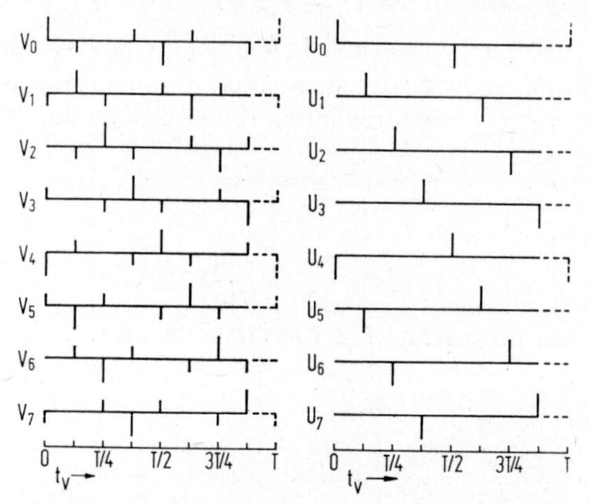

Fig. 187. Diagonalization of the cross-correlation function of Fig. 186.

proportionate to the rows or columns of Eq. (1) into eight voltages U_0 to U_7 that are proportionate to the rows or columns of Eq. (5). The values of the resistors and their connection to the adding and subtracting input terminals of the operational amplifiers represent the matrix (4).

Let us now turn to the block diagram of a synchronous receiver for electromagnetic Walsh waves shown in Fig. 189. The input voltage from the antenna is fed to the antenna switch A 1, which is operated according to the gating function $g(t)$ of Fig. 184, and to an integrator I 1. The integrated Dirac pulses are fed to the analog shift register ASR with 16 stages. The shift register is recycled via a feedback loop, an attenuator C 1 and the adder AD 1. This recycling causes the periodic signal to be added coherently while noise is added incoherently. The attenuator prevents an infinite increase of the recycled voltage. One may readily recognize the similarity between a frequency band-filter and the recycled shift register; the attenuator C 1 replaces the losses of the filter.

The feedback loop of the shift register ASR is connected to a counter COU. The wanted carrier is also fed to this counter. If the sequency of cal$(3, t/T)$ is not the same as that of the received signal, an inhibit pulse is sent to the gate IG which causes the gating function fed through PS

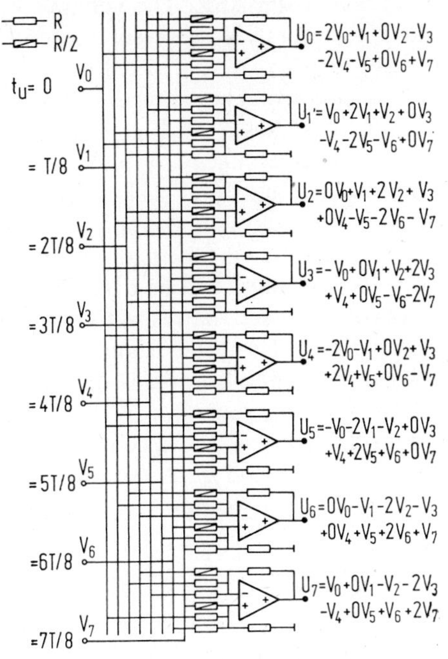

Fig. 188. Diagonalizing circuit for the transformation of the voltages V_i into U_i.

to A 1 to slip and to lock onto the next Dirac pulse. If the sequency of the received signal is correct nothing happens and the gating function stays locked.

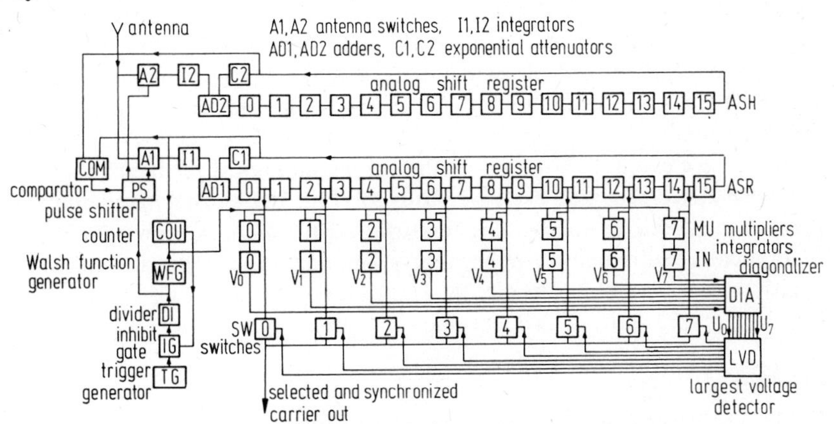

Fig. 189. Block diagram of a synchronous receiver for electromagnetic Walsh waves.

The function cal$(3, t/T)$ is fed from the generator WFG to the multipliers MU. The 8 time shifted functions cal$(3, t/T)$ to cal$(3, t/T - 7/8)$ are also fed to these multipliers from ASR. The voltages V_0 to V_7 at the output terminals of the integrators IN represent the cross-correlation function at one of the times $t_v = 0, T/8, \ldots, 7T/8$ according to Fig. 187, but the value of t_v is not yet known. The voltages are fed to the diagonalizer DIA, the circuit of which is shown by Fig. 188. The largest one of the output voltages U_0 to U_8 determines the multiplier MU to which the received signal synchronized with cal$(3, t/T)$ has been fed. The circuit LVD detects which voltage is the largest and applies the voltage $+1$ to the respective output terminal, while all others are at voltage 0. Circuits that detect the largest of several voltages will be discussed in section 6.2.2. The output voltage $+1$ of LVD closes the respective switch SW and applies the received and properly synchronized signal to the output terminal. It can be demodulated by multiplication with the function cal$(3, t/T)$ delivered by the Walsh function generator WFG.

It remains to be explained how the receiver of Fig. 189 can lock to a train of Dirac pulses with period T. According to Fig. 190 the gating function $g(t)$ is replaced by two functions $g_1(t)$ and $g_2(t)$ that are time

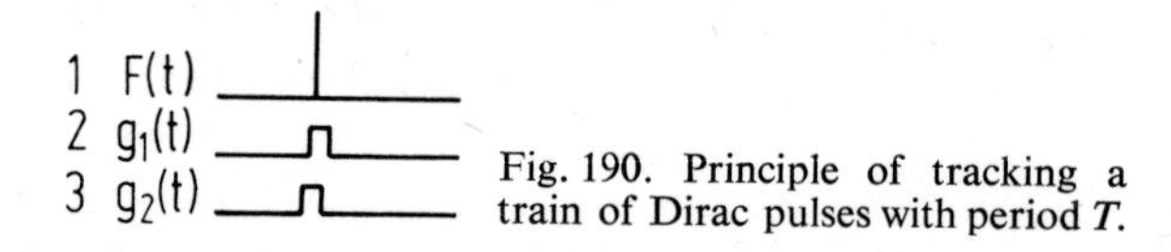

Fig. 190. Principle of tracking a train of Dirac pulses with period T.

shifted relative to each other but overlap. $g_1(t)$ operates the gate A 1 in Fig. 189 while $g_2(t)$ operates the gate A 2. If the signal $F(t)$ is lined up with $g_1(t)$ and $g_2(t)$ as shown in Fig. 190 the output voltages of the two analog shift registers ASR and ASH will be equal and the comparator COM will do nothing. If the pulse $F(t)$ is delayed or advanced so much that it passes the gate A 1 but not the gate A 2 or vice versa, the output voltages of ASR and ASH will not be equal and the comparator COM will shift the gating functions $g_1(t)$ and $g_2(t)$ so that the output voltages become equal.

There are evidently a large number of technological problems that have not been discussed and a closer study of Fig. 189 shows many improvements that could be made. The important point here is, however, that all theoretical requirements for asynchronous and synchronous reception of Walsh waves are met by the circuit, and that no part of the circuit is utopian for the present state of semiconductor technology.

Problems

542.1 The generation of the cross-correlation function and its quasi diagonalization are done in two steps in Fig. 189. Combine them into one.

542.2 The average received power of the carrier $\mathrm{cal}(3, t/T)$ is 1 mW, T is 100 microseconds and the switching time (the duration of the Dirac pulses) is 1 nanosecond. How large is the power flowing through the antenna switches A 1 and A 2 when they are closed?

542.3 What must be changed in Fig. 189 if the carrier $\mathrm{cal}(2, t/T)$ is to be received?

542.4 Repeat 542.3 for asynchronous reception.

542.5 Compare the matrix (1) with the auto-correlation function of $\mathrm{cal}(3, \theta)$ in Fig. 131. Write then the corresponding matrices for $\mathrm{cal}(2, \theta)$, $\mathrm{cal}(5, \theta)$, $\mathrm{cal}(6, \theta)$ and $\mathrm{cal}(7, \theta)$ [2].

542.6 Determine the matrices corresponding to Eq. (4) for the four functions of 542.5, write the matrices corresponding to Eq. (5) and check the position of the entries -1 with the values of θ_1 in Table 9 in section 4.1.2.

6. APPLICATION OF ORTHOGONAL FUNCTIONS TO STATISTICAL PROBLEMS

6.1 *Series Expansion of Stochastic Functions*

6.1.1 Thermal Noise

Consider a set of time functions $g_\lambda(\theta)$, $\lambda = 1, 2, \ldots$, which do not have to be orthogonal. Each function shall be expanded into a series of the complete orthonormal system $\{f(j, \theta)\}$ in the interval $-\frac{1}{2}\Theta \leqq \theta \leqq \frac{1}{2}\Theta$:

$$g_\lambda(\theta) = \sum_{j=0}^{\infty} a_\lambda(j) f(j, \theta)$$
$$a_\lambda(j) = \int_{-\Theta/2}^{\Theta/2} g_\lambda(\theta) f(j, \theta)\, d\theta \tag{1}$$

The coefficients $a_\lambda(j)$ have certain values for a fixed value of $j = j_0$ and variable values of λ. ι functions $g_\lambda(\theta)$ yield ι coefficients $a_\lambda(j)$. Let q_1 of them be in the interval $0 < A < \Delta A$, q_2 in the interval $\Delta A < A < 2\Delta A$, etc. The fractions $q_1/\iota, q_2/\iota, \ldots$ shall be plotted over the intervals 0 to ΔA, ΔA to $2\Delta A$, etc. The result is a step function. Assume that it can be approximated for small values of ΔA by a continuous density function. This density function can be different for each value of j. One calls $a_\lambda(j)$ equally distributed with reference to j, if the density functions are identical for all values of j. Furthermore, let the coefficients $a_\lambda(j)$ and $a_\lambda(k)$ be statistically independent for $j \neq k$. The set of time functions $g_\lambda(\theta)$ is called a sample of white noise with reference to the orthogonal system $\{f(j, \theta)\}$.

$a_\lambda(j_0)$ is called Gaussian distributed, if its density function is the derivative of the error function. The set of functions $g_\lambda(\theta)$ is called white Gaussian noise or thermal noise[1], if the $a_\lambda(j)$ are equally distributed with reference to j, statistically independent and Gaussian distributed for a certain $j = j_0$.

For the practical measurement of the coefficients $a_\lambda(j)$ consider a generator for the functions $f(j, \theta)$. The index j cannot run from zero to infi-

---

[1] Use of these terms is not uniform in the literature. Thermal noise is frequently called Johnson noise [10] or resistor noise. The noise generated by thermal agitation of electrons in an ohmic resistor is thermal noise, if the electrons are described by Boltzmann statistics rather than Fermi statistics.

nity as in Eq. (1); j can only assume a finite number m of values $0, \ldots, m-1$. Time is divided into non-overlapping intervals of duration Θ. The function $g_\lambda(\theta)$ in the first time interval is denoted by $g_1(\theta)$, the function in the time interval λ by $g_\lambda(\theta)$. A finite number ι of intervals is possible only; λ runs from 1 to ι. Let the m functions $f(j, \theta)$ be available simultaneously and let there be m multipliers and integrators. The m coefficients $a_1(j)$, $j = 0, \ldots, m-1$, can be measured in the first interval. These coefficients are represented by the integrator output voltages at the end of the first time interval of duration Θ. Repetition of these measurements for all ι time intervals yields the $m\iota$ coefficients $a_\lambda(j)$; $j = 0, \ldots, m-1$, $\lambda = 1, \ldots, \iota$.

Assume the set of functions $g_\lambda(\theta)$ is thermal noise. Let us plot the fraction q_r/ι of measurements yielding a value of $a_\lambda(j)$ in the interval $(r-1)\Delta A < A < r\Delta A$. The resulting m step functions may be approximated by continuous density functions $w_a(j, A)$ as shown in Fig. 191, if

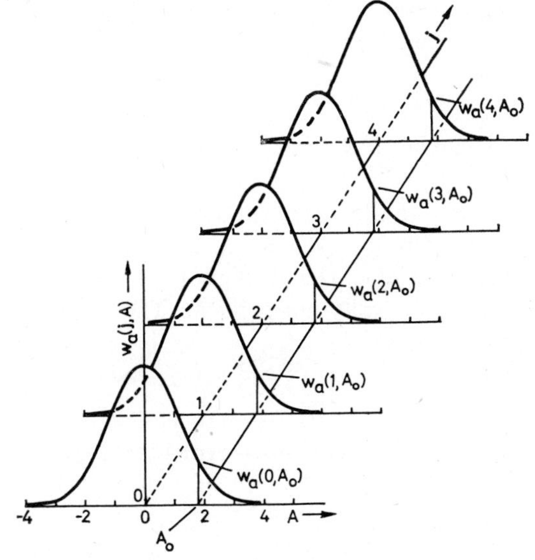

Fig. 191. Density functions $w_a(j, A)$ of thermal noise. $j = 0 \ldots m-1$; A denotes the normalized output voltages of the m integrators.

ΔA is sufficiently small and ι sufficiently large. The equal distribution with reference to j causes the following relation to hold for a certain $A = A_0$:

$$w_a(0, A_0) = w_a(1, A_0) = w_a(2, A_0) = \cdots \tag{2}$$

The distribution of the coefficients $a_\lambda(j)$ generally depends not only on the set of functions $g_\lambda(\theta)$, but also on the system $\{f(j, \theta)\}$. However, it is independent of the system $\{f(j, \theta)\}$ under very general assumptions for thermal noise.

For a proof of this statement let us replace the complete orthonormal system $\{f(j, \theta)\}$ by another system $\{h(j, \theta)\}$ that is also complete and orthonormal in the interval $-\frac{1}{2}\Theta \leqq \theta \leqq \frac{1}{2}\Theta$. The functions $f(j, \theta)$ and $h(j, \theta)$ shall be bounded. Let the functions $h(j, \theta)$ be expanded into a series

$$h(j, \theta) = \sum_{k=0}^{\infty} c_j(k) f(k, \theta), \quad c_j(k) = \int_{-\Theta/2}^{\Theta/2} h(j, \theta) f(k, \theta)\, d\theta \tag{3}$$

The sum $\sum c_j(k)$ shall converge absolutely. The series (3) then converges uniformly.

$g_\lambda(\theta)$ is expanded into a series of the system $\{h(j, \theta)\}$:

$$g_\lambda(\theta) = \sum_{j=0}^{\infty} b_\lambda(j)\, h(j, \theta), \quad b_\lambda(j) = \int_{-\Theta/2}^{\Theta/2} g_\lambda(\theta)\, h(j, \theta)\, d\theta \tag{4}$$

Using Eqs. (1) and (3) one obtains:

$$\begin{aligned} b_\lambda(j) &= \int_{-\Theta/2}^{\Theta/2} g_\lambda(\theta) \sum_{k=0}^{\infty} c_j(k) f(k, \theta) \\ &= \sum_{k=0}^{\infty} c_j(k) \int_{-\Theta/2}^{\Theta/2} g_\lambda(\theta) f(k, \theta)\, d\theta = \sum_{k=0}^{\infty} c_j(k)\, a_\lambda(k) \end{aligned} \tag{5}$$

The last sum converges absolutely, if all $a_\lambda(k)$ are bounded.

The sum of statistically independent, Gaussian distributed variables is a Gaussian distributed variable. Hence, the $b_\lambda(j)$ have a Gaussian distribution, if the $a_\lambda(k)$ are statistically independent. The mean of the $a_\lambda(k)$ and of the $b_\lambda(j)$ is zero.

The density function $w_a(k, A)$ of the $a_\lambda(k)$ reads for thermal noise as follows:

$$w_a(k, A) = \frac{1}{\sqrt{(2\pi)}\, m\, \sigma_a} \exp(-A^2/2\sigma_a^2) \tag{6}$$

Integration over A yields $1/m$,

$$w_a(k) = \int_{-\infty}^{\infty} w_a(k, A)\, dA = \frac{1}{m}$$

and the sum of m terms $w_a(k)$ yields 1:

$$\sum_{k=0}^{m-1} w_a(k) = 1$$

The variance σ_a^2 in Eq. (6) is defined by:

$$\sigma_a^2 = \lim_{\iota \to \infty} \frac{1}{\iota} \sum_{\lambda=1}^{\iota} a_\lambda^2(k) \tag{7}$$

The density function of $c_j(k)\, a_\lambda(k)$ equals:

$$w_c[k, c_j(k)\, A] = \frac{1}{\sqrt{(2\pi)}\, m\, \sigma_{ka}} \exp \frac{-c_j^2(k)\, A^2}{2\sigma_{ka}^2}$$

$$\sigma_{ka}^2 = \lim_{\iota \to \infty} \frac{1}{\iota} \sum_{\lambda=1}^{\iota} c_j^2(k)\, a_\lambda^2(k) = c_j^2(k) \lim_{\iota \to \infty} \frac{1}{\iota} \sum_{\lambda=1}^{\iota} a_\lambda^2(k) = c_j^2(k)\, \sigma_a^2 \tag{8}$$

The density function of the variable $b_\lambda(j)$ follows from Eqs. (5), (7) and (8):

$$w_b(j, A) = \frac{1}{\sqrt{(2\pi)}\, m\, \sigma_b} \exp \frac{-A^2}{2\sigma_b^2} \tag{9}$$

$$\sigma_b^2 = \lim_{\iota \to \infty} \frac{1}{\iota} \sum_{\lambda=1}^{\iota} b_\lambda^2(j) = \sum_{k=0}^{m-1} \sigma_{ka}^2 = \sigma_a^2 \sum_{k=0}^{m-1} c_j^2(k) \tag{10}$$

The last step in Eq. (10) makes use of the initial assumption that the distribution of $a_\lambda(k)$ and thus σ_a^2 does not depend on k.

If Parseval's theorem (1.1.2-6) is satisfied or, putting it differently, if the functions $h(j, \theta)$ may be represented with arbitrary accuracy in the sense of a vanishing mean square deviation by the system $\{f(j, \theta)\}$, one obtains from Eq. (3):

$$\int_{-\Theta/2}^{\Theta/2} h_j^2(j, \theta)\, d\theta = 1 = \int_{-\Theta/2}^{\Theta/2} \left[\sum_{k=0}^{\infty} c_j(k) f(k, \theta)\right]^2 d\theta = \sum_{k=0}^{\infty} c_j^2(k) \tag{11}$$

It follows from Eqs. (10) and (11), that the condition

$$\sigma_b^2 = \sigma_a^2(1 + \varepsilon) \tag{12}$$

is satisfied and that ε approaches zero for sufficiently large values of m. The variables $b_\lambda(j)$ and $a_\lambda(j)$ then have the same variance. The density functions of Fig. 191 remain unchanged, if the samples $g_\lambda(\theta)$ of thermal noise are expanded in a series of the system $\{h(j, \theta)\}$ instead of $\{f(j, \theta)\}$.

Thermal noise is usually defined in the literature by a Fourier series rather than by the general orthogonal series (1). One may substitute in Eq. (1) the sine and cosine pulses that vanish outside the interval $-\frac{1}{2}\Theta \leq \theta \leq \frac{1}{2}\Theta$ for the system $\{f(j, \theta)\}$. According to the results of this section, there is no difference whether thermal noise is defined by a Fourier series or by a series of functions $\{h(j, \theta)\}$ that can be expanded in a Fourier series as shown by Eq. (3).

It has been stated in section 3.3.1 that audio signals were found to have sequency formants, if decomposed by Walsh functions, just as they have frequency formants, if decomposed by sine-cosine functions. Furthermore, audio signals filtered by sequency filters could hardly be distinguished from signals filtered by frequency filters, if the information flow was the same. If audio signals had the distribution of thermal noise, and if the ear could decompose these signals into an infinite number of components according to Eqs. (1) or (4) one should expect such results. The experimental results show that audio signals are sufficiently similar to thermal noise and that the ear decomposes them into sufficiently many components to make the results of this section applicable.

Problems

611.1 What does uniform convergence of a series mean?

611.2 What is the difference between Boltzmann, Bose and Fermi statistics?

611.3 What is the difference between thermal noise of an ohmic resistor caused by electrons described by Fermi and by Boltzmann statistics?

6.1.2 Statistical Independence of the Components of an Orthogonal Expansion

It has been assumed in the preceding section that the coefficients $a_\lambda(h)$ and $a_\lambda(k)$ are statistically independent for $h \neq k$. It remains to be shown that this independence also holds for the coefficients $b_\lambda(j)$ and $b_\lambda(l)$ when $j \neq l$. These coefficients have a Gaussian distribution. They are statistically independent, if the correlation coefficient ϱ or the covariance σ_{jl}^2 vanish. Using the absolute convergence of the series in (6.1.1-5) one obtains:

$$\left.\begin{aligned} \sigma_{jl}^2 = \langle b_\lambda(j)\, b_\lambda(l)\rangle &= \lim_{\iota\to\infty} \frac{1}{\iota} \sum_{\lambda=1}^{\iota} b_\lambda(j)\, b_\lambda(l) \\ &= \lim_{\iota\to\infty} \frac{1}{\iota} \sum_{\lambda=1}^{\iota} \left[\sum_{h=0}^{\infty} c_j(h)\, a_\lambda(h) \sum_{k=0}^{\infty} c_l(k)\, a_\lambda(k)\right] \\ &= \lim_{\iota\to\infty} \frac{1}{\iota} \sum_{\lambda=1}^{\iota} \left[\sum_{h=0}^{\infty} \sum_{k=0}^{\infty} c_j(h)\, c_l(k)\, a_\lambda(h)\, a_\lambda(k)\right] \\ &= \sum_{h=0}^{\infty} \sum_{k=0}^{\infty} c_j(h)\, c_l(k) \lim_{\iota\to\infty} \frac{1}{\iota} \sum_{\lambda=1}^{\iota} a_\lambda(h)\, a_\lambda(k) \end{aligned}\right\} \quad (1)$$

Denote by ε the largest sum $\frac{1}{\iota} \sum_{\lambda=1}^{\iota} a_\lambda(h)\, a_\lambda(k)$ for any pair h, k and a finite value of ι; it follows:

$$\frac{1}{\iota} \sum_{\lambda=1}^{\iota} a_\lambda(h)\, a_\lambda(k) \leqq \varepsilon \quad (2)$$

$$\sigma_{jl}^2 \leqq \varepsilon \sum_{h=0}^{\infty} \sum_{k=0}^{\infty} c_j(h)\, c_l(k) \quad (3)$$

The double sum converges absolutely, since the sums of $c_j(h)$ and $c_l(k)$ converge absolutely:

$$\sum_{h=0}^{\infty} \sum_{k=0}^{\infty} c_j(h)\, c_l(k) = K \quad (4)$$

Equations (3) and (4) yield:

$$\sigma_{jl}^2 \leqq \varepsilon K \tag{5}$$

ε approaches zero for large values of ι by definition and the covariance σ_{jl}^2 vanishes.

6.2 *Additive Disturbances*

6.2.1 Least Mean Square Deviation of a Signal from Sample Functions

Let a time function $F_\chi(\theta)$ be composed of the first m functions of the orthogonal system $\{f(j,\theta)\}$:

$$F_\chi(\theta) = \sum_{j=0}^{m-1} a_\chi(j) f(j,\theta) \qquad -\tfrac{1}{2}\Theta \leqq \theta \leqq \tfrac{1}{2}\Theta \tag{1}$$

$F_\chi(\theta)$ is called a character of an alphabet. There is only a finite number of such characters, if the coefficients $a_\chi(j)$ are not arbitrary but can assume a finite number of values only. The teletype alphabet, e.g., contains 32 characters; m equals 5 and the coefficients $a_\chi(j)$ may assume two values.

Let $F_\chi(\theta)$ be transmitted. A disturbance $g_\lambda(\theta)$ is added during transmission and the signal

$$F(\theta) = F_\chi(\theta) + g_\lambda(\theta) \tag{2}$$

is received. Let us assume that $F(\theta)$ can be expanded in a series:

$$\begin{gathered} F(\theta) = \sum_{j=0}^{\infty} a(j) f(j,\theta) = \sum_{j=0}^{\infty} [a_\chi(j) + a_\lambda(j)] f(j,\theta) \\ a(j) = \int_{-\Theta/2}^{\Theta/2} F(\theta) f(j,\theta)\, d\theta; \quad a_\chi(j) = 0 \quad \text{for} \quad j \geqq m \end{gathered} \tag{3}$$

j runs from 0 to infinity and not from 0 to $m-1$. $a_\lambda(j)$ is defined by Eq. (6.1.1-1).

It must be decided at the receiver which character $F_\psi(\theta), \psi = 1 \ldots \chi \ldots$ is the one which most probably caused the signal $F(\theta)$. The probability of a transformation of $F_\psi(\theta)$ into $F(\theta)$ depends on the probability that $F_\psi(\theta)$ was transmitted. Let us assume all characters are transmitted with equal probability. The decision depends then only on the disturbances $g_\lambda(\theta)$. No decision is possible for a single character, if nothing is known about the set $g_\lambda(\theta)$. However, it is known in many cases that a disturbance $g_\lambda(\theta)$ with large energy is received less often than one with little energy. Putting it differently, the probability of receiving a disturbance $g_\lambda(\theta)$ with energy between W and $W + \Delta W$ decreases monotonically with increasing W. The signal $F(\theta)$ is most likely produced by a character $F_\psi(\theta)$ that may be transformed additively with the least energy into $F(\theta)$. The energy[1] ΔW_ψ required for this transformation is given by the integral

$$\Delta W_\psi = \int_{-\Theta/2}^{\Theta/2} [F(\theta) - F_\psi(\theta)]^2 \, d\theta = \int_{-\Theta/2}^{\Theta/2} [F^2(\theta) - 2F(\theta) F_\psi(\theta) + F_\psi^2(\theta)] \, d\theta \quad (4)$$

The integral of $F^2(\theta)$ yields the energy of the received signal, the integral of $F_\psi^2(\theta)$ the energy of the character $F_\psi(\theta)$ with which the signal is compared. The integral of $F(\theta) F_\psi(\theta)$ is the correlation integral or the correlation of the signal $F(\theta)$ and the character $F_\psi(\theta)$.

The contribution to ΔW_ψ by $F^2(\theta)$ is the same for all characters $F_\psi(\theta)$ and may be ignored. If, furthermore, the energy of all characters is the same,

$$W_\psi = \int_{-\Theta/2}^{\Theta/2} F_\psi^2(\theta) \, d\theta = W, \quad (5)$$

one may ignore $F_\psi^2(\theta)$ too. The smallest value ΔW_ψ is determined by the correlation integral alone in this case:

$$\Delta W_\psi = \text{minimum for} \int_{-\Theta/2}^{\Theta/2} F(\theta) F_\psi(\theta) \, d\theta = \text{maximum} \quad (6)$$

The transmitted character $F_\chi(\theta)$ will be detected correctly if ΔW_ψ has its minimum for $\psi = \chi$.

Signal detection by means of Eqs. (4) and (6) is called detection by the criterion of least mean square deviation. Samples $g_\lambda(\theta)$ of thermal noise satisfy the conditions for which such a detection is proper. There are many types of additive disturbances for which the conditions are not satisfied, such as pulse-type disturbances or so-called intelligent interference.

Using adders, multipliers and integrators, one may determine in principle the most probably transmitted character from Eq. (4) or Eq. (6). The effort required, however, is usually too great. Let an alphabet have n characters. n energies ΔW_ψ or n correlation integrals have to be computed according to Eq. (4) or Eq. (6). These computations should be done simultaneously. Hence, n or $n/2$ adders, multipliers and integrators are required.

Less expensive methods can be obtained by substituting Eqs. (1) and (3) into Eq. (4):

$$\begin{aligned} \Delta W_\psi &= \sum_{j=0}^{\infty} a^2(j) - 2\sum_{j=0}^{m-1} a(j) a_\psi(j) + \sum_{j=0}^{m-1} a_\psi^2(j) \\ &= \sum_{j=0}^{m-1} [a(j) - a_\psi(j)]^2 + \sum_{j=m}^{\infty} a^2(j) \end{aligned} \quad (7)$$

$$\Delta W_\psi = \text{minimum for} \sum_{j=0}^{m-1} [a(j) - a_\psi(j)]^2 = \text{minimum}$$

[1] The term energy is used for the definite integral of the square of a function. Its meaning is the same as the one generally used in electrical engineering, if the function represents the voltage across or the current through a unit resistor.

or

$$\Delta W_\psi = \text{minimum for } 2\sum_{j=0}^{m-1} a(j)\, a_\psi(j) - \sum_{j=0}^{m-1} a_\psi^2(j) = \text{maximum}$$

The sums $\sum_{j=m}^{\infty} a_\psi^2(j)$ or $\sum_{j=0}^{\infty} a^2(j)$ may be ignored, since they yield the same value for each ψ. One obtains from Eq. (7) for characters with equal energy:

$$\Delta W_\psi = \text{minimum for } \sum_{j=0}^{m-1} a(j)\, a_\psi(j) = \text{maximum} \tag{8}$$

Equations (3), (7) and (8) show that only the coefficients $a_\lambda(j), j < m$, of the noise sample $g_\lambda(\theta)$ affect the decision which character $F_\psi(\theta)$ was the most likely to produce the received signal $F(\theta)$.

m multipliers and integrators rather than n or $n/2$ are required for the practical implementation of Eqs. (7) and (8). This means a reduction from 32 or 16 to 5 in the case of the teletype alphabet.

Let us substitute the sum $a_\chi(j) + a_\lambda(j)$ from Eq. (3) for $a(j)$ in Eq. (7):

$$\Delta W_\psi = \text{minimum for } 2\sum_{j=0}^{m-1} [a_\chi(j) + a_\lambda(j)]\, a_\psi(j) - \sum_{j=0}^{m-1} a_\psi^2(j) = \text{maximum}$$

The effect of the disturbances $g_\lambda(\theta)$ on the signal decision is due to the sum $2\sum_{j=0}^{m-1} a_\lambda(j)\, a_\psi(j)$. The probability of a wrong decision depends solely on the statistical properties of the coefficients $a_\lambda(j) = \int_{-\Theta/2}^{\Theta/2} g_\lambda(\theta) \times$ $\times f(j, \theta)\, d\theta$.

Let $g_\lambda(\theta)$ be a sample of thermal noise. The statistical properties of the coefficients $a_\lambda(j)$ are then—under very general conditions—independent of the orthogonal system $\{f(j, \theta)\}$ used. The transmitted signal $F_\chi(\theta)$ is composed of these functions according to Eq. (1). Hence, it is quite unimportant for the probability of a wrong decision which functions $f(j, \theta)$ are used to compose the signal, if the disturbances are additive thermal noise.

Problems

621.1 In which communication links is thermal noise the most important cause of errors?

621.2 What are the most important causes of errors of digital signals transmitted through the switched telephone plant?

621.3 Give examples of practically used alphabets and state whether the characters have all the same energy.

6.2.2 Examples of Circuits

Let us discuss some circuits that use Eqs. (6.2.1-7) and (6.2.1-8) for signal detection. Figure 192 shows how the coefficients $a(j)$ are obtained from the received signal $F(\theta)$ by means of sample functions $f(j, \theta)$. This circuit is basically the same as the one of Fig. 117, except that the disturbed coefficients $a(j)$ instead of the undisturbed coefficients $a_\chi(j)$ are obtained.

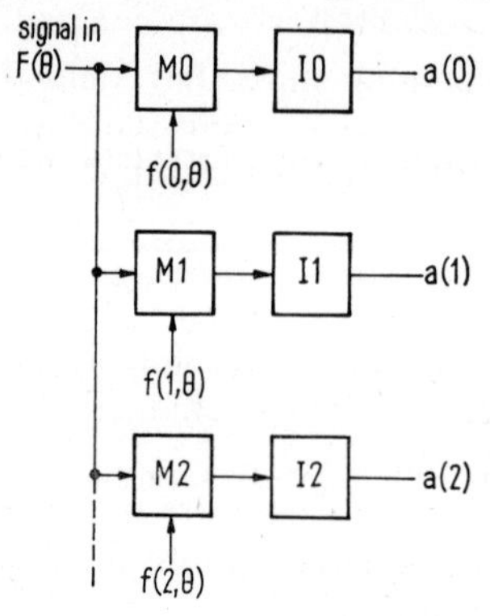

Fig. 192. Extraction of the coefficients $a(j)$ from the received signal $F(\theta)$. M, multiplier; I, integrator.

The sums of the products $a(j)\,a_\psi(j)$ according to Eq. (6.2.1-8) are produced by the circuit of Fig. 193. The characters are composed of three functions, $m = 3$. Hence, three coefficients $a_\psi(0)$, $a_\psi(1)$ and $a_\psi(2)$ occur

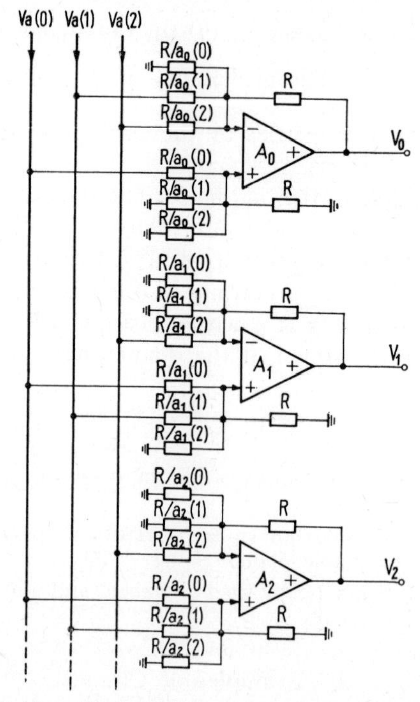

Fig. 193. Signal detection by the largest sum. All characters have equal energy.

$$V_0 = V[a(0)\,a_0(0) - a(1)\,a_0(1) - a(2)\,a_0(2)];$$
$$V_1 = V[a(0)\,a_1(0) + a(1)\,a_1(1) - a(2)\,a_1(2)];$$
$$V_2 = V[a(0)\,a_2(0) + a(1)\,a_2(1) - a(2)\,a_2(2)].$$

that are represented by voltages. The coefficients $a_\psi(0)$, $a_\psi(1)$ and $a_\psi(2)$, $\psi = 1, 2, \ldots$ are represented by resistors. The operational amplifiers A have differential inputs. The inverting input terminals are denoted by $(-)$, the noninverting ones by $(+)$.

For the implementation of Eq. (6.2.1-7) let us note that the sum of $a^2(j)$ is the same for all ΔW_ψ. This sum may be disregarded, if the smallest ΔW_ψ shall be determined without any need to know the value of ΔW_ψ. The sum of $a(j)\, a_\psi(j)$ is produced as before, except that the sign must be reversed. Hence, one may use the circuit of Fig. 193, but the inverting and noninverting input terminals of the operational amplifiers must be interchanged as in Fig. 194. The sum of $a_\psi^2(j)$ is produced by an additional line with constant voltage $+V$ and resistors of proper value.

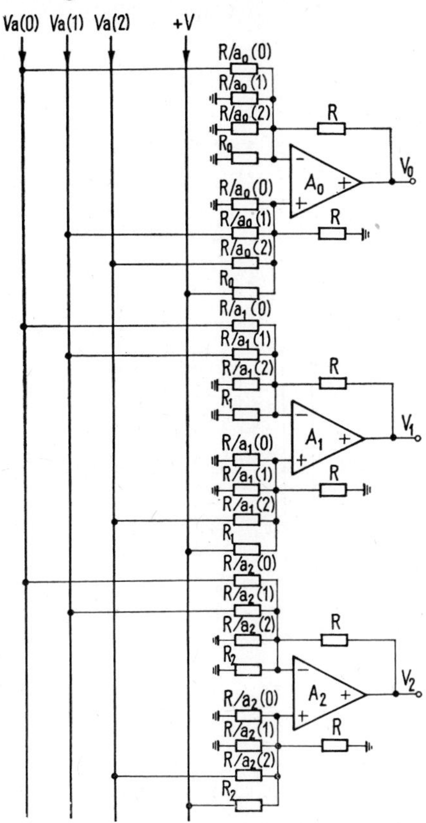

Fig. 194. Signal detection by the smallest sum. The characters do not have to have equal energy.

$$V_0 = V[a_0^2(0) + a_0^2(1) + a_0^2(2) - a(0)\, a_0(0) + a(1)\, a_0(1) + a(2)\, a_0(2)];$$
$$V_1 = V[a_1^2(0) + a_1^2(1) + a_1^2(2) - a(0)\, a_1(0) - a(1)\, a_1(1) + a(2)\, a_1(2)];$$
$$V_2 = V[a_2^2(0) + a_2^2(1) + a_2^2(2) - a(0)\, a_2(0) - a(1)\, a_2(1) + a(2)\, a_2(2)];$$
$$R_j = R/[a_j^2(0) + a_j^2(1) + a_j^2(2)]; \quad j = 0, 1, 2.$$

Circuits are required to determine which output voltage $V_0, V_1, V_2, \ldots$ in Fig. 193 is largest and which output voltage $V_0, V_1, \ldots$ in Fig. 194 is smallest. One type of circuit that determines the largest or smallest of n voltages uses a ramp voltage that is compared via n comparators with the n voltages. The first comparator to fire determines the smallest voltage in case of an increasing ramp voltage; the largest voltage is determined by the first comparator to fire in case of a decreasing ramp voltage. An advantage of this type of circuit is that the ramp voltage does not have to vary linearly with time and that voltage fluctuations are fairly unimportant. The drawback is the non-instantaneous operation.

An instantaneous comparator is shown in Fig. 195. The voltage at the common point of each group of 4 diodes equals the largest applied positive voltage. Let V_3 be the largest voltage. The voltage at the noninverting

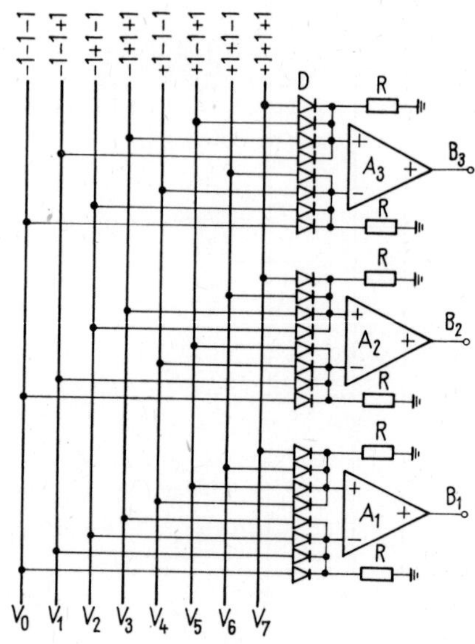

Fig. 195. Detection of the largest positive voltage V_0 to V_7. The largest voltage is determined by the values of B_1, B_2 and B_3 shown.

	V_0	V_1	V_2	V_3	V_4	V_5	V_6	V_7
B_1	-1	-1	-1	-1	$+1$	$+1$	$+1$	$+1$
B_2	-1	-1	$+1$	$+1$	-1	-1	$+1$	$+1$
B_3	-1	$+1$	-1	$+1$	-1	$+1$	-1	$+1$

input terminal $(-)$ of amplifier A_1 is larger than at the noninverting input terminal $(+)$. Assuming sufficient amplification the output voltage B_1 will be at negative saturation, which shall be indicated by $B_1 = -1$. The amplifiers A_2 and A_3 receive a larger voltage at the noninverting input terminal $(+)$ than at the reversing one $(-)$. Both are driven to positive saturation, denoted by $B_2 = B_3 = +1$. The output voltages B_1, B_2 and B_3 indicate the largest voltage $V_j, j = 0, \ldots, 7$, by representing j as binary number. The diode characteristics must be very similar for good results. r amplifiers are required for comparison of 2^r voltages. Variations of the circuit can detect which of several voltages has the largest or the smallest magnitude.

Figure 196 shows another circuit for determination of the largest voltage. The three amplifiers A_1, A_2 and A_3 are driven to positive or negative saturation by the differences between the three voltages V_0, V_1 and V_2. The $3! = 6$ possible permutations of the output voltages are shown in the table of the figure caption. They denote not only the largest but also the second and third largest—that is the smallest—voltage. The voltages

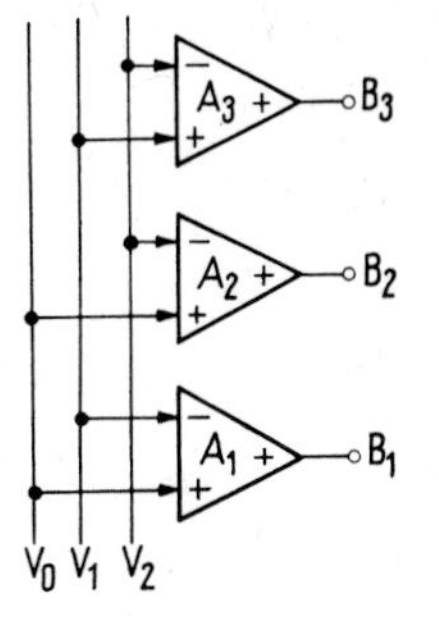

Fig. 196. Detection of the relative values of voltages.

largest voltage	V_0	V_1	V_2	V_0	V_1	V_2
second largest voltage	V_1	V_0	V_0	V_2	V_2	V_1
third largest voltage	V_2	V_2	V_1	V_1	V_0	V_0
B_1	+1	−1	+1	+1	−1	−1
B_2	+1	+1	−1	+1	−1	−1
B_3	+1	+1	−1	−1	+1	−1

V_0, V_1 and V_2 may be positive or negative. This circuit is much more sensitive than the one of Fig. 195, since the voltages are fed directly to the amplifiers rather than through diodes. Its drawback is the large number of amplifiers required. A comparison of n voltages requires measurement of $(n-1) + (n-2) + \cdots + 1 = \frac{1}{2}n(n-1)$ voltage differences. Hence, a total of $\frac{1}{2}n(n-1)$ differential amplifiers are needed. The circuit of Fig. 195, on the other hand, requires for $n = 2^r$ voltages $\log_2 n$ differential amplifiers only.

Problems

622.1 Devise a logic circuit that transforms the output voltages of the circuit of Fig. 195 into a voltage +1 at the output terminal i, and 0 at the other seven terminals if the input voltage V_i is the largest.

622.2 Modify the circuit of Fig. 195 for the detection of the voltage with the largest magnitude.

622.3 How can the circuit of Fig. 195 be made as sensitive as the one of Fig. 196?

6.2.3 Matched Filters

It has been assumed so far that the coefficients $a(j)$ are obtained by multiplication of the signal $F(\theta)$ with $f(j, \theta)$ and integration of the product. A mathematically equivalent but technically very different method uses matched filters. It is customary to use the pulse response rather than the frequency response of attenuation and phase shift to characterize matched filters. Consider a narrow block pulse $D(\theta)$ having the amplitude $1/\varepsilon$ inside the interval $-\frac{1}{2}\varepsilon \leqq \theta \leqq \frac{1}{2}\varepsilon$ and the amplitude 0 outside. This pulse approaches the delta function $\delta(\theta)$ for vanishing values of ε. Consider further a bank of filters. Let the pulse $\delta(\theta + \frac{1}{2})$ be applied at time $\theta = -\frac{1}{2}$ to the input of the filter j. The output function $f(j, \theta)$, $-\frac{1}{2} \leqq \theta \leqq +\frac{1}{2}$, of Eqs. (6.2.1-1) and (6.2.1-3) shall be produced. $f(j, \theta)$ is the pulse response of filter j.

The time function $F_\chi(\theta)$ of Eq. (6.2.1-1) can be produced by applying the pulses $a_\chi(j)\,\delta(\theta + \frac{1}{2})$ to m filters with pulse response $f(j, \theta)$ and summing the outputs. These filters are denoted as transmitter filters.

The receiver filters invert the process. The functions $F_\chi(\theta)$ or $F(\theta)$ are applied to their inputs during the time interval $-\frac{1}{2} \leqq \theta \leqq \frac{1}{2}$, and the coefficients $a_\chi(j)$ or $a(j)$ in Eq. (6.2.1-3) are obtained at the output of filter j at the time $\theta = +\frac{1}{2}$. Let the functions $f(j, \theta)$ be represented by the orthonormal system of pulses $D(\theta - k\,\varepsilon)$; $k = 0, \pm 1, \pm 2, \ldots$:

$$\left.\begin{aligned}
&a(j) f(j, \theta) \doteqdot a(j) \sum_k d_j(k)\, D(\theta - k\,\varepsilon) \\
&d_j(k) = \int_{-1/2}^{1/2} f(j, \theta)\, D(\theta - k\,\varepsilon)\, d\theta = \int_{k\varepsilon-\varepsilon/2}^{k\varepsilon+\varepsilon/2} f(j, \theta)\, d\theta \doteqdot f(j, k\,\varepsilon)\,\varepsilon \\
&k = 0, \pm 1, \ldots, \pm 1/2\varepsilon
\end{aligned}\right\} \tag{1}$$

The functions $f(j, \theta)$ are generally not represented exactly by the system $\{D(\theta - k\,\varepsilon)\}$, since the sum in Eq. (1) represents a step function. However, if ε becomes sufficiently small the mean square deviation between $f(j, \theta)$ and the step function becomes arbitrarily small for those functions that can be generated.

Let the receiver filter j produce the output $h(j, \theta)$, $-\frac{1}{2} \leqq \theta \leqq \frac{1}{2}$, if the input $\delta(\theta + \frac{1}{2})$ is applied. The input function $D(\theta - k\,\varepsilon) = D(\theta' + \frac{1}{2})$ produces the output $h(j, \theta') = h(j, \theta - \frac{1}{2} - k\,\varepsilon)$ if ε is sufficiently small. Hence, the function $a(j) f(j, \theta)$ applied to the input produces the following output signal:

$$a(j) \sum_k d_j(k)\, h(j, \theta - \tfrac{1}{2} - k\,\varepsilon), \qquad k = 0, \pm 1, \ldots, \pm 1/2\varepsilon \tag{2}$$

This signal has the value

$$a(j) \sum_k d_j(k)\, h(j, -k\,\varepsilon) \tag{3}$$

at time $\theta = +\frac{1}{2}$. Let us substitute $f(j, k\,\varepsilon)\,\varepsilon$ from Eq. (1) for $d_j(k)$ in Eq. (3). The sum yields the area of stripes of width ε and height $f(j, k\,\varepsilon) \times$ $\times\, h(j, -k\,\varepsilon)$. This sum may be replaced by an integral, if ε is sufficiently small:

$$a(j) \int_{-1/2}^{1/2} f(j, \theta)\, h(j, -\theta)\, d\theta, \quad \theta = \lim_{\substack{k \to \infty \\ \varepsilon \to 0}} k\,\varepsilon \quad d\theta = \lim_{\varepsilon \to 0} \varepsilon \tag{4}$$

This integral equals 1 for

$$h(j, \theta) = f(j, -\theta) \tag{5}$$

The coefficient $a(j)$ is obtained at the output of the receiver filter at the time $\theta = \frac{1}{2}$. On the other hand, the output 0 is obtained, if the function $a(l) f(l, \theta)$, $l \neq j$, is applied to the input of receiver filter j:

$$a(l) \int_{-1/2}^{1/2} f(l, \theta)\, h(j, -\theta)\, d\theta = a(l) \int_{-1/2}^{1/2} f(l, \theta) f(j, \theta)\, d\theta = 0 \qquad j \neq l. \tag{6}$$

The pulse response of the receiver filter j must be $f(j, -\theta)$ if the pulse response of the transmitter filter j is $f(j, \theta)$. Transmitter and receiver filters are identical for even functions $f(j, \theta) = f(j, -\theta)$, and for odd functions $f(j, \theta) = -f(j, -\theta)$.

Matched filters do not need multipliers to determine the coefficient $a(j)$ in Eq. (6.2.1-3). This is frequently an advantage over correlator circuits. In general, one cannot say whether correlators or matched filters are superior. Multipliers for Walsh functions, e.g., are very accurate. Matched filters, on the other hand, do not have to be constructed from coils and capacitors, but may be circuits like the one shown in Fig. 122.

Problems

623.1 Compare advantages and drawbacks of matched filters and correlators for Walsh functions. In which sections of this book were they discussed?

623.2 Devise a circuit using coils, capacitors and switches that can detect sine and cosine pulses like the circuit of Fig. 122.

623.3 Characterize a matched filter for the return signal of a chirp radar. Try to apply the principle to Walsh waves with varying sequency.

6.2.4 Compandors for Sequency Signals

It is well known that instantaneous compression of a frequency limited signal produces a signal that is not frequency limited anymore. The reason for this is that compression of sine functions always generates harmonics. This is not so for sequency limited functions composed of Walsh functions. Figure 197a shows as an example two characters $F_\lambda(\theta)$ and $F_\chi(\theta)$:

$$F_\lambda(\theta) = \mathrm{wal}(0, \theta) + \sum_{i=1}^{8} (-1)^i \,\mathrm{sal}(i, \theta) - \sum_{i=1}^{4} \mathrm{cal}(i, \theta) + \sum_{i=5}^{7} \mathrm{cal}(i, \theta)$$

$$F_\chi(\theta) = -\mathrm{wal}(0, \theta) + \sum_{i=1}^{8} (-1)^i \,\mathrm{sal}(i, \theta) + \sum_{i=1}^{7} \mathrm{cal}(i, \theta)$$

Sending these characters through a compressor having the characteristic shown by Fig. 197b produces the signals $F'_\lambda(\theta)$ and $F'_\chi(\theta)$ of Fig. 197c. These signals contain exactly the same Walsh functions as the characters $F_\lambda(\theta)$ and $F_\chi(\theta)$; they are only multiplied by different coefficients.

Consider a compressor characteristic $\eta = E \operatorname{erf}[\zeta/\sqrt{(2)}\,\sigma]$. Let $W_1(x) = W_1(-\infty < \zeta \leqq x)$ be the amplitude distribution function of a signal before compression. The function $W_2(y) = W_2(-\infty < \eta \leqq y)$ is obtained[1]:

$$\zeta = \sqrt{(2)}\,\sigma \operatorname{erf}^{-1} \frac{\eta}{E} \leqq \sqrt{(2)}\,\sigma \operatorname{erf}^{-1} \frac{y}{E}$$

$$W_2(y) = W_1\left(-\infty < \zeta \leqq \sqrt{(2)}\,\sigma \operatorname{erf}^{-1} \frac{y}{E}\right) = W_1\left(\sqrt{(2)}\,\sigma \operatorname{erf}^{-1} \frac{y}{E}\right)$$

$$w_2(y) = \sqrt{(2\pi)}\,\frac{\sigma}{2E} \exp\left(\operatorname{erf}^{-1} \frac{y}{E}\right)^2 w_1\left(\sqrt{(2)}\sigma \operatorname{erf}^{-1} \frac{y}{E}\right)$$

Consider further a signal composed of the 16 Walsh functions $\operatorname{wal}(0, \theta)$ to $\operatorname{wal}(15, \theta)$ of Fig. 2. All 16 functions equal $+1$ in the interval $0 < \theta < 1/16$. Among the 2^{16} binary characters that can be produced from the 16 functions there is $1 = \binom{16}{0}$ character with amplitude $16(+1) = 16$ in this interval; $16 = \binom{16}{1}$ characters have the amplitude $15(+1) + 1(-1) = 14$; $120 = \binom{16}{2}$ characters have the amplitude $14(+1) + 2(-1) = 12$; etc. The same result holds for any other time interval. Hence, binary characters composed of Walsh functions have a Bernoullian amplitude distribution. Let a character be composed of m Walsh functions with amplitude $+a$ or $-a$. The probability $p_B[(m-2h)\,a]$ of sampling an amplitude $(m-2h)\,a$ equals:

$$p_B[(m-2h)\,a] = \left(\frac{1}{2}\right)^m \binom{m}{h}; \qquad h = 0, 1, \ldots, m$$

The distribution function is $W_B(x)$,

$$W_B(x) = \left(\frac{1}{2}\right)^m \sum_{h=0}^{[x]} \binom{m}{h}$$

where $[x]$ denotes the largest integer smaller or equal to x.

$W_B(x)$ can be approximated for large values of m by the error function:

$$W_B(x) \doteqdot \frac{1}{2}\left(1 + \operatorname{erf} \frac{x}{\sqrt{(2)}\,E}\right) = W_1(x); \qquad E^2 = m\,a^2$$

The derivative $w_1(x)$ is shown in Fig. 198b. Compressor characteristics $\eta = E \operatorname{erf}[\zeta/\sqrt{(2)}\sigma]$ are shown for $\sigma = 0.5E$, E and $2E$ in Fig. 198a. The

[1] Equation (11) in chapter 4 of [3].

corresponding density functions $w_2(y)$ are shown in Fig. 198c:

$$w_2(y) = \frac{1}{2}\frac{\sigma}{E}\exp\left[(1-\sigma^2)\left(\operatorname{erf}^{-1}\frac{y}{E}\right)^2\right]$$

$$W_2(y) = \frac{1}{2}\left[1 + \operatorname{erf}\left(\sigma\operatorname{erf}^{-1}\frac{y}{E}\right)\right]$$

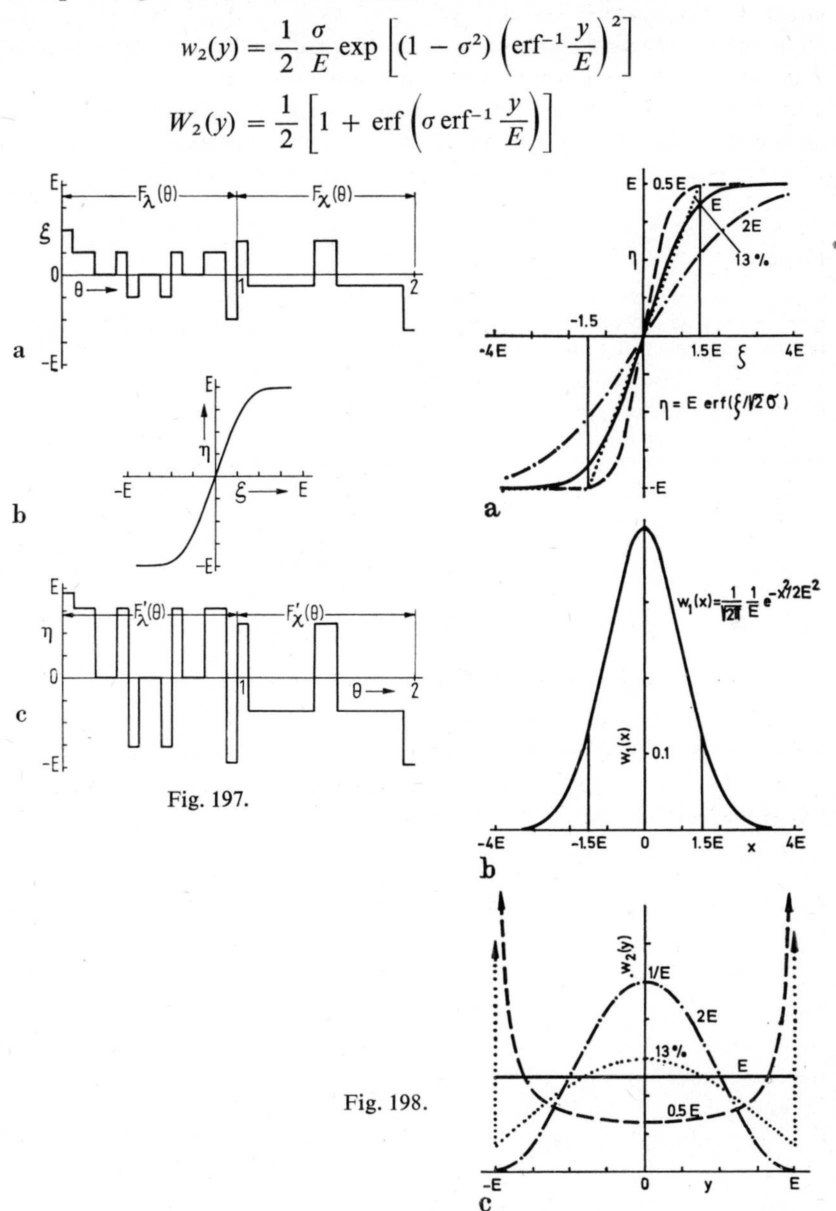

Fig. 197. Compression of sequency multiplex signals. (a) original signal; (b) compressor characteristic; (c) compressed signal.

Fig. 198. Compression of sequency multiplex signals. (a) compressor characteristics; (b) density function of the statistical variable ξ; (c) density functions after compression.

Note that the Gaussian distribution of Fig. 198b is transformed into an equal distribution for $\sigma = E$.

Figures 198a and c also show lines denoted by "13%". They hold for a non-reversible compressor with the characteristic $\eta = \zeta$ for $|\zeta| \leqq 1.5E$ and $\eta = \pm 1.5E$ for $|\zeta| \geqq 1.5E$. This compressor clips all amplitudes absolutely larger than $1.5E$, which are 13% of the amplitudes in the case of a Gaussian distribution. This clipper will be discussed in more detail in section 7.2.1.

Problems

624.1 Can you devise a compander for a modulated sinusoidal function that does not create any new harmonics? Try the sinusoidal carriers of Fig. 144 modulated by the step function.

624.2 What practical applications have companders?

624.3 Replace the curves denoted "13%" in Fig. 198a and c by curves that hold for a clipper that clips the 32% absolutely largest amplitudes of a Gaussian distribution.

6.3 Multiplicative Disturbances

6.3.1 Interference Fading

Let a radio signal be transmitted via several paths. The samples of the same signal interfere with one another at the receiver. Consider as example a sine wave transmitted via two paths. The samples $A_1 \cos 2\pi \nu_0 \theta$ and $A_2 \cos 2\pi \nu_0 (\theta - \theta_v)$ with a delay difference θ_v are received. The sum of the two samples may be written in two forms:

$$\begin{aligned}
&A_1 \cos 2\pi \nu_0 \theta + A_2 \cos 2\pi \nu_0 (\theta - \theta_v) \\
&= (A_1 + A_2 \cos 2\pi \nu_0 \theta_v) \cos 2\pi \nu_0 \theta + A_2 \sin 2\pi \nu_0 \theta \sin 2\pi \nu_0 \theta_v \\
&= [A_1^2 + 2A_1 A_2 \cos 2\pi \nu_0 \theta_v + A_2^2]^{1/2} \cos(2\pi \nu_0 \theta - \alpha)
\end{aligned} \tag{1}$$

A phase sensitive receiver receives one of the two terms of the second line. The amplitude of the signal received varies between $A_1 + A_2$ and $A_1 - A_2$ or between A_2 and $-A_2$. A phase insensitive receiver determines the amplitude of the oscillation in the third line of Eq. (1). It varies between $A_1 + A_2$ and 0.

The mathematical reason for this variation of the amplitudes is evidently that a time shift $\theta_v = \pi$ of an oscillation $\cos 2\pi \nu_0 \theta$ has the same effect as an amplitude reversal, $\cos 2\pi \nu_0 (\theta - \pi) = -\cos 2\pi \nu_0 \theta$. It appears reasonable to use other functions for which the equivalence between time shift and amplitude reversal does not hold or does hold for large values of θ_v only. A general theoretical investigation of useful functions is mathematically very complicated. It is, however, obvious that a superposition of time shifted, differentiated Walsh functions according to Fig. 163 would not cancel by interference.

A simpler application of orthogonal functions for the transmission through an interference-fading medium follows from the narrow bandwidth of the system discussed in section 3.2.2 for teletype transmission. The concept is as follows: Frequency diversity is a well known method for improving transmission reliability under the influence of interference fading. Signals are modulated onto several carriers rather than one. More than 2 or 3 carriers can generally not be used due to bandwidth limitations. The narrow bandwidth required by the system discussed in section 3.2.2 makes it possible to use 10 or more carriers instead of 2 or 3 without excessive requirements for bandwidth. The question is, whether it is worthwhile to spread a fixed transmitter power over so many carriers. A short digression into known results of diversity transmission is necessary before an answer can be given.

Let a harmonic oscillation with frequency ν_0 be radiated. Using the Rayleigh fading model, one obtains at the receiver input a voltage $e(\theta)$:

$$e(\theta) = v(\theta) \cos [2\pi \nu_0 \theta + \alpha(\theta)] \tag{2}$$

$v(\theta)$ is a slowly fluctuating envelope, which is practically constant during an interval $\theta_0 - \frac{1}{2}\theta_k \leqq \theta \leqq \theta_0 + \frac{1}{2}\theta_k$, and which has a Rayleigh distribution with the following density function:

$$\begin{aligned} w(v) &= \frac{2v}{\delta^2} \exp\left(-\frac{v^2}{\delta^2}\right) \qquad v \geqq 0 \\ &= 0 \qquad\qquad\qquad\qquad v < 0 \end{aligned} \tag{3}$$

δ^2 equals the expectation $\mathrm{E}(\zeta^2)$:

$$\delta^2 = \langle v^2 \rangle = \langle v^2(\theta) \rangle \tag{4}$$

The phase angle $\alpha(\theta)$ also fluctuates slowly. It shall have a constant density function:

$$\begin{aligned} w(\alpha) &= \frac{1}{2\pi} \qquad -\pi \leqq \alpha \leqq +\pi \\ &= 0 \quad \alpha < -\pi, \quad \alpha > +\pi \end{aligned} \tag{5}$$

An improvement in transmission reliability requires that two or more statistically independent "copies" of the signal are received. Hence, the density function of the joint distribution of the amplitudes shall be a product of density functions as defined by Eq. (3).

A number of methods are known for the reception of statistically independent copies of a signal. Space diversity uses several antennas spaced sufficiently far apart. Angle diversity obtains copies by means of directional antennas with narrow beams. Two polarized antennas discriminating between right and left circularly polarized waves provide fairly independent copies in the short wave region. Frequency diversity uses several sinusoidal carriers and time diversity transmits the signal repeatedly.

Having obtained several independent copies of a signal the problem of making best use of them arises. There are basically three methods available. a) The copy is used which has the largest average power during a time interval θ_k (optimal selection). b) All copies are added (equal gain summation). c) All copies are multiplied before summation by factors that depend on their average power during an interval θ_k (maximal ratio summation).

For a comparison of the three methods let q statistically independent copies of the signal $F(\theta)$ be available. Let fading transform copy l from $F(\theta)$ into $G_l(\theta)$. A sample $g_\lambda(\theta)$ of thermal noise is added to $G_l(\theta)$. Hence, the following is received as copy l:

$$H_l(\theta) = G_l(\theta) + g_\lambda(\theta) \tag{6}$$

$G_l(\theta)$ is represented during a short time θ_k by the following equation according to Eq. (2):

$$G_l(\theta) = v_l(\theta_0)\cos[2\pi\,\nu_0\,\theta + \alpha_l(\theta_0)], \quad \theta_0 - \tfrac{1}{2}\theta_k \leqq \theta \leqq \theta_0 + \tfrac{1}{2}\theta_k \tag{7}$$

v_l and α_l are assumed to be constant in the interval $\theta_0 - \frac{1}{2}\theta_k \leqq \theta \leqq$ $\leqq \theta_0 + \frac{1}{2}\theta_k$.

The probability $p(v_l < v_g)$ of v_l being smaller than a threshold v_g or, putting it differently, the fraction of time which v_l is smaller than v_g follows from Eq. (3):

$$p(v_l < v_g) = W(v_g) = \int_0^{v_g} \frac{2v}{\delta^2}\exp\left(-\frac{v^2}{\delta^2}\right)dv = 1 - \exp\left(-\frac{v^2}{\delta^2}\right) \tag{8}$$

Let q statistically independent copies be received, all having the same distribution. $p_q(v_l < v_g)$ is the probability that the amplitudes v_l of all q copies are smaller than v_g:

$$p_q(v_l < v_g) = W_q(v_g) = [1 - \exp(-v_g^2/\delta^2)]^q \tag{9}$$

The average power of the copy $G_l(\theta)$ in a time interval of duration θ_k, that is an integer multiple of $1/\nu_0$, follows from Eq. (7):

$$\frac{1}{\theta_k}\int_{\theta_0-\theta_k/2}^{\theta_0+\theta_k/2} G_l^2(\theta)\,d\theta = \frac{1}{2}\,v_l^2(\theta_0) = P_l(\theta_0) = \frac{1}{2}\,v_l^2 = P_l \tag{10}$$

Let P_r denote the average noise power received with copy l. The signal-to-noise power ratio,

$$P_l/P_r = v_l^2/2P_r \tag{11}$$

is a quantity that fluctuates due to the fading only. The probability of P_l being below a threshold P_g follows from Eqs. (8) and (11):

$$\left.\begin{aligned} P_l/P_r &= v_l^2/2P_r < P_g/P_r = v_g^2/2P_r \\ W(P_g) &= p(P_l/P_r < P_g/P_r) = p(v_l < v_g) = W(v_g) \\ W(P_g) &= 1 - \exp(-v_g^2/\delta^2) = 1 - \exp(-2P_g/\delta^2) \end{aligned}\right\} \qquad (12)$$

Let the copy with the largest value P_l/P_r be selected from the q available copies. The probability that P_l/P_r is smaller than P_g/P_r for all copies follows from Eq. (9), if all copies are statistically independent:

$$W_q(P_g) = [1 - \exp(-2P_g/\delta^2)]^q \qquad (13)$$

The mean of this distribution was calculated by Brennan:

$$\left\langle \frac{2P_l}{\delta^2} \right\rangle = \int_0^\infty \frac{2P_l}{\delta^2}\, dW_q(P_l) = \int_0^\infty q\, y(1 - e^{-y})^{q-1} e^{-y}\, dy = \sum_{l=1}^{q} \frac{1}{l} \qquad y = \frac{2P_l}{\delta^2} \qquad (14)$$

Let us denote the average signal-to-noise power ratio of each copy by P_s/P_r:

$$P_s/P_r = \langle P_l/P_r \rangle = \langle P_l \rangle / P_r$$

The average signal-to-noise power ratio P_{sq}/P_r of the best copy is obtained with the help of the relation $\langle P_l \rangle = \frac{1}{2}\delta^2 = P_s$:

$$\frac{P_{sq}}{P_r} = \left(\frac{P_s}{P_r}\right) \sum_{l=1}^{q} \frac{1}{l} \qquad (15)$$

The ratio $(P_{sq}/P_r)/(P_s/P_r)$ is shown in Fig. 199 by the points denoted by a. One may readily see that the average signal-to-noise power ratio increases insignificantly if more than three or four copies are used for optimal selection.

Replacing optimal selection by equal gain summation of q copies yields, according to Brennan, the following relation:

$$P_{sq}/P_r = (P_s/P_r)\,[1 + \tfrac{1}{4}\pi(q - 1)] \qquad (16)$$

P_{sq}/P_r stands now for the average signal-to-noise power ratio of the sum of all q copies of the signal.

The ratio $(P_{sq}/P_r)/(P_s/P_r)$ is shown in Fig. 199 by the points denoted by b. Optimal selection and equal gain summation differ only slightly if 2 copies are used ($q = 2$). However, equal gain summation yields an improvement of 4.5 dB over optimal selection if $q = 10$ copies are used.

For maximal ratio summation the amplitudes of copy l in a time interval of duration θ_k is multiplied by a weighting factor which is proportionate to the rms-value of copy l and inversely proportionate to the rms-value of the noise of that copy. Brennan derived the following expression replacing

Eqs. (15) and (16):

$$P_{sq}/P_r = (P_s/P_r)\,q \tag{17}$$

P_{sq}/P_r now denotes the average signal-to-noise power ratio of the weighted sum of q copies of the signal.

The ratio $(P_{sq}/P_r)/(P_s/P_r)$ is shown in Fig. 199 by the points denoted by c. Maximal ratio summation is somewhat better than equal gain summation. The difference is less than 1 dB for the range of values of q shown and approaches 1.05 dB for infinite values of q.

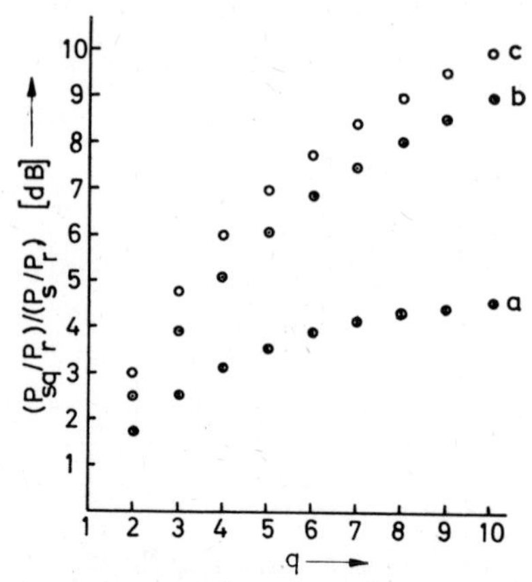

Fig. 199. Increase of the average signal-to-noise power ratio by diversity reception according to Brennan. q, number of received copies of the signal; $(P_{sq}/P_r)/(P_s/P_r)$ = (average signal-to-noise power ratio of q copies)/(average signal-to-noise power ratio for 1 copy). a, optimal selection; b, equal gain summation; c, maximal ratio summation.

The increase of the average signal-to-noise power ratio provides a good means for comparing the various methods for utilization of copies of the signal. The fraction of time during which transmission is possible is, however, a better measure for the reliability of a link. Equations (8) and (12) yield such a measure. The first gives the fraction of time during which a voltage v_l is below a threshold v_g, the second the fraction of time during which the average signal-to-noise power ratio P_l/P_r is below a threshold P_g/P_r. Let us now rewrite Eq. (12) by introducing the median $P_g = P_M$ for which $W(P_g)$ equals $\frac{1}{2}$:

$$W(P_M) = \tfrac{1}{2} = 1 - \exp(-2P_M/\delta^2) \tag{18}$$

It follows:

$$\begin{aligned} 2P_M/\delta^2 &= \ln 2 \doteq 0.693 \\ 2P_g/\delta^2 &= (P_g/P_M)\ln 2 \doteq 0.693\, P_g/P_M \end{aligned} \tag{19}$$

Equation (12) may be rewritten:

$$W(P_q) \doteq 1 - \exp(-0.693\, P_g/P_M) \tag{20}$$

The probability of P_l/P_r being larger than P_g/P_r becomes:

$$p(P_l/P_r > P_g/P_r) = 1 - W(P_g) \doteqdot \exp(-0.693\,P_g/P_M) \tag{21}$$

$p(P_l/P_r > P_g/P_r)$ is shown in Fig. 200 by the curve $q = 1$. The ordinate of that figure shows the percentage of the time during which P_l is larger than a threshold P_g. Here P_l and P_g are divided by the median P_M for normalization.

If q copies are received one obtains from Eq. (13) the probability that P_l/P_r is larger than P_g/P_r for at least one copy:

$$p_q(P_l/P_r > P_g/P_r) \doteqdot 1 - [1 - \exp(-2P_g/\delta^2)]^q$$

One may rewrite this equation using Eq. (19):

$$p_q(P_l/P_r > P_g/P_r) \doteqdot 1 - [1 - \exp(-0.693\,P_g/P_M)]^q \tag{22}$$

$p_q(P_l/P_r > P_g/P_r)$ is shown in Fig. 200 by the solid lines for $q = 2$, 4 and 8. These curves give the percentage of the time during which diversity

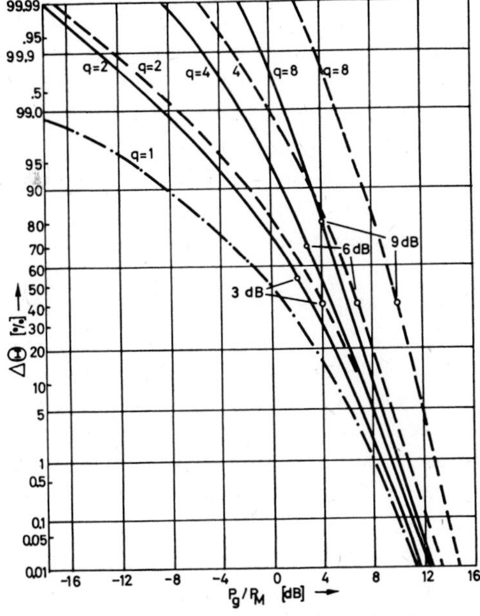

Fig. 200. Relative time $\Delta\Theta$ during which the normalized signal power of a diversity transmission exceeds a threshold P_g/P_M (derived from figures due to Brennan). q, number of received copies; solid lines: optimal selection; dashed lines: equal gain summation; dashed-dotted line: reception without diversity.

transmission is possible if optimal selection is used with 2, 4 or 8 copies and if a ratio P_l/P_r larger than P_g/P_M is required.

The dashed lines in Fig. 200 show the percentage of the time during which P_l of a sum of 2, 4 or 8 copies is larger than the threshold P_g. Hence, they give the fraction of time during which diversity transmission is possible if equal gain summation is used with 2, 4 or 8 copies and if a ratio P_l/P_r larger than P_g/P_M is required[1].

Problems

631.1 Verify the relation in Eq. (4) of the Rayleigh distribution.

631.2 Discuss qualitatively the effect of multipath transmission on Walsh dipole waves.

631.3 A time shift is equivalent to an amplitude reversal for Walsh functions as well as for sinusoidal functions. What can one do in the case of Walsh functions to make the necessary time shift very large?

6.3.2 Diversity Transmission Using Many Copies

The methods discussed in the previous section for obtaining statistically independent copies of a signal usually provide only a few copies. Polarization diversity cannot yield more than two copies. Space and angle diversity could theoretically yield many copies. However, consideration of cost and the space required limit this number in practice. For instance, antennas have to be spaced several hundred meters apart for space diversity in the short wave region. Frequency and time diversity are the only practical methods that can provide many copies of the signal.

In order to apply the curves of Fig. 200 to frequency and time diversity, one must keep in mind that q "equal" signals are radiated while only one signal is radiated for space and angle diversity. Given a certain average transmitter power, the average power radiated per signal is smaller by a factor $1/q$ for frequency and time diversity than for space and angle diversity. This drawback of frequency and time diversity may, of course, be compensated. Instead of using q antennas for reception as in space and angle diversity, one may use one antenna with q-times the gain; this would just compensate the reduced signal power of each copy.

Let the transmitter power and the receiver antenna be fixed. Replacing ordinary transmission ($q = 1$) by q-fold frequency diversity will bring an improvement only if the average signal-to-noise power ratio at the receiver input is increased, despite the decrease of the signal-to-noise power ratio of each copy by $1/q$. Reduction of the average signal power per copy by $1/q$ reduces the median P_M in Fig. 200 to P_M/q. Given a certain threshold P_g the ratio P_g/P_M becomes $q\,P_g/P_M$. Using q-fold frequency diversity, the fraction of time during which the average signal power exceeds $q\,P_g/P_M$ must be larger than the fraction of time during which the average signal power exceeds P_g/P_M for ordinary transmission. Consider an example: The curve $q = 1$ in Fig. 200 yields $\Delta\Theta = 95\%$ for $10 \log P_g/P_M = -11$ dB, while the curves for $q = 2$ yield for $10 \log 2 P_g/P_M = -8$ dB

[1] These curves cannot be represented by tabulated functions for $q > 2$ but have to be computed numerically [2].

the values $\Delta\Theta = 99\%$ and 99.3%. Hence, twofold diversity increases the fraction of time during which the average signal power is larger than P_g from 95% to 99% or 99.3%. One may readily see that such an improvement is possible if the curves $q = 1$ and $q = 2$ are separated horizontally by at least $10 \log q = 10 \log 2 \doteq 3$ dB. The points denoted by "3 dB" show where this separation is just 3 dB. Evidently twofold diversity with equal gain summation is worthwhile if satisfactory operation occurs for more than 40% of the time, while optimal selection will be worthwhile only if satisfactory operation is possible for more than 55% of the time.

Based on the same considerations the separation between the curves $q = 1$ and $q = 4$ must be at least $10 \log 4 \doteq 6$ dB and between $q = 1$ and $q = 8$ at least $10 \log 8 \doteq 9$ dB in order to make fourfold or eightfold diversity worthwhile. The points in Fig. 200 denoted by "6 dB" and "9 dB" indicate where these separations are just 6 and 9 dB. Optimal selection is worthwhile only if satisfactory operation is possible at least 70% or 80% of the time. For equal gain summation all three points "3 dB", "6 dB" and "9 dB" are located on the line $\Delta\Theta = 40\%$. Frequency diversity using equal gain summation is worthwhile if satisfactory operation is possible at least 40% of the time and is then best with the largest number of copies.

It has been discussed in section 3.2.2 that a certain frequency bandwidth may be well utilized by sine and cosine pulses. Six teletype circuits require about 120 Hz bandwidth according to Table 8, twelve circuits about 240 Hz. A total of 2400 Hz bandwidth are required for tenfold frequency diversity transmission of these twelve circuits. The spacing of the ten copies by multiples of 240 Hz is sufficient in the short wave region.

Problems

632.1 What are suitable frequencies for the ten carriers required for tenfold frequency diversity transmission of six teletype channels in the telephony band from about 300 to 3600 Hz?

632.2 How can the unused bandwidth in problem 632.1 be exploited for additional synchronization signals spaced throughout the used frequency band and used for monitoring the time variation of the channel?

632.3 How are the results of problems 632.1 and 632.2 changed if teletype signals of 100 ms rather than 150 ms duration are used?

7. SIGNAL DESIGN FOR IMPROVED RELIABILITY

7.1 Transmission Capacity

7.1.1 Measures of Bandwidth

It was recognized very early during the development of communications that the possible transmission rate of symbols of a communication channel depended on its frequency response of attenuation and phase shift. For instance, the famous theorem by Nyquist [1] states that one independent symbol may be transmitted per time interval of duration τ through an idealized frequency low-pass filter of bandwidth Δf, where

$$\tau = 1/2\Delta f \tag{1}$$

The transmission rate of symbols is defined in this case by the number $1/\tau$ of independent symbols transmitted per unit time [2, 3]. Shannon took into account that the possible transmission rate of information depends on statistical disturbances as well as on the frequency response of attenuation and phase shift [4, 5]. He obtained the celebrated formula for the possible transmission rate of information through an idealized frequency low-pass filter under the influence of additive thermal noise,

$$C = \Delta f \log(1 + P/P_{\Delta f}) \tag{2}$$

where C is the transmission capacity stated, e.g., in bits per second. Δf is the frequency bandwidth of the idealized low-pass filter and $P/P_{\Delta f}$ is the quotient (average signal power)/(average noise power in the band $0 \leqq f \leqq \leqq \Delta f$).

It is important for the present purpose that Eq. (1) as well as Eq. (2) contain the frequency bandwidth Δf. A consistent theory of communication based on complete systems of orthogonal functions requires a definition of transmission capacity that does not need the concept of frequency.

First, frequency is eliminated from the average noise power $P_{\Delta f}$ in the frequency band $0 \leqq f \leqq \Delta f$. To do so let the noise, represented by a voltage, be applied through an input resistance R_a to an integrator that integrates this voltage over a time interval ΔT. A total of ι integrations is performed. The integrator output voltage at the end of the λ-th integration will be denoted by V_λ. For thermal noise the mean voltage $\langle V_\lambda \rangle$ is equal to zero. The mean square deviation from zero, multiplied by $1/R_a$, is an average power and may characterize the noise just like $P_{\Delta f}$ does:

$$\left\langle \frac{V_\lambda^2}{R_a} \right\rangle = \lim_{\iota\to\infty} \frac{1}{\iota} \sum_{\lambda=1}^{\iota} \frac{V_\lambda^2}{R_a} \tag{3}$$

Using section 6.1.1 this result may be generalized. Let $g_\lambda(\theta)$ of Eq. (6.1.1-1) be a voltage across a resistor R_a which is caused by thermal noise. The notation $V_\lambda(t)$ will be used instead of $g_\lambda(\theta)$ and the functions $f(j, \theta)$ in the same equation are replaced by the normalized voltages $[V(j, t)]/V$, where V is defined as follows:

$$\int_{-\Theta/2}^{\Theta/2} f(j, \theta) f(k, \theta)\, d\theta = \frac{1}{T V^2} \int_{-T'/2}^{T'/2} V(j, t)\, V(k, t)\, dt = 1; \quad \frac{T'}{T} = \Theta \tag{4}$$

The coefficients of (6.1.1-1) are represented by normalized voltages using the notation $V^{-1} V_\lambda(j)$. Equation (6.1.1-1) then assumes the following form:

$$\begin{aligned} g_\lambda(\theta) &= V^{-1} V_\lambda(t) = \sum_{j=0}^{\infty} a_\lambda(j) f(j, \theta) = V^{-2} \sum_{j=0}^{\infty} V_\lambda(j)\, V(j, t) \\ a_\lambda(j) &= V^{-1} V_\lambda(j) = \int_{-\Theta/2}^{\Theta/2} g_\lambda(\theta) f(j, \theta)\, d\theta = \frac{1}{T V^2} \int_{-T'/2}^{T'/2} V_\lambda(t)\, V(j, t)\, dt \end{aligned} \tag{5}$$

Let the voltages $V^{-1} V_\lambda(t)\, V(j, t)$ be applied to an integrator and integrated from $-\frac{1}{2}T'$ to $+\frac{1}{2}T'$. The output voltage at the time $\frac{1}{2}T'$ equals $-V_\lambda(j)$ if the time constant of the integrator is chosen equal to the unit of time T. The quantity $V_\lambda^2(j)/R_a$, with dimension of power, may be derived from the output voltage. Let $V_\lambda(t)$ in Eq. (5) be squared, divided by TR_a, and then integrated from $-\frac{1}{2}T'$ to $\frac{1}{2}T'$:

$$\frac{1}{T} \int_{-T'/2}^{T'/2} V_\lambda^2(t)\, R_a^{-1}\, dt = \sum_{j=0}^{\infty} V_\lambda^2(j)\, R \tag{6}$$

Since the left side is the average power of the noise sample $g_\lambda(\theta)$, the right side must have the same meaning. A certain term $V_\lambda^2(j)/R_a$ in the sum represents the average power of the component j, or $f(j, \theta)$, of the noise sample $g_\lambda(\theta)$. Averaging the term $V_\lambda^2(j)/R$ over ι samples of noise $g_\lambda(\theta)$,

$$P_j = \langle V_\lambda^2(j)\, R_a^{-1} \rangle = \lim_{\iota\to\infty} \frac{1}{\iota} \sum_{\lambda=1}^{\iota} V_\lambda^2(j)\, R_a^{-1} \tag{7}$$

yields the average power P_j of the component j of the noise samples or of "the noise". The distribution of $V_\lambda(j)$ is the same for any j in the case of thermal noise. Hence it makes no difference which component is averaged. In this case one may replace[1] the average over λ by the average over j.

[1] This exchange of time and ensemble average requires that the ergodic hypothesis is satisfied.

Furthermore, the average of m components equals m times the average of one component:

$$\left\langle \sum_{j=0}^{m-1} V_\lambda^2(j) R_a^{-1} \right\rangle = m\langle V_\lambda^2(j) R_a^{-1}\rangle = m P_j \tag{8}$$

The value of P_j is quite independent of the orthogonal system $\{f(j, \theta)\}$. Multiplication of the noise samples $g_\lambda(\theta)$ by the functions of an orthogonal system $\{h(j, \theta)\}$, which have the same orthogonality interval as the functions $f(j, \theta)$ and can be expanded into a series according to (6.1.1-3), yields voltages $V_\lambda'(j)$ instead of $V_\lambda(j)$. It follows, however, from Eqs. (6.1.1-4) to (6.1.1-12):

$$\langle V_\lambda^2(j) R_a^{-1}\rangle = \langle V_\lambda'^2(j) R_a^{-1}\rangle \tag{9}$$

This finishes the investigation about the replacement of $P_{\Delta f}$ in Eq. (2). Let us now turn to the replacement of Δf in that equation by a parameter that is independent of sine and cosine functions.

Assume the $m = 2l + 1$ orthogonal functions $f(0, \theta), f_c(1, \theta), f_s(1, \theta), \ldots, f_c(l, \theta), f_s(l, \theta)$ may be transmitted through a communication channel during the orthogonality interval $-\frac{1}{2} \leqq \theta \leqq \frac{1}{2}$. Consider as a special case the functions of the Fourier series:

$$\begin{gathered} f(0, \theta) = 1, \quad f_c(i, \theta) = \sqrt{(2)} \cos 2\pi\, i\, \theta, \quad f_s(i, \theta) = \sqrt{(2)} \sin 2\pi\, i\, \theta, \\ -\tfrac{1}{2} \leqq \theta \leqq \tfrac{1}{2}, \quad i = 1, \ldots, l, \quad \theta = t/T \end{gathered} \tag{10}$$

These sine and cosine *elements* are orthonormal in the interval $-\frac{1}{2} \leqq \theta \leqq \frac{1}{2}$ and undefined outside. Let them be stretched by the substitution $\theta' = \theta/\xi$ as in section 1.2.1:

$$f(0, \theta') = f(0/\xi, \theta) \tag{11}$$

$$f_c(i, \theta') = \sqrt{(2)} \cos 2\pi\, i(\theta/\xi) = \sqrt{(2)} \cos 2\pi(i/\xi)\, \theta = f_c(i/\xi, \theta)$$

$$f_s(i, \theta') = \sqrt{(2)} \sin 2\pi\, i(\theta/\xi) = \sqrt{(2)} \sin 2\pi(i/\xi)\, \theta = f_s(i/\xi, \theta)$$

$$-\tfrac{1}{2} \leqq \theta' \leqq \tfrac{1}{2}, \quad -\tfrac{1}{2}\xi \leqq \theta \leqq \tfrac{1}{2}\xi$$

The duration of the orthogonality interval has been increased from 1 to ξ. The number of functions transmitted per unit of time shall remain constant. $\xi(2l + 1)$ functions must be transmitted in the interval ξ-times as large. The index i runs from 1 to k, where k is defined by the equation

$$(2l + 1)\,\xi = 2k + 1, \quad k = \xi(l + \tfrac{1}{2} + 1/2\xi) \tag{12}$$

Let ξ approach infinity. The time limited sine and cosine elements become the periodic sine and cosine functions with the frequencies $i/\xi = \nu = fT$. The frequency ν runs from $\nu_1 = 1/\xi$ to $\nu_k = k/\xi$ since i runs from 1 to k. The difference $\nu_k - \nu_1$, denoted as the frequency bandwidth

$\Delta\nu$, is given by

$$\Delta\nu = \Delta f T = \lim_{\xi\to\infty} (\nu_k - \nu_1) = \lim_{\xi\to\infty} k/\xi = \tfrac{1}{2}(2l+1) \tag{13}$$

$$\Delta f = \frac{2l+1}{2T} = \frac{1}{2}\frac{m}{T} \tag{14}$$

m/T is the number of orthogonal functions transmitted per unit time T. The bandwidth $\Delta\nu$ or Δf is a measure of the number k of orthogonal sine or cosine elements transmitted during the interval of orthogonality, if the number of elements and their orthogonality interval $-\frac{1}{2}\xi \leqq \theta \leqq \frac{1}{2}\xi$ approaches infinity. According to Eq. (14) one may use m/T, instead of Δf, which is the number of sine and cosine elements transmitted per unit time T.

The frequency bandwidth Δf is only a measure of the number of sine and cosine functions that can be transmitted. On the other hand, m/T may be interpreted as a measure of the number of orthogonal functions that can be transmitted per unit of time, without reference to sine and cosine functions. Hence, m/T is a generalization of the concept of frequency bandwidth.

The difference between Δf and m/T goes beyond the greater generality of m/T. It is often cumbersome for theoretical investigations that every function occupies an infinite section of the time-frequency-domain. The hatched section in Fig. 201a shows the section of the time-frequency-domain occupied by a function that differs from zero in the interval

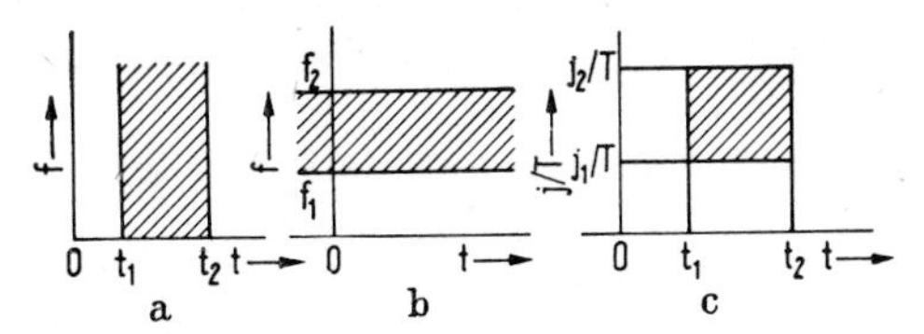

Fig. 201. Time-frequency-domain and time-function-domain. (a) section of the time-frequency-domain occupied by a time limited signal; (b) section of the time-frequency-domain occupied by a frequency limited signal; (c) section of the time-function-domain occupied by a time and function limited signal. $f_2 - f_1 = \Delta f$; $t_2 - t_1 = T$; $j_2 - (j_1 - 1) = m$.

$t_1 \leqq t \leqq t_2$ only. Figure 201b shows the section occupied by a frequency-limited function that is non-zero in the interval $f_1 \leqq f \leqq f_2$. The hatched areas can be made finite only by truncating them arbitrarily at some value of f or t since there are no time- and frequency-limited functions.

It has been shown in section 1.3.3 that there is a class of time and sequency limited functions. This makes it tempting to replace the time-frequency-domain by a time-sequency-domain. But this would unnecessarily distinguish the system of Walsh functions. It is better to introduce a "time-function-domain." Consider a system of functions $\{f(j, \theta)\}$, which are orthogonal in the finite interval $t_1 \leqq t \leqq t_2$ and zero outside; $j = 0, 1, \ldots, j_1, \ldots, j_2, \ldots$ Let signals be composed of functions with the index j running from j_1 to j_2. According to Fig. 201c, the time is plotted along the abscissa and the indices j or $j/T = j/(t_2 - t_1)$ along the ordinate of a cartesian coordinate system. The signals considered occupy the hatched section of this time-function-domain. These signals are exactly time and "function" limited.

Let us investigate the connection between sequency bandwidth $\Delta\phi$ and m/T. The system of functions

$$\{f(0,\theta), f_c(i,\theta), f_s(i,\theta)\}, \quad i = 1, \ldots, l; \quad -\tfrac{1}{2} \leqq \theta \leqq \tfrac{1}{2} \tag{15}$$

shall be orthogonal and $2i$ shall equal the number of zero crossings in the orthogonality interval. The same considerations apply as for the sine and cosine elements. Equations (13) and (14) are obtained again, but the normalized sequency μ has to be substituted for the normalized frequency ν. Furthermore, $\Delta\nu$ is replaced by $\Delta\mu$:

$$\Delta\mu = \lim_{\xi\to\infty} (\mu_k - \mu_1) = \lim_{\xi\to\infty} k/\xi = \tfrac{1}{2}(2l + 1) \tag{16}$$

$$\frac{\Delta\mu}{T} = \Delta\phi = \frac{2l+1}{2T} = \frac{1}{2}\frac{m}{T} \tag{17}$$

Comparison of Eqs. (13) and (14) with Eqs. (16) and (17) shows:

a) The normalized frequency bandwidth $\Delta\nu$ is a measure of the number of functions of the system $\{\sqrt{(2)}\cos 2\pi\nu\theta, \sqrt{(2)}\sin 2\pi\nu\theta\}$ that can be transmitted in a normalized time interval of duration 1.

b) The normalized sequency bandwidth $\Delta\mu$ is a measure of the number of functions of the more general system $\{f_c(\mu,\theta), f_s(\mu,\theta)\}$ that can be transmitted in a normalized time interval of duration 1.

c) The frequency bandwidth $\Delta f = \Delta\nu/T$ is a special case of the sequency bandwidth $\Delta\phi = \Delta\mu/T$, but $m/2T$ is an even more general measure of bandwidth since it applies to all complete systems of orthogonal functions including those to which the concept of sequency in its present definition is not applicable. $m/2T$ equals "one half the average number of functions transmitted per unit time T".

Problems

711.1 What does the ergodic hypothesis state? Its experimental verification requires that many sources of statistical (time) variables are available simultaneously. Can you give an example occurring in communications?

711.2 How can one check by measurements whether a noise generator delivers noise with Gaussian amplitude distribution in a certain frequency band?

711.3 What section of the time-frequency-domain is occupied by the parabolic cylinder functions of Fig. 11?

7.1.2 Transmission Capacity of Communication Channels

Consider signals $F_\chi(\theta)$ that are composed of the system of functions $\{f(0,\theta), f_c(i,\theta), f_s(i,\theta)\}$ orthogonal in the interval $-\frac{1}{2} \leqq \theta \leqq \frac{1}{2}$.

$$F_\chi(\theta) = a_\chi(0) f(0,\theta) + \sum_{i=1}^{\infty} [a_{c\chi}(i) f_c(i,\theta) + a_{s\chi}(i) f_s(i,\theta)] \qquad \chi = 1, 2, \ldots \tag{1}$$

Let $F_\chi(\theta)$ be transmitted through a communication channel. Then assume for the time being that the functions $f(0,\theta)$, $f_c(i,\theta)$ and $f_s(i,\theta)$ are only attenuated and delayed by the time $\theta(0)$ during transmission. Using the attenuation coefficients $K(0)$, $K_c(i)$ and $K_s(i)$ of section 1.3.2 one obtains for the signal at the receiver:

$$F_{\chi E}(\theta) = b_\chi(0) f[0, \theta - \theta(0)] + \sum_{i=1}^{\infty} \{b_{c\chi}(i) f_c[i, \theta - \theta(0)] + b_{c\chi}(i) f_s[i, \theta - \theta(0)]\} \tag{2}$$

$$b_\chi(0) = K(0)\, a_\chi(0), \qquad b_{c\chi}(i) = K_c(i)\, a_{c\chi}(i), \qquad b_{s\chi}(i) = K_s(i)\, a_{s\chi}(i)$$

The receiver shall determine which one of the possible characters $F_\psi(\theta)$ was transmitted. The least-mean-square-deviation criterion shall be used for the decision. Sample signals $F_{\psi M}(\theta)$ must be produced at the receiver which are as similar as possible to the received signals $F_{\chi E}(\theta)$. It is then necessary to decide which integral $I(\psi, \chi)$ is smallest:

$$I(\psi,\chi) = \int_{\theta_0 - 1/2}^{\theta_0 + 1/2} [F_{\chi E}(\theta) - F_{\psi M}(\theta)]^2 \, d\theta \qquad \psi = 1, 2, \ldots, \chi, \ldots; \chi = 1, 2, \ldots \tag{3}$$

Let us assume the sample functions $F_{\chi M}(\theta)$ could be made exactly equal to the received signals:

$$F_{\psi M}(\theta) = F_{\chi E}(\theta) \quad \text{for} \quad \psi = \chi \tag{4}$$

The integral $I(\psi, \psi)$ is then zero. The integral $I(\psi, \chi)$ for $\psi \neq \chi$ must differ from zero by at least ΔI. The minimum difference ΔI cannot be arbitrarily small since it is only possible to determine a finite difference. It follows from Eqs. (2) and (3) due to the orthonormality of the system $\{f(0, \theta)$, $f_c(i, \theta), f_s(i, \theta)\}$:

$$I(\psi,\chi) = [b_\psi(0) - b_\chi(0)]^2 + \sum_{i=1}^{\infty} \{[b_{c\psi}(i) - b_{c\chi}(i)]^2 + [b_{s\psi}(i) - b_{s\chi}(i)]^2\} \geqq \Delta I \tag{5}$$

Consider those signals $F_{\chi E}(\theta)$ that differ from $F_{\psi E}(\theta)$ in only one of the coefficients $a_\chi(0)$, $a_{c\chi}(i)$ or $a_{s\chi}(i)$; one of the following conditions must hold:

$$[b_{\psi}(0) - b_{\chi}(0)]^2 \geqq \Delta I,$$
$$[b_{c\psi}(i) - b_{c\chi}(i)]^2 \geqq \Delta I, \quad [b_{s\psi}(i) - b_{s\chi}(i)]^2 \geqq \Delta I \tag{6}$$

The minimal values $\Delta a(0)$, $\Delta a_c(i)$ and $\Delta a_s(i)$ by which the coefficients of two signals must differ at the transmitter follow from Eqs. (2) and (6):

$$|a_{\psi}(0) - a_{\chi}(0)| \geqq \Delta a(0) = (\Delta I)^{1/2}/K(0)$$
$$|a_{c\psi}(i) - a_{c\chi}(i)| \geqq \Delta a_c(i) = (\Delta I)^{1/2}/K_c(i) \tag{7}$$
$$|a_{s\psi}(i) - a_{s\chi}(i)| \geqq \Delta a_s(i) = (\Delta I)^{1/2}/K_s(i)$$

Let $a_{\chi}(0)$, $a_{c\chi}(i)$ and $a_{s\chi}(i)$ be restricted to values between $+A$ and $-A$. The number of possible coefficients is then given by r_o, r_{ci} and r_{si}:

$$r_o \leqq 2A\, K(0)/(\Delta I)^{1/2} + 1$$
$$r_{ci} \leqq 2A\, K_c(i)/(\Delta I)^{1/2} + 1, \quad r_{si} \leqq 2A\, K_s(i)/(\Delta I)^{1/2} + 1 \tag{8}$$

The "ones" on the right sides take into account the possibility that the coefficients may have the value zero. The largest integers that satisfy the inequalities (8) must be taken for r_o, r_{ci} and r_{si}. The permissible values of the coefficient $a_{\chi}(0)$ are $0, \pm\Delta a(0), \pm 2\Delta a(0), \ldots$ if r_o is odd; for even r_o they are $\pm\frac{1}{2}\Delta a(0), \pm\frac{3}{2}\Delta a(0), \ldots$

Let $K_c(i)$, for $i > l_c$, and $K_s(i)$, for $i > l_s$, be so small that the following relations hold:

$$r_{ci} = 1 \quad \text{for} \quad i > l_c, \qquad r_{si} = 1 \quad \text{for} \quad i > l_s \tag{9}$$

No information can be transmitted with a single function $f_c(i, \theta)$, for $i > l_c$, or $f_s(i, \theta)$, for $i > l_s$. For simplification let us put

$$l_c = l_s = l \tag{10}$$

where l is called the band limit. Information can be transmitted beyond the band limit but the process is different. The coefficients of at least two functions $f_c(i, \theta)$ and/or $f_s(i, \theta)$, $i > l$, must be changed to obtain a different signal. This type of transmission is impossible if the attenuation increases so rapidly beyond the band limit that the condition

$$\sum_{i=l+1}^{\infty} \{[b_{c\psi}(i) - b_{c\chi}(i)]^2 + [b_{s\psi}(i) - b_{s\chi}(i)]^2\} < \Delta I \tag{11}$$

is satisfied for any pair χ and ψ.

The number of distinguishable signals that can be transmitted during a time interval of duration T is then given by the product

$$r_o \prod_{i=1}^{l} r_{ci}\, r_{si}$$

The information transmitted per unit of time, or the transmission capacity of the channel, is the logarithm of this product divided by T:

$$C = \frac{1}{T}\left[\log r_o + \sum_{i=1}^{l} (\log r_{ci} + \log r_{si})\right] \tag{12}$$

Let us consider a special case of Eq. (12). It follows from Eq. (8) for $r_o = r_{ci} = r_{si} = r$:

$$K(0) = K_c(i) = K_s(i), \quad i = 1, \ldots, l$$

All functions $f(0, \theta)$, $f_c(i, \theta)$ and $f_s(i, \theta)$, $i \leqq l$, are attenuated equally. It follows from Eq. (12):

$$C = \frac{2l + 1}{T} \log r \tag{13}$$

If the system of functions used are the periodic sine and cosine functions, one may substitute Δf from Eq. (7.1.1-14) and obtain:

$$C = 2\Delta f \log r = \Delta f \log r^2 \tag{14}$$

This formula has the structure of Shannon's formula (7.1.1-2) although it was derived under different assumptions. It will be shown in the remainder of this section that r in Eqs. (13) and (14) is replaced by $(1 + P/P_{\Delta f})^{1/2}$ if the same assumptions are made as in the derivation of Eq. (7.1.1-2).

Consider signals $F_\chi(\theta)$ composed of r functions $f(j, \theta)$. The orthogonality interval is $-\frac{1}{2} \leqq \theta \leqq \frac{1}{2}$ or $-\frac{1}{2}T \leqq t \leqq \frac{1}{2}T$.

$$F_\chi(\theta) = \sum_{j=0}^{r-1} a_\chi(j) f(j, \theta)$$
$$\{f(j, \theta)\} = \{f(0, \theta), f_c(i, \theta), f_s(i, \theta)\}; \quad r = 2k + 1. \tag{15}$$

The integral of $F_\chi^2(\theta)$ yields the average power of the signal:

$$\int_{-1/2}^{1/2} F_\chi^2(\theta)\, d\theta = \frac{1}{T} \int_{-T/2}^{T/2} F_\chi^2(t/T)\, dt = P_\chi$$
$$= \frac{1}{T} \int_{-T/2}^{T/2} \sum_{j=0}^{r-1} a_\chi^2(j) f^2(j, t/T)\, dt = \sum_{j=0}^{r-1} a_\chi^2(j) \tag{16}$$

Instead of representing a signal by a time function $F_\chi(\theta)$, one may represent it by a point in a r-dimensional cartesian signal space, according to section 3.1.1. Let the unit vectors $\mathbf{e}_j$, $j = 0, \ldots, r - 1$, point in the direction of the r coordinate axes. The square of the length of these unit vectors equals the integral of the square of $f(j, \theta)$.

$$\int_{-1/2}^{1/2} f^2(j, \theta)\, d\theta = \mathbf{e}_j^2 = 1, \quad \int_{-T/2}^{T/2} f^2(j, t/T)\, dt = T = T\mathbf{e}_j^2 \tag{17}$$

A signal is represented by the following sum:

$$\mathbf{F}_\chi = \sum_{j=0}^{r-1} a_\chi(j)\,\mathbf{e}_j \tag{18}$$

$\mathbf{F}_\chi$ rather than $F_\chi(\theta)$ is written in vector representation, and $\mathbf{F}_\chi$ represents a certain point in the r-dimensional space. Its distance from the origin is D_χ:

$$D_\chi = \left[T \sum_{j=0}^{r-1} a_\chi^2(j)\,\mathbf{e}_j^2\right]^{1/2} = \left[T \sum_{j=0}^{r-1} a_\chi^2(j)\right]^{1/2} = (T\,P_\chi)^{1/2} \tag{19}$$

A sample of thermal noise,

$$g_\lambda(\theta) = \sum_{j=0}^{\infty} a_\lambda(j) f(j, \theta) \tag{20}$$

may also be represented by a vector:

$$\mathbf{g}_\lambda = \sum_{j=0}^{\infty} a_\lambda(j)\,\mathbf{e}_j \tag{21}$$

According to Eqs. (6.2.1-7) and (6.2.1-8) only the r components $f(j, \theta)$ or $\mathbf{e}_j$ that occur in the signal are important in Eqs. (20) and (21). Hence, $g_\lambda(\theta)$ is divided into two parts $g'_\lambda(\theta)$ and $g''_\lambda(\theta)$; the part $g''_\lambda(\theta)$ may be ignored:

$$\begin{aligned} g'_\lambda(\theta) &= \sum_{j=0}^{r-1} a_\lambda(j) f(j, \theta) \quad & \mathbf{g}'_\lambda &= \sum_{j=0}^{r-1} a_\lambda(j)\,\mathbf{e}_j \\ g''_\lambda(\theta) &= \sum_{j=r}^{\infty} a_\lambda(j) f(j, \theta) \quad & \mathbf{g}''_\lambda &= \sum_{j=r}^{\infty} a_\lambda(j)\,\mathbf{e}_j \end{aligned} \tag{22}$$

The distance of the point $\mathbf{g}'_\lambda$ from the origin equals D'_λ:

$$D'_\lambda = \left[T \sum_{j=0}^{r-1} a_\lambda(j)\right]^{1/2} \tag{23}$$

The average power of many noise samples $\mathbf{g}'_\lambda$ is denoted by $P_{r,T}$; the indices r and T indicate the number of orthogonal components of the noise sample and the duration of the orthogonality interval:

$$P_{r,T} = \lim_{\iota\to\infty} \frac{1}{\iota} \sum_{\lambda=1}^{\iota} \sum_{j=0}^{r-1} a_\lambda^2(j) \tag{24}$$

It has been shown in sections 6.1.1 and 6.1.2 under very general assumptions, that the distribution of the coefficients $a_\lambda(j)$ is the same for all j, if the $g_\lambda(\theta)$ are samples of thermal noise. Equation (24) may thus be rewritten as follows:

$$P_{r,T} = \lim_{\iota\to\infty} \frac{r}{\iota} \sum_{\lambda=1}^{\iota} a_\lambda^2(j) = r\,P_{1,T} \tag{25}$$

The average over λ for fixed j may be replaced by the average over j for fixed λ:

$$P_{r,T} = \lim_{\iota\to\infty} \frac{r}{\iota} \sum_{j=1}^{\iota} a_\lambda^2(j) \tag{26}$$

The substitution $\iota = r$ yields:

$$\lim_{r\to\infty} P_{r,T} = \lim_{r\to\infty} \sum_{j=1}^{r} a_\lambda^2(j) \tag{27}$$

Comparison of Eqs. (23) and (27) shows that the distance D'_λ of all points $\mathbf{g}'_\lambda$ from the origin approaches $(TP_{r,T})^{1/2}$ for large values of r. The points representing thermal noise are located in signal space arbitrarily close to the surface of a r-dimensional sphere with radius $(TP_{r,T})^{1/2}$.

The average power of ι signals $\mathbf{F}_\chi$ follows from Eq. (16):

$$P = \lim_{\iota\to\infty} \frac{1}{\iota} \sum_{\chi=1}^{\iota} P_\chi = \lim_{\iota\to\infty} \frac{1}{\iota} \sum_{\chi=1}^{\iota} \sum_{j=0}^{r-1} a_\chi^2(j) \tag{28}$$

This equation may be rewritten, if the coefficients $a_\chi(j)$ have the same distribution for all j and if they are statistically independent:

$$P = \lim_{\iota\to\infty} \frac{r}{\iota} \sum_{\chi=1}^{\iota} a_\chi^2(j) = \lim_{\iota\to\infty} \frac{r}{\iota} \sum_{j=1}^{\iota} a_\chi^2(j) \tag{29}$$

The substitution $\iota = r$ yields:

$$\lim_{r\to\infty} P = \lim_{r\to\infty} \sum_{j=1}^{r} a_\chi^2(j) \tag{30}$$

Comparison with Eq. (19) shows that all points $\mathbf{F}_\chi$ are located arbitrarily close to the surface of a r-dimensional sphere with radius $(TP)^{1/2}$ for large values of r.

A signal with an additive noise sample $\mathbf{g}'_\lambda$ superimposed is represented by the point

$$\mathbf{F}_\chi + \mathbf{g}'_\lambda = \sum_{j=0}^{r-1} [a_\chi(j) + a_\lambda(j)]\, \mathbf{e}_j.$$

The points $\mathbf{F}_\chi + \mathbf{g}'_\lambda$, $\chi = 1, 2, \ldots$ are located arbitrarily close to the surface of a sphere with radius $\sqrt{(T)}\,(P + P_{r,T})^{1/2}$ for large values of r:

$$\lim_{r\to\infty} \left\{T \sum_{j=0}^{r-1} [a_\chi(j) + a_\lambda(j)]^2\right\}^{1/2} = \lim_{r\to\infty} \left\{T \sum_{j=0}^{r-1} [a_\chi^2(j) + a_\chi^2(j)]\right\}^{1/2}$$

$$= \lim_{r\to\infty} \sqrt{(T)}\,(P + P_{r,T})^{1/2} \tag{31}$$

One may decide unambiguously which signal $\mathbf{F}_\chi$ caused the disturbed signal $\mathbf{F}_\chi + \mathbf{g}'_\lambda$ if the distance between any two signal points is at least $2(TP_{r,T})^{1/2}$. The possible number of points having this minimum distance from one another is equal to the possible number of signals. To determine this number consider the volume V of a r-dimensional sphere having the radius R [3, 4]:

$$V = \frac{\pi^{r/2}}{\Gamma(\frac{1}{2}r + 1)} R^r \tag{32}$$

The volume V_ε between two concentric spheres with radius R and $R - \varepsilon$ approaches for large numbers r the volume V:

$$V_\varepsilon = \frac{\pi^{r/2}}{\Gamma(\frac{1}{2}r + 1)} [R^r - (R - \varepsilon)^r] = \frac{\pi^{r/2}}{\Gamma(\frac{1}{2}r - 1)} R^r \left[1 - \left(1 - \frac{\varepsilon}{R}\right)^r\right] \doteqdot V \tag{33}$$

Hence, most of the volume of the r-dimensional sphere is close to its surface. A good estimate of the possible number of signal points is obtained by dividing the volume of the sphere with radius $\sqrt{(T)}\,(P + P_{r,T})^{1/2}$ by that of a sphere with radius $(T P_{r,T})^{1/2}$:

$$\frac{T^{r/2}(P + P_{r,T})^{r/2}}{(T\,P_{r,T})^{1/2}} = \left(1 + \frac{P}{P_{r,T}}\right)^{r/2} \tag{34}$$

Each signal $F_\chi(\theta)$ has the duration T. The limit of the error-free transmission rate thus becomes:

$$C = \lim_{r\to\infty} \frac{1}{T} \log\left(1 + \frac{P}{P_{r,T}}\right)^{r/2} = \lim_{r\to\infty} \frac{r}{2T} \log\left(1 + \frac{P}{P_{r,T}}\right) \tag{35}$$

One may see that r, which is the number of orthogonal functions in a signal $F_\chi(\theta)$, must approach infinity. The interval of orthogonality $-\frac{1}{2}T \leqq t \leqq \frac{1}{2}T$ may be finite or infinite. The ratio $P/P_{r,T}$ of the average signal to the average noise power may also be finite or infinite. Equation (27) shows that the average noise power $P_{r,T}$ is infinite for finite T; according to Eq. (30) the same holds true for the average signal power P. The transmission capacity grows beyond all bounds if T is finite and $P/P_{r,T}$ is not zero.

Consider two special cases of Eq. (35) for which the orthogonality interval approaches infinity. For the first example let us use a system of m orthogonal functions $\{f(j, \theta)\}$ that vanish outside the interval $-\frac{1}{2}k\,T' \leqq \leqq t \leqq -\frac{1}{2}k\,T' + T'$, where $k = T/T'$ is an integer. Consider, furthermore, a system of m orthogonal functions $\{f(j, \theta - 1)\}$ that are shifted by -1 and which may have the same shape as the functions $f(j, \theta)$. These functions vanish outside the interval $-\frac{1}{2}k\,T' + T' \leqq t \leqq -\frac{1}{2}k\,T' + 2T'$. Continue this way until the system $\{f(j, \theta - k + 1)\}$ is reached which vanishes outside the interval $\frac{1}{2}k\,T' - T' \leqq t \leqq \frac{1}{2}k\,T'$. The total number of orthogonal functions produced is then:

$$m\,k = r; \quad m, k, r = \text{integers} \tag{36}$$

The factor $r/2T$ in Eq. (35) becomes $m\,k/2k\,T' = m/2T'$. The average noise power $P_{r,T}$ becomes $P_{m,T'}$, because m functions only are non-zero in any one of the k time intervals and because T' is the duration of the orthogonality interval. One obtains from Eq. (35):

$$C = \lim_{k\to\infty} \frac{m\,k}{2k\,T'} \log\left(1 + \frac{P}{P_{mk,\,kT'}}\right) = \frac{m}{2T'} \log\left(1 + \frac{P}{P_{m,T'}}\right) \qquad (37)$$

The derivation of this formula shows that one does not have to wait infinitely long to obtain the information in the signal $F_\chi(\theta)$. Part of the information is available at the ends of the k time intervals.

As a second example consider $m = 2l + 1$ sine and cosine elements in the interval $-\frac{1}{2}T \leqq t \leqq \frac{1}{2}T$. These elements are stretched by a factor $\xi > 1$. According to Eq. (7.1.1-12) one has to substitute $k = \xi(l + \frac{1}{2} + 1/2\xi)$ for l in order to keep the number of functions transmitted per unit of time constant. The number of orthogonal functions in the interval $-\frac{1}{2}\xi\,T \leqq$ $\leqq t \leqq \frac{1}{2}\xi\,T$ is given by:

$$r = 2k + 1 = (2l + 1)\,\xi = m\,\xi \qquad (38)$$

It follows from Eq. (7.1.1-14) that the factor $r/2T$ in Eq. (35) is replaced by $m\,\xi/2\xi\,T = m/2T = \Delta f$. The average noise power $P_{r,T}$ is replaced by $P_{m\xi,\xi T}$. It follows that $P_{m\xi,\xi T} = P_{\Delta f}$ since the signals occupy the frequency band $0 \leqq f \leqq \Delta f$ and all sine and cosine components of the noise samples with frequencies in this band are received. Shannon's formula is thus obtained from Eq. (35):

$$C = \lim_{\xi\to\infty} \frac{m\,\xi}{2\xi\,T} \log\left(1 + \frac{P}{P_{m\xi,\,\xi T}}\right) = \Delta f \log\left(1 + \frac{P}{P_{\Delta f}}\right)$$

$$\Delta f = \frac{1}{2}\,\frac{m}{T}; \qquad P_{m\xi,\xi T} = P_{\Delta f} \qquad (39)$$

Some care must be exercised in interpreting the formulas (13), (14), (35), (37) and (39). They hold for the transmission of orthogonal functions with the one independent variable time. This corresponds to the transmission of signals represented by voltages or currents. An electromagnetic wave travelling in free space in the z-direction also has the independent variable t only, but has two orthogonal positions for the polarization vector, and the formulas apply to each of them. However, in a wave travelling in a wave guide in direction z, the variables x and y may appear in addition to t as independent variables. These additional degrees of freedom show up as modes and the formulas apply to each of them. Hence, Shannon's formula should not only be viewed as the limit of what existing communication channels can transmit, but as a guide to better channels. Equations (13) and (35) show that the number of transmittable orthogonal functions is the principal factor determining transmission capacity. A possible way to increase this number is to use channels that transmit signals which are variables of time and space coordinates. Optical telescopes are used in this way.

Problems

712.1 A communication receiver for time signals connected to a parabolic antenna receives all the noise collected by the antenna from a solid (spatial) angle. A photographic film in the image plane of a parabolic mirror receives a spatial signal. Does a particular point in the image plane receive all the noise collected by the parabolic mirror?

712.2 Why is it theoretically possible but practically difficult to use the parabolic antenna as efficiently as the parabolic mirror?

712.3 Can you find a theoretical way to overcome the difficulty of 712.2 by using Walsh waves with short switching time instead of sinusoidal waves?

7.1.3 Signal Delay and Signal Distortions

Several simplifying assumptions have been made for the derivation of the transmission capacity in Eq. (7.1.2-12). The elimination of these assumptions will be investigated in this section.

Let the functions $f_c(i, \theta)$ and $f_s(i, \theta)$ in Eq. (7.1.2-2) be delayed by $\theta_c(i)$ and $\theta_s(i)$ rather than by a common delay time $\theta(0)$. The functions in Eq. (7.1.2-2) are then no longer orthogonal and $\theta(0)$ is no longer the delay time of the signal. For a more general definition of a signal delay time let $f(0, \theta)$, $f_c(i, \theta)$ and $f_s(i, \theta)$ be transmitted individually. The functions $K(0)f[0, \theta - \theta(0)]$, $K_c(i)f_c[i, \theta - \theta_c(i)]$ and $K_s(i)f_s[i, \theta - \theta_s(i)]$ are then received. Let them be cross-correlated with sample functions $f(0, \theta)$, $f_c(i, \theta)$ and $f_s(i, \theta)$. The time difference between $\theta = 0$ and the absolute maxima of the cross-correlation functions yield the delays $\theta(0)$, $\theta_c(i)$ and $\theta_s(i)$. The values of the maxima yield the attenuation coefficients $K(0)$, $K_c(i)$ and $K_s(i)$. Using these coefficients one may derive a sample function $F'_{\chi M}(\theta)$ from $F_\chi(\theta)$ in Eq. (18):

$$F'_{\chi M}(\theta) = K(0)\,a_\chi(0) f(0,\theta) + \sum_{i=1}^{\infty} [K_c(i)\,a_{c\chi}(i) f_c(i,\theta) + K_s(i)\,a_{s\chi}(i) f_s(i,\theta)] \quad (1)$$

The received signal $F_{\chi E}(\theta)$ has the same shape, but θ must be replaced by $\theta - \theta(0)$, $\theta - \theta_c(i)$ or $\theta - \theta_s(i)$ on the right side. Let the cross-correlation function of $F'_{\chi M}(\theta)$ and $F_{\chi E}(\theta)$,

$$\int_{-\infty}^{\infty} F_{\chi E}(\theta)\, F'_{\chi M}(\theta - \theta')\, d\theta = f(\theta') \quad (2)$$

yield an absolute maximum for a certain value of $\theta' = \theta_\chi$. This value is defined as the delay time or the propagation time of the signal $F_\chi(\theta)$. Since it is not known at the receiver which signal is going to arrive, it is advantageous to define a propagation time independent of χ. One may, for instance, average the values θ_χ, if there are R different signals $F_\chi(\theta)$, to define a propagation time θ_L:

$$\theta_L = \frac{1}{R} \sum_{\chi=1}^{R} \theta_\chi \quad (3)$$

The propagation time of a signal carrying information is a statistical variable which can be identified in first approximation only with the concepts of group delay or signal delay originally defined in optics [2].

The received signal $F_{\chi E}(\theta)$ has the shape of Eq. (1) if θ is replaced by $\theta - \theta(0)$, $\theta - \theta_c(i)$ or $\theta - \theta_s(i)$ on the right side. The sample function $F_{\chi M}(\theta)$ has this shape too, but θ must be replaced by $\theta - \theta_L$ on the right side. Let the functions $f[0, \theta - \theta(0)]$, $f_c[i, \theta - \theta_c(i)]$ and $f_s[i, \theta - \theta_s(i)]$ be expanded in a series of the system $\{f(0, \theta - \theta_L), f_c(i, \theta - \theta_L), f_s(i, \theta - \theta_L)\}$. One obtains in analogy to Eq. (3.3.2-1) the following equations, in which $v = \theta - \theta_L$, $f_0 = f[0, \theta - \theta(0)]$, $f_c = f_c[i, \theta - \theta_c(i)]$ and $f_s = f_s[i, \theta - \theta_s(i)]$ is written for abbreviation:

$$\left.\begin{aligned} f_0 &= K(0,0) f(0,v) + \sum_{k=1}^{\infty} [K(0,ck) f_c(k,v) + K(0,sk) f_s(k,v)] \\ f_c &= K(ci,0) f(0,v) + \sum_{k=1}^{\infty} [K(ci,ck) f_c(k,v) + K(ci,sk) f_s(k,v)] \\ f_s &= K(si,0) f(0,v) + \sum_{k=1}^{\infty} [K(si,ck) f_c(k,v) + K(si,sk) f_s(k,v)] \end{aligned}\right\} \tag{4}$$

Let these series be substituted into the formula for $F_{\chi E}(\theta)$. The first term of $F_{\chi E}(\theta)$ has the following form:

$$\left\{K(0)\, a_\chi(0)\, K(0,0) + \sum_{i=1}^{\infty} [K_c(i)\, a_{c\chi}(i)\, K(ci,0) + K_s(i)\, a_{s\chi}(i)\, K(si,0)]\right\} \times \\ \times f(0, \theta - \theta_L) \tag{5}$$

There is mutual interference or crosstalk between the coefficients. It is possible in principle to devise distortion correcting circuits that compensate the crosstalk within the accuracy of measurement, so that $K(0)\, a_\chi(0) \times \times f(0, \theta - \theta_L)$ is obtained in place of (5). $F_{\chi E}(\theta)$ and $F_{\chi M}(\theta)$ are then identical.

Let us further assume that $f(0, \theta)$, $f_c(i, \theta)$ and $f_s(i, \theta)$ are not only attenuated and delayed during transmission but also suffer a linear, time invariant distortion. $f(0, \theta)$, $f_c(i, \theta)$ and $f_s(i, \theta)$ are transformed into $g(0, \theta)$, $g_c(i, \theta)$ and $g_s(i, \theta)$ according to section 3.3.2. Let the functions $f(0, \theta)$, $f_c(i, \theta)$ and $f_s(i, \theta)$ be transmitted individually. The correlation functions of the received functions $g(0, \theta)$, $g_c(i, \theta)$ and $g_s(i, \theta)$ with sample functions $f(0, \theta)$, $f_c(i, \theta)$ and $f_s(i, \theta)$ are produced. The time shift between their absolute maxima and $\theta = 0$ yields the delays $\theta(0)$, $\theta_c(i)$ and $\theta_s(i)$. The values of the maxima yield the attenuation coefficients $K(0) = K(0, 0)$, $K_c(i) = K(ci, ci)$ and $K_s(i) = K(si, si)$. Sample functions $F'_{\chi M}(\theta)$ of Eq. (1) may be constructed with these coefficients. Equation (2) yields θ_χ and Eq. (3) defines a propagation time θ_L. Now let the distorted functions $g(0, \theta)$, $g_c(i, \theta)$ and $g_s(i, \theta)$ be expanded in a series of the system

$\{f(0, \theta - \theta_L), f_c(i, \theta - \theta_L), f_s(i, \theta - \theta_L)\}$. The resulting expressions are formally the same as those in Eq. (4) and the same conclusions apply.

Problems

731.1 How are phase velocity and group velocity defined?

713.2 Under which conditions can phase velocity or group velocity become infinite?

713.3 Using the results of problem 713.2 and considering the restrictions on the velocity of energy transport imposed by the special theory of relativity, specify when concepts like group velocity and group delay can be used in communication to characterize the velocity with which signals are transmitted.

7.2 Error Probability of Signals

7.2.1 Error Probability of Simple Signals due to Thermal Noise

Consider the transmission of teletype characters in the presence of thermal noise. The probability of error will be computed for several methods of transmission and detection. The general form of such characters represented by time functions is:

$$F_\chi(\theta) = \sum_{j=0}^{4} a_\chi(j) f(j, \theta), \quad \chi = 1, \ldots, 32, \quad \theta = t/T \tag{1}$$

The functions $f(j, \theta)$ are orthonormal in the interval $-\frac{1}{2} \leqq \theta \leqq \frac{1}{2}$. T is the duration of a teletype character which is usually 100, 150 or 167 ms. The coefficients $a_\chi(j)$ have the values $+1$ and -1, or $+a$ and $-a$, for a balanced system; they are $+1$ and 0 for an on-off system. A sample $g_\lambda(\theta)$ of additive thermal noise transforms the character $F_\chi(\theta)$ into the signal $F(\theta)$:

$$\left.\begin{aligned} &F(\theta) = F_\chi(\theta) + g_\lambda(\theta) \\ &g_\lambda(\theta) = \sum_{j=0}^{\infty} a_\lambda(j) f(j, \theta), \qquad a_\lambda(j) = \int_{-1/2}^{1/2} g_\lambda(\theta) f(j, \theta)\, d\theta \\ &F(\theta) = \sum_{j=0}^{\infty} a(j) f(j, \theta), \qquad a(j) = a_\chi(j) + a_\lambda(j) \end{aligned}\right\} \tag{2}$$

The energy of all characters is the same in a balanced system. Using the least-mean-square-deviation criterion from sample functions $F_\psi(\theta)$,

$$F_\psi(\theta) = \sum_{j=0}^{4} a_\psi(j) f(j, \theta) \tag{3}$$

one may decide, according to Eq. (6.2.1-8), which value of ψ will give

$$\sum_{j=0}^{4} a(j)\, a_\psi(j) \tag{4}$$

its maximum value. All coefficients $a(j)$ must have the same sign as the coefficients $a_\chi(j)$ if the maximum is to occur for $\psi = \chi$. The sum (4) then has the following value:

$$a \sum_{j=0}^{4} |a(j)|, \quad a_\chi(j) = +a \quad \text{or} \quad -a; \quad \chi = \psi$$

If, for example, $a(0)$ had the opposite sign of $a_\chi(0)$, the sum (4) would be larger for the character $F_\eta(\theta)$ with the coefficients $a_\eta(0) = -a_\chi(0)$, $a_\eta(k) = a_\chi(k)$, $k = 1, \ldots, 4$ than for $F_\chi(\theta)$:

$$\frac{1}{a} \sum_{j=0}^{4} a(j)\, a_\psi(j) = \begin{cases} +|a(0)| + |a(1)| + |a(2)| + |a(3)| + |a(4)|, & \psi = \eta \\ -|a(0)| + |a(1)| + |a(2)| + |a(3)| + |a(4)|, & \psi = \chi \end{cases}$$

The following two conditions must be satisfied, according to Eq. (2), in order to have different signs for $a(j)$ and $a_\chi(j)$:

a) $\operatorname{sig} a_\chi(j) \neq \operatorname{sig} a_\lambda(j)$

b) $|a_\chi(j)| < |a_\lambda(j)|$, equivalent $\dfrac{a_\lambda(j)}{|a_\chi(j)|} > 1$ or $\dfrac{a_\lambda(j)}{|a_\chi(j)|} < -1$ (5)

$\operatorname{sig} a_\chi(j)$ means "sign of $a_\chi(j)$".

In the case of thermal noise the probability of $a_\lambda(j)$ being positive is $\frac{1}{2}$ and the probability of being negative is also $\frac{1}{2}$. Hence, the probability of condition (*a*) being satisfied equals $\frac{1}{2}$, independent of the sign of $a_\chi(j)$.

The distribution of $x = a_\lambda(j)/|a_\chi(j)|$ is needed for the computation of the probability of condition (*b*) being satisfied. Since $a_\chi(j)$ can be $+a$ or $-a$ only, $|a_\chi(j)|$ is a constant. Therefore, x has the same distribution as $a_\lambda(j)$. The density function $w_a(k, x)$ is obtained from Eq. (6.1.1-6) by substituting x for A. From $w_a(k, x)$ follows the conditional density function $w(x)$ for the condition $k = j$. The probability that k equals one of the $m = 5$ values of j is $1/m$, since the coefficients $a_\lambda(j)$ have the same distribution for all j in case of thermal noise. Thus the density function $w(x)$ follows from Eq. (6.1.1-6):

$$\left.\begin{aligned} w(x) &= \frac{w_a(k, x)}{m^{-1}} = \frac{1}{\sqrt{(2\pi)}\,\sigma_a} \exp(-x^2/2\sigma_a^2) \\ &= \lim_{\iota\to\infty} \frac{1}{\iota} \sum_{\lambda=1}^{\iota} \frac{a_\lambda^2(j)}{|a_\chi(j)|^2} = a^{-2} \lim_{\iota\to\infty} \frac{1}{\iota} \sum_{\lambda=1}^{\iota} a_\lambda^2(j) \\ x &= a_\lambda(j)/|a_\chi(j)| = a_\lambda(j)/a \end{aligned}\right\} \quad (6)$$

Each coefficient $a_\chi(j)$ in Eq. (1) is transmitted with equal energy. Hence, the average signal power P equals:

$$P = \frac{1}{T} \int_{-T/2}^{T/2} F_\chi^2(\theta)\, dt = \sum_{j=0}^{4} a_\chi^2(j) = 5a^2$$

This result may be generalized and solved for a^2:

$$a^2 = P/m \tag{7}$$

Equation (7.1.2-25) yields for $r = \xi = 1$:

$$\lim_{\iota \to \infty} \frac{1}{\iota} \sum_{\lambda=1}^{\iota} a_\lambda^2(j) = \langle a_\lambda^2(j) \rangle = P_{1,T} \tag{8}$$

The mean square deviation becomes

$$\sigma_a^2 = m\, P_{1,T}/P = P_{m,T}/P, \quad P_{m,T} = m\, P_{1,T} \tag{9}$$

where $P_{m,T}$ is the average power of m orthogonal components of thermal noise in an orthogonality interval of duration T.

Using Eq. (7.1.2-39) one may rewrite σ_a^2:

$$\sigma_a^2 = P_{\Delta f}/P, \quad \Delta f = m/2T \tag{10}$$

$P_{\Delta f}$ is the average power of thermal noise in a frequency band of width Δf.

The probability $p(x > 1) + p(x < -1)$ that x is larger than $+1$ or smaller than -1 follows from Eq. (6) by integration:

$$p(x > 1) + p(x < -1) = \frac{2}{\sqrt{(2\pi)}\,\sigma_a} \int_1^\infty \exp\left(\frac{-x^2}{2\sigma_a^2}\right) dx$$

$$= 1 - \operatorname{erf}\left(\frac{1}{\sqrt{(2)}\,\sigma_a}\right) = 1 - \operatorname{erf}\left(\sqrt{\frac{P}{2P_{\Delta f}}}\right) \tag{11}$$

The probability p_1 that conditions (a) as well as (b) of Eq. (5) are satisfied becomes:

$$p_1 = \tfrac{1}{2}\left[1 - \operatorname{erf}\left(\sqrt{\frac{P}{2P_{\Delta f}}}\right)\right] \tag{12}$$

The probability that the conditions of Eq. (5) are not satisfied is $1 - p_1$; the probability that they are not satisfied for any of the $m = 5$ coefficients $a_\chi(j)$ is $(1 - p_1)^m$; the probability that they are satisfied for at least one of the m coefficients equals p_m:

$$p_m = 1 - (1 - p_1)^m = 1 - \left(\frac{1}{2}\right)^m \left[1 + \operatorname{erf}\left(\sqrt{\frac{P}{2P_{\Delta f}}}\right)\right]^m \tag{13}$$

The probability of error p_m does not depend on the system of functions $\{f(j, \theta)\}$ used, provided these functions satisfy the conditions of sections 6.1.1 and 6.1.2.

The numerical values $m = 5$ and $\Delta f = \dfrac{m}{2T} = \dfrac{5}{0.3} = 16.7$ Hz apply to the much used teletype standard of 150 ms per character. Curve a of Fig. 202 shows $p_m = p_5$ of Eq. (13) as function of $P/P_{\Delta f}$ for these values of m and Δf. The measured points a were obtained with an early version of the equipment shown in Fig. 117, with the system $\{f(j, \theta)\}$ consisting of sine and cosine pulses according to Fig. 1.

Let the system $\{f(j, \theta)\}$ consist of the functions

$$f(j, \theta) = \frac{\sin \pi(m\,\theta - j)}{\pi(m\,\theta - j)} = \frac{\sin \pi(\theta' - j)}{\pi(\theta' - j)} \tag{14}$$

$$j = 0, 1, 2, \ldots, m = 5, \quad \theta = t/T, \quad \theta' = t/(T/m)$$

Equation (13) applies to this system too. The energy of these functions is concentrated in the frequency band $-\frac{1}{2} \leqq \nu = fT/m \leqq \frac{1}{2}$ with the bandwidth $\Delta f = m/2T = 16.7$ Hz.

According to section 3.1.3 the same values are obtained for the coefficients $a(j)$ whether $F(\theta)$ is multiplied by the functions (14) and the product is integrated, or whether $F(\theta)$ is passed through an ideal frequency low-pass filter 16.7 Hz wide and the amplitudes are sampled. Hence, Eq. (13) also holds for filtering and amplitude sampling of the pulses of Eq. (14). A low-pass filter 120 Hz wide increases the average noise power in Eq. (13) by $120/16.7 \doteqdot 7.2$. This means a shift of the curve a in Fig. 202 by $10 \log 7.2 \doteqdot 8.58$ dB; the shifted curve is denoted by b.

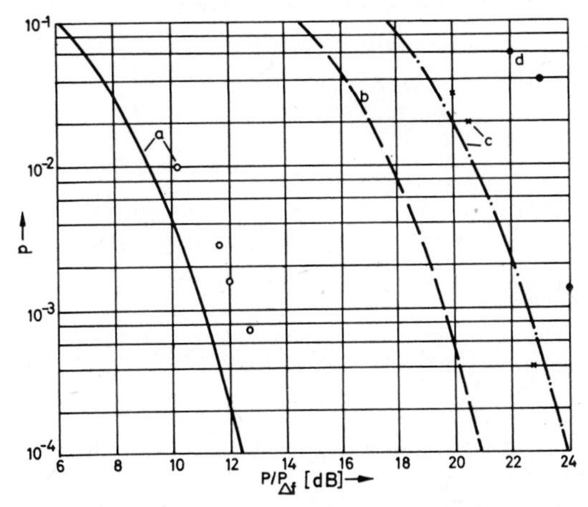

Fig. 202. Error probability p for the reception of teletype signals superimposed by additive thermal noise. $P/P_{\Delta f}$ = average signal power/average noise power in a 16.67 Hz wide band. (*a*) balanced system, detection by cross-correlation; (*b*) balanced system, filtering by a 120 Hz wide ideal low-pass filter, detection by amplitude sampling; (*c*) same as (*b*) but on-off system; (*d*) same as (*c*) but start-stop synchronization disturbed by the noise.

Consider an on-off system. The coefficient $a_\chi(j)$ may assume the values $+b$ or 0 instead of $+a$ or $-a$. The following conditions must be satisfied in order for a coefficient $a_\chi(j)$ to be detected as $+b$ instead of 0, or as 0 instead of $+b$:

a) $\operatorname{sig}[a_\chi(j) - \frac{1}{2}b] \neq \operatorname{sig} a_\lambda(j)$

b) $\left|a_\chi(j) - \dfrac{1}{2}\,b\right| < \left|a_\lambda(j)\right|$, equivalent $\dfrac{a_\lambda(j)}{|a_\chi(j) - b/2|} \begin{cases} > +1 \\ < -1 \end{cases}$ (15)

$a_\chi(j) - \frac{1}{2}b$ may be $+\frac{1}{2}b$ or $-\frac{1}{2}b$, since $a_\chi(j)$ may be $+b$ or 0. The conditions in Eqs. (5) and (15) are thus the same, but $+a$ and $-a$ have to be replaced by $+\frac{1}{2}b$ and $-\frac{1}{2}b$. The average power of the m coefficients with values $+b$ or 0 equals $P = \frac{1}{2}m\,b^2$ and it follows:

$$(\tfrac{1}{2}b)^2 = P/2m \tag{16}$$

Comparison of Eqs. (7) and (16) shows that P has to be replaced by $\frac{1}{2}P$ in the equations holding for a balanced system to get the equations for an on-off system. This means a shift of curves a and b in Fig. 202 by $10 \log 2 \doteqdot 3$ dB. The shifted curve b is denoted by c. The measured points c were obtained by adding thermal noise to the block pulses of teletype characters, after which the disturbed signals were filtered by a 120 Hz wide low-pass filter and then fed to the receiving magnet of a teletype receiver. The measured points agree fairly well with curve c, although the block pulses did not have the shape of the pulses in Eq. (14), the low-pass filter was not ideal, and the magnet of a teletype receiver works only very roughly as an amplitude sampler. The measured points d hold for the same teletype transmission, but start-stop pulses were transmitted through the noisy channel for synchronization. The points c and d depend strongly on the care taken in adjusting the teletype receiver.

Problems

721.1 Determine the first two terms of a series expansion of Eq. (13) for small values of p_1 and specify when the first term can be used to represent p_m.

721.2 The concept of conditional probability was used in the derivation of Eq. (6). How is it defined?

721.3 How are the received digits sampled in a mechanical teletype receiver? Discuss what quantity is actually sampled.

7.2.2 Peak Power Limited Signals

It has been assumed so far that the average signal power is the determining factor in the error probability. However, power amplifiers generally limit the peak power rather than the average power. Consider an amplifier that clips amplitudes at $\pm E$ and delivers a peak power P_E. Average power P and peak power P_E of a signal consisting of binary block pulses, having positive or negative amplitudes, are the same. The error probability p_1 of Eq. (7.2.1-12) for one digit is plotted in Fig. 203 as a function of $P/P_{\Delta f} = P_E/P_{\Delta f}$ and denoted "Theoretical limit".

Let these pulses be amplitude modulated onto a carrier. The curve "Theoretical limit" would still apply if the carrier is a Walsh carrier. The peak power of a sinusoidal carrier would have to be 3 dB larger to yield the same average power; the curve denoted by $m/n = 1$ in Fig. 203 holds for a sinusoidal carrier, amplitude modulated by binary block pulses.

Only about one quarter of the channels in a telephony multiplex system are busy during peak traffic. Using block pulses for PCM transmission, the amplifiers are used $\frac{1}{4}$ of the time only, while no signals, or at least not very useful signals, are transmitted $\frac{3}{4}$ of the time. The peak power must be increased by $10 \log 4 \doteqdot 6$ dB to obtain the same average signal power that the amplifier would deliver if useful signals would be amplified all the time. The resulting curve is denoted by $m/n = 0.25$ in Fig. 203. The ratio

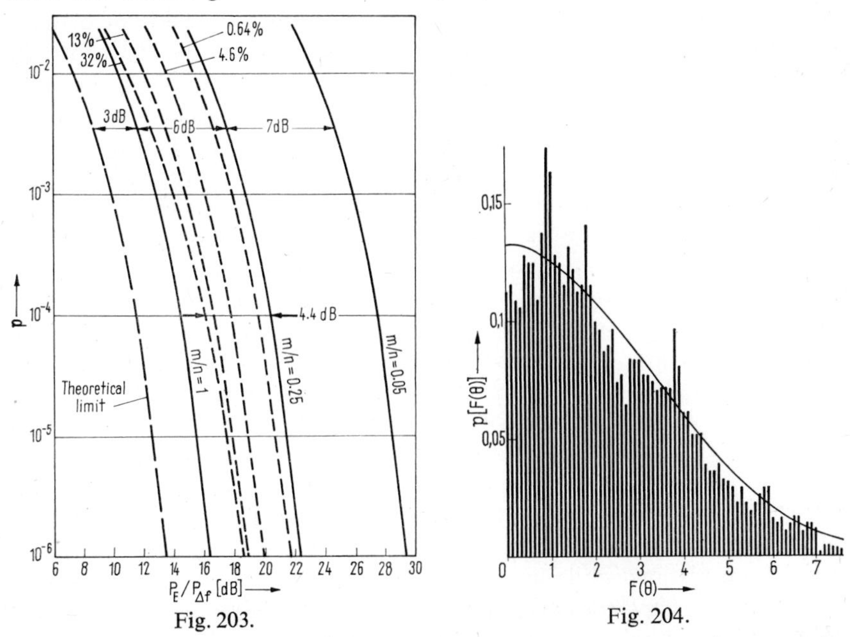

Fig. 203. Fig. 204.

Fig. 203. Error probability p as function of $P_E/P_{\Delta f}$ = peak signal power/average noise power in a band of width $\Delta f = m/2T$. Solid lines: time division, sine carrier, activity factors 1, 0.25 and 0.05; dashed lines: 4 sine and 4 cosine pulses, percentage of clipped amplitudes shown.

Fig. 204. Probability $p[F(\theta)]$ of the amplitudes of the 512 signals $F_\chi(\theta)$ being in intervals of width 0.1. Gaussian density function with equal mean and mean square deviation shown for comparison.

$$a_\chi(0),\ a_\chi(i),\ b_\chi(i) = \pm 1;\ F_\chi(\theta) = a_\chi(0) + \sqrt{(2)} \sum_1^4 [a_\chi(i) \cos 2\pi i\,\theta + b_\chi(i) \sin 2\pi i \theta].$$

m/n is the activity factor, m being the number of busy channels and n the number of available channels. Very low activity factors occur in the ground stations of multiple access satellite systems, since the sum of the activity factors of the ground stations is equal to the activity factor of the satellite transponder. A representative curve for $m/n = 0.05$ is shown in Fig. 203.

Consider now the transmission of binary digits by sine-cosine or Walsh pulses. The resulting signals $F(\theta)$ have very large peaks although most amplitudes are much smaller than the peaks. Figure 204 shows the probability $p[F(\theta)]$ of such a signal having an amplitude within an interval 0.1 wide. Superimposed is a Gaussian density function having the same mean and mean square deviation. According to the results of section 6.2.4 this density function approximates very accurately the probability function of the amplitudes of signals consisting of a sum of Walsh pulses. The plots of Fig. 204 are symmetrical for negative values of $F(\theta)$.

The average power of the signals would be very small if the large but rare peaks would be transmitted. The large peaks must be limited to increase the average signal power. The dashed lines in Fig. 203 show the results of amplitude clipping for sine-cosine pulses in the presence of additive thermal noise. The parameters 0.64%, 4.6%, 13% and 32% indicate the percentage of amplitudes clipped. The curves hold for dc transmission or for transmission by a Walsh carrier. They also hold approximately for single sideband modulation of sine carriers. Walsh pulses yield very similar curves [1]. Little energy is transferred to adjacent frequency bands by clipping of sums of sine and cosine pulses [2], while no energy is transferred to adjacent sequency bands by clipping of sums of Walsh pulses.

The following conclusions may be drawn from Fig. 203. Serial transmission of binary block pulses produces the lowest error rates, if the activity factor is close to 1. Parallel transmission by sine-cosine or Walsh pulses yields lower error rates, if the activity factor is 0.5 or less. The exact percentage of clipped amplitudes is not critical. The clipper characteristic and the density function of a clipped Gaussian amplitude distribution is shown in Fig. 198 for 13% of the amplitudes clipped. Note that a decrease of the activity factor keeps the energy of a pulse unchanged in the case of serial transmission, while in the case of parallel transmission the average power will be kept constant, and the energy of a pulse will be increased if an automatic gain control amplifier is used.

An increase of the required peak power is needed for equal error rates if the block pulses are replaced by other pulse shapes used in serial transmission. Table 19 shows the increase ΔP_E required for some typical pulse shapes. The solid curves in Fig. 203 have to be shifted to the right by ΔP_E

Table 19. *Increase ΔP_E of peak signal power over the peak power of a block pulse for equal error probability*

	Pulse shape	ΔP_E [dB]
1	dc block pulse, E for $0 < t < T/n$, 0 otherwise	0
2	raised cosine pulse in frequency domain; roll-off factor $r = 1$	1.8
3	same, $r = 0.75$	2.7
4	same, $r = 0.5$	4.1
5	raised cosine pulse in time domain, Fig. 125	1.2
6	triangular pulse, $E(1 + n\,t/T)$, $-T/n < t < 0$ $E(1 - n\,t/T)$, $0 < t < T/n$	1.8

to apply to these pulses. The "raised cosine pulse in frequency domain" is defined by the equation

$$f(t/T) = \mathrm{E}\frac{\sin\pi\, n\, t/T}{\pi\, n\, t/T}\,\frac{\cos\pi\, r\, n\, t/T}{1-(2r\, n\, t/T)^2}$$

r is the so-called roll-off factor of the low-pass filter used for pulse shaping [3], n is the number of channels, and T/n is the duration of a block pulse if n of them have to be transmitted during the time T.

Problems

722.1 What causes the conspicuous peaks around $F(\theta) = 1, 2, 4$ in Fig. 204?

722.2 The so-called TASI system (Time Assignement Speech Interpolation) is used to improve the activity factor in submarine cables. What prevents its use in multiple access satellite communication? On which link is the activity factor most important and what makes its use there difficult?

722.3 Substitute Walsh pulses for sine-cosine pulses and compute curves like those in Fig. 203.

7.2.3 Pulse-Type Disturbances

The error probability of digital signals is independent of the particular system of orthogonal functions used for their transmission if the disturbances are caused by additive thermal noise. This is not so for pulse-type disturbances which are more important than thermal noise on telephone lines.

Let us assume that the amplitude of a disturbing pulse is much larger than the largest amplitude of the undisturbed signal. Then let these pulses pass through an amplitude limiter. If the rise and fall times of the pulses are sufficiently short, block pulses of various length but equal amplitude will be obtained at its output. Let these pulses be observed during R time intervals of duration T; there shall be one pulse in r intervals. r/R is the probability for the occurrence of a pulse during an interval of duration T, if r and R are very large. $W_1(T)$ is written for r/R as r and R approach infinity; $W_1(T)$ is the distribution function for the occurrence of a pulse.

Let the duration $\Delta\tau_s$ of the pulses be observed and let q out of Q have a duration $\Delta\tau_s \leqq \tau_s$. The limit q/Q for infinite values of q and Q is denoted by $W_2(\tau_s)$, the distribution function for the length of the pulses.

Let the occurrence and the length of the pulses be statistically independent. The distribution function $W(T, \tau_s)$ of the joint distribution is then defined by the product

$$W(T, \tau_s) = W_1(T)\, W_2(\tau_s) \tag{1}$$

$W(T, \tau_s)$ cannot be determined by separate measurement of $W_1(T)$ and $W_2(\tau_s)$ if statistical independence does not hold. A total of $R\,Q$ rather than $R + Q$ measurements would then be required.

The distribution function $W(T, \tau_s)$ applies when only one pulse occurs in an interval of duration T. If more pulses occur, computations get very involved. Hence, it is assumed that more than one pulse occurs very infrequently.

Denote by p the probability that a signal of duration T is changed beyond recognition by a pulse of duration $\Delta\tau_s \leqq \tau_s$. The conditional probability of an error equals p_b, under the condition that a pulse of duration $\Delta\tau_s \leqq \tau_s$ is received:

$$p_b = \frac{p}{p(t \leqq T, \Delta\tau_s \leqq \tau_s)}; \quad p(t \leqq T, \Delta\tau_s \leqq \tau_s) = W(T, \tau_s) \tag{2}$$

The conditional probability p_b may be calculated for various pulse shapes and detection methods. p can be computed if $W(T, \tau_s)$ is known from measurements. The knowledge of p_b suffices for a comparison of the susceptibility of various pulse shapes and detection methods to disturbances.

Let the transmitted character consist of m block pulses as shown in Fig. 3 for $m = 5$. Each pulse has the duration T/m. A positive or negative amplitude shall be detected by amplitude sampling. A disturbing pulse with duration $\Delta\tau_s \geqq T/m$ causes an error with probability $p_b \geqq \frac{1}{2}$ since half of the disturbing pulses change the sign of at least one of the m sampled amplitudes. The probability p_b increases linearly with $\Delta\tau_s$ in the interval $0 \leqq \Delta\tau_s \leqq T/m$, as shown by curve *1* in Fig. 205.

Let the sign of the amplitudes of the block pulses be determined by cross-correlation. This means that the integral of the pulses is sampled. The amplitudes of the received signal can be limited at $+a$ and $-a$ if the undisturbed signal has the amplitude $+a$ or $-a$. A disturbing pulse with positive amplitude superimposed on a signal pulse with amplitude $+a$ will be suppressed completely. On the other hand, the amplitude of a negative disturbing pulse would be limited to $-2a$ since $+a - 2a$ is $-a$, the smallest amplitude the limiter would allow. On the average, one half of the disturbing pulses have an amplitude $+2a$ or $-2a$, the others have an amplitude zero. No error will occur if the duration $\Delta\tau_s$ of the disturbing pulses is so short that the following relation holds:

$$\begin{aligned} 2a\,\Delta\tau_s &< a\,T/m \\ \Delta\tau_s &< T/2m \end{aligned} \tag{3}$$

The conditional probability p_b depends for

$$a\,T/m \leqq 2a\,\Delta\tau_s \leqq 2a\,T/m \quad \text{or} \quad T/2m \leqq \Delta\tau_s \leqq T/m \tag{4}$$

on the position in time of the disturbing pulse. p_b jumps from 0 to $\frac{1}{4}$ at $\Delta\tau_s = T/2m$ and increases for larger values of $\Delta\tau_s$ linearly to $\frac{1}{2}$ as shown by curve *2* in Fig. 205. There is a strong threshold effect at $\Delta\tau_s = T/2m$.

Consider the transmission of characters composed of m Walsh functions. Let each function have the amplitude $+a/m$ or $-a/m$. The largest and the smallest amplitudes of a sum of m such functions is $+a$ and $-a$. An amplitude limiter may thus clip at $+a$ and $-a$ without changing the undisturbed signal.

Let m be a power of 2. At a certain moment a character has the amplitude a_k if $m-k$ Walsh functions have the amplitude $+a/m$ and k have the amplitude $-a/m$:

$$a_k = (1 - 2k/m)\, a, \qquad k = 0, 1, \ldots, m \tag{5}$$

The probability of a_k ocurring is denoted by $r(k)$:

$$r(k) = 2^{-m} \binom{m}{k} \tag{6}$$

The amplitude b_k of a disturbing pulse superimposed on the amplitude a_k of the signal may have one of the two following values after amplitude limiting at $\pm a$:

$$b_k = a - (1 - 2k/m)\, a = 2k\, a/m \tag{7}$$

or

$$b_k = -a - (1 - 2k/m)\, a = -2(1 - k/m)\, a \tag{8}$$

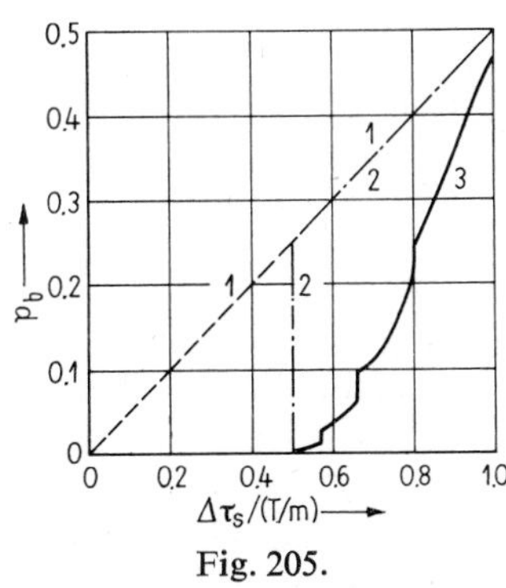

Fig. 205.

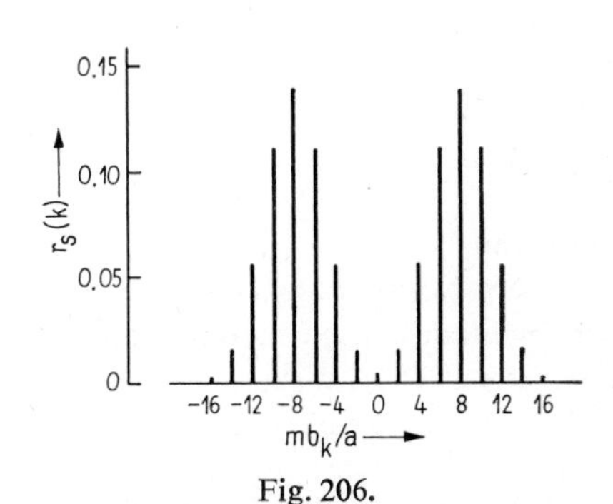

Fig. 206.

Fig. 205. Probability p_b of an error caused by a disturbing pulse of duration $\Delta\tau_s/(T/m)$. *1*, block pulses of Fig. 3, amplitude sampling; *2*, same block pulses, correlation; *3*, Walsh pulses, correlation.

Fig. 206. Probability $r_s(k)$ of the amplitudes b_k of disturbing pulses after amplitude limiting; signals consisting of $m = 8$ Walsh pulses.

The probability $r_s(k)$ of having an amplitude b_k between $-2a$ and $+2a$ follows from Eq. (6):

$$r_s(k) = \frac{1}{2}\, 2^{-m} \binom{m}{k} \tag{9}$$

An example of $r_s(k)$ is shown for $m = 8$ in Fig. 206. Positive as well as negative disturbing pulses have a Bernoulli distribution after amplitude limiting.

The cross-correlation of a binary signal $F_\chi(\theta)$ composed of Walsh pulses $\mathrm{wal}(j, \theta)$,

$$F_\chi(\theta) = \frac{a}{m} \sum_{j=0}^{m-1} a_\chi(j) \,\mathrm{wal}(j, \theta), \quad a_\chi(j) = \pm 1$$

yields:

$$\int_{-T/2}^{T/2} F_\chi(\theta) \,\mathrm{wal}(l, \theta)\, dt = \frac{aT}{m}, \quad \theta = \frac{t}{T}, \quad l = 0, \ldots, m-1 \tag{10}$$

According to Eqs. (7) and (8), the absolute value of the amplitude of the disturbing pulse cannot be larger than $2a$ and no error can occur if its duration $\Delta\tau_s$ is so small that the relation $2a\,\Delta\tau_s < a\,T/m$ holds. Hence, p_b is zero for

$$\Delta\tau_s < T/2m \tag{11}$$

This is the same value as for block pulses.

The calculation of the conditional error probability p_b is very tedious for longer disturbing pulses [5]. The result of the calculation is shown by curve *3* of Fig. 205. The error probability is somewhat lower than for block pulses and has several thresholds.

Better results are obtained if the signal composed of Walsh or sine-cosine pulses is amplitude limited at the transmitter as discussed in section 7.2.2, since the amplitude limiter at the receiver may then be set to lower levels.

Problem

723.1 Investigate errors caused by impulse disturbances on signals consisting of amplitude clipped superpositions of Walsh pulses according to problem 722.3.

7.3 Coding

7.3.1 Coding with Binary Elements

It has been discussed in section 3.1.1 that a signal may be represented by a time function $F_\chi(\theta)$, a vector $\mathbf{F}_\chi$ or a set of coefficients $a_\chi(j)$. A set of R different signals is called an alphabet. A certain function $F_\chi(\theta)$ is a character of the alphabet. Some problems of designing the characters will be discussed here for which orthogonality may be used to advantage.

A disturbance of a character may cause it to be mistaken for a different character at the receiver. A suitable choice of the R characters of an alphabet may reduce the probability of this happening for certain types of disturbances. Some methods for making a suitable choice will be investigated. Let the R characters be represented by m coefficients:

$$a_\chi(0), a_\chi(1), \ldots, a_\chi(m-1); \quad \chi = 1, \ldots, R \tag{1}$$

Such alphabets are called block codes. Using a system of functions $\{f(j, \theta)\}$, orthogonal in the interval $-\frac{1}{2} \leqq \theta \leqq \frac{1}{2}$, one obtains the representation by time functions:

$$F_\chi(\theta) = \sum_{j=0}^{m-1} a_\chi(j) f(j, \theta) \tag{2}$$

Generally, the signal at the input of a receiver may be a time dependent electric or magnetic field strength, in other instances a time dependent voltage or current. It appears reasonable to use the representation by time functions when looking for alphabets with low error probability. However, it has been shown in section 6.2.1 that the functions $f(j, \theta)$ are unimportant and the coefficients $a_\chi(j)$ alone decide the probability of error in case of additive thermal noise. Different systems of functions $\{f(j, \theta)\}$ require different frequency bandwidths for transmission and the practical difficulties for their generation and detection are different, but they do not influence the error rate. One may represent the characters by the coefficients $a_\chi(j)$ in this special case.

A further simplification is achieved by restricting the coefficients $a_\chi(j)$ to two values which are usually denoted by $+1$ and -1, or by 1 and 0. One often makes the additional assumption that a disturbance leaves a coefficient unchanged or changes it to the other permitted value[1]. This means a character with coefficients $a_\chi(0) = +1$ and $a_\chi(1) = +1$, written in short notation as the character $+1 + 1$, can be changed by a disturbance into one of the four forms $+1 + 1$, $+1 - 1$, $-1 + 1$ or $-1 - 1$ only and not, e.g., into $+\frac{1}{4} + \frac{1}{2}$. It has been shown in section 6.2.1 that the coefficient $a_\chi(j)$ is changed by a disturbance into $a(j)$ which may have any value even though $a_\chi(j)$ can be only $+1$ or -1. There are a number of reasons why only the values $+1$ and -1 are often permitted for $a(j)$. At the beginning of development of coding theory it was usually assumed that the functions $f(j, \theta)$ were block pulses and were decoded by amplitude sampling. A positive amplitude was interpreted as $+1$ and a negative one as -1. This quantization changes the sums in Eqs. (6.2.1-7) and (6.2.1-8) and generally increases the error rate.

Disturbances from sources other than additive thermal noise require, in principle, an investigation of their effects on the time functions $F_\chi(\theta)$ of $a_\chi(j) f(j, \theta)$ rather than on the coefficients $a_\chi(j)$. It has been shown in the previous section 7.2.3 that the effect of pulse-type disturbances depends on the shape of the functions $f(j, \theta)$ and on the clipping amplitude. Despite these results, it is customary to consider only the coefficients

[1] The characteristic function, which is the Fourier transform of the density function of a statistical variable, can be used to advantage in the investigation of sums of statistical variables. Pearl has shown that a dyadic characteristic function, which is the Walsh-Fourier transform of the density function, has very similar properties in the investigation of modulo 2 sums [12]. The type of error discussed here may be considered to be caused by the modulo 2 addition of a disturbing signal with values 0 or 1 to the transmitted signal with values 0 or 1.

of code alphabets that are designed for resistance to pulse-type disturbances, and to distinguish only between positive and negative values of the coefficients. The reasons for this are the requirement of simple implementation and the compatibility with existing equipment.

The theory of coding by binary elements is based on the assumption that the undisturbed coefficients $a_\chi(j)$ as well as the disturbed coefficients $a(j)$ can be $+1$ and -1, or 1 and 0, only. The coefficients $a_\chi(j)$ and $a(j)$ are usually called elements in this case. Alphabets consisting of characters with equal number of elements are called binary block-alphabets or binary block-codes. Since the undisturbed as well as the disturbed characters contain only the elements $+1$ and -1, one may consider them to be binary numbers. Number theory applied to binary numbers may then be used in the investigation of coding problems. Binary coding has been treated in a large number of publications starting with Hamming [1–5]. An excellent summary was given in a book by Peterson [6, 7]. Non-binary alphabets have also been investigated using number theory [8, 9].

The value of a code alphabet for communications depends on the error rate that can be achieved. Computation of this error rate is often very difficult. Hence, it is common to use the "Hamming distance" for judging the quality of an alphabet in the theory of coding by binary elements. It denotes the number of binary elements in which two characters differ. For instance, the characters $+1+1+1+1+1$ and $+1+1+1+1-1$ or 11111 and 11110 have the Hamming distance 1. The probability of decoding a disturbed character χ into the wrong character ψ often decreases with increasing Hamming distance between the two characters. Consider, for instance, characters consisting of a sequence of block pulses. The larger the number of pulses in which the characters differ, the larger may be the number of disturbed pulses without an error occurring. The Hamming distance is particularly useful if the peak power rather than the energy of the transmitted signal is limited.

The following example shows that a large Hamming distance does not necessarily mean a low probability of error. 2^m characters can be constructed from m elements $+1$ or -1. The smallest Hamming distance d between two characters is 1. One may increase d by constructing the 2^m characters from $m' > m$ elements. The energy of each transmitted character is increased by the factor m'/m if the energy per element is kept constant. The decrease of the error probability is partly derived from the construction of the characters and partly by their larger energy. It is often reasonable to base the comparison of two alphabets on equal energy of the characters or on equal average energy. A character with $m' > m$ elements must then contain m/m' times the energy per element. Hence, the Hamming distance is increased, but the probability of error for one element is also increased, e.g., if the errors are caused by additive thermal noise. It cannot be decided without calculation which effect dominates.

Alphabets with one parity check digit are an example of a reduction of the error rate under the influence of additive thermal noise by increasing the smallest Hamming distance. Consider the $2^m = 32$ characters of the teletype alphabet:

$$
\left.\begin{array}{lrrrrr}
1. & 1 & 1 & 1 & 1 & 1 \\
2. & 1 & 1 & 1 & 1 & -1 \\
3. & 1 & 1 & 1 & -1 & 1 \\
4. & 1 & 1 & 1 & -1 & -1 \\
 & \text{etc.} & & & &
\end{array}\right\} \tag{3}
$$

The smallest Hamming distance equals 1. Let a parity check digit $+1$ be added to all characters having an odd number of elements 1 and a check digit -1 to all characters with an even number of elements 1:

$$
\left.\begin{array}{lrrrrrr}
1. & 1 & 1 & 1 & 1 & 1 & 1 \\
2. & 1 & 1 & 1 & 1 & -1 & -1 \\
3. & 1 & 1 & 1 & -1 & 1 & -1 \\
4. & 1 & 1 & 1 & -1 & -1 & 1 \\
 & \text{etc.} & & & & &
\end{array}\right\} \tag{4}
$$

The smallest Hamming distance has thus been increased to 2. The energy per element must be reduced to 5/6 or generally to $m/(m+1)$. The factor $m/(m+1)$ approaches 1 for large values of m, while the Hamming distance is still doubled.

The smallest Hamming distance between the characters of an alphabet may be made 3, 4, ... or generally d, by adding sufficiently many check digits. These alphabets are called systematic alphabets. Making $d = 2l + 1$ one may decode all characters correctly, if no more than l elements have been reversed by disturbances. $d = 2l$ permits the correction of $l - 1$ reversals and the detection without correction of l reversals. Hence, one distinguishes between l-errors-correcting and l-errors-detecting alphabets. This distinction is necessary only if the disturbed coefficients $a(j)$ are limited to the values $+1$ or -1. According to Eqs. (6.2.1-7) and (6.2.1-8) the relation $\Delta W_\chi = \Delta W_\psi$ would have to hold in order to make an error detection without correction possible. The probability that ΔW_χ and ΔW_ψ are equal is zero if the disturbances are due to thermal noise. There are, however, disturbances for which this probability is not zero.

The 2^m characters constructed from m binary elements 1 and 0 form a group under addition modulo 2. Note that the Walsh functions have the same feature [11]. An alphabet is called a binary group alphabet or a binary group code if its characters are a subgroup of this group. A systematic group code is a systematic code whose characters form a group.

A special class of binary group codes are the Reed-Muller codes [5, 10]. Their characters contain m elements, m being a power of 2. The number of check elements is $m - k$ and the number of characters is 2^k; k has the value

$$k = \sum_{i=0}^{r} \binom{m}{i}, \qquad r < m \tag{5}$$

The smallest Hamming distance is $d = 2^{m-r}$.

Consider an example where $m = 2^4 = 16$ and $r = 1$ (Reed-Muller alphabet of first order). It follows $d = 2^{4-1} = 8$ and $k = 1 + 4 = 5$. This alphabet contains $2^5 = 32$ characters constructed from 16 elements, $16 - 5 = 11$ of which are check elements. This alphabet is denoted as (16, 5)-alphabet or generally as a (m, k)-alphabet. Table 20 shows the characters of this (16, 5)-alphabet with the elements represented by $+1$ and -1. Compare the signs of the elements of the first 16 characters with the Walsh functions of Fig. 2. The signs correspond to the positive and negative amplitudes of the Walsh functions. The signs of the elements of characters 17 through 32 are obtained by reversing the signs of the characters 16 through 1. One may thus construct a Reed-Muller alphabet with m characters as follows: The $\frac{1}{2}m$ Walsh functions $\mathrm{wal}(j, \theta), j = 0 \ldots \frac{1}{2}m - 1$, represented by $\frac{1}{2}m$ plus and minus signs yield one half of the characters; the other half of the characters are represented by the functions $-\mathrm{wal}(j, \theta)$. Thus the Reed-Muller alphabets belong to the class of orthogonal alphabets.

Problems

731.1 Indicate which houses of Diadicville in Fig. 49 have to be eliminated to obtain the smallest Hamming distance 2.

731.2 How many different answers to problem 731.1 exist?

731.3 What is changed if the check digits in Eq. (4) are not added at the end of the character but inserted between two columns?

7.3.2 Orthogonal, Transorthogonal and Biorthogonal Alphabets

To save space let us consider a (4, 3)-alphabet instead of a (16, 5)-alphabet. It contains $2^3 = 8$ characters. The first four are the first four Walsh functions of Fig. 2:

$$\left.\begin{array}{llrrrr} 1. & & +1 & +1 & +1 & +1 \\ 2. & & -1 & -1 & +1 & +1 \\ 3. & & -1 & +1 & +1 & -1 \\ 4. & & +1 & -1 & +1 & -1 \end{array}\right\} \tag{1}$$

The elements of Eq. (1) form a Hadamard matrix **X**. Interchanging the rows and columns yields the transposed matrix **X***:

$$\mathbf{X}^* = \begin{bmatrix} +1 & -1 & -1 & +1 \\ +1 & -1 & +1 & -1 \\ +1 & +1 & +1 & +1 \\ +1 & +1 & -1 & -1 \end{bmatrix} \tag{2}$$

The product **X X*** yields the unit matrix **1** multiplied by 4:

$$\begin{bmatrix} +1 & +1 & +1 & +1 \\ -1 & -1 & +1 & +1 \\ -1 & +1 & +1 & -1 \\ +1 & -1 & +1 & -1 \end{bmatrix} \begin{bmatrix} +1 & -1 & -1 & +1 \\ +1 & -1 & +1 & -1 \\ +1 & +1 & +1 & +1 \\ +1 & +1 & -1 & -1 \end{bmatrix} = \\ = 4 \begin{bmatrix} +1 & 0 & 0 & 0 \\ 0 & +1 & 0 & 0 \\ 0 & 0 & +1 & 0 \\ 0 & 0 & 0 & +1 \end{bmatrix} \tag{3}$$

A matrix is called orthogonal if its product with its transposed matrix yields the unit matrix multiplied by a constant. An alphabet is called orthogonal if its elements can be written as the elements of an orthogonal matrix. The alphabet (1) is an orthogonal alphabet; the characters 1 to 16 of Table 20 form an orthogonal alphabet, as do the characters 17 to 32.

Let us omit the third element of all characters in Eq. (1). An alphabet with three elements and four characters is obtained:

$$\begin{array}{ll} 1. & +1 \quad +1 \quad +1 \\ 2. & -1 \quad -1 \quad +1 \\ 3. & -1 \quad +1 \quad -1 \\ 4. & +1 \quad -1 \quad -1 \end{array} \tag{4}$$

The product of the matrix **Y** and the transposed matrix **Y***

$$\mathbf{Y} = \begin{bmatrix} +1 & +1 & +1 \\ -1 & -1 & +1 \\ -1 & +1 & -1 \\ +1 & -1 & -1 \end{bmatrix} \qquad \mathbf{Y}^* = \begin{bmatrix} +1 & -1 & -1 & +1 \\ +1 & -1 & +1 & -1 \\ +1 & +1 & -1 & -1 \end{bmatrix}$$

yields

$$\mathbf{Y}\,\mathbf{Y}^* = 3 \begin{bmatrix} +1 & -\frac{1}{3} & -\frac{1}{3} & -\frac{1}{3} \\ -\frac{1}{3} & +1 & -\frac{1}{3} & -\frac{1}{3} \\ -\frac{1}{3} & -\frac{1}{3} & +1 & -\frac{1}{3} \\ -\frac{1}{3} & -\frac{1}{3} & -\frac{1}{3} & +1 \end{bmatrix} \tag{5}$$

Table 20. *The coefficients $a_\chi(j)$ of the characters of a (16,5)-alphabet according to Reed-Muller.* $\chi = 1 \ldots 32,\ j = 0 \ldots 15$

χ	0	1	2	3	4	5	6	7	8	9	10	11	12	13	14	15
1	+1	+1	+1	+1	+1	+1	+1	+1	+1	+1	+1	+1	+1	+1	+1	+1
2	−1	−1	−1	−1	−1	−1	−1	−1	+1	+1	+1	+1	+1	+1	+1	+1
3	−1	−1	−1	−1	+1	+1	+1	+1	+1	+1	+1	+1	−1	−1	−1	−1
4	+1	+1	+1	+1	−1	−1	−1	−1	+1	+1	+1	+1	−1	−1	−1	−1
5	+1	+1	−1	−1	−1	−1	+1	+1	+1	+1	−1	−1	−1	−1	+1	+1
6	−1	−1	+1	+1	+1	+1	−1	−1	+1	+1	−1	−1	−1	−1	+1	+1
7	−1	−1	+1	+1	−1	−1	+1	+1	+1	+1	−1	−1	+1	+1	−1	−1
8	+1	+1	−1	−1	+1	+1	−1	−1	+1	+1	−1	−1	+1	+1	−1	−1
9	+1	−1	−1	+1	+1	−1	−1	+1	+1	−1	−1	+1	+1	−1	−1	+1
10	−1	+1	+1	−1	−1	+1	+1	−1	+1	−1	−1	+1	+1	−1	−1	+1
11	−1	+1	+1	−1	+1	−1	−1	+1	+1	−1	−1	+1	−1	+1	+1	−1
12	+1	−1	−1	+1	−1	+1	+1	−1	+1	−1	−1	+1	−1	+1	+1	−1
13	+1	−1	+1	−1	−1	+1	−1	+1	+1	−1	+1	−1	−1	+1	−1	+1
14	−1	+1	−1	+1	+1	−1	+1	−1	+1	−1	+1	−1	−1	+1	−1	+1
15	−1	+1	−1	+1	−1	+1	−1	+1	+1	−1	+1	−1	+1	−1	+1	−1
16	+1	−1	+1	−1	+1	−1	+1	−1	+1	−1	+1	−1	+1	−1	+1	−1
17	−1	+1	−1	+1	−1	+1	−1	+1	−1	+1	−1	+1	−1	+1	−1	+1
18	+1	−1	+1	−1	+1	−1	+1	−1	−1	+1	−1	+1	−1	+1	−1	+1
19	+1	−1	+1	−1	−1	+1	−1	+1	−1	+1	−1	+1	+1	−1	+1	−1
20	−1	+1	−1	+1	+1	−1	+1	−1	−1	+1	−1	+1	+1	−1	+1	−1
21	−1	+1	+1	−1	+1	−1	−1	+1	−1	+1	+1	−1	+1	−1	−1	+1
22	+1	−1	−1	+1	−1	+1	+1	−1	−1	+1	+1	−1	+1	−1	−1	+1
23	+1	−1	−1	+1	+1	−1	−1	+1	−1	+1	+1	−1	−1	+1	+1	−1
24	−1	+1	+1	−1	−1	+1	+1	−1	−1	+1	+1	−1	−1	+1	+1	−1
25	−1	−1	+1	+1	−1	−1	+1	+1	−1	−1	+1	+1	−1	−1	+1	+1
26	+1	+1	−1	−1	+1	+1	−1	−1	−1	−1	+1	+1	−1	−1	+1	+1
27	+1	+1	−1	−1	−1	−1	+1	+1	−1	−1	+1	+1	+1	+1	−1	−1
28	−1	−1	+1	+1	+1	+1	−1	−1	−1	−1	+1	+1	+1	+1	−1	−1
29	−1	−1	−1	−1	+1	+1	+1	+1	−1	−1	−1	−1	+1	+1	+1	+1
30	+1	+1	+1	+1	−1	−1	−1	−1	−1	−1	−1	−1	+1	+1	+1	+1
31	+1	+1	+1	+1	+1	+1	+1	+1	−1	−1	−1	−1	−1	−1	−1	−1
32	−1	−1	−1	−1	−1	−1	−1	−1	−1	−1	−1	−1	−1	−1	−1	−1

The difference between the elements on the principal diagonal and the others is larger for the matrix (5) than for the unit matrix (3). For this reason the alphabet (4) is called transorthogonal. The practical meaning of transorthogonality is evident from the alphabets (1) and (4). Both contain four characters, and the Hamming distance between any two characters equals 2. However, the alphabet (1) requires four elements and the alphabet (4) only three.

Let the characters of the alphabet (4) be represented by vectors:

1. $\mathbf{F}_0 = +\mathbf{e}_0 \quad +\mathbf{e}_1 \quad +\mathbf{e}_2$

2. $\mathbf{F}_1 = -\mathbf{e}_0 \quad -\mathbf{e}_1 \quad +\mathbf{e}_2$

3. $\mathbf{F}_2 = -\mathbf{e}_0 \quad +\mathbf{e}_1 \quad -\mathbf{e}_2$

4. $\mathbf{F}_3 = +\mathbf{e}_0 \quad -\mathbf{e}_1 \quad -\mathbf{e}_2$

The end points of these four vectors are the corners of a tetrahedron, as shown in Fig. 113a, if the origin of the coordinate system is placed at the center of the tetrahedron and the coordinate system is rotated into a proper position.

The terms off the principal diagonal of the matrix $\mathbf{Y}\mathbf{Y}^*$ are close to zero for transorthogonal alphabets with more than four characters.

Let the orthogonal alphabet (1) be supplemented by the characters obtained by changing the signs of the elements:

$$\left.\begin{array}{lrrrr} 5. & -1 & +1 & -1 & +1 \\ 6. & +1 & -1 & -1 & +1 \\ 7. & +1 & +1 & -1 & -1 \\ 8. & -1 & -1 & -1 & -1 \end{array}\right\} \tag{6}$$

The (4, 3)-alphabet consisting of the characters in Eqs. (1) and (6) is called biorthogonal. The (16, 5)-alphabet of Table 20 is also biorthogonal. Any character of a biorthogonal alphabet has the Hamming distance d from any other except for one which has the distance $2d$. An example of a biorthogonal alphabet that is not a Reed-Muller alphabet is the one shown by the octahedron in Fig. 113b.

Let the representation of characters by elements or coefficients be replaced by the representation by time functions. Consider a system of 16 orthogonal functions $f(j, \theta)$. Each function is multiplied by one of the 16 coefficients of a character in Table 20 and the products are added. If the functions $f(j, \theta)$ are block pulses, the first 16 characters are represented by the Walsh functions of Fig. 2, the second 16 characters by the same Walsh functions multiplied by -1.

Instead of multiplying the 16 block pulses by $+1$ or -1 and adding the products, one could just as well multiply one Walsh function by $+1$ or -1 and the other fifteen by 0 and add the products. The characters are then represented by the coefficients $+1$, -1 and 0 as shown in Table 21 where the first row lists the index j of $\mathrm{wal}(j, \theta)$ and the first column lists the number χ of the character. The functions $\mathrm{wal}(j, \theta)$ are multiplied by the coefficients $+1$, -1 or 0. The summation of the products is trivial since one product only is unequal zero for each character.

One has the curious result that the ternary alphabet of Table 21 and the binary alphabet of Table 20 yield the same signals. Both alphabets must have the same error rate under the influence of any kind of disturbance.

Instead of representing the characters of the ternary alphabet of Table 21 by 16 Walsh pulses, one may use the constant $f(0, \theta)$, 8 sine and 7 cosine pulses according to Fig. 9. The frequency power spectra of the first 5 pulses are shown in Fig. 47 by the curves a, b and c. The sixteenth character would be $F_{16}(\theta) = \sqrt{(2)}\sin(16\pi\,\theta + \frac{1}{4}\pi)$. Its power spectrum would be

Table 21. *The coefficients $a_\chi(j)$ of the characters of a ternary biorthogonal alphabet.* $\chi = 1 \ldots 32$, $j = 0 \ldots 15$

χ	0	1	2	3	4	5	6	7	8	9	10	11	12	13	14	15
1	+1	0	0	0	0	0	0	0	0	0	0	0	0	0	0	0
2	0	+1	0	0	0	0	0	0	0	0	0	0	0	0	0	0
3	0	0	+1	0	0	0	0	0	0	0	0	0	0	0	0	0
4	0	0	0	+1	0	0	0	0	0	0	0	0	0	0	0	0
5	0	0	0	0	+1	0	0	0	0	0	0	0	0	0	0	0
6	0	0	0	0	0	+1	0	0	0	0	0	0	0	0	0	0
7	0	0	0	0	0	0	+1	0	0	0	0	0	0	0	0	0
8	0	0	0	0	0	0	0	+1	0	0	0	0	0	0	0	0
9	0	0	0	0	0	0	0	0	+1	0	0	0	0	0	0	0
10	0	0	0	0	0	0	0	0	0	+1	0	0	0	0	0	0
11	0	0	0	0	0	0	0	0	0	0	+1	0	0	0	0	0
12	0	0	0	0	0	0	0	0	0	0	0	+1	0	0	0	0
13	0	0	0	0	0	0	0	0	0	0	0	0	+1	0	0	0
14	0	0	0	0	0	0	0	0	0	0	0	0	0	+1	0	0
15	0	0	0	0	0	0	0	0	0	0	0	0	0	0	+1	0
16	0	0	0	0	0	0	0	0	0	0	0	0	0	0	0	+1
17	0	0	0	0	0	0	0	0	0	0	0	0	0	0	0	−1
18	0	0	0	0	0	0	0	0	0	0	0	0	0	0	−1	0
19	0	0	0	0	0	0	0	0	0	0	0	0	0	−1	0	0
20	0	0	0	0	0	0	0	0	0	0	0	0	−1	0	0	0
21	0	0	0	0	0	0	0	0	0	0	0	−1	0	0	0	0
22	0	0	0	0	0	0	0	0	0	0	−1	0	0	0	0	0
23	0	0	0	0	0	0	0	0	0	−1	0	0	0	0	0	0
24	0	0	0	0	0	0	0	0	−1	0	0	0	0	0	0	0
25	0	0	0	0	0	0	0	−1	0	0	0	0	0	0	0	0
26	0	0	0	0	0	0	−1	0	0	0	0	0	0	0	0	0
27	0	0	0	0	0	−1	0	0	0	0	0	0	0	0	0	0
28	0	0	0	0	−1	0	0	0	0	0	0	0	0	0	0	0
29	0	0	0	−1	0	0	0	0	0	0	0	0	0	0	0	0
30	0	0	−1	0	0	0	0	0	0	0	0	0	0	0	0	0
31	0	−1	0	0	0	0	0	0	0	0	0	0	0	0	0	0
32	−1	0	0	0	0	0	0	0	0	0	0	0	0	0	0	0

centered at $\nu = 8$ in Fig. 47. Choosing $T = 150$ ms, which is a much used standard for teletype signals, one obtains the unnormalized frequencies shown there. The signal $F_{16}(\theta)$ would have its energy centered at about 53.33 Hz and there would be practically no energy above 60 Hz. One should not conclude from this narrow bandwidth that the alphabet of Table 21 is better than that of Table 20. One may multiply pulses according to Fig. 9 by the coefficients +1 and −1 of Table 20 and add the products. The resulting 32 signals have almost no energy above 60 Hz.

One may construct 2^{16} characters from 16 binary coefficients. The (16, 5)-alphabet of Table 20 uses 2^5 of them. It is usual to say that this alphabet contains 5 information digits and 11 check digits or – better – that each character contains 5 bits of information and 11 bits of redundancy. A total of 3^{16} characters may be constructed from 16 ternary coefficients. The alphabet of Table 21 uses 2^5 of them; one may assign the information 5 bits to each character. One will, however, be reluctant to assign the redundancy $\log_2(3^{16} - 2^5)$ to them. The concept of redundancy is useful, if alphabets of a certain order are considered. Without this restriction there is no reason why the characters of the (16, 5)-alphabet should not be considered to be derived from the r^{16} characters of an alphabet of order r rather than from the 2^{16} characters of an alphabet of order 2.

The concept of distance has also proven useful in the general theory of coding, no longer restricted to binary elements. For a generalization of the Hamming distance consider two characters represented by time functions $F_\chi(\theta)$ and $F_\psi(\theta)$ in the interval $-\frac{1}{2} \leqq \theta \leqq \frac{1}{2}$. The energy required to transform $F_\chi(\theta)$ into $F_\psi(\theta)$ is $W_{\chi\psi}$:

$$W_{\chi\psi} = \int_{-1/2}^{1/2} [F_\chi(\theta) - F_\psi(\theta)]^2 \, d\theta \tag{7}$$

The energy of the character $F_\chi(\theta)$ is W_χ:

$$W_\chi = \int_{-1/2}^{1/2} F_\chi^2(\theta) \, d\theta \tag{8}$$

The average energy of all R characters of an alphabet is W,

$$W = \sum_{\chi=1}^{R} p_\chi W_\chi \tag{9}$$

where p_χ is the probability of transmission of character χ.

The energy distance[1] $d_{\chi\psi}$ of the characters $F_\chi(\theta)$ and $F_\psi(\theta)$ is defined by normalization of the energy $W_{\chi\psi}$:

$$d_{\chi\psi} = W_{\chi\psi}/W \tag{10}$$

Let $F_\chi(\theta)$ and $F_\psi(\theta)$ be constructed from m orthogonal functions $f(j, \theta)$:

$$F_\chi(\theta) = \sum_{j=0}^{m-1} a_\chi(j) f(j, \theta), \quad F_\psi(\theta) = \sum_{j=0}^{m-1} a_\psi(j) f(j, \theta) \tag{11}$$

One obtains for $W_{\chi\psi}$ and W_χ:

$$W_{\chi\psi} = \sum_{j=0}^{m-1} [a_\chi(j) - a_\psi(j)]^2 \qquad W_\chi = \sum_{j=0}^{m-1} a_\chi^2(j) \tag{12}$$

Let all characters have the same energy $W = W_\chi$. It follows:

$$\frac{d_{\chi\psi}}{2} = \frac{W_{\chi\psi}}{2W} = 1 - \frac{1}{W} \int_{-1/2}^{1/2} F_\chi(\theta) F_\psi(\theta) \, d\theta = 1 - \frac{\sum_{j=0}^{m-1} a_\chi(j) \, a_\psi(j)}{\sum_{j=0}^{m-1} a_\chi^2(j)} \tag{13}$$

---

[1] The term "normalized nonsimilarity" has been used for energy distance if the integration interval is infinite [7].

It holds for the characters of Table 21:

$$\sum_{j=0}^{m-1} a_\chi^2(j) = \sum_{j=0}^{15} a_\chi^2(j) = 1$$

$$\sum_{j=0}^{m-1} a_\chi(j)\, a_\psi(j) = \sum_{j=0}^{15} a_\chi(j)\, a_\psi(j) = \begin{cases} -1 & \text{for} \quad \chi = 32 - \psi + 1 \\ +1 & \text{for} \quad \chi = \psi \\ 0 & \text{for} \quad \chi \neq \psi,\ 32 - \psi + 1 \end{cases}$$

The following energy distances are thus obtained for the characters of Table 21:

$$\left.\begin{aligned} d_{\chi\psi} &= 4 \quad \text{for} \quad \chi = 32 - \psi + 1 \\ &= 0 \quad \text{for} \quad \chi = \psi \\ &= 2 \quad \text{for} \quad \chi \neq \psi,\ 32 - \psi + 1 \end{aligned}\right\} \tag{14}$$

The characters of Table 20 yield:

$$\sum_{j=0}^{m-1} a_\chi^2(j) = \sum_{j=0}^{15} a_\chi^2(j) = 16 \tag{15}$$

$$\sum_{j=0}^{m-1} a_\chi(j)\, a_\psi(j) = \sum_{j=0}^{15} a_\chi(j)\, a_\psi(j) = \begin{cases} -16 & \text{for} \quad \chi = 32 - \psi + 1 \\ +16 & \text{for} \quad \chi = \psi \\ 0 & \text{for} \quad \chi \neq \psi,\ 32 - \psi + 1 \end{cases}$$

$$\begin{aligned} d_{\chi\psi} &= 4 \quad \text{for} \quad \chi = 32 - \psi + 1 \\ &= 0 \quad \text{for} \quad \chi = \psi \\ &= 2 \quad \text{for} \quad \chi \neq \psi,\ 32 - \psi + 1 \end{aligned} \tag{16}$$

The distances $d_{\chi\psi}$ of the characters of Table 20 would have the values 16, 0 or 8, if $W_{\chi\psi}$ in Eq. (10) were divided by $W/\log_2 m = W/4$ rather than by W. This is just the number of elements in which the characters differ, i.e., their Hamming distance.

The energy distance $d_{\chi\psi}$ of two characters is equal to the square of the vector connecting their signal points in signal space. These vectors are represented by the rods between the signal points in Fig. 113. The term distance has an evident meaning in the vector representation. Due to the normalization of $d_{\chi\psi}$ one must require, for the vector representation, that the signal points have the average distance 1 from their common center of gravity.

Let the R characters of a biorthogonal alphabet be listed in such a sequence that the relation

$$F_\chi(\theta) = -F_{R-\chi+1}(\theta) \tag{17}$$

is satisfied. It follows:

$$\frac{1}{W} \int_{-1/2}^{1/2} F_\chi(\theta)\, F_\psi(\theta)\, d\theta = \begin{cases} 1 & \text{for} \quad \chi = \psi \\ -1 & \text{for} \quad \chi = R - \psi + 1 \\ 0 & \text{for} \quad \chi \neq \psi,\ R - \psi + 1 \end{cases} \tag{18}$$

It follows from Eq. (13) that the character χ of a biorthogonal alphabet has an energy distance 4 from the character $R - \chi + 1$ and an energy distance 2 from all other characters; $\chi = 1, \ldots, R$.

Problems

732.1 Find a transorthogonal alphabet with 8 characters based on a Hadamard matrix.

732.2 Using the result of problem 732.1, for which number of characters can you readily find transorthogonal alphabets?

732.3 Find a biorthogonal alphabet with 8 characters based on a Hadamard matrix but different from Eqs. (1) and (6).

732.4 Give three examples of orthogonal alphabets with 16 characters based on Hadamard matrices.

7.3.3 Coding for Error-Free Transmission

Shannon's formula for the transmission capacity of a communication channel proves that an error-free transmission is possible as a limiting case. From the derivation of that formula in section 7.1.2 it is evident how alphabets may be obtained which approach the transmission capacity of the channel and which have vanishing error rates in the presence of additive thermal noise.

Consider a system of Fourier expandable orthogonal functions $f(j, \theta)$ in the interval $-\frac{1}{2} \leqq \theta \leqq \frac{1}{2}$. Random numbers $a_0(j)$ with a Gaussian distribution are taken from a table and the character $F_0(\theta)$ is constructed [1–3]:

$$F_0(\theta) = \sum_{j=0}^{m-1} a_0(j) f(j, \theta) \tag{1}$$

One may assume that the numbers $a_0(j)$ represent voltages. $F_0(\theta)$ is then a time variable voltage. $F_0(\theta)$ cannot be distinguished from a sample of thermal noise if m grows beyond all bounds.

Using another set of m random numbers $a_1(j)$, one may construct a second character $F_1(\theta)$. The general character $F_\chi(\theta)$ can be constructed by means of m Gaussian distributed random numbers $a_\chi(j)$. The unnormalized duration of these characters equals T. The transmission capacity of the channel of Eq. (7.1.2-37) follows from m, T and the average signal-to-noise power ratio $P/P_{m,T}$:

$$C = \frac{m}{2T} \log_2 \left(1 + \frac{P}{P_{m,T}}\right) \tag{2}$$

Let n be the largest integer smaller than 2^{CT} and let n characters $F_\chi(\theta)$ be constructed:

$$\chi = 0, 1, \ldots, n - 1 \tag{3}$$

These n characters form the first alphabet. Now let L alphabets with n characters each be constructed in this way and pick one alphabet at random. If n and L approach infinity, the probability is arbitrarily close to 1 that this alphabet yields an error rate approaching zero.

These "random alphabets" are very satisfying from the theoretical point of view. There are, however, practical drawbacks. It is not only interesting to see how good the alphabet is in the limit, but what the probability of error is for a finite amount of information per character. Elias found the first non-random alphabet approaching the error probability zero for finite energy per bit of information [4, 5]. The transmission rate of information was, however, much smaller than Shannon's limit. The so-called combination alphabets also yield vanishing error probabilities and come very close to Shannon's limit.

Problem

733.1 Random numbers are usually generated with equal distribution, e.g., by drawing numbers in a lottery or by computing irrational numbers. How are random numbers with Gaussian distribution obtained from them?

7.3.4 Ternary Combination Alphabets

m orthogonal functions $f(j, \theta)$ can transmit m coefficients $a_\chi(j)$. A total of $R = 3^m$ characters can be constructed if $a_\chi(j)$ may assume the three values $+1$, 0 and -1. Writing $(1 + 2)^m$ instead of 3^m yields the following expansion:

$$R = (1 + 2)^m = 2^0\binom{m}{0} + 2^1\binom{m}{1} + \cdots + 2^h\binom{m}{h} + \cdots + 2^m\binom{m}{m} \quad (1)$$

This decomposition divides the set of R characters into subsets of characters containing equally many functions $f(j, \theta)$. There is $1 = 2^0\binom{m}{0}$ character containing no function, because all coefficients $a_\chi(j)$ are zero. Furthermore, there are $2^1\binom{m}{1} = 2m$ characters, consisting of one function each, because only one coefficient $a_\chi(j)$ equals $+1$ or -1. These characters form the biorthogonal alphabets. In general, there are $2^h\binom{m}{h}$ characters, each containing h functions $a_\chi(j)f(j, \theta)$, where $a_\chi(j)$ equals $+1$ or -1. Since $\binom{m}{h}$ is the number of combinations of h out of m functions, these alphabets are called ternary combination alphabets for $h \neq 0$, 1 or m. Table 22 shows the number $2\binom{m}{h}$ of characters in such alphabets.

Table 22. *Number of characters in ternary combination alphabets. According to Kasack* [2], *the numbers above the line drawn through the table belong to "good" alphabets*

h \ m	2	3	4	5	6	7	8	9	10
1	4	6	8	10	12	14	16	18	20
2	4	12	24	40	60	84	112	144	180
3		8	32	80	160	280	448	672	960
4			16	80	240	560	1,120	2,016	3,360
5				32	192	672	1,792	4,032	8,064
6					64	448	1,792	5,376	13,440
7						128	1,024	4,608	15,360
8							256	2,304	11,520
9								512	5,120
10									1,024

Equation (1) yields, for $h = m$, the $2^m\binom{m}{m} = 2^m$ characters that contain all m functions $a_\chi(j)f(j,\theta)$ with $a_\chi(j)$ equal $+1$ or -1. These are the characters of the binary alphabets.

Consider an alphabet with characters containing h functions $f(j,\theta)$. Each character contains h coefficients $a_\chi(j)$ equal to $+a_0$ or $-a_0$ and $m-h$ coefficients equal to zero. Let these characters be transmitted. Cross-correlation of the received signal with the functions $f(j,\theta)$ yields the coefficients $a_\chi(j)$. Let additive thermal noise be superimposed on the signal. The coefficients $a(j)$ are obtained, which have a Gaussian distribution with a mean either $+a_0$, 0 or $-a_0$, denoted by $a^{(+1)}(j)$, $a^{(0)}(j)$ and $a^{(-1)}(j)$:

$$\left\langle \frac{a^{(+1)}(j)}{a_0} \right\rangle = \langle a^{(+1)} \rangle = +1, \quad \left\langle \frac{a^{(-1)}(j)}{a_0} \right\rangle = \langle a^{(-1)} \rangle = -1$$
$$\left\langle \frac{a^{(0)}(j)}{a_0} \right\rangle = \langle a^{(0)} \rangle = 0 \tag{2}$$

The variance σ^2 of these distributions follows from Eqs. (7.2.1-7) to (7.2.1-10):

$$\sigma^2 = \left\langle \frac{a_\lambda^2(j)}{a_0^2} \right\rangle = \frac{P_{1,T}}{P/h} = \frac{P_{h,T}}{P} = \frac{h\,P_{\Delta f}}{n\,P} \tag{3}$$

h number of coefficients $a_\chi(j)$ with value $+a_0$ or $-a_0$;

$n = \log_2 2^h\binom{m}{h}$ information per character in bits, if all characters are transmitted with equal probability;

$P_{n,T}$ average power of n orthogonal components of thermal noise in an orthogonality interval of duration T;

$P = h\,a_0^2$ average signal power; $\Delta f = n/2T$;

$P_{\Delta f}$ average power of thermal noise in a frequency band of width Δf.

The average noise power $P_{n,T}$ rather than $P_{h,T}$ or $P_{m,T}$ is used as a reference in order to facilitate comparison between binary and ternary alphabets.

The $2^m\binom{m}{h}$ sums

$$S_\psi = \sum_{j=0}^{m-1} a(j)\, a_\psi(j) \tag{4}$$

must be produced from the m coefficients $a(j)$ received, and the largest one must be determined for decoding according to Eq. (6.2.1-8). $m - h$ of the coefficients $a_\psi(j)$ are 0 for any ψ. Consider those sums for which certain coefficients $a_\psi(j)$ are 0, for instance those for $j = 0, \ldots, m - h - 1$. The remaining h coefficients $a_\psi(j)$ equal $+a_0$ or $-a_0$ and yield 2^h different sums S_ψ. The largest of these 2^h sums will contain h positive terms $a(j)\, a_\psi(j)$, while the remaining $m - h$ terms are 0. The largest of all $2^h\binom{m}{h}$ sums S_ψ will be the sum whose non-vanishing terms contain the h coefficients $a(j)$ with the largest magnitude. The sum will be largest for the transmitted character $F_\chi(\theta)$ when the absolute value of the h coefficients $a^{(+1)}(j)$ and $a^{(-1)}(j)$ is larger than that of the $m - h$ coefficients $a^{(0)}(j)$, and if in addition $a^{(+1)}(j)$ is larger and $a^{(-1)}(j)$ is smaller than zero. Hence, the following two conditions must be satisfied for error-free decoding (*see* Fig. 207):

1. All coefficients $a^{(+1)} = a^{(+1)}(j)/a_0$ and $-a^{(-1)} = -a^{(-1)}(j)/a_0$ are non-negative:

$$0 \leqq a^{(+1)}, -a^{(-1)} < \infty \tag{5}$$

2. None of the h coefficients $+a^{(+1)}$ and $-a^{(-1)}$ is smaller than the absolute value of one of the $m - h$ coefficients $a^{(0)} = a^{(0)}(j)/a_0$. This condition needs to be satisfied only if condition 1 is satisfied:

$$\left.\begin{aligned} +a^{(+1)} - |a^{(0)}| &\geqq 0 \\ -a^{(-1)} - |a^{(0)}| &\geqq 0 \end{aligned}\right\} \quad \text{for} \quad 0 \leqq a^{(+1)}, -a^{(-1)} < \infty \tag{6}$$

The density functions $w_1(x)$ of $a^{(+1)}$ and $w_2(y)$ of $|a^{(0)}|$ are given by:

$$w_1(x) = \frac{1}{\sqrt{(2\pi)}\,\sigma} \exp\left(\frac{-(x-1)^2}{2\sigma^2}\right) \qquad -\infty < x < \infty$$

$$w_2(y) = \frac{2}{\sqrt{(2\pi)}\,\sigma} \exp\left(\frac{-y^2}{2\sigma^2}\right) \qquad 0 \leqq y < \infty$$

$$w_2(y) = 0 \qquad y < 0$$

The probability of $p(a^{(+1)} < 0) = W_1(0)$ of the condition (5) not being satisfied equals:

$$p(a^{(+1)} < 0) = W_1(0) = \frac{1}{\sqrt{(2\pi)}\,\sigma} \int_{-\infty}^{0} \exp\left(\frac{-(x-1)^2}{2\sigma^2}\right) dx$$
$$= \frac{1}{2}\left[1 - \operatorname{erf}\left(\frac{1}{\sqrt{(2)}\,\sigma}\right)\right] \tag{7}$$

The probability $p(-a^{(-1)} < 0)$ has the same value:

$$p(-a^{(-1)} < 0) = \tfrac{1}{2}[1 - \operatorname{erf}(1/\sqrt{(2)}\,\sigma)] \tag{8}$$

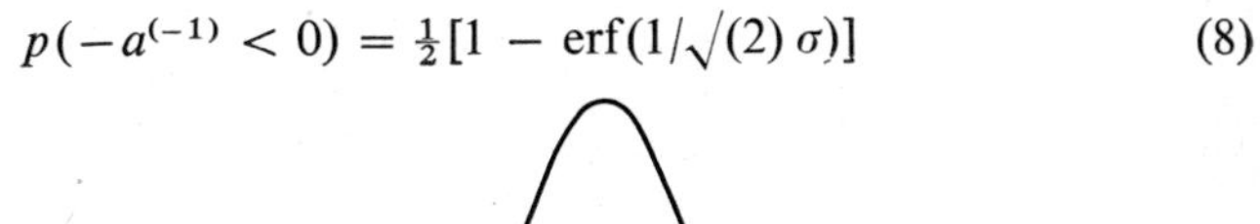

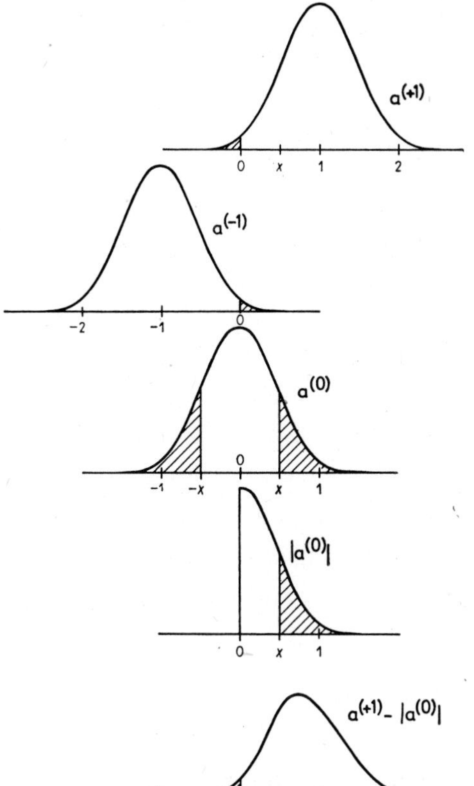

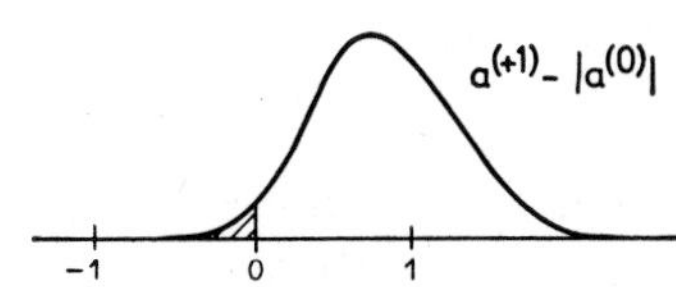

Fig. 207. Density functions of $a^{(+1)}$, $a^{(-1)}$, $a^{(0)}$, $|a^{(0)}|$ and $a^{(+1)} - |a^{(0)}|$ for a ternary combination alphabet. The hatched areas indicate errors.

$p_h^{(3)}$ denotes the probability that the condition (5) is not satisfied for at least one of the h coefficients $a^{(+1)}$ and $a^{(-1)}$:

$$p_h^{(3)} = 1 - [1 - W_1(0)]^h = 1 - 2^{-h}[1 + \operatorname{erf}(1/\sqrt{(2)}\,\sigma)]^h \tag{9}$$

Consider the distribution of $a^{(+1)} - |a^{(0)}|$, $0 < a^{(+1)} < \infty$. Its density function is given by[1]:

$$w(z) = \frac{1}{\pi C \sigma^2} \int_{-z}^{\infty} \exp\left(-\frac{(z+y-1)^2}{2\sigma^2}\right) \exp\left(\frac{-y^2}{2\sigma^2}\right) dy$$

$$= \frac{1}{2\sqrt{(\pi)}\, C \sigma} \exp\left(-\frac{(z-1)^2}{4\sigma^2}\right) \left[1 - \operatorname{erf}\left(\frac{-z-1}{2\sigma}\right)\right],$$

$$z = x - y \leqq 0$$

$$C = \tfrac{1}{2}[1 + \operatorname{erf}(1/\sqrt{(2)}\,\sigma)]$$

The probability that the condition (6) is not satisfied for a certain one of the $h(m-h)$ differences $a^{(+1)} - |a^{(0)}|$ and $-a^{(-1)} - |a^{(0)}|$ is

$$\frac{p(a^{(+1)} - |a^{(0)}| < 0)}{p(a^{(+1)} > 0)} = W(0) = \int_{-\infty}^{0} w(z)\, dz$$

$$= \frac{1}{2\sqrt{(\pi)}\, C\sigma} \int_{-\infty}^{0} \exp\left(-\frac{(z-1)^2}{4\sigma^2}\right) \left[1 - \operatorname{erf}\left(\frac{-z-1}{2\sigma}\right)\right] dz$$

$$= \frac{1 + 2\operatorname{erf}(1/\sqrt{(2)}\,\sigma) - 2\operatorname{erf}(1/2\sigma) - \operatorname{erf}^2(1/2\sigma)}{2[1 + \operatorname{erf}(1/\sqrt{(2)}\,\sigma)]} \tag{10}$$

$p^{(3)}_{h,m-h}$ denotes the probability that the condition (6) is not satisfied for all $h(m-h)$ differences $a^{(+1)} - |a^{(0)}|$ and $-a^{(-1)} - |a^{(0)}|$:

$$p^{(3)}_{h,m-h} = 1 - [1 - W(0)]^{h(m-h)} \tag{11}$$

Equations (9) and (11) yield the error probability $p^{(3)}_{m,h}$ of ternary combination alphabets, biorthogonal alphabets and binary (m, m)-alphabets:

$$p^{(3)}_{m,h} = 1 - (1 - p^{(3)}_h)(1 - p^{(3)}_{h,m-h})$$

$$= 1 - 2^{-h}\left[1 + \operatorname{erf}\left(\frac{1}{\sqrt{(2)}\,\sigma}\right)\right]^h \left\{\frac{1}{2}\,\frac{[1 + \operatorname{erf}(1/2\sigma)]^2}{1 + \operatorname{erf}(1/\sqrt{(2)}\sigma)}\right\}^{h(m-h)} \tag{12}$$

$$\sigma^2 = h P_{\Delta f}/n P$$

Equation (12) yields for $h = m$ the error probability of a binary alphabet with m coefficients $a_\chi(j)$ and 2^m characters, which is the same as Eq. (7.2.1-13):

$$p^{(3)}_{m,m} = 1 - (1 - p^{(3)}_m)(1 - p^{(3)}_{m,0}) = 1 - \left(\frac{1}{2}\right)^m \left[1 + \operatorname{erf}\left(\frac{1}{\sqrt{(2)}\,\sigma}\right)\right]^m$$

$$\sigma^2 = \frac{m P_{\Delta f}}{n P} = \frac{P_{\Delta f}}{P}, \quad n = \log_2 2^m \binom{m}{m} = m \tag{13}$$

[1] *See* pages 198 and 199 of [3].

The error probability of biorthogonal alphabets follows for $h = 1$:

$$p_{m,1}^{(3)} = 1 - (1 - p_1^{(3)})(1 - p_{1,m-1}^{(3)})$$
$$= 1 - \frac{1}{2}\left[1 + \operatorname{erf}\left(\frac{1}{\sqrt{(2)}\,\sigma}\right)\right]\left(\frac{1}{2}\,\frac{[1 + \operatorname{erf}(1/2\sigma)]^2}{1 + \operatorname{erf}(1/\sqrt{(2)}\,\sigma)}\right)^{m-1} \tag{14}$$
$$\sigma^2 = \frac{P_{1,T}}{P} = \frac{P_{\Delta f}}{n\,P}, \qquad n = \log_2 2^1 \binom{m}{1} = \log_2 2m$$

Figure 208 shows the error probability for some biorthogonal alphabets. The error probability of the binary (5, 5)-alphabet (curve $n = 5$, $m = 5$) and the (16, 16)-alphabet (curve $n = 16$, $m = 16$) are shown for comparison. The curve $n = 5$, $m = 5$ is the same as curve a in Fig. 202.

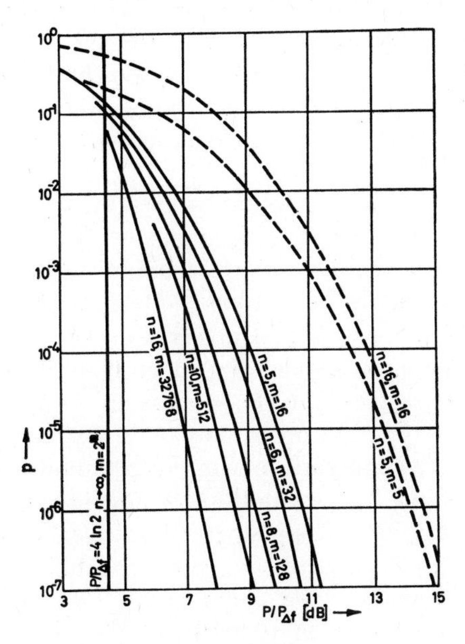

Fig. 208. Error probability p of biorthogonal alphabets. P, average signal power; $P_{\Delta f}$, average power of thermal noise in a frequency band of width $\Delta f = n/2T$; n, information of the characters in bit; T, duration of the characters; m, number of orthogonal functions in the alphabet. Solid lines: biorthogonal alphabets; dashed lines: binary alphabets (5, 5) and (16, 16).

The choice of the average signal-to-noise power ratio plotted along the abscissa requires explanation. The meaning of the average signal power P is evident. The average noise power $P_{1,T}$ of one orthogonal component of thermal noise in an orthogonality interval of duration T is used as reference. Plotting $P/P_{1,T}$ would give a false impression, since the characters of the various alphabets transmit different amounts of information. It is better to use P/n, the average signal power per bit of information, rather than P. This gives $P/P_{\Delta f}$, which is used in Fig. 208:

$$(P/n)/P_{1,T} = P/n\,P_{1,T} = P/P_{n,T} = P/P_{\Delta f}; \quad \Delta f = n/2T \tag{15}$$

Consider the transmission of characters with $n = 5$ bits of information with an error probability of 10^{-3}. According to Fig. 208, the binary alphabet ($n = 5$, $m = 5$) requires a ratio $P/P_{\Delta f}$ of 11 dB and the biorthogonal alphabet ($n = 5$, $m = 16$) one of 8 dB. Hence, the biorthogonal alphabet requires $11 - 8 = 3$ dB less signal power. The price paid for this gain is an increase in the number of orthogonal functions required from $m = 5$ to $m = 16$; a 16/5-times larger section of the time-function-domain or, somewhat less precise, a 16/5-times wider frequency band is required. Consider further the transmission of characters with $n = 16$ bits of information with an error probability of 10^{-3}. The binary alphabet ($n = 16$, $m = 16$) requires a ratio $P/P_{\Delta f}$ of 11.7 dB; the biorthogonal alphabet ($n = 16$, $m = 32768$) one of 5.8 dB. Thus the biorthogonal alphabet requires only about one quarter of the signal power of the binary alphabet ($11.7 - 5.8 = 5.9$ dB). The number of functions required increases, however, from 16 to 32768.

The smaller the required error probability the more justified is the use of a biorthogonal alphabet. For example, it requires a ratio $P/P_{\Delta f}$ of 11.2 dB for an error probability of 10^{-7} (curve $n = 5$, $m = 16$ in Fig. 208), while the binary alphabet ($n = 5$, $m = 5$) requires a ratio of 14.8 dB, a possible reduction of the signal power by $14.8 - 11.2 = 3.6$ dB. This same difference amounts to somewhat more than 7 dB for the alphabets $n = 16$, $m = 32768$ and $n = 16$, $m = 16$ at an error probability of 10^{-7}.

Figure 209 shows the error probability of ternary combination alphabets according to Eq. (12). A comparison with Fig. 208 shows that these particular ones need a larger ratio $P/P_{\Delta f}$ than the biorthogonal alphabet but a smaller one than the binary (m, m)-alphabets. For instance, the biorthogonal alphabet $n = 10$, $m = 512$ yields an error probability of 10^{-5} for a ratio $P/P_{\Delta f}$ of 8 dB; the comparable combination alphabet $n = 9.9$, $m = 8$, $h = 3$ requires a ratio of about 10.5 dB.

Figure 209 shows that there are alphabets which transmit more information n with the same number m of functions than the binary (m, m)-alphabet and nevertheless yield a lower error probability. These alphabets do more than exchange "more functions" for "less signal power". Consider the curve $n = 8.8$, $m = 8$, $h = 3$. A character of a binary $(8, 8)$-alphabet transmits with $m = 8$ functions the information $n = 8$ bits, which is less than the $n = 8.8$ bits of the ternary combination alphabet that requires also $m = 8$ functions. The error probability of the binary $(8, 8)$-alphabet is represented by a curve that lies between the curves $n = 5$, $m = 5$ and $n = 16$, $m = 16$ in Fig. 209. This curve is about 3 dB to the right of the curve $n = 8.8$, $m = 8$, $h = 3$ for error probabilities between 10^{-4} and 10^{-7}.

Consider the error probability $p^{(3)}_{m,h}$ of Eq. (12) for large values of m and n. Using the approximations

$$\operatorname{erf}(x) \cong 1 - \frac{1}{\sqrt{(\pi)}\,x}\, e^{-x^2}, \quad x \gg 1, \quad \text{and} \quad 1 - y \doteqdot e^{-y}, \quad y \ll 1,$$

one obtains:

$$p^{(3)}_{m,h} \cong 1 - \exp(-\xi\, e^{n\eta}), \quad m \gg 1$$

$$\xi = \frac{2}{\sqrt{(\pi)}} \sqrt{\frac{P_{\Delta f}}{P}}, \quad \eta = \frac{1}{n} \ln\left(h(m-h)\sqrt{\frac{h}{n}}\right) - \frac{P}{4h\,P_{\Delta f}} \tag{16}$$

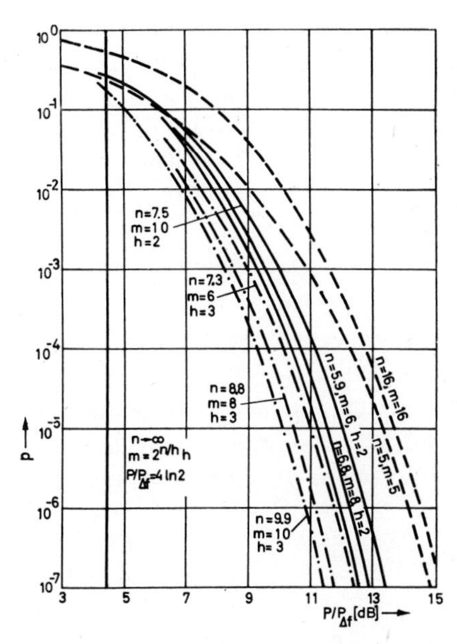

Fig. 209. Error probability p of ternary combination alphabets; P, $P_{\Delta f}$, n and m defined in the caption of Fig. 208. h, number of orthogonal functions in a character. Dashed lines show the error probabilities of the binary alphabets (5, 5) and (16, 16).

Let n and m approach infinity:

$$\lim_{m\to\infty} p^{(3)}_{m,h} = 1 \quad \text{for} \quad \eta > 0$$

$$\lim_{m\to\infty} p^{(3)}_{m,h} = 0 \quad \text{for} \quad \eta < 0 \tag{17}$$

Thus, error-free transmission is achieved in the limit $n = \infty$ for $\eta < 0$. Using the relation

$$n = \log_2 2^h \binom{m}{h} \cong h \log_2 \frac{m}{h}, \quad m \gg h \tag{18}$$

one may transform the condition $\eta < 0$ into the following condition, holding for a constant value of h:

$$P/P_{\Delta f} > 4\ln 2 \tag{19}$$

A ratio $P/P_{\Delta f}$ larger than $4\ln 2$ yields error-free transmission for infinitely large m and n, and finite h; the error probability is 1 if $P/P_{\Delta f}$ is smaller than $4\ln 2$. The limit function $\lim p^{(3)}_{m,h}$ for $m \to \infty$, $n \to \infty$ is shown in Figs. 208 and 209. This limit is the same for the biorthogonal alphabets ($h = 1$) and the combination alphabets ($h > 1$). Hence, the combination alphabets are the superior ones from the standpoint of functions or bandwidth required, since the number m of functions required increases proportionate to 2^n for biorthogonal alphabets but only proportionate to $2^{n/h}h$ for combination alphabets.

Let h not remain constant as m and n approach infinity, but let it increase proportionate to m^α, $0 \leqq \alpha < 1$. The condition $m \gg h$ of Eq. (18) is still satisfied for large m. The condition $\eta < 0$ then yields the following condition in place of Eq. (19):

$$\frac{P}{P_{\Delta f}} > 4\frac{1+\alpha}{1-\alpha}\ln 2, \qquad h = m^\alpha, \qquad 0 \leqq \alpha < 1 \tag{20}$$

Error-free transmission is possible if α is smaller than 1.

Let us investigate how Shannon's limit of the transmission capacity in the form of Eq. (7.1.2-37) is approached by ternary combination alphabets. The average noise power $P_{\Delta f}$ must be replaced by $P_{m,T}$ in Eq. (16). It follows:

$$P_{\Delta f} = P_{n,T} = n\,P_{1,T} = \frac{n}{m}\,P_{m,T} \tag{21}$$

The condition $\eta < 0$ becomes:

$$\frac{1}{n}\ln\left[h(m-h)\sqrt{\frac{h}{n}}\right] < \frac{m}{n\,h\,P_{m,T}} \tag{22}$$

The approximation $n = \log_2 2\binom{m}{h} \cong h\log_2\frac{m}{h}$, $m \gg h$, is substituted on the left side and the terms are reordered:

$$n < \frac{m}{4\ln 2}\left[\frac{\ln\dfrac{m}{h}}{\ln m\,h - \dfrac{1}{2}\ln\left(\log_2\dfrac{m}{h}\right)}\right]\frac{P}{P_{m,T}} \tag{23}$$

The term in the brackets becomes 1 when m becomes infinite and h remains finite; it becomes $(1-\alpha)/(1+\alpha)$ for $h = m^\alpha$, $0 \leqq \alpha < 1$. The information transmitted error-free per unit time is equal to n/T since n is the information of each character transmitted during an orthogonality interval of duration T:

$$C = \frac{n}{T} < \frac{1}{2} \frac{m}{2T} \frac{1}{\ln 2} \frac{P}{P_{m,T}}, \quad h = \text{constant} \tag{24}$$

$$C = \frac{n}{T} < \frac{1}{2} \frac{1-\alpha}{1+\alpha} \frac{m}{2T} \frac{1}{\ln 2} \frac{P}{P_{m,T}}, \quad h = m^{\alpha}, \quad 0 \leqq \alpha < 1 \tag{25}$$

The logarithm to the base 2 must be used in Eq. (7.1.2-37) if the transmission capacity is to be obtained in bits per unit time:

$$C = \frac{m}{2T} \log_2 \left(1 + \frac{P}{P_{m,T}}\right) \doteqdot \frac{m}{2T} \frac{1}{\ln 2} \frac{P}{P_{m,T}}, \quad \frac{P}{P_{m,T}} \ll 1 \tag{26}$$

The right side of Eq. (26) is larger by a factor 2 or $2(1+\alpha)/(1-\alpha)$ than the right hand sides of Eqs. (24) and (25). Hence, a ternary combination alphabet with h = constant transmits half as much information error-free as permitted by Shannon's limit, provided the signal-to-noise power ratio $P/P_{m,T}$ is small. The physical meaning of the condition $P/P_{m,T} \ll 1$ is evident; the coefficients of a ternary alphabet have only the three values $+1$, 0 and -1. An increase of the average signal-to-noise power ratio $P/P_{m,T}$ is worthless once the error probability has reached zero. Use could be made of an increased ratio $P/P_{m,T}$ only if the coefficients could assume more than the three values $+1$, 0 and -1. The ternary combination alphabets must be replaced by alphabets of higher order.

A more detailed investigation of ternary combination alphabets was published by Kasack [2].

Problems

734.1 Compute some curves with $m = 16$ and various values of h according to Fig. 209. What computational difficulty arises for large values of m?

734.2 Investigate the rate of approach of the curves in Figs. 208 and 209 to the theoretical limit for various values of h.

734.3 The curves of Figs. 208 and 209 show that coding becomes increasingly worthwhile with decreasing error probability. Which communication links exist with a requirement for error probability in the order of 10^{-6} and errors mainly caused by thermal noise? Which links may be anticipated?

7.3.5 Combination Alphabets of Order $2r + 1$

Let characters $F_\chi(\theta)$ be composed of m orthogonal functions $f(j, \theta)$, $-\frac{1}{2} \leqq \theta \leqq \frac{1}{2}$, multiplied by coefficients $a_\chi(j)$. These coefficients may assume $2r + 1$ values rather than 3 as for ternary alphabets. A total of $(1 + 2r)^m$ characters $F_\chi(\theta)$ can be produced. Let $(1 + 2r)^m$ be expanded in a binomial series:

$$(1 + 2r)^m = 1 + 2r\binom{1}{m} + \cdots + (2r)^h\binom{m}{h} + \cdots + (2r)^m\binom{m}{m} \tag{1}$$

$(2r)^h\binom{m}{h}$ is the number of characters in the alphabet containing h of the m functions $f(j, \theta)$. This means that h of the coefficients $a_\chi(j)$ are non-zero; χ runs from 1 to $(2r)^h\binom{m}{h}$. These characters form a combination alphabet of order $2r + 1$. Let all these characters be transmitted with equal probability. The information per character in bits equals:

$$n = \log_2 (2r)^h \binom{m}{h} \cong h \log_2 \frac{2r\,m}{h}, \quad m \gg h \tag{2}$$

Each of the h coefficients $a_\chi(j) \neq 0$ may assume $2r$ values. They are denoted by a_ϱ, $\varrho = \pm 1, \ldots, \pm r$. The probability of a coefficient $a_\chi(j)$ assuming the value a_ϱ is denoted by $p(\varrho)$. Let $p(\varrho)$ be independent of j. The average power of the functions $f(j, \theta)$ is P_j:

$$\begin{gathered} P_j = \sum_{\substack{\varrho=-r \\ \neq 0}}^{+r} p(\varrho) \frac{1}{T} \int_{-T/2}^{T/2} a_\varrho^2 f^2(j, \theta)\, dt = \sum_{\substack{\varrho=-r \\ \neq 0}}^{+r} p(\varrho)\, a_\varrho^2 \\ \sum_{\substack{\varrho=-r \\ \neq 0}}^{+r} p(\varrho) = 1 \end{gathered} \tag{3}$$

The average power of the characters composed of h functions $f(j, \theta)$ is P:

$$P = h\, P_j \tag{4}$$

The following assumptions are made:

a) The probability of a coefficient $a_\chi(j)$ having the value a_ϱ is independent of ϱ: $p(\varrho) = 1/2r$.

b) The difference $|a_\varrho - a_{\varrho-1}|$ is independent of ϱ. $|a_\varrho - a_{\varrho-1}| = a_0$. This condition is satisfied if a_ϱ is a multiple of a_0: $a_\varrho = \varrho\, a_0$, $\varrho = \pm 1, \ldots, \pm r$.

The average power P_j of a function $f(j, \theta)$ follows from Eqs. (3) and (4):

$$\begin{gathered} P_j = \sum_{\substack{\varrho=-r \\ \neq 0}}^{+r} \varrho^2 \frac{a_0^2}{2r} = \left(\frac{a_0^2}{r}\right) \sum_{\varrho=1}^{r} \varrho^2 = \frac{(r+1)(2r+1)}{6} a_0^2 = \frac{P}{h} \\ a_0^2 = \frac{6P}{h(r+1)(2r+1)} \end{gathered} \tag{5}$$

Let a character $F_\chi(\theta)$ be transmitted. Cross-correlation with the functions $f(j, \theta)$ yields the coefficients $a_\chi(j)$ at the receiver. Superimposed additive thermal noise changes these coefficients into $a(j)$. They have a Gaussian distribution with mean $|\varrho|\, a_0$, $-|\varrho|\, a_0$ or 0; $|\varrho| = 1, \ldots, r$. These coefficients are denoted by $a^{(+\varrho)}(j)$, $a^{(-\varrho)}(j)$ and $a^{(0)}(j)$:

$$\left\langle \frac{a^{(+\varrho)}(j)}{a_0} \right\rangle = \langle a^{(+\varrho)} \rangle = \varrho, \qquad \left\langle \frac{a^{(-\varrho)}(j)}{a_0} \right\rangle = \langle a^{(-\varrho)} \rangle = -\varrho$$

$$\left\langle \frac{a^{(0)}(j)}{a_0} \right\rangle = \langle a^{(0)} \rangle = 0 \tag{6}$$

The variance of these distributions follows in analogy to Eq. (7.3.4-3):

$$\begin{aligned} \sigma^2 &= \langle a_\lambda^2(j)/a_0 \rangle = h(r+1)(2r+1)\, P_{1,T}/6P \\ &= h(r+1)(2r+1)\, P_{n,T}/6n\, P = h(r+1)(2r+1)\, P_{\Delta f}/6n\, P \end{aligned} \tag{7}$$

h number of non-zero coefficients $a_\chi(j)$;

$n = \log_2 (2r)^h \binom{m}{h}$ information per character in bits, if all characters are transmitted with equal probability;

$2r$ number of non-zero values which the coefficients $a_\chi(j)$ may assume;

$P_{n,T}$ average power of n orthogonal components of thermal noise in an orthogonality interval of duration T;

$P = h\, a_0^2 (r+1)(2r+1)/6$ average signal power; $\Delta f = n/2T$;

$P_{\Delta f}$ average power of thermal noise in a frequency band of width Δf.

The characters of combination alphabets of higher than third order are not transmitted with equal energy. One must determine the smallest energy ΔW_ψ according to Eq. (6.2.1-7) for the detection of the signal. This means that the $(2r)^h \binom{m}{h}$ sums

$$S_\psi = \sum_{j=0}^{m-1} [a(j) - a_\psi(j)]^2 \tag{8}$$

must be computed and the one with the smallest value determined. An error occurs if S_ψ is not smallest for $\psi = \chi$, where χ denotes the transmitted character $F_\chi(\theta)$.

The smallest value of S_ψ is obtained if the h smallest terms $[a(j) - a_\psi(j)]^2$ are added. The h terms, for which $a(j)$ is equal to $a_\chi(j) \neq 0$ in the noise-free case, will be the h smallest terms in the presence of additive thermal noise if the following conditions are satisfied (*see* Fig. 210):

1. None of the h coefficients $a^{(+\varrho)} = a^{(+\varrho)}(j)/a_0$ and $-a^{(-\varrho)} = -a^{(-\varrho)}(j)/a_0$ is farther from its correct mean $|a_\varrho/a_0| = |\varrho|$ than from the other means $|\varrho'| = 1, \ldots, r, \neq |\varrho|$.

2. None of the h coefficients $a^{(+\varrho)}$ and $-a^{(-\varrho)}$ is farther from one of the means $1, \ldots, r$ than the absolute value of one of the $m-h$ coefficients $a^{(0)}$. This condition must be satisfied only if condition 1 is satisfied.

These two conditions are essentially equal to the conditions (7.3.4-5) and (7.3.4-6) for ternary combination alphabets. The calculation of the

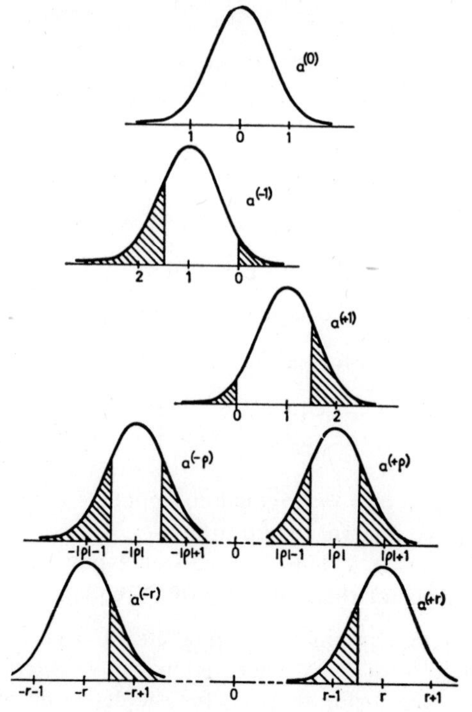

Fig. 210. Density functions of $a^{(0)}$, $a^{(-1)}$, $a^{(+1)}$, $a^{(-\varrho)}$, $a^{(+\varrho)}$, $a^{(-r)}$ and $a^{(+r)}$. The hatched areas indicate errors.

error probability is much more complicated. Only the results will be stated here.

Let $p_{m,h}^{(2r+1)}$ denote the error probability due to thermal noise of a combination alphabet of order $2r + 1$, using h out of m functions. For large values m, h, r and small values of σ^2,

$$m \gg h \gg 1, \quad r \gg 1, \quad \sigma^2 \doteq h\, r^2\, P_{\Delta f}/3n\, P \ll 1, \tag{9}$$

$$n = \log_2 (2r)^h \binom{m}{h} \cong h \log_2 \left(\frac{r\, m}{h}\right)$$

one obtains the following formula:

$$p_{m,h}^{(2r+1)} \cong 1 - \exp(-e^{n\eta_1}) \exp(-e^{n\eta_2}) \tag{10}$$

$$\eta_1 = \frac{1}{n} \ln h - \frac{3P}{8h\, r^2\, P_{\Delta f}}, \quad \eta_2 = \frac{1}{n} \ln (m - h)\, h - \frac{3P}{4h\, r^2\, P_{\Delta f}}$$

Let the information n per character grow beyond all bounds:

$$\lim_{n\to\infty} p_{m,h}^{(2r+1)} = 1 \quad \text{for} \quad \eta_2 > 0, \quad \eta_1 \text{ arbitrary}$$
$$\lim_{n\to\infty} p_{m,h}^{(2r+1)} = 0 \quad \text{for} \quad \eta_2 < 0, \quad \eta_1 < 0 \tag{11}$$

The case $\eta_2 < 0$, $\eta_1 > 0$ is not possible because it holds:

$$\eta_2 = \left[\frac{1}{n}\ln(m-h) - \frac{3P}{8hr^2 P_{\Delta f}}\right] + \left[\frac{1}{n}\ln h - \frac{3P}{8hr^2 P_{\Delta f}}\right] \tag{12}$$

The term in the second bracket is equal to η_1. It follows from $\eta_1 > 0$ that η_2 must be larger than 0. Hence, $\eta_2 < 0$ yields the error probability 0 and $\eta_2 > 0$ yields the error probability 1 for $n \to \infty$. Rewriting η_2 yields the following condition for error-free transmission:

$$\frac{P}{P_{\Delta f}} > \frac{4h}{3n} r^2 \ln(m-h)\,h \tag{13}$$

Substitution of n from Eq. (9) yields:

$$\frac{P}{P_{\Delta f}} > \frac{4}{3} r^2 \ln \frac{2\ln(m-h)\,h}{\ln(r\,m/h)} \tag{14}$$

Let us investigate how Shannon's limit in the form of Eq. (7.1.2-37) can be approached by combination alphabets of order $2r+1$. The average noise power $P_{\Delta f}$ must be replaced by $P_{m,T}$ in Eq. (10) for η_2:

$$P_{\Delta f} = P_{n,T} = n\,P_{1,T} = \frac{n}{m} P_{m,T}$$

One obtains from Eq. (11):

$$\lim_{n\to\infty} p_{m,h}^{(2r+1)} = 0 \quad \text{for} \quad \frac{P}{P_{m,T}} > \frac{4h}{3m} r^2 \ln(m-h)\,h \tag{15}$$

Using the relation

$$r \cong 2^{n/h} \frac{h}{m} \tag{16}$$

which follows from Eq. (9) for $r \gg 1$, one obtains from Eq. (15):

$$n < \frac{h}{2} \log_2 \left\{\left[\frac{3m^3}{4h^3 \ln(m-h)\,h}\right] \frac{P}{P_{m,T}}\right\} \tag{17}$$

One must choose $h = h(m)$ so that the right side of this inequality becomes as large as possible for a certain value of m and a fixed ratio $P/P_{m,T}$. The expression is too complicated to find a maximum by differentiation. One may see, however, that the factor h in front of the logarithm should be as large as possible. If h becomes too large, the term in the brackets becomes smaller than 1. This term would then become arbitrarily small with increasing m. Hence, h is chosen so that the equation

$$\frac{3m^3}{4h^3 \ln(m-h)\, h} = K = \text{constant} \tag{18}$$

is satisfied. This suggests the choice:

$$h = m/\sqrt[3]{(\ln m)} \tag{19}$$

It follows from Eq. (17):

$$n < \frac{m}{2} \frac{1}{(\ln m)^{1/3}} \log_2 \left(\frac{KP}{P_{m,T}} \right) \tag{20}$$

The information n/T transmitted per unit time becomes:

$$C = \frac{n}{T} < \frac{m}{2T} \frac{1}{(\ln m)^{1/3}} \left[\log_2 \left(\frac{P}{P_{m,T}} \right) + \log_2 K \right] \tag{21}$$

This formula differs from Shannon's limit in Eq. (7.1.2-37) for large values of $P/P_{m,T}$ only by the factor $(\ln m)^{-1/3}$. This small difference is probably accounted for by having chosen an equal distribution for $p(\varrho)$ rather than a Gaussian distribution. The physical meaning of the condition $P/P_{m,T} \gg K$ is readily understandable. $r \gg 1$ had been assumed in Eq. (9); many different values for the coefficients $a_\chi(j)$ will permit an error-free transmission only if the average signal-to-noise power ratio is large.

REFERENCES

ordered by sections[1]

Introduction

1. Rademacher, H.: Einige Sätze von allgemeinen Orthogonalfunktionen. *Math. Ann.* 87 (1922) 122–138.
2. Fowle, F. F.: The transposition of conductors. *Trans. AIEE* 23 (1905) 659 to 687.
3. Osborne, H. S.: The design of transpositions for parallel power and telephone line circuits. *Trans. AIEE* 37 (1918) 897–936.
4. Pinkert, H. S.: Induktionsschutz für Fernsprechleitungen. *Telegraphen- und Fernsprechtechnik*, 3. Sonderheft (1919) 108–119.
5. Klein, W.: *Die Theorie des Nebensprechens auf Leitungen.* Berlin/Göttingen/ Heidelberg: Springer, 1955.
6. Walsh, J. L.: A closed set of orthogonal functions. *Amer. J. Math.* 45 (1923) 5–24.
7. Bernal, I., Groth, I.: *Ancient Mexico in color.* New York: McGraw-Hill, 1968, p. 127.
8. Mann, P. A.: Der Zeitablauf von Rauschspannungen. *Elektr. Nachrichten-Technik* 20 (1943) 183–189.
9. Stumpers, F. L.: Theory of frequency modulation noise. *Proc. IRE* 36 (1948) 1081–1092.
10. Voelcker, H. B.: Toward a unified theory of modulation. *Proc. IEEE* 54 (1966) 340–353, 735–755.
11. Andrews, H. C.: *Computer techniques in image processing.* New York: Academic Press, 1970.
12. Weiser, F. E.: Walsh function analysis of instantaneous nonlinear stochastic problems. Thesis, Department of Electrical Engineering, Polytechnic Institute of Brooklyn (1964).
13. Jain, P. C.: Problems of synchronization and nonlinear distortion, and their effect on the detection of Walsh functions. *Proc. 1971 Walsh Functions Symposium*, 192–198.
14. Decker, J. A.: Hadamard-transform exhaust-analysis spectrometer. *Appl. Opt.* 10 (1971) 24–27.
15. Nelson, E. D., Fredman, M. L.: Hadamard spectroscopy. *J. Opt. Soc. Amer.* 60 (1970) 1664–1669.

---

[1] References showing a number of the form AD 000-000 may be ordered by this number from Clearinghouse, National Bureau of Standards, Springfield, Virginia 22151. The order number for the Proceedings of the 1970 Walsh Functions Symposium is AD 707-431, the one for the Proceedings of the 1971 Walsh Functions Symposium is AD 727-000. The Proceedings of the 1971 Symposium where also published as special issue of the IEEE Transactions on Electromagnetic Compatibility EMC-13 (August 1971). It can be ordered under number 71 C 14-EMC from IEEE Headquarters, 345 East 47 Street, New York, New York 10017.

16. Gibbs, J. E., Gebbie, H. A.: Application of Walsh functions to transform spectroscopy. *Nature* 224 (1969) 1012–1013.
17. Bell, A. G.: Researches in telephony. *Proc. American Academy of Arts and Sciences* 12 (1876) 1–10.
18. Marland, E. A.: *Early electrical communication.* New York: Abelard-Schuman, 1964.
19. Decker, J. A.: Experimental realization of the multiplex advantage with a Hadamard transform spectrometer. *Appl. Opt.* 10 (1971) 510–514.
20. Soroko, L. M.: The Hadamard transform scintillation counter. 16 pages; Pseudo noise sequences applicable to multiplex systems of particle detection. 15 pages; Multiplex targets. 7 pages (all three papers in Russian); Communications of the Joint Institutes for Nuclear Research R 13-5696, R 13-5722, R 13-5699 (1971). Available from Joint Institutes of Nuclear Research (Dubna), Head Post Office Box 79, Moscow, USSR.
21. Canada: *see* reference [11], section 4.2.1; England: *see* reference 4, section 4.2.2; Netherlands: *see* reference [14], section 4.2.1; Switzerland: *see* reference [12], section 4.2.1; U.S.A.: D. I. Durst, Grumman Aerospace Corp., Bethpage, N. Y. (personal communication); H. E. Jones, Westinghouse Electric Corporation, Baltimore, Md. (personal communication); West Germany: *see* references [10] and [13], section 4.2.1.
22. Hübner, H.: Symposium über Anwendungen von Walshfunktionen Washington 1971. *Internationale Elektronische Rundschau* 25 (August 1971) 208–212.
23. Bath, M., Burman, S.: Walsh spectroscopy of Rayleigh waves. *Geophysics* (J. Soc. Explor. Geophys., Tulsa, Oklahoma), (in press).
24. Bößwetter, C.: Stand und Entwicklung der Anwendung von Walsh-Funktionen in der Nachrichtentechnik und verwandten Gebieten. *NTZ* 24 (1971) 478–480.
25. Lackey, R. B.: So what's a Walsh function. *Proc. IEEE Fall Electronics Conf., Chicago, 18–20 Oct., 1971,* 368–371.
26. Gubbins, D., Scollar, I., Wisskirchen, P.: Two dimensional digital filtering with Haar and Walsh transforms. *Ann. Géophys.* vol. 27/2 (1971) 85–104.

1.11

1. Tricomi, F.: *Vorlesungen über Orthogonalreihen.* 2nd ed. Berlin/Heidelberg/New York: Springer, 1970.
2. Sansone, G.: *Orthogonal functions.* New York: Interscience, 1959.
3. Lense, J.: *Reihenentwicklungen in der mathematischen Physik.* Berlin: de Gruyter, 1953.
4. Milne-Thomson, J. M.: *The calculus of finite differences.* London: Macmillan, 1951.
5. Nörlund, N. E.: *Vorlesungen über Differenzenrechnung.* Berlin: Springer, 1924.
6. Gradshtein, I. S., Ryzhik, I. M.: *Tables of integrals, series and products.* New York: Academic Press, 1966.

1.1.2

1. Courant, R., Hilbert, D.: *Methoden der mathematischen Physik.* Berlin: Springer, 1931.
2. Morse, P. M., Feshbach, H.: *Methods of theoretical physics.* New York: McGraw-Hill, 1953.
3. Whittaker, E. T., Watson, G. N.: *A course of modern analysis.* London: Cambridge University Press, 1952, Chapter 9.

4. Titchmarsh, E. C.: *Theory of the Fourier-integral.* London: Oxford University Press, 1937.
5. Alexits, G.: *Konvergenzprobleme der Orthogonalreihen.* Berlin: Deutscher Verlag der Wissenschaften, 1960.
6. Davis, H. F.: *Fourier series and orthogonal functions.* Boston: Allyn & Bacon, 1963.

1.1.3

1. Smirnow, W. I.: *Lehrgang der höheren Mathematik.* Teil 2, Berlin: Deutscher Verlag der Wissenschaften, 1961.
2. Titchmarsh, E. C.: *Theory of the Fourier-integral.* London: Oxford University Press, 1937.
3. Bracewell, R.: *The Fourier-transform and its applications.* New York: McGraw-Hill, 1965.
4. Bennet, W. R., Davey, J. R.: *Data transmission.* New York: McGraw-Hill, 1959.
5. Wiener, N.: *The Fourier-integral and certain of its applications.* London: Cambridge University Press, 1933.
6. Hartley, R. V. L.: A more symmetrical Fourier analysis. *Proc. IRE* 30 (1942) 144–150.
7. Boesswetter, C.: Analog sequency analysis and synthesis of voice signals. *Proc. 1970 Walsh Functions Symposium,* 220–229.
8. —: The mutual spectral representation of trigonometric functions and Walsh functions. *Proc. 1971 Walsh Functions Symposium,* 43–46.
9. Filipowsky, R. F.: On the correlation matrices of trigonometric product functions. *IBM J. Research and Development* 14 (1970) 680–685.
10. Simmons, R. E.: A short review of Walsh functions and the Walsh-Hadamard transform. Naval Electronics Laboratory Center Technical Note 1850, 6-11-01N/ZR011-01-01/NELC Z155. San Diego, Calif. 92152.
11. Bößwetter, C.: Die Eigenschaften der Sequenz-Analyse und -Synthese von Signalen. *NTZ* 24 (1971) 193–201.

1.1.4

1. Walsh, J. L.: A closed set of orthogonal functions. *Amer. J. of Mathematics* 45 (1923) 5–24.
2. Rademacher, H.: Einige Sätze von allgemeinen Orthogonalfunktionen. *Math. Annalen* 87 (1922) 122–138.
3. Henderson, K. W.: Some notes on the Walsh functions. *IEEE Trans. Electronic Computers* EC-13 (1964) 50–52.
4. Liedl, R.: Über eine spezielle Klasse von stark multiplikativ orthogonalen Funktionensystemen. *Monatshefte für Mathematik* 68 (1964) 130–137.
5. —: Walsh-Funktionen und endlichdimensionale Hilberträume. *Monatshefte für Mathematik* 70 (1966) 342–348.
6. —: Über gewisse Funktionale im Raum $C^{(v)}[0, 1]$ und Walsh-Fourier-Koeffizienten. *Monatshefte für Mathematik* 72 (1968) 38–44.
7. Weiss, P.: Zusammenhang von Walsh-Fourier-Reihen mit Polynomen. *Monatshefte für Mathematik* 71 (1967) 165–179.
8. Pichler, F.: Synthese linearer periodisch zeitvariabler Filter mit vorgeschriebenem Sequenzverhalten. *Arch. elektr. Übertragung* 22 (1968) 150–161.
9. —: Das System der sal- und cal-Funktionen als Erweiterung des Systems der Walsh-Funktionen und die Theorie der sal- und cal-Fouriertransformation. Thesis, Dept. of Mathematics, Innsbruck Univ., Austria 1967.

10. Vilenkin, N. W.: On a class of complete orthogonal systems (in Russian). *Izv. Akad. Nauk SSSR, Ser. Math.* 11 (1947) 363–400.
11. Fine, N. J.: On the Walsh-functions. *Trans. Amer. Math. Soc.* 65 (1949) 372–414.
12. —: The generalized Walsh functions. *Trans. Amer. Math. Soc.* 69 (1950) 66–77.
13. Paley, R. E.: A remarkable series of orthogonal functions. *Proc. London Math. Soc.* (2) 34 (1932) 241–279.
14. Selfridge, R. G.: Generalized Walsh transforms. *Pacific J. of Math.* 5 (1955) 451–480.
15. Toni, S.: Su un notevole sistema orthogonale di funzioni. *Atti Accad. Sci. Ist. Bologna, Cl. Sci. fis.*, Ann. 246 Rend. XI, Ser. 5, No. 1 (1958) 225–230.
16. Morgenthaler, G. W.: On Walsh-Fourier series. *Trans. Amer. Math. Soc.* 84 (1957) 472–507.
17. Wiener, N.: *Nonlinear problems in random theory.* New York: MIT Press and Wiley, 1958, p. 21.
18. Fowle, F. F.: The transposition of conductors. *Trans. AIEE* 23 (1905) 659 to 687.
19. Peterson, W. W.: *Error-correcting codes.* New York: MIT Press and Wiley, 1961.
20. Loomis, L. H.: *An introduction to abstract harmonic analysis.* New York: Van Nostrand, 1953.
21. Hammond, J. L., Johnson, R. S.: A review of orthogonal square wave functions and their application to linear networks. *J. Franklin Inst.* 273 (1962) 211–225.
22. Vilenkin, N. W.: On the theory of Fourier integrals on topologic groups (in Russian). *Mat. Sbornik* (*N. S.*) 30 (72) (1952) 233–244.
23. Fine, N. J.: The Walsh functions, in *Encyclopaedic Dictionary of Physics.* Oxford: Pergamon Press (in press).
24. Polyak, B., Shreider, Yu. A.: The application of Walsh functions in approximate calculations. *Voprosy Teorii Matematicheskikh Mashin* 2, Bazilevskii (Editor), Moscow: Fizmatgiz 1962.
25. Boulton, P. I.: Smearing techniques for pattern recognition (Hadamard-Walsh transformation). Thesis, Univ. of Toronto, Canada (1968).
26. Kokilashvili, V. M.: Approximation of functions by means of Walsh-Fourier-series (in Russian). *Soobshcheniya Akademii Nauk Gruzunskoy SSR* 55 (1967) 305–310.
27. Shneyder, A. A.: The uniqueness of expansions in Walsh functions (in Russian). *Mat. Sbornik* 24 (66) (1949) 279–300.
28. —: Series of Walsh functions with monotonic coefficients (in Russian). *Izv. Akad. Nauk SSSR, Ser. Mat.* 12 (1948) 179–192.
29. Koenig, M., Zolesio, J.: Problems numeriques relatifs aus fonctions de Walsh. *Comp. Ren. Acad. Sc.* Paris 271 (23 Nov. 1970) 1074–1077.
30. Franks, L. E.: *Signal theory.* Englewood Cliffs, N.J.: Prentice Hall 1969, p. 56.
31. Maxwell, J. C.: *A treatise on electricity and magnetism*, reprint of the third edition of 1891. New York: Dover 1954, Vol. 1, p. 470.
32. Adams, R. N.: Properties and applications of Walsh functions. Thesis, University of Pennsylvania, Moore School of Electrical Engineering, Philadelphia (1962).
33. Ohnsorg, F.: Properties of complex Walsh functions. *Proc. IEEE Fall Electronics Conf., Chicago, 18–20 Oct. 1971*, 383–385.

1.1.5

1. Crittenden, R. B.: On Hadamard matrices and complete orthonormal systems. *IEEE Trans. Computers*, submitted.
2. Paley, R. E. A.: On orthogonal matrices. *J. Math. Phys.* 12 (1933) 311–320.
3. Hadamard, M. J.: Resolution d'une question relative aux determinants. *Bull. Sci. Math.* A 17 (1893) 240–246.
4. Golomb, S. W., Baumert, L. D.: The search for Hadamard matrices. *Amer. Math. Monthly* 70 (1) (1963) 12–17.
5. Sylvester, J. J.: Thoughts on inverse orthogonal matrices, simultaneous sign-successions, and tessalated pavements in two or more colors, with applications to Newton's rule, ornamental tile-work, and the theory of numbers. *Phil. Mag.* 34 (1867) 461–475.
6. Welch, L. R.: Walsh functions and Hadamard matrices. *Proc. 1970 Walsh Functions Symposium*, 163–165.
7. Henderson, K. W.: Previous investigation of realization of an arbitrary switching function with a network of threshold and parity elements. *IEEE Trans. Computers* C-20 (1971) 941–942.

1.1.6

1. Haar, A.: Zur Theorie der orthogonalen Funktionensysteme. *Math. Ann.* 69 (1910) 331–371; 71 (1912) 38–53.

1.1.7

1. Meltzer, B., Searle, N. H., Brown, R.: Numerical specification of biological form. *Nature* 216 (7 Oct. 1967) 32–36.

1.2.1

1. Fourier, J.: *Théorie analytique de la chaleur* (1822). English edition: *The analytical theory of heat* (1878), reprinted New York: Dover 1955.

1.2.2

1. Morse, P. M., Feshbach, H.: *Methods of theoretical physics*, Vol. 1. New York: McGraw-Hill, 1953, pp. 942–945.
2. Bracewell, R.: *The Fourier-transform and its applications*. New York: McGraw-Hill, 1965.
3. Kantorowitsch, L. W., Akilow, G. P.: Funktionalanalysis in normierten Räumen. Berlin: Akademie-Verlag, 1964, Chapter 8, section 1.
4. Ahmed, N., Bates, R. M.: A power spectrum and related physical interpretation for the multidimensional BIFORE Transform. *Proc. 1971 Walsh Functions Symposium*, 47–50.
5. Ahmed, N., Rao, K. R.: Generalized transform. *Proc. 1971 Walsh Functions Symposium*, 60–67.
6. Kak, S. C.: Classification of random binary sequences using Walsh-Fourier analysis. *Proc. 1971 Walsh Functions Symposium*, 74–77.

1.2.4

1. Harmuth, H.: Verallgemeinerung des Fourier-Integrales und des Begriffes Frequenz. *Arch. elektr. Übertragung* 18 (1964) 439–451.
2. Pichler, F.: Das System der sal- und cal-Funktionen als Erweiterung des Systems der Walsh-Funktionen und die Theorie der sal- und cal-Fourier-transformation. Thesis, Dept. of Mathematics, Innsbruck Univ., Austria 1967.

3. Bowyer, D.: Walsh functions, Hadamard matrices and data compression. *Proc. 1971 Walsh Functions Symposium*, 33–37.
4. Andrews, H. C.: Walsh functions in image processing, feature extraction and pattern recognition. *Proc. 1971 Walsh Functions Symposium*, 26–32.
5. Alexandridis, N., Klinger, A.: Walsh orthogonal functions in geometrical feature extraction. *Proc. 1971 Walsh Functions Symposium*, 18–25.
6. Moharir, P. S., Sama, K. R., Prasada, B.: Amplitude bounds and quantization schemes in the Walsh-Fourier domain. *Proc. 1971 Walsh Functions Symposium*, 142–150.
7. Sandy, G. F.: Some Walsh functions analysis techniques. *Proc. 1971 Walsh Functions Symposium*, 151–157.
8. Carl, J. W., Kabrisky, M.: Playing the identification game with Walsh functions. *Proc. 1971 Walsh Functions Symposium*, 203–209.
9. Ahmed, N., Rao, K. R.: Transform properties of Walsh functions. *Proc. IEEE Fall Electronics Conf., Chicago, 18–20 Oct. 1971*, 378–382.

1.2.5

1. Green, R. R.: A serial orthogonal decoder. Space Programs Summary, Jet Propulsion Laboratory, Pasadena, Calif., No. 37–39, vol. 4 (1966) 247–251.
2. Posner, E. C.: Combinatorial structures in planetary reconnaissance. *Symposium on error-correcting codes.* Math. Research Center of the US Army, Univ. of Wisconsin, 1968.
3. Welch, L. R.: Computation of finite Fourier series. *Space Programs Summary*, Jet Propulsion Laboratory, Pasadena, Calif., No. 37–39, Vol. 4 (1966) 295–297.
4. Pratt, W. K., Kane, J., Andrews, H. C.: Hadamard transform image coding. *Proc. IEEE* 57 (1969) 58–68.
5. Whelchel, J. E., Guinn, D. F.: Fast Fourier-Hadamard transform and its use in signal representation and classification. *EASCON' 68 Record* (1968) 561–573.
6. Shanks, J. L.: Computation of the fast Walsh-Fourier transform. *IEEE Trans. Computers* C-18 (1969) 457–459.
7. Agarwal, V. K.: A new approach to the fast Hadamard transform algorithm. *Proc. 1971 Hawaii Int. Conf. System Sciences*, 408–410.
8. Kabrisky, M., Carl, J. W.: Sequency filtered densely connected transforms for pattern recognition. *Proc. 1971 Hawaii Int. Conf. System Sciences*, 223–228.
9. Parkyn, W. A.: Sequency and periodicity concepts in imagery analysis. *Proc. 1971 Hawaii Int. Conf. System Sciences*, 229–232.
10. Kennedy, J. D.: Experimental Walsh-transform image processing. *Proc. 1971 Hawaii Int. Conf. System Sciences*, 233–235.
11. Schmidt, R. O.: A collection of Walsh function programs. *Proc. 1971 Walsh Functions Symposium*, 88–94.
12. Pratt, W. K.: An algorithm for a fast Hadamard matrix transform of order twelve. *IEEE Trans. Computers* C-18 (1969) 1131–1132.
13. Carl, J. W.: Comment on 'An algorithm for a fast Hadamard matrix transform of order twelve'. *IEEE Trans. Computers* C-20 (1971) 587–588.
14. Webb, S. R.: Some comments on the fast Hadamard transform of order twelve. *IEEE Trans. Computers* C-20 (1971) 588–589.
15. Henderson, K. W.: Comments on 'Computation of the fast Walsh-Fourier transform'. *IEEE Trans. Computers* C-19 (1970) 850–851.
16. Lechner, R. J.: Comment on 'Computation of the fast Walsh-Fourier transform' *IEEE Trans. Computers* C-19 (1970) 174.

17. Pearl, J.: Basis-restricted transformations and performance measures for spectral representations. *IEEE Trans. Information Theory* IT-17 (1971) 751–752.

1.2.6

1. Andrews, H. C., Pratt, W. K.: Digital image transform processing. *Proc. 1970 Walsh Functions Symposium*, 183–194.
2. Pearl, J.: Basis-restricted transformations and performance measures for spectral representations. *Proc. 1971 Hawaii Int. Conf. System Sciences*, 321–323.

1.2.8

1. Gibbs, E.: Walsh spectroscopy, a form of spectral analysis well suited to binary digital computation (1967). This paper is now accepted as fundamental but Gibbs never succeeded in publishing it.
2. Gibbs, E., Millard, M. J.: Walsh functions as solutions of a logical differential equation. DES Report 1 (1969); Some methods of solution of linear ordinary logical differential equations. DES Report 2 (1969). National Physical Laboratory, Teddington, Middlesex, England.
3. Gibbs, E.: Some properties of functions on the nonnegative integers less than 2^n. DES Report 3 (1969); Functions that are solutions of a logical differential equation. DES Report 4 (1970). National Physical Laboratory, Teddington, Middlesex, England.
4. Gibbs, E.: Discrete complex Walsh functions; Sine and Walsh waves in physics. *Proc. 1970 Walsh Functions Symposium*, 106–122, 260–274.
5. Pichler, F.: Some aspects of a theory of correlation with respect to Walsh harmonic analysis (1970), AD 714–596.
6. Pearl, J.: Application of Walsh transform to statistical analysis. *Proc. 1971 Hawaii Int. Conf. System Sciences*, 406–407.
7. Pichler, F.: Walsh-Fourier-Synthese optimaler Filter. *Arch. elektr. Übertragung* 24 (1970) 350–360.
8. Pichler, F.: Walsh functions and optimal linear systems; Walsh functions and linear system theory. *Proc. 1970 Walsh Functions Symposium*, 17–22 and 175–182.
9. Simmons, G. J.: Correlation properties of pseudo Rademacher-Walsh codes. *Proc. 1971 Walsh Functions Symposium*, 84–87.
10. Pitassi, D. A.: Fast convolution using Walsh transforms. *Proc. 1971 Walsh Functions Symposium*, 130–133.
11. Gethoefer, H.: Sequency analysis using correlation and convolution. *Proc. 1971 Walsh Functions Symposium*, 118–123.
12. Frank, T. H.: Implementation of dyadic correlation. *Proc. 1971 Walsh Functions Symposium*, 111–117.
13. Woodward, P. M.: *Probability and information theory with applications to radar*. New York: McGraw-Hill, 1953.
14. Rihaczek, A. W.: *Principles of high-resolution radar*. New York: McGraw-Hill, 1969.
15. Erickson, C. W.: Clutter cancelling in auto-correlation functions by binary sequency pairing (1961) AD 446-146.
16. Golay, M. J.: Complementary series. *IRE Trans. Inform. Theory* IT-7 (1961) 82–87.

1.3.1

1. Stumpers, F. L.: Theory of frequency modulation noise. *Proc. IRE* 36 (1948) 1081–1092.
2. Mann, P. A.: Der Zeitablauf von Rauschspannungen. *Elektr. Nachr.-Technik* 20 (1943) 183–189.
3. Panter, P. F.: *Modulation, noise and spectral analysis*. New York: McGraw-Hill, 1965.
4. Harmuth, H.: A generalized concept of frequency and some applications. *IEEE Trans. Inform. Theory* IT-14 (1968) 375–382.
5. Alexandridis, N. A.: Relations among sequency, axis symmetry, and period of Walsh functions. *IEEE Trans. Inform. Theory* IT-17 (1971) 495–497.

1.3.2

1. Oppenheim, A. V., Schafer, R. W., Stockham, T. G.: Nonlinear filtering of multiplied and convolved signals. *Proc. IEEE* 56 (1968) 1264–1291.
2. Pichler, F., Gibbs, J. E.: Comments on transformation of Fourier power spectra into Walsh power spectra. *Proc. 1971 Walsh Functions Symposium*, 51–54.
3. Ohnsorg, F. R.: Spectral modes of the Walsh-Hadamard transform. *Proc. 1971 Walsh Functions Symposium*, 55–59.
4. Yuen, C. K.: Upper bounds on Walsh transforms, Report No. 66 (1971), Basser Computing Department, University of Sidney, Australia.

1.3.3

1. Claire, E. J., Farber, S. M., Green, R. R.: Practical techniques for transform data compression/image coding. *Proc. 1971 Walsh Functions Symposium*, 2–6.
2. Yuen, C. K.: Walsh functions and Gray code. *Proc. 1971 Walsh Functions Symposium*, 68–73.
3. Claire, E. J., Farber, S. M., Green, R. R.: Acoustic signature detection and classification techniques. *Proc. 1971 Walsh Functions Symposium*, 124–129.

1.3.4

1. Rosenbloom, J. H.: Physical representation of the dyadic group. *Proc. 1971 Walsh Functions Symposium*, 158–165.
2. Pichler, F.: On space state description of linear dyadic invariant systems. *Proc. 1971 Walsh Functions Symposium*, 166–170.

2.1.1

1. Lebert, F. J.: Walsh function generator for a million different functions. *Proc. 1970 Walsh Functions Symposium*, 52–54.
2. Peterson, H. L.: Generation of Walsh functions. *Proc. 1970 Walsh Functions Symposium*, 55–57.
3. Lee, J. S.: Generation of Walsh functions as binary group codes. *Proc. 1970 Walsh Functions Symposium*, 58–61.
4. Manoli, S. H.: Walsh function generator. *Proc. IEEE* 59 (1971) 93–94.
5. Elliott, A. R.: A programmable Walsh function generator. *1971 Int. Electrical, Electronics Conference and Exposition*, Toronto, Digest 144–145.
6. Yuen, C. K.: A new Walsh function generator. *Electronics Letters* (in press).
7. Redinbo, G. R.: An implementation technique for Walsh functions. *IEEE Trans. Computers* C-20 (1971) 706–707.

2.1.2

1. Wagner, K. W.: Spulen- und Kondensatorleitungen. *Arch. Elektrotechn.* 8 (1919) 61–92, received by the publisher on 7 January 1915.
2. Campbell, G. A.: Physical theory of the electric wave filter. *Bell System Tech. J.* 1 (1922) 1–32; US-patent applied for on 15 July 1915.
3. Zobel, O. J.: Theory and design of uniform and composite electric wave filters. *Bell System Tech. J.* 2 (1923) 1–46.
4. Bartlett, A. C.: *The theory of electrical artificial lines and filters.* New York: Wiley, 1930.
5. Guillemin, E. A.: *Communication Networks.* Vol. 1: The classical theory of lumped constant networks. Vol. 2: The classical theory of long lines, filters, and related networks. New York: Wiley, 1931.
6. Cauer, W.: *Theorie der linearen Wechselstromschaltungen.* Leipzig: Akademische Verlagsgesellschaft, 1941. English edition: *Synthesis of linear communication networks.* New York: McGraw-Hill, 1958.
7. Pichler, F.: Synthese linearer periodisch zeitvariabler Filter mit vorgeschriebenem Sequenzverhalten. *Arch. elektr. Übertragung* 22 (1968) 150–161.
8. Harmuth, H.: Sequency filters based on Walsh functions. *IEEE Trans. Electromagnetic Compatibility* EMC-10 (1968) 293–295.
9. —: Sequency filters. *Proc. of the Summer School on Circuit Theory 1968*, Czechoslovak Academy of Science, Prague.

2.1.3

1. Vandivere, E. F.: A flexible Walsh filter design for signals for moderately low sequency. *Proc. 1970 Walsh Functions Symposium*, 3–6.
2. Lee, T.: Hardware approach to Walsh function sequency filters. *Proc. 1970 Walsh Functions Symposium*, 7–11.
3. Harmuth, H.: Survey of analog sequency filters based on Walsh functions. *Proc. 1970 Walsh Functions Symposium*, 208–219.
4. Pratt, W. K.: Linear and nonlinear filtering in the Walsh domain. *Proc. 1971 Walsh Functions Symposium*, 38–42.
5. Walsh, D. M.: Walsh domain filter techniques. Thesis, Dept. Electrical Engineering, University of South Florida (1970).

2.1.4

1. Roth, D.: Special filters based on Walsh functions. *Proc. 1970 Walsh Functions Symposium*, 12–16.
2. Harmuth, H.: Grundzüge einer Filtertheorie für die Mäanderfunktionen $A_n(\theta)$. *Arch. elektr. Übertragung* 18 (1964) 544–554. Meander functions and Walsh functions are identical: $A_n(\theta) = \text{wal}(n, \theta)$.
3. Clark, B. R.: Convergence of the Walsh expansion of x^2, x^3 and x^4 for $-\frac{1}{2} < x < \frac{1}{2}$. *Proc. 1971 Walsh Functions Symposium*, 155–157.

2.2.1

1. Haard, H. B., Svala, C. G.: US Patent No. 2718621.
2. Harmuth, H.: Resonance filters based on Walsh functions. *Proc. 1970 Kyoto Int. Conf. Circuit and System Theory*, 195–198; Sequenzfilter für Signale mit zwei Raumvariablen und LCS-Filter. *NTZ* 23 (1970) 377–383.
3. Milne-Thomson, L. M.: *The calculus of finite differences.* London: Macmillan, 1951.
4. Nörlund, N. E.: *Vorlesungen über Differenzenrechnung.* Berlin: Springer, 1924.

5. Gelfond, A. O.: *Differenzenrechnung*. Berlin: VEB Deutscher Verlag der Wissenschaften, 1958.
6. Golden, J. P., James, S. N.: Implementation of Walsh function resonant filters. *Proc. 1971 Walsh Functions Symposium*, 106–110.
7. Golden, J. P., James, S. N.: LCS resonant filters for Walsh functions. *Proc. IEEE Fall Electronics Conf., Chicago, 18–20 Oct. 1971*, 386–390.

2.2.3

1. Tucker, D. G.: *Circuits with periodically varying parameters*. London: Macmillan, 1964.
2. Manley, J. M., Rowe, H. E.: Some general properties of nonlinear elements, part 1. General energy relations. *Proc. IRE* 44 (1956) 904–913.
3. Walker, J. E.: Parametric amplifier based on Walsh functions. *Proc. 1970 Walsh Functions Symposium*, 62–64.
4. Ries, R. P., Satterthwaite, C. B.: Superconducting parametric amplifier for the measurement of small voltages. *Rev. Scientific Instruments* 38 (1967) 1203–1209.

2.3.1

1. Harmuth, H.: Survey of analog sequency filters based on Walsh functions. *Proc. 1970 Walsh Functions Symposium*, 208–219.

2.3.2

1. Andrews, H. C., Pratt, W. K.: Digital image transform processing. *Proc. 1970 Walsh Functions Symposium*, 183–194.
2. Pratt, W. K., Kane, J., Andrews, H. C.: Hadamard transform image coding. *Proc. IEEE* 57 (1969) 58–68.
3. Parkyn, W. A.: Digital image processing aspects of the Walsh transform. *Proc. 1970 Walsh Functions Symposium*, 152–156.
4. Harmuth, H.: Sequency filters based on Walsh functions for signals with two space variables. *Proc. 1971 Hawaii Int. Conf. System Sciences*, 414–416.
5. Habibi, A., Wintz, P. A.: Image coding by linear transformations and block quantization. *IEEE Trans. Communication Technology* COM-19 (1971) 50–62.
6. Andrews, H. C.: Multidimensional rotations in feature selection. *IEEE Trans. Computers* C-20 (1971) 1045–1050.

2.3.3

1. Boesswetter, C.: Analog sequency analysis and synthesis of voice signals. *Proc. 1970 Walsh Functions Symposium*, 220–229 (particulary p. 224).
2. Enomoto, H., Shibata, K.: Features of a Hadamard transformed television signal. *Proc. 1965 National Conf. of the Institute of Electrical and Communications Engineers of Japan*, paper 881, 1 page. Television signal coding method by orthogonal transformations. *Proc. 1966 Joint Convention of Electrical and Electronics Engineers of Japan*, paper 1436, 2 pages. Television signal coding method by orthogonal transformation. *Papers of the 6th Research Group on Television Transmission, Institute of Television Engineers of Japan* (1968), 24 pages. Experiment on television signal PCM system by orthogonal transformation. *Proc. 1969 Joint Convention of Electrical and Electronics Engineers of Japan*, paper 2219, 2 pages. Orthogonal transform coding system for television signals. *Television* (Journal of the Institute of Television Engineers of Japan) 24 (1970), No. 2, 99–108. All papers in Japanese.

3. Shibata, K., Ohira, T.: PCM CODEC for orthogonal transformed television signals (in Japanese). *Proc. 1969 Joint Convention of Electrical and Electronics Engineers*, paper 2619, 1 page.
4. Shibata, K., Ohira, T., Terauchi, S.: Color television signal orthogonal transformation PCM terminal equipment (in Japanese). *Papers of the Technical Group on Communication Systems* (1970), Institute of Electrical and Communications Engineers of Japan; order No. CS 70-47 (1970-07), 25 pages.
5. Shibata, K.: On PCM of color television signals using an orthogonal transformation (in Japanese). *Proc. 1970 National Conference of the Institute of Electrical and Communications Engineers of Japan*, paper S. 9—11, 2 pages.
6. Shibata, K.: Television signal PCM by orthogonal transformation (internal report in English) (1968). Available from KDD Research Laboratory, 1—23, 2-chome, Nekameguro, Meguro-ku, Tokyo.
7. Taki, Y., Hatori, M. *et al.*: On the band compression of television signals by the *E*-sequence transformation technique (in Japanese). *Papers of the Technical Group on Information Theory* (1970), Institute of Electrical and Communications Engineers of Japan; order No. IT 70-13 (1970-05).
8. Enomoto, H., Shibata, K.: Orthogonal transform coding system for television signals. *Proc. 1971 Walsh Functions Symposium*, 11—17.
9. Skolnik, M. I.: *Radar handbook*. New York: McGraw-Hill, 1970, p. 11—66.

2.4.1

1. Lackey, R. B., Meltzer, D.: A simplified definition of Walsh functions. *IEEE Trans. Computers* (in press).

2.4.3

1. Oppenheim, A. V., Schafer, R. W., Stockham, T. G.: Nonlinear filtering of multiplied and convolved signals. *Proc. IEEE* 56 (1968) 1264—1291.
2. Boyle, W. S., Smith, G. E.: Charge coupled devices — A new approach to MIS device structure. *IEEE Spectrum* 8, No. 7 (July 1971) 18—27.

2.5.1

1. Nowak, D. J., Schmid, P. E.: Introduction to digital filters. *IEEE Trans. Electromagnetic Compatibility* EMC-10 (1968) 210—220.
2. Robinson, G. S., Granger, R. L.: A design procedure for nonrecursive digital filters based on Walsh functions. *Proc. 1971 Walsh Functions Symposium*, 95—100.
3. Murray, G. G.: Digital Walsh filter design. *Proc. 1971 Walsh Functions Symposium*, 101—105.
4. Walsh, D. M.: Design considerations for digital Walsh filters. *Proc. IEEE Fall Electronics Conf., Chicago, 18—20 Oct. 1971*, 372—377.

3.1.1

1. Lerner, R. M.: Representation of signals; design of signals. *Lectures on Communication System Theory*. New York: McGraw-Hill, 1961.
2. Akiyama, M.: Orthogonal PCM transmission with weighted bit length. *J. Inst. Electr. Comm. Engrs. Japan* 49 (1966) 1153—1159.
3. Schmid, P. E., Dudley, H. S., Skinner, S. E.: Partial response signal formats for parallel data transmission. *1968 IEEE Int. Conf. on Communications, Record*, 811—816.

4. Chang, R. W., Gibby, R. A.: A theoretical study of performance of an orthogonal multiplexing data transmission scheme. *1968 IEEE Int. Conf. on Communications, Record,* 833–837.
5. Lange, F. H.: *Signale und Systeme.* Vol. 1. Braunschweig: Vieweg, 1966.
6. Blachman, N. M., Spectral analysis with sinusoids and Walsh functions. *IEEE Trans. Aerospace and Electronic Systems* AES-7 (1971) 900–905.

3.1.3

1. Whittaker, J. M.: *Interpolatory function theory.* Cambridge Tracts in Mathematics and Mathematical Physics 33. London: Cambridge University Press, 1935.
2. Levinson, N.: *Gap and density theorems.* Amer. Math. Soc. Coll. Publ. 26 (1940).
3. Shannon, C. E.: Communication in the presence of noise. *Proc. IRE* 37 (1949) 10–21.
4. Goldmann, S.: *Information theory.* Englewood Cliffs, N.J.: Prentice Hall, 1953.
5. Linden, D. A.: A discussion of sampling theorems. *Proc. IRE* 47 (1959) 1219–1226.
6. Kohlenberg, A.: Exact interpolation of band-limited functions. *J. Appl. Phys.* 24 (1953), 1432–1436.
7. Kluvanec, I.: Sampling theorem in abstract harmonic analysis. *Matematicko-fyzikálny Casopis Sloven. Akad. Vied* 15 (1965) 43–48.

3.2.1

1. Lange, F. H.: *Korrelationselektronik.* Berlin: Verlag Technik, 1959.
2. BURR-BROWN Research Corp.: *Handbook of operational amplifier applications.* Tucson, Ariz., 1963.
3. PHILBRICK RESEARCHES, Inc.: *Applications manual for computing amplifiers.* Dedham, Mass., 1966.
4. Johnson, C. K.: *Analog computer techniques.* New York: McGraw-Hill, 1963.
5. Korn, G. A., Korn, T. M.: *Electronic analog and hybrid computers.* New York: McGraw-Hill, 1964.
6. Schmid, P., Nowak, D., Harmuth, H.: Detection of orthogonal sine and cosine pulses by linear active RC networks. *Proc. 1967 Int. Telemetering Conf., Washington, D.C.,* 210–220.
7. Swick, D. A.: Walsh-function generation. *IEEE Trans. Inform. Theory* IT-15 (1969) 167.

3.2.2

1. Kawai, K., Michishita, H., Yamauchi, K., Sasaki, H.: A new carrier multiplex telegraph system using phase modulation. *J. Inst. Electr. Comm. Engrs. Japan* 48, 8 (1965) 1369–1377.
2. Sanders, R. W.: The digilock orthogonal modulation system. *Advances in Communication Systems* 1 (1965) 57–75.
3. Kuhn, B. G., Morey, K. H., Smith, W. B.: The orthomatch data transmission system. *IEEE Trans. Space Electronics and Telemetry* SET-9 (1963) 63–66.
4. Viterbi, A. J.: On coded phase-coherent communications. *IRE Trans. Space Electronics and Telemetry,* SET-7 (1961) 3–14.

5. Mosier, R. R., Clabaugh, R. G.: Kineplex, a bandwidth-efficient binary transmission system. *Trans. AIEE, Communication and Electronics* 76 (1957) 723–727.
6. Jaffe, R. M.: Digilock telemetry system for the Air Force special weapons center's Blue Scout Jr. *IRE Trans. Space Electronics and Telemetry,* SET-8 (1962) 44–50.
7. Sanders, R. W.: Communication efficiency comparison of several communication systems. *Proc. IRE* 48 (1960) 575–588.
8. Doelz, M. L., Heald, E. T., Martin, D. L.: Binary data transmission techniques for linear systems. *Proc. IRE* 45 (1957) 656–661.
9. Wier, J. M.: Digital data communication techniques. *Proc. IRE* 49 (1961) 169–209.
10. Filipowsky, R. F., Muehldorf, E. I.: Space communications systems; Space communications techniques. Englewood Cliffs, N.J.: Prentice Hall, 1965.
11. Katsumaru, K., Hayashi, T., Takada, Y., Ogawa, K.: Super multichannel carrier telegraph system by phase modulation (Rectiplex system). *Fujitsu Sci. Techn. J.* 1 (1965) 261–279.
12. Harmuth, H.: On the transmission of information by orthogonal time functions. *Trans. AIEE Communication and Electronics* 79 (1960) 248–255.
13. —: Radio communication with orthogonal time functions. *Trans. AIEE Communication and Electronics* 79 (1960) 221–228.
14. Osatake, T., Kirisawa, K.: An orthogonal pulse code modulation system. *Electronics and Communications in Japan* 50 (1967) 35–43. Translated by Scripta Electronica Inc.
15. Michishita, H., Kawai, K.: Rectiplex. *J. Inst. Electr. Comm. Engrs. Japan* 53, 1 (1970) 22–32.

3.2.3

1. Sunde, E. D.: Ideal binary pulse transmission by AM and FM. *Bell System Tech. J.* 38 (1959) 1357–1426.
2. Nyquist, H.: Certain topics in telegraph transmission theory. *Trans. AIEE* 47 (1928) 617–644.
3. Schreiber, H. H.: Bandwidth requirements for Walsh functions. *IEEE Trans. Information Theory* IT-16 (1970) 491–493.
4. Kak, S. C.: Sampling theorem in Walsh-Fourier analysis. *Electronics Letters* 6 (1970) 447–448.
5. DeBuda, R., Harmuth, H. F.: Conversion of sequency limited signals into frequency limited signals and vice versa. *IEEE Trans. Information Theory* IT-17 (1971) 343–344.

3.3.1

1. Wagner, K. W.: *Elektromagnetische Wellen.* Basel: Birkhäuser, 1953.
2. Smirnow, W. I.: *Lehrgang der höheren Mathematik,* Part 2. Berlin: Deutscher Verlag der Wissenschaften, 1961.
3. Boesswetter, C.: Analog sequency analysis and synthesis of voice signals. *Proc. 1970 Walsh Functions Symposium,* 220–229.
4. Rathbun, D. K., Jensen, H. J.: Nuclear test instrumentation with miniature superconductive cables. *IEEE Spectrum* 5 (1968) 91–99.
5. Allen, R. J., Nahmann, N. S.: Analysis and performance of superconductive coaxial transmission lines. *Proc. IEEE* 52 (1964) 1147–1154.
6. Sandy, G. F.: Square wave (Rademacher-Walsh functions) analysis. Mitre Corp., Working Paper WP-1585 (1968).

7. Robinson, G.: Walsh-Hadamard transform speech compression. *Proc. 1971 Hawaii Int. Conf. System Sciences*, 411–413.
8. Robinson, G., Campanella, S. J.: Digital sequency decomposition of voice signals. *Proc. 1970 Walsh Functions Symposium*, 230–237.
9. Campanella, S. J., Robinson, G. S.: A comparison of Walsh and Fourier transformations for application to speech. *Proc. 1971 Walsh Functions Symposium*, 199–202.
10. Wien, R. A.: Walsh functions. USAECOM-CADPL Report No. 132 (1970), 24 pages.

3.3.2

1. Siebert, W. M.: Signals in linear time invariant systems. *Lectures on communication system theory*. New York: McGraw-Hill, 1961.

4.1.1

1. Costas, J. P.: Synchronous communication. *Proc. IRE* 44 (1956) 1713–1718.
2. Black, H. S.: *Modulation theory*. New York: Van Nostrand, 1953.
3. Schwartz, M.: *Information transmission, modulation and noise*. New York: McGraw-Hill, 1959.
4. Panter, P. F.: *Modulation, noise and spectral analysis*. New York: McGraw-Hill, 1965.
5. Viterbi, A. J.: *Principles of coherent communication*. New York: McGraw-Hill, 1966.

4.1.2

1. Baghdady, E. J.: Analog modulation systems. *Lectures on communication system theory*. New York: McGraw-Hill, 1961.
2. Viterbi, A. J.: *Principles of coherent communication*. New York: McGraw-Hill, 1966.
3. Pichler, F.: Das System der sal- und cal-Funktionen als Erweiterung des Systems der Walsh-Funktionen und die Theorie der sal- und cal-Fourier-transformation. Thesis, Dept. of Mathematics, Innsbruck University, Austria 1967, p. 47.
4. Wetscher, J.: Korrelationsfunktionen der ursprünglichen Walshfunktionen. Thesis, Dept. of Mathematics, Innsbruck University, Austria 1970.

4.1.3

1. Honey, J. F., Weaver, D. K.: An introduction to single sideband communication. *Proc. IRE* 44 (1956) 1667–1675.
2. Norgaard, D. E.: The phase-shift method of single-sideband signal generation; The phase-shift method of single-sideband signal reception. *Proc. IRE* 44 (1956) 1718–1743.
3. Saraga, W.: Single-sideband generation. *Electronic Technology* 39 (1962) 168–171.
4. Weaver, D. K.: A third method of generation and detection of single-sideband signals. *Proc. IRE* 44 (1956) 1703–1705.
5. Nyquist, H.: Certain topics in telegraph transmission theory. *Trans. AIEE* 47 (1928) 617–644.
6. Becker, F. K., Davey, J. R., Saltsberg, B. R.: An AM vestigial sideband data transmission set using synchronous detection for serial transmission up to 3000 bits per second. *Trans. AIEE, Communication and Electronics* 81 (1962) 97–101.

7. Bennett, W. R., Davey, J. R.: *Data transmission.* New York: McGraw-Hill, 1965.
8. Harmuth, H., Schmid, P. E., Nowak, D. L.: Transposed sideband modulation for data transmission. *IEEE Trans. Communication Technology* COM-15 (1967) 868–870.

4.2.1

1. Crowley, T. H., Harris, G. G., Miller, S. E., Pierce, J. R., Runyon, J. P.: *Modern communications.* New York: Columbia University Press, 1962.
2. Flood, J. E.: Time division multiplex systems. *Electronic Engr.* 25 (1953) 2–5, 58–63, 101–106, 146–150.
3. Roberts, F. F., Simmonds, J. C.: Multichannel communication systems. *Wireless Engr.* 22 (1945) 538–549, 576–589.
4. Bennett, W. R.: Time division multiplex systems. *Bell System Tech. J.* 20 (1941) 199–221.
5. Technical Staff Bell Telephone Laboratories: *Transmission systems for communications.* Winston-Salem, N.C.: Western Electric Co. Technical Publications, 1965.
6. Schwartz, M., Bennett, W. R., Stein, S.: *Communication systems and techniques.* New York: McGraw-Hill, 1966.
7. Landon, V. D.: Theoretical analysis of various systems of multiplex transmission. *RCA Review* 9 (1948) 287–351, 438–482.
8. Harmuth, H.: Sequenz-Multiplexsysteme für Telephonie- und Datenübertragung. 1. Quadraturmodulation, 2. Einseitenband-Modulation. *Arch. elektr. Übertragung* 22 (1968) 27–35, 102–108.
9. Pichler, F.: Das Sequenzvielfach, ein neues Sprechwegenetz für vollelektronische Fernsprechvermittlungsämter. *XII. Int. Wiss. Kolloquium der Technischen Hochschule Ilmenau* 7 (1967) 15–20.
10. Hübner, H.: On the transmission of Walsh multiplexed signals; Analog and digital multiplexing by means of Walsh functions. *Proc. 1970 Walsh Functions Symposium,* 41–45, 238–247.
11. Davidson, I. A.: The use of Walsh functions for multiplexing signals. *Proc. 1970 Walsh Functions Symposium,* 23–25.
12. Bagdasarjanz, F., Loretan, R.: Theoretical and experimental studies of a sequency multiplex system. *Proc. 1970 Walsh Functions Symposium,* 36–40.
13. Lüke, H. D., Maile, R.: Telephonie-Multiplexsystem mit orthogonalen Walshfunktionen — Aufbau und Meßergebnisse. Report FE/FI Nr. 2/69, AEG-Telefunken, Research Institute, Ulm, West Germany.
14. Roefs, H. F. A.: Een code division multiplex system met behulp van orthogonale Hadamard codes (1971); Van Ittersum, F.: Orthogonale code detektie door middel van de "fast Walsh-Fourier" transformatie (1971). Theses, Dept. Electrical Engineering, Technological University Delft, Netherlands.
15 Lopez de Zavalia, J., Moro, S. M. (Universidad Nacional de Tucumán, Instituto de Ingeniera Eléctrica, San Miguel de Tucumán, Argentina): Trabajos sobre funciones de Walsh. *Supplemento de Revista Telegrafica Electronica* (August 1971), p. S10 (Summary of three papers presented at the XXIII Semana de la Ingeniera Eléctrica y Electronica, 13 to 17 September 1971, Buenos Aires).

4.2.2

1. Sunde, E. D.: Ideal binary pulse transmission by AM and FM. *Bell System Tech. J.* 38 (1959) 1357–1426.

2. Bayless, J. W., Jones, R. A., Gupta, S. C.: A comparison of time division and Walsh multiplexing for a multiuser communication satellite link. *Proc. 1971 Hawaii Int. Conf. System Sciences*, 318–320.
3. Hübner, H.: Multiplex systems using sums of Walsh functions as carriers. *Proc. 1971 Walsh Functions Symposium*, 180–191.
4. Gordon, J. A., Barrett, R. A.: Majority multiplexing using Walsh functions. *Proc. 1971 Walsh Functions Symposium*, 171–176.
5. Barrett, R., Karran, J.: Correlation-multiplex data-transmission system. *Electronics Letters* 4 (29 Nov. 1968), 538–539.

4.2.3

1. Searle, N. H.: A 'logical' Walsh-Fourier transform. *Proc. 1970 Walsh Functions Symposium*, 95–98.
2. Helm, H. A.: An application of coding algebra to the design of a digital multiplexing system using linear sequential circuits. *Proc. IEEE Seventh Annual Symposium on Switching and Automata Theory* (1966), 14 pages.

4.3.1

1. Van der Pol, B.: Frequency modulation. *Proc. IRE* 18 (1930) 1194–1205.
2. Hund, A.: *Frequency modulation.* New York: McGraw-Hill, 1942.
3. Cuccia, C. L.: *Harmonics, sidebands and transients in communication engineering.* New York: McGraw-Hill, 1952.
4. Black, H. S.: *Modulation theory.* Princeton, N.J.: Van Nostrand, 1953.

4.3.2

1. Crosby, M. G.: Communication by phase modulation. *Proc. IRE* 27 (1939) 126–136.
2. Black, H. S.: *Modulation theory.* Princeton, N.J.: Van Nostrand, 1953.
3. Panter, P. F.: *Modulation, noise and spectral analysis.* New York: McGraw-Hill, 1965.

4.3.3

1. Oliver, B. M., Pierce, J. R., Shannon, C. E.: The philosophy of PCM. *Proc. IRE* 36 (1948) 1324–1331.
2. Mayer, H. F.: *Principles of pulse code modulation.* Advan. Electron. New York: Academic Press, 1951.
3. Flood, J. E.: Time division multiplex systems, part 4. *Electronic Engr.* 25 (1953) 146–150.
4. Goodall, W. M.: Telephony by pulse-code-modulation. *Bell System Tech. J.* 26 (1947) 395–409.
5. de Jager, F.: Delta modulation: A method of PCM transmission using the 1-unit-code. *Philips Research Report* 7 (1952) 442–466.
6. Bennett, W. R.: Spectra of quantized signals. *Bell System Tech. J.* 27 (1948) 446–472.

5.1.1

1. Hertz, H.: Die Kräfte elektrischer Schwingungen behandelt nach der Maxwellschen Theorie. *Ann. Phys., Neue Folge* 36 (1889) 1–22.
2. Hertz, H.: *Electric waves.* New York: Dover, 1962 (originally published by Macmillan in 1893).
3. Slater, J. C., Frank, H.: *Electromagnetism.* New York: Mc-Graw-Hill, 1947.

4. Ware, L. A.: *Elements of electromagnetic waves.* New York: Pitman, 1949.
5. Becker, R., Sauter, F.: *Theorie der Elektrizität*, vol. 1, 18th ed. Stuttgart: Teubner, 1964.
6. Landau, L. D., Lifschitz, E. M.: *Lehrbuch der theoretischen Physik*, vol. 2, *Klassische Feldtheorie*. Berlin: Akademie-Verlag, 1964.
7. Kraus, J.: *Antennas.* New York: McGraw-Hill, 1950.
8. Schelkunoff, S. A.: *Electromagnetic waves.* New York: Van Nostrand, 1943.
9. —: *Advanced antenna theory.* New York: Wiley, 1952.

5.1.4

1. Rumsey, V. H.: *Frequency independent antennas.* New York: Academic Press, 1966.

5.3.1

1. Perlman, J.: Radiation patterns for antennas with Walsh current inputs. *Proc. 1970 Walsh Functions Symposium*, 65–69.
2. Harmuth, H.: Electromagnetic Walsh waves in communication. *Proc. 1970 Walsh Functions Symposium*, 248–259.

5.3.3

1. Harmuth, H.: Grundlagen und mögliche Anwendungen der Sequenztechnik. *Bull. schweizer. elektrotechn. Ver.* 59 (1968) 1196–1203.
2. Rihaczek, A. W.: *Principles of high-resolution radar.* New York: McGraw-Hill 1969.

5.4.1

1. Harmuth, H. F.: Asynchronous filters and mobile radio communication based on Walsh functions. *Proc. 1971 Walsh Functions Symposium*, 210–218.
2. Harmuth, H. F., Frank, T. H.: Theoretical study of the synchronous demodulation problem for a Walsh wave receiver (1971) AD 725–755.

6.1.1

1. Doob, J. L.: *Stochastic processes.* New York: Wiley, 1953.
2. van der Ziel, A.: *Noise.* Englewood Cliffs, N.J.: Prentice Hall, 1954.
3. Rice, S. O.: Mathematical analysis of random noise. *Bell System Tech. J.* 23 (1944) 282–332, 24 (1945) 46–156.
4. Smullin, D., Haus, H. A.: *Noise in electron devices.* New York: Wiley, 1959.
5. Bennett, W. R.: *Electrical noise.* New York: McGraw-Hill, 1960.
6. Davenport, W. B., Jr., Root, W. L.: *An introduction to the theory of random signals and noise.* New York: McGraw-Hill, 1958.
7. Schwartz, M.: *Information transmission, modulation and noise.* New York: McGraw-Hill, 1959.
8. Root, W. L., Pitcher, T. S.: On the Fourier series expansion of random functions. *Ann. Math. Statistics* 26 (1955) 313–318.
9. Haus, H. A., *et al.*: IRE standards of methods of measuring noise in linear twoports. *Proc. IRE* 48 (1960) 60–68.
10. Johnson, J. B.: Thermal agitation of electricity in conductors. *Phys. Rev.* 32 (1928) 97–109.
11. Johnson, J. B.: Electronic noise: The first two decades. *IEEE Spectrum* 8, No. 2 (1971) 42–46.

6.2.1

1. Kotel'nikov, V. A.: *The theory of optimum noise immunity* (translation of the Russian original, published in 1947, by R. A. Silverman). New York: McGraw-Hill, 1959.
2. Siebert, W. M., Root, W. L.: Statistical decision theory and communications, in *Lectures on communication system theory*. New York: McGraw-Hill, 1961.
3. Middleton, D.: *An introduction to statistical communication theory*. New York: McGraw-Hill, 1960.
4. Wainstain, L. A., Zubakov, V. D.: *Extraction of signals from noise*. Englewood Cliffs, N.J.: Prentice Hall, 1962.
5. Harman, W. A.: *Principles of the statistical theory of communication*. New York: McGraw-Hill, 1963.
6. Wiener, N.: *Extrapolation, interpolation and smoothing of stationary time series*. New York: MIT Press and Wiley, 1949.
7. Hancock, J. C.: *Signal detection theory*. New York: McGraw-Hill, 1966.
8. Levinson, N.: The Wiener RMS error criterion in filter design and prediction. *J. Math. Phys.* 25 (1947) 261–278.
9. Kolmogoroff, A.: Interpolation and extrapolation of stationary random sequencies. *Bulletin de l'academie des sciences de USSR, Ser. Math.* 5 (1941) 3–14.
10. Sherman, S.: Non-mean square error criteria. *IRE Trans. Information Theory* IT-4 (1959) 125–126.
11. Bode, H. W.: A simplified derivation of linaer least-square smoothing and prediction theory. *Proc. IRE* 38 (1950) 417–426.
12. Arthurs, E., Dym, H.: On the optimum detection of digital signals in the presence of white Gaussian noise. *IRE Trans. Communication Systems* CS-10 (1962) 336–372.

6.2.2

1. North, D. O.: An analysis of the factors which determine signal/noise discrimination in pulsed-carrier systems. Reprinted in *Proc. IEEE* 51 (1963) 1016 to 1027.
2. Turin, G. L.: An introduction to matched filters. *IRE Trans. Information Theory* IT-6 (1960) 311–329.
3. Zadeh, L. A., Ragazzini, I. R.: Optimum filters for the detection of signals in noise. *Proc. IRE* 40 (1952) 1123–1131.
4. Peterson, E. L.: *Statistical analysis and optimization of systems*. New York: Wiley, 1961.

6.2.4

1. Corrington, M. S., Adams, R. N.: Advanced analytical and signal processing techniques. Applications of Walsh functions to nonlinear analysis (1962) AD 277–942.
2. Weiser, F. E.: Walsh function analysis of instantaneous nonlinear stochastic problems. Thesis, Polytechnic Institute of Brooklyn, 1964.
3. Harmuth, H.: *Transmission of information by orthogonal functions*. 1st ed. New York/Heidelberg/Berlin: Springer, 1969.
4. Ahmed, N., Rao, K. R., Abdussattar, A. L.: BIFORE or Hadamard transform. IEEE Trans. Audio and Electroacoustics AU-19 (1971) 225–234.

6.3.1

1. Baghdady, E. J.: Diversity techniques. *Lectures on communication 'system theory*. New York: McGraw-Hill, 1961.

2. Brennan, D. G.: Linear diversity combining techniques. *Proc. IRE* 47 (1959) 1075–1102.
3. Pierce, J. N., Stein, S.: Multiple diversity with non-independent fading. *Proc. IRE* 48 (1960) 89–104.
4. Price, R.: Optimum detection of random signals in noise with application to scatter multipath communications. *IRE Trans. Information Theory* IT-2 (1956) 125–135.
5. Price, R., Green, P. E.: A Communication technique for multipath channels, *Proc. IRE* 46 (1958) 555–570.
6. Glen, A. B.: Comparison of PSK vs FSK and PSK-AM vs FSK-AM binary coded transmission systems. *IEEE Trans. Communication Systems* CS-8 (1960) 87–100.
7. Ridout, P. N., Wheeler, L. K.: Choice of multi-channel telegraph systems for use on HF radio links. *Proc. IEE* 110 (1963) 1402–1410.
8. Turin, G. L.: On optimal diversity reception I. *IRE Trans. Information Theory* IT-7 (1961) 154–166.
9. —: On optimal diversity reception II. *IRE Trans. Communication Systems* CS-10 (1962) 22–31.
10. Law, H. B.: The detectability of fading radiotelegraph signals in noise. *Proc. IEE* 104 B (1957) 130–140.
11. Voelcker, H. B.: Phase shift keying in fading channels. *Proc. IRE* 107 B (1960) 31–38.
12. Pierce, J. N.: Theoretical diversity improvement in frequency-shift keying. *Proc. IRE* 46 (1958) 903–910.
13. Alnatt, J. W., Jones, E. D., Law, H. B.: Frequency diversity in the reception of selective fading binary frequency-modulated signals. *Proc. IEE* 104 B (1957) 98–110.
14. Bello, P. A., Nelin, B. D.: The effect of frequency selective fading on the binary error probabilities of incoherent and differentially coherent matched filter receivers. *IEEE Trans. Communication Systems* CS-11 (1963) 170–186.
15. Schwartz, M., Bennett, W. R., Stein, S.: *Communication systems and techniques*. New York: McGraw-Hill, 1966.

7.1.1

1. Nyquist, H.: Certain topics in telegraph transmission theory. *Trans. AIEE* 47 (1928) 617–644.
2. Hartley, R. V. L.: Transmission of information. *Bell System Tech. J.* 7 (1928) 535–563.
3. Küpfmüller, K.: *Die Systemtheorie der elektrischen Nachrichtenübertragung*. Stuttgart: Hirzel, 1952.
4. Shannon, C. E.: A mathematical theory of communication. *Bell System Tech. J.* 27 (1948) 379–423, 623–656.
5. —: Communication in the presence of noise. *Proc. IRE* 37 (1949) 10–21.

7.1.2

1. Fano, R. M.: *Transmission of information*. New York: MIT Press and Wiley, 1961.
2. Fey, P.: *Informationstheorie*. Berlin: Akademie-Verlag, 1963.
3. Sommerville, D. M. Y.: *An introduction to the geometry of N dimensions*. New York: Dutton, 1929.
4. Madelung, E.: *Die mathematischen Hilfsmittel des Physikers*. Berlin/Göttingen/Heidelberg: Springer, 1957.

7.1.3

1. Harmuth, H.: Die Übertragungskapazität von Nachrichtenkanälen nach der Verallgemeinerung des Begriffes Frequenz. *Arch. elektr. Übertragung* 19 (1965) 125–133.
2. Sommerfeld, A.: Über die Fortpflanzung des Lichtes in dispergierenden Medien. *Ann. Phys.* 44 (1914) 177–202.

7.2.1

1. Davenport, W. B., Jr., Root, W. L.: *An introduction to the theory of random signals and noise.* New York: McGraw-Hill, 1958.
2. Harman, W. W.: *Principles of the statistical theory of communication.* New York: McGraw-Hill, 1963.
3. Wainstein, L. A., Zubakov, V. D.: *Extraction of signals from noise.* Englewood Cliffs, N.J.: Prentice Hall, 1962.

7.2.2

1. Harmuth, H., Schmid, P. E., Dudley, H. S.: Multiple access communication with binary orthogonal sine and cosine pulses using heavy amplitude clipping. *1968 IEEE Int. Conf. on Communication, Record,* pp. 794–799.
2. van Vleck, J. H., Middleton, D.: The spectrum of clipped noise. *Proc. IEEE* 54 (1966) 2–19.
3. Sunde, E. D.: Ideal binary pulse transmission by AM and FM. *Bell System Tech. J.* 38 (1959) 1357–1426.
4. Harmuth, H., Schmid, P. E., Dudley, H. S.: Multiple access communication with binary orthogonal sine and cosine pulses using heavy amplitude clipping. *IEEE Trans. Communication Technology* COM-19 (1971) 1247–1252.

7.2.3

1. Aikens, A. J., Lewinski, D. A.: Evaluation of message circuit noise. *Bell System Tech. J.* 39 (1960) 879–909.
2. Smith, D. B., Bradley, W. E.: The theory of impulse noise in ideal frequency-modulation receivers. *Proc. IRE* 34 (1946) 743–751.
3. Bennett, W. R.: *Electrical noise.* New York: McGraw-Hill, 1960.
4. Stumpers, F. L.: On the calculation of impulse-noise transients in frequency-modulation receivers. *Philips Research Reports* 2 (1947) 468–474.
5. Harmuth, H.: Kodieren mit orthogonalen Funktionen. *Arch. elektr. Übertragung* 17 (1963) 429–437, 508–518.

7.3.1

1. Hamming, R. W.: Error detecting and error correcting codes. *Bell System Tech. J.* 29 (1950) 147–160.
2. Slepian, D.: A class of binary signaling alphabets. *Bell System Tech. J.* 35 (1956) 203–234.
3. Wozencraft, J. M., Reiffen, B.: *Sequential decoding.* New York: MIT Press and Wiley, 1961.
4. Gallager, R. G.: *Low-density parity-check codes.* Cambridge, Mass.: MIT Press, 1963.
5. Muller, D. E.: Application of Boolean algebra to switching circuit design and to error detection. *IRE Trans. Electronic Computers* EC-3 (1954) 6–12.
6. Peterson, W. W.: *Error correcting codes.* New York: MIT Press and Wiley, 1961.

7. —: Progress of information theory 1960—63. *IEEE Trans. Information Theory* IT-10 (1963) 221—264.
8. Lee, C. Y.: Some properties of non-binary error correcting codes. *IRE Trans. Information Theory* IT-4 (1958) 72—82.
9. Ulrich, W.: Non-binary error correcting codes. *Bell System Tech. J.* 36 (1957) 1341—1388.
10. Reed, I. S.: A class of multiple-error-correcting codes and the decoding scheme. *IRE Trans. Information Theory* IT-4 (1954) 38—49.
11. Weiss, P.: Über die Verwendung von Walshfunktionen in der Codierungstheorie. *Arch. elektr. Übertragung* 21 (1967) 255—258.
12. Pearl, J.: Walsh processing of random signals. *Proc. 1971 Walsh Functions Symposium*, 137—141.
13. Helm, H. A.: Group codes and Walsh functions. *Proc. 1971 Walsh Functions Symposium*, 78—83.
14. Pearl, J.: Application of Walsh transform to statistical analysis. *IEEE Trans. Systems, Man and Cybernetics* SMC-1 (1971) 111—119.

7.3.2

1. Golomb, S. W., Baumert, L. D., Easterling, M. F., Stiffler, J. J., Viterbi, A. J.: *Digital communications*. Englewood Cliffs, N.J.: Prentice Hall, 1964.
2. Harmuth, H.: Orthogonal codes. *Proc. IEE* 107 C (1960) 242—248.
3. Aronstein, R. H.: Comparison of orthogonal and block codes. *Proc. IEE* 110 (1963) 1965—1967.
4. Hsieh, P., Hsiao, M. Y.: Several classes of codes generated from orthogonal functions. *IEEE Trans. Information Theory* IT-10 (1964) 88—91.
5. Fano, R.: Communication in the presence of additive Gaussian noise. *Communication Theory*. New York: Academic Press, 1953.
6. Lachs, G.: Optimization of signal waveforms. *IEEE Trans. Information Theory* IT-9 (1963) 95—97.
7. Neidhardt, P.: *Informationstheorie und automatische Informationsverarbeitung*. Berlin: Verlag Technik, 1964.

7.3.3

1. Wood, H.: *Random normal deviates*. Tracts for Computers 25. London: Cambridge University Press, 1948.
2. US Department of Commerce: *Handbook of mathematical functions*. National Bureau of Standards, Applied Mathematical Series 55. Washington, D.C.: US Government Printing Office, 1964.
3. The Rand Corporation: *A Million random digits with 100000 normal deviates*. Glencoe, Ill.: The Free Press, 1955.
4. Peterson, W. W.: *Error correcting codes*. New York: MIT Press and Wiley, 1961.
5. Elias, P.: Error-free coding. *IRE Trans. Information Theory* IT-4 (1954) 29—37.

7.3.4

1. Harmuth, H.: Kodieren mit orthogonalen Funktionen, II. Kombinations-Alphabete und Minimum-Energie-Alphabete. *Arch. elektr. Übertragung* 17 (1963) 508—518.
2. Kasack, U.: Korrelationsempfang von Buchstaben in binärer bzw. ternärer Darstellung bei Bandbegrenzungen und gaußschem Rauschen. *Arch. elektr. Übertragung* 22 (1968) 487—493.
3. Harmuth, H.: *Transmission of information by orthogonal functions*. 1st ed. New York/Heidelberg/Berlin: Springer, 1969.

Additional references may be found in the first edition, second printing of this book and particularly in: Lee, J. D.: Review of recent work on applications of Walsh functions in communications. *Proc. 1970 Walsh Functions Symposium*, 26–35. The following papers were published in the *Proceedings of the Symposium on Theory and Applications of Walsh Functions*, 29 and 30 June 1971, at the Hatfield Polytechnic, Dept. of Electrical Engineering and Physics, P.O. Box 109, Hatfield, Hertfordshire, England:

1. Barrett, R., Gordon, J. A.: Walsh functions — an introduction. 2. Davies, A. C.: Some basic ideas about binary discrete signals. 3. Kennett, B. L. N.: Introduction to the finite Walsh transform and the theory of the fast Walsh transform. 4. Searle, N. H.: Walsh functions and information theory. 5. Kremer, H.: Algorithms for the Haar functions and the fast Haar transform. 6. Harmuth, H. F., Frank, T.: Multiplexing of digital signals for time division channels by means of Walsh functions. 7. Gordon, J. A., Barrett, B.: On coordinate transformations and digital majority multiplexing. 8. Hübner, H.: Methods for multiplexing and transmitting signals by means of Walsh functions. 9. Moss, G. C.: The use of Walsh functions in identification of systems with output nonlinearities. 10. Durrani, T. S., Stafford, E. M.: Control applications of Walsh functions. 11. Morgan, D. G.: The use of Walsh functions in the analysis of physiological signals. 12. Pichler, F.: On discrete dyadic systems. 13. Gethöffer, H.: Convolution and deconvolution with Walsh functions. 14. Hook, R. C.: Time variable recursive sequency filters. 15. Frank, T. H.: Circuitry for the reception of Walsh waves. 16. Georgi, K. H.: An analog Walsh vocoder using a matrix transformation. 17. Liedl, R., Pichler, F.: On harmonic analysis of switching functions. 18. Koenig, M., Zolésio, J.: Problèmes numériques relatifs aux fonctions de Walsh.

INDEX

721/14/72 — III/18/203